# Signals and Communication Technology

More information about this series at http://www.springer.com/series/4748

Marcus Greferath · Mario Osvin Pavčević
Natalia Silberstein · María Ángeles Vázquez-Castro
Editors

# Network Coding and Subspace Designs

 Springer

*Editors*
Marcus Greferath
Department of Mathematics
and Systems Analysis
Aalto University
Espoo
Finland

Mario Osvin Pavčević
Faculty of Electrical Engineering
and Computing, Department
of Applied Mathematics
University of Zagreb
Zagreb
Croatia

Natalia Silberstein
Yahoo Research
Haifa
Israel

María Ángeles Vázquez-Castro
Department of Telecommunications
and Systems Engineering
Universitat Autònoma de Barcelona
Cerdanyola del Vallès, Barcelona
Spain

ISSN 1860-4862         ISSN 1860-4870 (electronic)
Signals and Communication Technology
ISBN 978-3-319-88917-7         ISBN 978-3-319-70293-3 (eBook)
https://doi.org/10.1007/978-3-319-70293-3

Printed on acid-free paper

This Springer imprint is published by Springer Nature
The registered company is Springer International Publishing AG
The registered company address is: Gewerbestrasse 11, 6330 Cham, Switzerland

This article/publication is based upon work from COST Action IC1104 Random Network Coding and Designs over GF(q), supported by COST (European Cooperation in Science and Technology).

COST (European Cooperation in Science and Technology) is a funding agency for research and innovation networks. Our Actions help connect research initiatives across Europe and enable scientists to grow their ideas by sharing them with their peers. This boosts their research, career and innovation.

http://www.cost.eu

Funded by the Horizon 2020 Framework Programme of the European Union

# Foreword

The beautiful and the elegant are often inspired by the seemingly mundane. The chapters of this book follow from a line of work rooted in a practical engineering problem: namely, the efficient transmission of information in packet networks. Traditional approaches to the operation of packet networks treat information flow as commodity flow, emphasizing the efficient routing of the information along network pathways, while avoiding or cleverly resolving contention for transmission resources. *Network coding*, introduced in 2000 in the seminal paper of Ahlswede, Cai, Li, and Yeung, challenges this conventional approach.

Network coding is based on a simple, yet far-reaching, idea: Rather than simply routing packets, intermediate nodes are permitted to "mix" packets, transmitting functions of the data that they receive. At the network boundary, received packets are treated as evidence that must be gathered in sufficient quantity so as to permit decoding of the transmitted message.

A special case arises when the packets are interpreted as vectors of symbols drawn from a finite field, and the local mixing functions are chosen to be linear transformations. From the 2003 paper of Li, Yeung, and Cai, it is known that such *linear network coding* can achieve, when the underlying field is large enough, the so-called multicast capacity of a network, in which a single source wishes to communicate the same message to a number of different terminals. An elegant algebraic proof of this fact is given in the 2003 paper of Kötter and Médard via the existence of a nonzero for a particular multivariate polynomial arising as a product of matrix determinants. Indeed, as was demonstrated in the 2006 paper of Ho, Médard, Kötter, Karger, Effros, Shi, and Leong, the multicast capacity of a network is achieved, with high probability in a sufficiently large field, by a completely random choice of local mixing functions, obviating the need for a deliberate network-topology-dependent code design.

Such *random linear network coding* yields an interesting new type of data transmission model. In this model, to send a message the transmitter injects into the network a collection of vectors which propagate through intermediate nodes in the network, where they are randomly linearly combined, before arriving at any given terminal. A terminal observes a number of such vectors, from which the transmitted message must be inferred. Since the linear transformation by the channel of the transmitted vectors is not known in advance by the transmitter or any of the receivers, such a model is sometimes referred to as a *noncoherent* transmission model (in contrast with a so-called *coherent* model, in which knowledge of the channel transformation is assumed).

To achieve information transmission in this noncoherent model of random linear network coding, one might seek a communication invariant: some property of the transmitted vectors that is preserved by the operation of the channel. As observed in my 2008 paper with Ralf Kötter, an essential invariant—the key property preserved by the random action of the channel—is the vector space spanned by the transmitted vectors. No matter the action of the channel, we are guaranteed (in the absence of errors) that each of the vectors observed at any terminal belongs to the space spanned by the transmitted vectors. Thus, we are naturally led to consider information transmission not via the choice of the transmitted vectors themselves, but rather by the choice of the *vector space* that they span.

Such a transmission model also lends itself to a concise description of the effects of the injection (by an adversary, say) of erroneous vectors into the network: The erroneous vectors combine linearly with the transmitted ones, resulting in a vector space observed at any receiver that may differ from the transmitted one. Whenever a space $V$ is transmitted (by injection into the network of a basis for $V$), and a space $U$ is observed at a terminal (by observation of a basis for $U$), the transmitted $V$ is related to the received $U$ via the direct sum $U = \mathcal{H}(V) \oplus E$, where the operator $\mathcal{H}$ selects some subspace of $V$ and $E$ is a suitable error space intersecting trivially with $V$. This so-called *operator channel* takes in a vector space from the transmitter and produces some vector space at any given receiver, where the received space may suffer from erasures (deletion of vectors from the transmitted space) or errors (addition of vectors to the transmitted space).

A coding theory for the operator channel thus presents itself very naturally. The collection $\mathcal{P}(W)$ of subspaces of the ambient packet space $W$ plays the role of input and output alphabets. One may define metrics on $\mathcal{P}(W)$ to measure the adversarial effort required to convert a transmitted subspace $V$ to a received one $U$; intuitively, two spaces should be near each other if they intersect in a space of relatively large dimension. One natural measure, which equally weights erasures and errors, is the "subspace metric" $d_S(U, V) = \dim(U + V) - \dim(U \cap W)$. Another measure, introduced in my 2008 paper with Silva and Kötter, is the "injection distance" $d_I(U, V) = \max\{\dim(U), \dim(V)\} - \dim(U \cap V)$, which accounts for the possibility that a single injection of a packet by an adversary may simultaneously case the deletion of a dimension (an erasure) and the insertion of one (an error). These two metrics coincide (except for a factor of two) in the case of constant dimension, i.e., in the case when $\dim(U) = \dim(V)$.

A *subspace code* for the operator channel is then a nonempty subset of $\mathcal{P}(W)$, i.e., a nonempty codebook of vector spaces, each a subspace of the ambient packet space $W$. As in classical coding theory, the error- and erasure-correcting capability of such a code is determined by the minimum distance between codewords, measured according to either the subspace distance or the injection distance. In the important special case when the codewords all have the same dimension, a so-called *constant dimension* code, an analog of a "constant weight code" in classical coding theory, arises.

Many very interesting mathematical and engineering questions arise immediately; indeed, the chapters of this book aim to pose, study, and answer some of them.

As in classical coding theory, one can ask for extremal codes. For example, when the ambient packet space is $\mathbb{F}_q^n$, which codes with codewords of constant dimension $k$ have maximum codebook cardinality $M$ while preserving minimum injection distance $d$? Similarly, for a fixed codebook cardinality $M$, which codes of constant dimension $k$ have largest possible minimum injection distance $d$? One might also be interested in extremal codes with codewords not all of the same dimension, in which case the distinction between the subspace distance and the injection distance becomes important. Many classical coding theory bounds on extremal codes (e.g., the Hamming bound, the Singleton bound, the Gilbert bound) have analogs in the new setting of subspace codes.

One may also ask for code constructions that admit efficient decoding algorithms. In the case where $k \mid n$, the analog of a classical extremal code, the repetition code, is now an extremal object in finite geometry: a *spread*, a collection of pairwise trivially intersecting $k$-dimensional subspaces of $\mathbb{F}_q^n$ whose union is $\mathbb{F}_q^n$. In our 2008 papers, we showed that constant dimension codes of very large ("nearly" extremal) cardinality can be constructed by a "lifting" of extremal codes—such as the maximum rank-distance Delsarte–Gabidulin codes—constructed for the rank metric. These codes consist of the row spaces of matrices of the form $[I \mid M]$, where $I$ is the identity matrix and $M$ is a matrix codeword of a rank-metric code. Furthermore, we showed that efficient analogs of classical bounded distance decoding algorithms for Reed–Solomon codes can be developed for such lifted Delsarte–Gabidulin codes.

Closely related to the combinatorial questions of code construction, a large part of combinatorial mathematics and finite geometry deals with the existence, construction, and properties of so-called combinatorial designs: collections of sub*sets* of some ambient *set* exhibiting particular balance or symmetry properties. A prototypical example is a Steiner system: a collection of $k$-element subsets (called blocks) of $S = \{1, \ldots, n\}$ with the property that each $t$-element subset of $S$ is contained in exactly one block. In this context of subspaces of a fixed ambient space, one can ask for $q$-analogs of such designs: namely, collections of sub*spaces* of some ambient *space* over $\mathbb{F}_q$ exhibiting particular balance or symmetry properties. For example, the $q$-analog of a Steiner system would be a collection of $k$-dimensional subspaces (blocks) of $W = \mathbb{F}_q^n$ with the property that each $t$-dimensional subspace of $W$ is contained in exactly one block. That nontrivial $(t \geq 2)$ $q$-Steiner systems even exist was not known until the 2013 paper of Braun, Etzion, Östergård, Vardy, and Wassermann.

Written by leading experts in the field, this book is an exploration of the beautiful and elegant mathematical and engineering ideas that are related to network coding. Here, you will find new constructions of subspace codes of various types as well as interesting generalizations. You will find a deep discussion of rank-metric codes and their properties, and you will find connections made to finite geometry and the theory of combinatorial designs, including a description of state-of-the-art computational methods that can be used to search for designs. You will learn how network coding is related to problems of broadcasting with side information at the receivers, and how network coding can be applied in various wireless communication scenarios. Network coding ideas have also been used to determine bounds on the parameters of codes that are useful in distributed storage systems. In this book, you will find a nice overview of the desirable features—such as "locality"—that such codes should possess, along with new constructions for locally repairable codes, and codes that permit so-called private information retrieval from a distributed storage system.

That such a diversity of new ideas should arise from the problem of efficient transmission of information in a packet network suggests that the original problem may not have been so mundane after all.

Toronto, Canada                                                                               Frank R. Kschischang
                                                                      Dept. of Electrical & Computer Engineering
                                                                                     University of Toronto

# Preface

When the four editors authoring this short editorial came up with the plan for this book project, the initial idea was to present, disseminate, and advertise for a large collection of results of COST Action IC1104. This Action, named *Random Network Coding and Designs over* GF($q$), as granted by the European Science Foundation in 2011, supported 4 years of a joint endeavor in European research within the newly established field of network coding and subspace designs. When we look at the collection of material of this book today, we recognize that this book has developed to be much more: We feel that what we present here has become more significant than the Action itself in that it describes the state-of-the-art in many aspects of network coding and related topics.

Historically, the new concepts and inventive ideas of the seminal paper

R. Koetter and F. R. Kschischang: *Coding for errors and erasures in random network coding*, IEEE Transactions on Information Theory **54** (2008), 3579–3591.

attracted attention of scholars from many areas of coding theory and related scientific disciplines. They felt that this endeavor allowed them to make original contributions to their established knowledge and techniques. Algebraic coding theorists, combinatorialists, computers scientists, and engineers—they all saw their opportunities to apply their skills and experience to an interdisciplinary work on the introduced topics. So, it happened that the COST Action offered the perfect platform at the right time for enabling fruitful discussions and exchange resulting in an abundance of results by scientists all over the field.

With growing enthusiasm, a competent union of more than hundred and twenty scholars from 27 different countries joined actively the project illuminating different directions, opening new questions, and connecting more and more branches of science.

This book consists of 16 chapters by various protagonists of Action IC1104 and a foreword by Frank Kschischang that briefly provides conceptual framework, initial information on the contents of the chapters, and points the coherence between the interdisciplinary chapters. Authors are invited project members and internationally renowned experts who proved excellence in the field, not only within the 4 years of European support.

The chapters are as much self-contained as due and possible; they all start with basics but come soon to the frontiers of knowledge without feeling obliged to overdo in explaining too many details. Some of the chapters include respectable new results that have not been published yet elsewhere, along with a list of open problems and ideas for further research. This makes each of the chapters an excellent start for Ph.D. students or newcomers to the field who wish to commence research on a particular topic connected with network coding or combinatorial design theory. Since this research area is still growing and opening new problems and topics, we sincerely hope that this book will provide ample and appropriate guidance, when used by mathematicians as well as computer scientists and electrical engineers.

We greatly appreciate the kindness of all the authors to cooperate and present the chapters in such a useful review paper style. Our special thanks are addressed to the European Science Foundation and its COST framework for the financial support for Action IC1104 and in particular for helping this book to come to existence.

Espoo, Finland                                          Marcus Greferath
Zagreb, Croatia                                    Mario Osvin Pavčević
Haifa, Israel                                          Natalia Silberstein
Cerdanyola del Vallès, Barcelona, Spain        María Ángeles Vázquez-Castro

# Contents

# Part I
# Subspace Codes and Rank Metric Codes

# Codes Endowed with the Rank Metric

**Elisa Gorla and Alberto Ravagnani**

**Abstract** We review the main results of the theory of error-correcting codes with the rank metric, introducing combinatorial techniques for their analysis. We study their duality theory and MacWilliams identities, comparing in particular rank-metric codes in vector and matrix representation. We then investigate the structure of MRD codes and cardinality-optimal anticodes in the rank metric, describing how they relate to each other.

## Introduction

A $q$-ary rank-metric code is a set of matrices over $\mathbb{F}_q$ equipped with the rank distance, which measures the rank of the difference of a pair of matrices. Rank-metric codes were first studied in [3] by Delsarte for combinatorial interest.

More recently, codes endowed with the rank metric have been re-discovered for error correction in the context of linear network coding, and featured prominently in the coding theory literature.

In linear network coding, a source attempts to transmit information packets to multiple destinations via a network of intermediate nodes. The nodes compute and forward in the direction of the sinks *linear functions* of the received packets, rather than simply routing them. In [1, 9] it was shown that linear network coding achieves the optimal multicast throughput over sufficiently large alphabets.

Rank-metric codes were proposed in [8, 18] for end-to-end error correction in noisy and adversarial networks. In this context, as shown in [17], the correction capability of a rank-metric code is measured by a fundamental parameter, called the *minimum rank distance* of the code.

In this chapter, we survey the main results of the mathematical theory of rank-metric codes, with emphasis on their combinatorial structure.

E. Gorla
Institut de Mathématiques, Université de Neuchâtel, Neuchâtel, Switzerland
e-mail: elisa.gorla@unine.ch

A. Ravagnani (✉)
School of Mathematics and Statistics, University College Dublin,
Dublin, Ireland
e-mail: alberto.ravagnani@ucd.ie

© Springer International Publishing AG 2018
M. Greferath et al. (eds.), *Network Coding and Subspace Designs*,
Signals and Communication Technology,
https://doi.org/10.1007/978-3-319-70293-3_1

In Sect. 1 we introduce the most important parameters of a rank-metric code, namely, the minimum distance, the weight distribution, and the distance distribution. We then define the trace-dual of a linear rank-metric code, and compare the duality theories of codes in matrix and vector representation. In particular, we show that the former generalizes the latter.

Section 2 is devoted to the duality theory of codes endowed with the rank metric. We study how combinatorial properties of a linear code relate to combinatorial properties of the dual code. In particular, we show that the weight distribution of a linear code and the weight distribution of its dual code determine each other via a MacWilliams-type transformation. We also show an application of the MacWilliams identities for the rank metric to an enumerative combinatorics problem.

In Sect. 3 we study codes that have the largest possible cardinality for their parameters. These are called *Maximum Rank Distance* codes (MRD in short) and have very remarkable properties. We first show the existence of linear MRD codes for all choices of the parameters and of the field size. Then we prove that the dual of a linear MRD code is MRD. Finally, we show that the distance distribution of a (possibly non-linear) rank-metric code is completely determined by its parameters.

Section 4 is devoted to rank-metric anticodes, i.e., sets of matrices where the distance between any two of them is bounded from above by a given integer. We study how codes and anticodes relate to each other, deriving in particular an upper bound for the cardinality of any anticode of given parameters. We conclude the section showing that the dual of an optimal linear anticode is an optimal anticode.

# 1 Rank-Metric Codes

Throughout this chapter, $q$ denotes a fixed prime power, and $\mathbb{F}_q$ the finite field with $q$ elements. Moreover, $k$ and $m$ denote positive integers with $k \leq m$ without loss of generality, and $\mathbb{F}_q^{k \times m}$ is the space of $k \times m$ matrices over $\mathbb{F}_q$. Finally, for given integers $a, b \in \mathbb{N}$ we denote by

$$\begin{bmatrix} a \\ b \end{bmatrix}_q$$

the $q$-ary binomial coefficient of $a$ and $b$, which counts the number of $b$-dimensional subspaces of an $a$-dimensional space over $\mathbb{F}_q$. See e.g. [19, Sect. 1.7] for details.

**Definition 1** The **rank distance** is the function $d : \mathbb{F}_q^{k \times m} \times \mathbb{F}_q^{k \times m} \to \mathbb{N}$ defined by $d(M, N) = \text{rk}(M - N)$ for all $M, N \in \mathbb{F}_q^{k \times m}$.

It is easy to check that $d$ is indeed a distance function on $\mathbb{F}_q^{k \times m}$.

**Definition 2** A **(rank-metric) code** over $\mathbb{F}_q$ is a non-empty subset $\mathscr{C} \subseteq \mathbb{F}_q^{k \times m}$. When $|\mathscr{C}| \geq 2$, the **minimum distance** of $\mathscr{C}$ is the positive integer

$$d(\mathscr{C}) = \min\{d(M, N) \mid M, N \in \mathscr{C}, \ M \neq N\}.$$

A code $\mathscr{C}$ is **linear** if it is an $\mathbb{F}_q$-linear subspace of $\mathbb{F}_q^{k \times m}$. In this case its **dual code** is defined as

$$\mathscr{C}^{\perp} = \{N \in \mathbb{F}_q^{k \times m} \mid \mathrm{Tr}(MN^t) = 0 \text{ for all } M \in \mathscr{C}\} \subseteq \mathbb{F}_q^{k \times m},$$

where $\mathrm{Tr}(\cdot)$ denotes the trace of a square $k \times k$ matrix.

The map $(M, N) \rightarrow \mathrm{Tr}(MN^t) \in \mathbb{F}_q$ is a scalar product on $\mathbb{F}_q^{k \times m}$, i.e., it is symmetric, bilinear and non-degenerate. In particular, the dual of a linear code is a linear code of dimension

$$\dim(\mathscr{C}^{\perp}) = km - \dim(\mathscr{C}).$$

Other fundamental parameters of a rank-metric code are the following.

**Definition 3** The **weight distribution** and the **distance distribution** of a code $\mathscr{C}$ are the collections $\{W_i(\mathscr{C}) \mid i \in \mathbb{N}\}$ and $\{D_i(\mathscr{C}) \mid i \in \mathbb{N}\}$ respectively, where

$$W_i(\mathscr{C}) = |\{M \in \mathscr{C} \mid \mathrm{rk}(M) = i\}|, \quad D_i(\mathscr{C}) = 1/|\mathscr{C}| \cdot |\{(M, N) \in \mathscr{C}^2 \mid d(M, N) = i\}|$$

for all $i \in \mathbb{N}$.

If $\mathscr{C}$ is a linear code, then for all $P \in \mathscr{C}$ there are precisely $|\mathscr{C}|$ pairs $(M, N) \in \mathscr{C}^2$ such that $M - N = P$. Therefore

$$D_i(\mathscr{C}) = 1/|\mathscr{C}| \cdot \sum_{\substack{P \in \mathscr{C} \\ \mathrm{rk}(P)=i}} |\{(M, N) \in \mathscr{C}^2 \mid M - N = P\}| = W_i(\mathscr{C})$$

for all $i \in \mathbb{N}$. Moreover, if $|\mathscr{C}| \geq 2$ then $d(\mathscr{C}) = \min\{\mathrm{rk}(M) \mid M \in \mathscr{C}, M \neq 0\}$.

In [5], Gabidulin proposed independently a different notion of rank-metric code, in which the codewords are vectors with entries from an extension field $\mathbb{F}_{q^m}$ rather than matrices over $\mathbb{F}_q$.

**Definition 4** The **rank** of a vector $v = (v_1, ..., v_k) \in \mathbb{F}_{q^m}^k$ is the dimension of the linear spaces generated over $\mathbb{F}_q$ by its entries, i.e., $\mathrm{rk}_G(v) = \dim_{\mathbb{F}_q}\langle v_1, ..., v_k\rangle$. The **rank distance** between vectors $v, w \in \mathbb{F}_{q^m}^k$ is $d_G(v, w) = \mathrm{rk}_G(v - w)$.

One can check that $d_G$ is a distance function on $\mathbb{F}_{q^m}^k$.

**Definition 5** A **vector rank-metric code** over $\mathbb{F}_{q^m}$ is a non-empty subset $C \subseteq \mathbb{F}_{q^m}^k$. When $|C| \geq 2$, the **minimum distance** of $C$ is the positive integer

$$d_G(C) = \min\{d_G(v, w) \mid v, w \in C, v \neq w\}.$$

The code $C$ is **linear** if it is an $\mathbb{F}_{q^m}$-linear subspace of $\mathbb{F}_{q^m}^k$. In this case the **dual** of $C$ is defined as

$$C^{\perp} = \left\{ w \in \mathbb{F}_{q^m}^k \mid \sum_{i=1}^{k} v_i w_i = 0 \text{ for all } v \in C \right\} \subseteq \mathbb{F}_{q^m}^k.$$

The map $(v, w) \mapsto \sum v_i w_i$ is an $\mathbb{F}_{q^m}$-scalar product on $\mathbb{F}_{q^m}^k$. Therefore for all linear vector rank-metric codes $C \subseteq \mathbb{F}_{q^m}^k$ we have

$$\dim_{\mathbb{F}_{q^m}}(C^{\perp}) = k - \dim_{\mathbb{F}_{q^m}}(C).$$

**Definition 6** The **weight distribution** and the **distance distribution** of a vector rank-metric code $C$ are the integer vectors $(W_i(C) \mid i \in \mathbb{N})$ and $(D_i(C) \mid i \in \mathbb{N})$ respectively, where

$$W_i(C) = |\{v \in C \mid \mathrm{rk}_G(v) = i\}|, \quad D_i(C) = 1/|C| \cdot |\{(v, w) \in C^2 \mid d_G(v, w) = i\}|$$

for all $i \in \mathbb{N}$.

There exists a natural way to associate to a vector rank-metric code a code in matrix representation with the same cardinality and metric properties.

**Definition 7** Let $\Gamma = \{\gamma_1, ..., \gamma_m\}$ be a basis of $\mathbb{F}_{q^m}$ over $\mathbb{F}_q$. The matrix **associated** to a vector $v \in \mathbb{F}_{q^m}^k$ with respect to $\Gamma$ is the $k \times m$ matrix $\Gamma(v)$ with entries in $\mathbb{F}_q$ defined by

$$v_i = \sum_{j=1}^{m} \Gamma(v)_{ij} \gamma_j \quad \text{for all } i = 1, ..., k.$$

The rank-metric code **associated** to a vector rank-metric code $C \subseteq \mathbb{F}_{q^m}^k$ with respect to $\Gamma$ is

$$\Gamma(C) = \{\Gamma(v) \mid v \in C\} \subseteq \mathbb{F}_q^{k \times m}.$$

Notice that in the previous definition the $i$-th row of $\Gamma(v)$ is the expansion of the entry $v_i$ over the basis $\Gamma$.

The proof of the following result is standard and left to the reader.

**Proposition 1** For every $\mathbb{F}_q$-basis $\Gamma$ of $\mathbb{F}_{q^m}$ the map $v \mapsto \Gamma(v)$ is an $\mathbb{F}_q$-linear bijective isometry $(\mathbb{F}_{q^m}^k, d_G) \to (\mathbb{F}_q^{k \times m}, d)$.

In particular, if $C \subseteq \mathbb{F}_{q^m}^k$ is a vector rank-metric code, then $\Gamma(C)$ has the same cardinality, rank distribution and distance distribution as $C$. Moreover, if $|C| \geq 2$ then $d_G(C) = d(\Gamma(C))$.

In the remainder of the section we compare the duality theories of matrix and vector rank-metric codes, showing that the former generalizes the latter. The following results appear in [12].

Given an $\mathbb{F}_{q^m}$-linear vector rank-metric code $C \subseteq \mathbb{F}_{q^m}^k$ and a basis $\Gamma$ of $\mathbb{F}_{q^m}$ over $\mathbb{F}_q$, it is natural to ask whether the codes $\Gamma(C^{\perp})$ and $\Gamma(C)^{\perp}$ coincide or not. The answer is negative in general, as we show in the following example.

*Example 1* Let $q = 3$, $k = m = 2$ and $\mathbb{F}_{3^2} = \mathbb{F}_3[\eta]$, where $\eta$ is a root of the irreducible primitive polynomial $x^2 + 2x + 2 \in \mathbb{F}_3[x]$. Let $\xi = \eta^2$, so that $\xi^2 + 1 = 0$. Set $\alpha = (\xi, 2)$, and let $C \subseteq \mathbb{F}_{3^2}^2$ be the 1-dimensional vector rank-metric code generated by $\alpha$ over $\mathbb{F}_{3^2}$. Take $\Gamma = \{1, \xi\}$ as basis of $\mathbb{F}_{3^2}$ over $\mathbb{F}_3$. One can check that $\Gamma(C)$ is generated over $\mathbb{F}_3$ by the two matrices

$$\Gamma(\alpha) = \begin{bmatrix} 0 & 1 \\ 2 & 0 \end{bmatrix}, \qquad \Gamma(\xi\alpha) = \begin{bmatrix} -1 & 0 \\ 0 & 2 \end{bmatrix}.$$

Let $\beta = (\xi, 1) \in \mathbb{F}_{3^2}^2$. We have $\alpha_1\beta_1 + \alpha_2\beta_2 = 1 \neq 0$, and so $\beta \notin C^\perp$. It follows $\Gamma(\beta) \notin \Gamma(C^\perp)$. On the other hand,

$$\Gamma(\beta) = \begin{bmatrix} 0 & 1 \\ 1 & 0 \end{bmatrix},$$

and it is easy to see that $\Gamma(\beta)$ is trace-orthogonal to both $\Gamma(\alpha)$ and $\Gamma(\xi\alpha)$. Therefore $\Gamma(\beta) \in \Gamma(C)^\perp$, hence $\Gamma(C)^\perp \neq \Gamma(C^\perp)$.

Although the duality notions for matrix and vector rank-metric codes do not coincide, there is a simple relation between them via orthogonal bases of finite fields.

Let Trace : $\mathbb{F}_{q^m} \to \mathbb{F}_q$ be the map defined by $\text{Trace}(\alpha) = \alpha + \alpha^q + \cdots + \alpha^{q^{m-1}}$ for all $\alpha \in \mathbb{F}_{q^m}$. Bases $\Gamma = \{\gamma_1, ..., \gamma_m\}$ and $\Gamma' = \{\gamma_1', ..., \gamma_m'\}$ of $\mathbb{F}_{q^m}$ over $\mathbb{F}_q$ are called **orthogonal** if $\text{Trace}(\gamma_i'\gamma_j) = \delta_{ij}$ for all $i, j \in \{1, ..., m\}$. It is well-known that every basis $\Gamma$ of $\mathbb{F}_{q^m}$ over $\mathbb{F}_q$ has a unique orthogonal basis $\Gamma'$ (see [10], p. 54).

**Theorem 1** *Let* $C \subseteq \mathbb{F}_{q^m}^k$ *be an* $\mathbb{F}_{q^m}$-*linear vector rank-metric code, and let* $\Gamma$, $\Gamma'$ *be orthogonal bases of* $\mathbb{F}_{q^m}$ *over* $\mathbb{F}_q$. *We have*

$$\Gamma'(C^\perp) = \Gamma(C)^\perp.$$

*In particular,* $C$ *has the same weight distribution as* $\Gamma(C)$, *and* $C^\perp$ *has the same weight distribution as* $\Gamma(C)^\perp$.

*Proof* Write $\Gamma = \{\gamma_1, ..., \gamma_m\}$ and $\Gamma' = \{\gamma_1', ..., \gamma_m'\}$. Let $M \in \Gamma'(C^\perp)$ and $N \in \Gamma(C)$. There exist $\alpha \in C^\perp$ and $\beta \in C$ such that $M = \Gamma'(\alpha)$ and $N = \Gamma(\beta)$. By Definition 7 we have

$$0 = \sum_{i=1}^{k} \alpha_i\beta_i = \sum_{i=1}^{k}\sum_{j=1}^{m} M_{ij}\gamma_j' \sum_{t=1}^{m} N_{it}\gamma_t = \sum_{i=1}^{k}\sum_{j=1}^{m}\sum_{t=1}^{m} M_{ij}N_{it}\gamma_j'\gamma_t. \qquad (1)$$

Applying the function Trace : $\mathbb{F}_{q^m} \to \mathbb{F}_q$ to both sides of Eq. (10) we obtain

$$0 = \text{Trace}\left(\sum_{i=1}^{k}\sum_{j=1}^{m}\sum_{t=1}^{m} M_{ij}N_{it}\gamma_j'\gamma_t\right) = \sum_{i=1}^{k}\sum_{j=1}^{m}\sum_{t=1}^{m} M_{ij}N_{it}\text{Trace}(\gamma_j'\gamma_t) = \text{Tr}(MN^t).$$

Therefore $\Gamma'(C^{\perp}) \subseteq \Gamma(C)^{\perp}$. Proposition 1 implies that $\Gamma'(C^{\perp})$ and $\Gamma(C)^{\perp}$ have the same dimension over $\mathbb{F}_q$. Hence the two codes are equal. The second part of the statement follows from Proposition 1.

Theorem 1 shows that the duality theory of $\mathbb{F}_q$-linear rank-metric codes in matrix representation can be regarded as a generalization of the duality theory of $\mathbb{F}_{q^m}$-linear vector rank-metric codes. For this reason, in the sequel we only treat rank-metric codes in matrix representation.

Rank-metric codes can be used to construct several types of subspace codes, a class of error-correcting codes introduced in [8] in the context of random linear network coding. See chapters "Constructions of Constant Dimension Codes", "Constructions of Cyclic Subspace Codes and Maximum Rank Distance Codes", "Generalizing subspace codes to flag codes using group actions", and "Partial spreads and vector space partitions" for constructions of subspace codes, and chapter "Geometrical aspects of subspace codes" for their structural properties.

## 2  MacWilliams Identities for the Rank Metric

This section is devoted to the duality theory of codes endowed with the rank metric. We concentrate on linear rank-metric codes, and show that the weight distributions of a code $\mathscr{C}$ and its dual code $\mathscr{C}^{\perp}$ determine each other via a MacWilliams-type transformation. This result was established by Delsarte in [3, Theorem 3.3] using the machinery of association schemes, and may be regarded as the rank-analogue of a celebrated theorem by MacWilliams on the weight distribution of linear codes endowed with the Hamming metric (see [11]). In this section we present a lattice-theoretic proof inspired by [13, Theorem 27].

**Notation 1** *We denote by* $\mathrm{colsp}(M) \subseteq \mathbb{F}_q^k$ *the* $\mathbb{F}_q$-*space generated by the columns of a matrix* $M \in \mathbb{F}_q^{k \times m}$. *Given a code* $\mathscr{C} \subseteq \mathbb{F}_q^{k \times m}$ *and an* $\mathbb{F}_q$-*subspace* $U \subseteq \mathbb{F}_q^k$, *we let*

$$\mathscr{C}(U) = \{M \in \mathscr{C} \mid \mathrm{colsp}(M) \subseteq U\} \subseteq \mathbb{F}_q^{k \times m}$$

*be the set of matrices in* $\mathscr{C}$ *whose columnspace is contained in* $U$.

Note that for all $M, N \in \mathbb{F}_q^{k \times m}$ we have $\mathrm{colsp}(M + N) \subseteq \mathrm{colsp}(M) + \mathrm{colsp}(N)$. As a consequence, if $U \subseteq \mathbb{F}_q^k$ is an $\mathbb{F}_q$-linear subspace and $\mathscr{C} \subseteq \mathbb{F}_q^{k \times m}$ is a linear code, then $\mathscr{C}(U)$ is a linear code as well.

We start with a series of preliminary results. In the sequel we denote by $U^{\perp}$ the orthogonal (or dual) of an $\mathbb{F}_q$-vector space $U \subseteq \mathbb{F}_q^k$ with respect to the standard inner product of $\mathbb{F}_q^k$. It will be clear from context if by "$\perp$" we denote the trace-dual in $\mathbb{F}_q^{k \times m}$ or the standard dual in $\mathbb{F}_q^k$.

**Lemma 1** *Let* $U \subseteq \mathbb{F}_q^k$ *be a subspace. The following hold.*

1. $\dim(\mathbb{F}_q^{k \times m}(U)) = m \cdot \dim(U)$.

2. $\mathbb{F}_q^{k \times m}(U)^\perp = \mathbb{F}_q^{k \times m}(U^\perp)$.

*Proof*  1. Let $s = \dim(U)$ and $V = \{(x_1, ..., x_k) \in \mathbb{F}_q^k \mid x_i = 0 \text{ for } i > s\} \subseteq \mathbb{F}_q^k$.
There exists an $\mathbb{F}_q$-isomorphism $g : \mathbb{F}_q^k \to \mathbb{F}_q^k$ that maps $U$ to $V$. Let $G \in \mathbb{F}_q^{k \times k}$ be
the invertible matrix associated to $g$ with respect to the canonical basis $\{e_1, ..., e_k\}$
of $\mathbb{F}_q^k$, i.e.,

$$g(e_j) = \sum_{i=1}^k G_{ij} e_i \qquad \text{for all } j = 1, ..., k.$$

The map $M \mapsto GM$ is an $\mathbb{F}_q$-isomorphism $\mathbb{F}_q^{k \times m}(U) \to \mathbb{F}_q^{k \times m}(V)$. Property 1
of the lemma now directly follows from the definition of $\mathbb{F}_q^{k \times m}(V)$.

2. Let $N \in \mathbb{F}_q^{k \times m}(U^\perp)$ and $M \in \mathbb{F}_q^{k \times m}(U)$. Using the definition of trace-product one
sees that $\text{Tr}(MN^t) = \sum_{i=1}^m \langle M_i, N_i \rangle$, where $\langle \cdot, \cdot \rangle$ is the standard inner product of
$\mathbb{F}_q^k$, and $M_i, N_i$ denote the $i$-th column of $M$ and $N$ (respectively). Each column of
$N$ belongs to $U^\perp$, and each column of $M$ belongs to $U$. Therefore $\text{Tr}(MN^t) = 0$,
hence $\mathbb{F}_q^{k \times m}(U^\perp) \subseteq \mathbb{F}_q^{k \times m}(U)^\perp$. By property 1, the two spaces $\mathbb{F}_q^{k \times m}(U^\perp)$ and
$\mathbb{F}_q^{k \times m}(U)^\perp$ have the same dimension over $\mathbb{F}_q$. Therefore they are equal.

The following result is [12, Lemma 28].

**Proposition 2** *Let $\mathscr{C} \subseteq \mathbb{F}_q^{k \times m}$ be a linear code, and let $U \subseteq \mathbb{F}_q^k$ be a subspace of
dimension $u$ over $\mathbb{F}_q$. Then*

$$|\mathscr{C}(U)| = \frac{|\mathscr{C}|}{q^{m(k-u)}} |\mathscr{C}^\perp(U^\perp)|.$$

*Proof* We have $\mathscr{C}(U)^\perp = (\mathscr{C} \cap \mathbb{F}_q^{k \times m}(U))^\perp = \mathscr{C}^\perp + \mathbb{F}_q^{k \times m}(U)^\perp = \mathscr{C}^\perp + \mathbb{F}_q^{k \times m}(U^\perp)$, where
the last equality follows from part 2 of Lemma 1. Therefore

$$|\mathscr{C}(U)| \cdot |\mathscr{C}^\perp + \mathbb{F}_q^{k \times m}(U^\perp)| = q^{km}. \tag{2}$$

On the other hand, part 1 of Lemma 1 gives

$$\dim(\mathscr{C}^\perp + \mathbb{F}_q^{k \times m}(U^\perp)) = \dim(\mathscr{C}^\perp) + m \cdot \dim(U^\perp) - \dim(\mathscr{C}^\perp(U^\perp)).$$

As a consequence,

$$|\mathscr{C}^\perp + \mathbb{F}_q^{k \times m}(U^\perp)| = \frac{q^{km} \cdot q^{m(k-u)}}{|\mathscr{C}| \cdot |\mathscr{C}^\perp(U^\perp)|}. \tag{3}$$

Combining Eqs. (2) and (3) one obtains the proposition.

We will also need the following preliminary lemma, which is an explicit version
of the Möbius inversion formula for the lattice of subspaces of $\mathbb{F}_q^k$. We include a short
proof for completeness. See [19, Sects. 3.7–3.10] for details.

**Lemma 2** Let $\mathscr{P}(\mathbb{F}_q^k)$ be the set of all $\mathbb{F}_q$-subspaces of $\mathbb{F}_q^k$, and let $f : \mathscr{P}(\mathbb{F}_q^k) \to \mathbb{Z}$ be any function. Define $g : \mathscr{P}(\mathbb{F}_q^k) \to \mathbb{Z}$ by $g(V) = \sum_{U \subseteq V} f(U)$ for all $V \subseteq \mathbb{F}_q^k$. Then for all $i \in \{0, ..., k\}$ and for any subspace $V \in \mathscr{P}(\mathbb{F}_q^k)$ with $\dim(V) = i$ we have

$$f(V) = \sum_{u=0}^{i} (-1)^{i-u} q^{\binom{i-u}{2}} \sum_{\substack{U \subseteq V \\ \dim(U)=u}} g(U).$$

*Proof* Fix an integer $i \in \{0, ..., k\}$ and a vector space $V \in \mathscr{P}(\mathbb{F}_q^k)$ with $\dim(V) = i$. We inductively define a function $\mu : \{U \in \mathscr{P}(\mathbb{F}_q^k) \mid U \subseteq V\} \to \mathbb{Z}$ by $\mu(U) = 1$ if $U = V$, and $\mu(U) = -\sum_{U \subsetneq S \subseteq V} \mu(S)$ if $U \subsetneq V$. By definition of $g$ we have

$$\sum_{U \subseteq V} \mu(U) g(U) = \sum_{U \subseteq V} \mu(U) \sum_{S \subseteq U} f(S) = \sum_{S \subseteq V} f(S) \sum_{S \subseteq U \subseteq V} \mu(U) = f(V),$$

where the last equality immediately follows from the definition of $\mu$. Therefore it suffices to show that for all $U \subseteq V$ we have

$$\mu(U) = (-1)^{i-u} q^{\binom{i-j}{2}}, \tag{4}$$

where $u = \dim(U)$. We proceed by induction on $i - u$. If $i = u$ then Eq. (4) is trivial. Now assume $i > u$. By definition of $\mu$ and the induction hypothesis we have

$$
\begin{aligned}
\mu(U) &= -\sum_{U \subsetneq S \subseteq V} \mu(S) = -\sum_{s=u+1}^{i} (-1)^{i-s} q^{\binom{i-s}{2}} \begin{bmatrix} i-j \\ s-u \end{bmatrix}_q \\
&= -\sum_{s=u+1}^{i} (-1)^{i-s} q^{\binom{i-s}{2}} \begin{bmatrix} i-u \\ i-s \end{bmatrix}_q \\
&= -\sum_{s=0}^{i-u} (-1)^{s} q^{\binom{s}{2}} \begin{bmatrix} i-u \\ s \end{bmatrix}_q + (-1)^{i-u} q^{\binom{i-u}{2}} \\
&= (-1)^{i-u} q^{\binom{i-u}{2}},
\end{aligned}
$$

where the last equality follows from the $q$-Binomial Theorem (see [19], p. 74).

We can now prove the main result of this section, first established by Delsarte in [3, Theorem 3.3]. A proof for the special case of $\mathbb{F}_{q^m}$-linear vector rank-metric codes using different techniques can be found in [6].

**Theorem 2** (MacWilliams identities for the rank metric) Let $\mathscr{C} \subseteq \mathbb{F}_q^{k \times m}$ be an linear rank-metric code. For all $i \in \{0, ..., k\}$ we have

$$W_i(\mathscr{C}^\perp) = \frac{1}{|\mathscr{C}|} \sum_{j=0}^{k} W_j(\mathscr{C}) \sum_{u=0}^{k} (-1)^{i-u} q^{mu+\binom{i-u}{2}} \begin{bmatrix} k-u \\ k-i \end{bmatrix}_q \begin{bmatrix} k-j \\ u \end{bmatrix}_q.$$

*Proof* For all subspaces $V \subseteq \mathbb{F}_q^k$ define

$$f(V) = |\{M \in \mathscr{C}^\perp \mid \mathrm{colsp}(M) = V\}|, \qquad g(V) = \sum_{U \subseteq V} f(U) = |\mathscr{C}^\perp(V)|.$$

By Lemma 2, for any $i \in \{0, ..., k\}$ and for any vector space $V \subseteq \mathbb{F}_q^k$ of dimension $i$ we have

$$f(V) = \sum_{u=0}^{i} (-1)^{i-u} q^{\binom{i-u}{2}} \sum_{\substack{U \subseteq V \\ \dim(U)=u}} |\mathscr{C}^\perp(U)|$$

$$= \sum_{u=0}^{i} (-1)^{i-u} q^{\binom{i-u}{2}} \sum_{\substack{T \subseteq \mathbb{F}_q^k \\ T \supseteq V^\perp \\ \dim(T)=k-u}} |\mathscr{C}^\perp(T^\perp)|$$

$$= \frac{1}{|\mathscr{C}|} \sum_{u=0}^{i} (-1)^{i-u} q^{mu+\binom{i-u}{2}} \sum_{\substack{T \subseteq \mathbb{F}_q^k \\ T \supseteq V^\perp \\ \dim(T)=k-u}} |\mathscr{C}(T)|,$$

where the last equality follows from Proposition 2. Now observe that

$$W_i(\mathscr{C}^\perp) = \sum_{\substack{V \subseteq \mathbb{F}_q^k \\ \dim(V)=i}} f(V)$$

$$= \frac{1}{|\mathscr{C}|} \sum_{u=0}^{i} (-1)^{i-u} q^{mu+\binom{i-u}{2}} \sum_{\substack{V \subseteq \mathbb{F}_q^k \\ \dim(V)=i}} \sum_{\substack{T \subseteq \mathbb{F}_q^k \\ T \supseteq V^\perp \\ \dim(T)=k-u}} |\mathscr{C}(T)|$$

$$= \frac{1}{|\mathscr{C}|} \sum_{u=0}^{i} (-1)^{i-u} q^{mu+\binom{i-u}{2}} \sum_{\substack{T \subseteq \mathbb{F}_q^k \\ \dim(T)=k-u}} \sum_{\substack{V \subseteq \mathbb{F}_q^k \\ V \supseteq T^\perp \\ \dim(V)=i}} |\mathscr{C}(T)|$$

$$= \frac{1}{|\mathscr{C}|} \sum_{u=0}^{i} (-1)^{i-u} q^{mu+\binom{i-u}{2}} \begin{bmatrix} k-u \\ i-u \end{bmatrix}_q \sum_{\substack{T \subseteq \mathbb{F}_q^k \\ \dim(T)=k-u}} |\mathscr{C}(T)|. \qquad (5)$$

On the other hand,

$$\sum_{\substack{T\subseteq\mathbb{F}_q^k \\ \dim(T)=k-u}} |\mathscr{C}(T)| = \sum_{\substack{T\subseteq\mathbb{F}_q^k \\ \dim(T)=k-u}} \sum_{j=0}^{k-u} \sum_{\substack{S\subseteq T \\ \dim(S)=j}} |\{M \in \mathscr{C} \mid \mathrm{colsp}(M) = S\}|$$

$$= \sum_{j=0}^{k-u} \sum_{\substack{S\subseteq\mathbb{F}_q^k \\ \dim(S)=j}} \sum_{\substack{T\subseteq\mathbb{F}_q^k \\ T\supseteq S \\ \dim(T)=k-u}} |\{M \in \mathscr{C} \mid \mathrm{colsp}(M) = S\}|$$

$$= \sum_{j=0}^{k-u} \begin{bmatrix} k-j \\ u \end{bmatrix}_q W_j(\mathscr{C}). \qquad (6)$$

Combining Eqs. (5) and (6) one obtains the desired result.

*Example 2* Let $q = 2$, $k = 2$, $m = 3$. Let $\mathscr{C} \subseteq \mathbb{F}_q^{k\times m}$ be the 2-dimensional linear code generated over $\mathbb{F}_5 \cong \mathbb{Z}/5\mathbb{Z}$ by the matrices

$$\begin{bmatrix} 1 & 0 & 2 \\ 0 & 2 & 4 \end{bmatrix}, \quad \begin{bmatrix} 2 & 3 & 0 \\ 1 & 4 & 0 \end{bmatrix}.$$

We have $W_0(\mathscr{C}) = 1$, $W_1(\mathscr{C}) = 8$ and $W_2(\mathscr{C}) = 16$. Using Theorem 2 one computes $W_0(\mathscr{C}^\perp) = 1$, $W_1(\mathscr{C}^\perp) = 65$ and $W_2(\mathscr{C}) = 560$. Observe that $\mathscr{C}^\perp$ has dimension $6 - 2 = 4$, and that $1 + 64 + 560 = 625 = 5^4$, as expected.

We now present a different formulation of the MacWilliams identities for the rank metric. The following result is [12, Theorem 31].

**Theorem 3** *Let $\mathscr{C} \subseteq \mathbb{F}_q^{k\times m}$ be a linear code. For all $0 \le v \le k$ we have*

$$\sum_{i=0}^{k-v} W_i(\mathscr{C}) \begin{bmatrix} k-i \\ v \end{bmatrix}_q = \frac{|\mathscr{C}|}{q^{mv}} \sum_{j=0}^{v} W_j(\mathscr{C}^\perp) \begin{bmatrix} k-j \\ v-j \end{bmatrix}_q.$$

*Proof* Proposition 2 gives

$$\sum_{\substack{U\subseteq\mathbb{F}_q^k \\ \dim(U)=k-v}} |\mathscr{C}(U)| = \frac{|\mathscr{C}|}{q^{mv}} \sum_{\substack{U\subseteq\mathbb{F}_q^k \\ \dim(U)=v}} |\mathscr{C}^\perp(U)|. \qquad (7)$$

Observe that

$$\sum_{\substack{U\subseteq\mathbb{F}_q^k \\ \dim(U)=k-v}} |\mathscr{C}(U)| = |\{(U, M) \mid U \subseteq \mathbb{F}_q^k, \dim(U) = k - v, M \in \mathscr{C}, \mathrm{colsp}(M) \subseteq U\}|$$

$$= \sum_{M \in \mathscr{C}} |\{U \subseteq \mathbb{F}_q^k, \ \dim(U) = k - v, \ \text{colsp}(M) \subseteq U\}|$$

$$= \sum_{i=0}^{k} \sum_{\substack{M \in \mathscr{C} \\ \text{rk}(M)=i}} |\{U \subseteq \mathbb{F}_q^k, \ \dim(U) = k - v, \ \text{colsp}(M) \subseteq U\}|$$

$$= \sum_{i=0}^{k} \sum_{\substack{M \in \mathscr{C} \\ \text{rk}(M)=i}} \begin{bmatrix} k-i \\ k-v-i \end{bmatrix}_q = \sum_{i=0}^{k-v} W_i(\mathscr{C}) \begin{bmatrix} k-i \\ v \end{bmatrix}_q. \tag{8}$$

Using the same argument with $\mathscr{C}^\perp$ and $k - v$ one shows that

$$\sum_{\substack{U \subseteq \mathbb{F}_q^k \\ \dim(U)=v}} |\mathscr{C}^\perp(U)| = \sum_{j=0}^{k-v} W_j(\mathscr{C}^\perp) \begin{bmatrix} k-j \\ v-j \end{bmatrix}_q. \tag{9}$$

The result now follows combining Eqs. (7), (8) and (9).

*Remark 1* The two formulations of the MacWilliams identities for the rank metric given in Theorems 2 and 3 are equivalent. See [6, Corollary 1 and Proposition 3] and [12, Theorem 61] for details.

The next theorem is [2, Theorem 27], and shows that the weight distribution of a linear code is determined by its parameters, together with the number of codewords of small weight. We state it without proof. An application of this result will be given in Sect. 3 (see Corollary 3).

**Theorem 4** *Let $\mathscr{C} \subseteq \mathbb{F}_q^{k \times m}$ be a linear code with $1 \leq \dim(\mathscr{C}) \leq km - 1$, minimum distance $d = d(\mathscr{C})$, and dual minimum distance $d^\perp = d(\mathscr{C}^\perp)$. Let $\varepsilon = 1$ if $\mathscr{C}$ is MRD, and $\varepsilon = 0$ otherwise. For all $1 \leq i \leq d^\perp$ we have*

$$W_{k-d^\perp+i}(\mathscr{C}) = (-1)^i q^{\binom{i}{2}} \sum_{u=d^\perp}^{k-d} \begin{bmatrix} u \\ d^\perp - i \end{bmatrix}_q \begin{bmatrix} u - d^\perp + i - 1 \\ i - 1 \end{bmatrix}_q W_{k-u}(\mathscr{C})$$

$$+ \begin{bmatrix} k \\ d^\perp - i \end{bmatrix}_q \sum_{u=0}^{i-1-\varepsilon} (-1)^u q^{\binom{u}{2}} \begin{bmatrix} k - d^\perp + i \\ u \end{bmatrix}_q \left( q^{\dim(\mathscr{C}) - m(d^\perp - i + u)} - 1 \right).$$

*In particular, $k$, $m$, $t$, $d$, $d^\perp$ and $W_d(\mathscr{C}), \ldots, W_{k-d^\perp}(\mathscr{C})$ completely determine the weight distribution of $\mathscr{C}$.*

We conclude this section showing how MacWilliams identities for the rank metric can be employed to solve certain enumerative problems of matrices over finite fields. The following result is [13, Corollary 52].

**Corollary 1** *Let $I \subseteq \{(i,j) \in \{1, ..., k\} \times \{1, ..., m\} \mid i = j\}$ be a set of diagonal entries. For all $0 \leq r \leq k$ the number of $k \times m$ matrices $M$ over $\mathbb{F}_q$ having rank $r$ and $M_{ij} = 0$ for all $(i,j) \in I$ is*

$$q^{-|I|} \sum_{t=0}^{k} \binom{|I|}{t} (q-1)^t \sum_{u=0}^{k} (-1)^{r-u} q^{mu+\binom{r-u}{2}} \begin{bmatrix} k-u \\ k-r \end{bmatrix}_q \begin{bmatrix} k-t \\ u \end{bmatrix}_q.$$

*Proof* Define the linear code $\mathscr{C} = \{M \in \mathbb{F}_q^{k \times m} \mid M_{ij} = 0 \text{ for all } (i,j) \notin I\} \subseteq \mathbb{F}_q^{k \times m}$. Then $\dim(\mathscr{C}) = |I|$, $W_t(\mathscr{C}) = 0$ for $|I| < t \le k$, and

$$W_t(\mathscr{C}) = \binom{|I|}{t}(q-1)^t$$

for $0 \le t \le |I|$. Moreover, $\mathscr{C}^\perp = \{M \in \mathbb{F}_q^{k \times m} \mid M_{ij} = 0 \text{ for all } (i,j) \in I\}$. Therefore the number of matrices $M \in \mathbb{F}_q^{k \times m}$ having rank $r$ and $M_{ij} = 0$ for all $(i,j) \in I$ is precisely $W_r(\mathscr{C}^\perp)$. The corollary now follows from Theorem 2. $\qquad \blacksquare$

## 3   MRD Codes

In this section we study rank-metric codes that have the largest possible cardinality for their parameters. We start with a Singleton-type bound for the cardinality of a rank-metric code of given minimum distance. A code is called MRD if it attains the bound. We then show that for any admissible choice of the parameters there exists a linear MRD code with those parameters.

In the second part of the section we study general structural properties of MRD codes. We first prove in Theorem 7 that the dual of a linear MRD code is MRD. Then we show in Theorem 8 that the weight distribution of a possibly non-linear MRD code $\mathscr{C} \subseteq \mathbb{F}_q^{k \times m}$ with $0 \in \mathscr{C}$ is determined by $k$, $m$ and $d(\mathscr{C})$. As a corollary, we prove that these three parameters completely determine the distance distribution of any MRD code. Our proofs are inspired by the lattice-theory approach to the weight functions of coding theory proposed in [13, 14].

**Theorem 5** (Singleton-like bound) *Let* $\mathscr{C} \subseteq \mathbb{F}_q^{k \times m}$ *be a rank-metric code with* $|\mathscr{C}| \ge 2$ *and minimum distance* $d$. *Then* $|\mathscr{C}| \le q^{m(k-d+1)}$.

*Proof* Let $\pi : \mathscr{C} \to \mathbb{F}_q^{(k-d+1) \times m}$ denote the projection on the last $k - d + 1$ rows. Since $\mathscr{C}$ has minimum distance $d$, the map $\pi$ is injective. Therefore

$$|\mathscr{C}| = |\pi(\mathscr{C})| \le q^{m(k-d+1)}. \qquad \blacksquare$$

A code is MRD if its parameters attain the Singleton-like bound.

**Definition 8** We say that $\mathscr{C} \subseteq \mathbb{F}_q^{k \times m}$ is an **MRD** code if $|\mathscr{C}| = 1$, or $|\mathscr{C}| \ge 2$ and $|\mathscr{C}| = q^{m(k-d+1)}$, where $d = d(\mathscr{C})$.

We now prove that for any choice of $q, k, m$ and $d$ there exists a linear rank-metric code $\mathscr{C} \subseteq \mathbb{F}_q^{k \times m}$ that attains the bound of Theorem 5. This result was first shown by

Delsarte in [3], and rediscovered independently by Gabidulin in [5] and by Kötter and Kschischang in [8] in the context of linear network coding.

**Theorem 6** *For all $1 \leq d \leq k$ there exists an $\mathbb{F}_{q^m}$-linear vector rank-metric code $C \subseteq \mathbb{F}_{q^m}^k$ with $d_G(C) = d$ and $\dim_{\mathbb{F}_{q^m}}(C) = k - d + 1$. In particular, there exists a linear MRD code $\mathscr{C} \subseteq \mathbb{F}_q^{k \times m}$ with $d(\mathscr{C}) = d$.*

We include an elegant proof for Theorem 6 from [8]. Recall that a **linearized polynomial** $p$ over $\mathbb{F}_{q^m}$ is a polynomial of the form

$$p(x) = \alpha_0 x + \alpha_1 x^q + \alpha_2 x^{q^2} + \cdots + \alpha_s x^{q^s}, \qquad \alpha_i \in \mathbb{F}_{q^m}, \quad i = 0, ..., s.$$

The **degree** of $p$, denoted by $\deg(p)$, is the largest integer $i \geq 0$ such that $\alpha_i \neq 0$. The $\mathbb{F}_{q^m}$-vector space of linearized polynomials over $\mathbb{F}_{q^m}$ of degree at most $s$ is denoted by $\mathrm{Lin}_q(m, s)$. It is easy to see that $\dim_{\mathbb{F}_{q^m}}(\mathrm{Lin}_q(m, s)) = s + 1$.

*Remark 2* The roots of a linearized polynomial $p$ over $\mathbb{F}_{q^m}$ form an $\mathbb{F}_q$-vector subspace of $\mathbb{F}_{q^m}$ (see [10], Theorem 3.50), which we denote by $V(p) \subseteq \mathbb{F}_{q^m}$ in the sequel. Clearly, for any non-zero linearized polynomial $p$ we have $\dim_{\mathbb{F}_q} V(p) \leq \deg(p)$ by the Fundamental Theorem of Algebra.

*Proof (of Theorem 6)* Let $E = \{\beta_1, ..., \beta_k\} \subseteq \mathbb{F}_{q^m}$ be a set of $\mathbb{F}_q$-independent elements. These elements exist as $k \leq m$ by assumption. Define the $\mathbb{F}_{q^m}$-linear map

$$\mathrm{ev}_E : \mathrm{Lin}_q(m, k - d) \rightarrow \mathbb{F}_{q^m}^k, \quad \mathrm{ev}_E(p) = (p(\beta_1), ..., p(\beta_k)) \text{ for } p \in \mathrm{Lin}_q(m, k - d).$$

We claim that $C = \mathrm{ev}_E(\mathrm{Lin}_q(m, k - d)) \subseteq \mathbb{F}_{q^m}^k$ is a vector rank-metric code with the desired properties.

Clearly, $C$ is $\mathbb{F}_{q^m}$-linear. Now let $p \in \mathrm{Lin}_q(m, k - d)$ be a non-zero linearized polynomial, and let $W \subseteq \mathbb{F}_{q^m}$ denote the space generated over $\mathbb{F}_q$ by the evaluations $p(\beta_1), ..., p(\beta_k)$. The polynomial $p$ induces an $\mathbb{F}_q$-linear evaluation map $p : \langle \beta_1, ..., \beta_k \rangle_{\mathbb{F}_q} \rightarrow \mathbb{F}_{q^m}$. The image of $p$ is $W$, and therefore by the rank-nullity theorem we have $\dim_{\mathbb{F}_q}(W) = k - \dim_{\mathbb{F}_q} V(p)$. By Remark 2 we conclude $\dim_{\mathbb{F}_q}(W) \geq k - (k - d) = d$. This shows that $d_G(C) \geq d$. In particular, as $d \geq 1$, the map $\mathrm{ev}_E$ is injective, and the dimension of $C$ is $\dim_{\mathbb{F}_{q^m}}(C) = k - d + 1$. Combining Proposition 1 and Theorem 5 we obtain $d_G(C) = d$.

The second part of the theorem immediately follows from Proposition 1.

The MRD code construction in the proof of Theorem 6 was later generalized by Sheekey in [15], introducing a new class of MRD codes. See chapter "Constructions of Cyclic Subspace Codes and Maximum Rank Distance Codes" for other constructions of MRD codes.

The reminder of the section is devoted to the structural properties of MRD codes. We start with a preliminary result from [14, Chap. 7].

**Lemma 3** *Let $\mathscr{C} \subseteq \mathbb{F}_q^{k \times m}$ be an MRD code with $|\mathscr{C}| \geq 2$ and minimum distance $d$. For all subspaces $U \subseteq \mathbb{F}_q^k$ with $u = \dim(U) \geq d - 1$ we have*

$$|\mathscr{C}(U)| = q^{m(u-d+1)}.$$

*Proof* As in Lemma 1, define $V = \{(x_1, ..., x_k) \in \mathbb{F}_q^k \mid x_i = 0 \text{ for } i > u\} \subseteq \mathbb{F}_q^k$. Let $g : \mathbb{F}_q^k \to \mathbb{F}_q^k$ be an $\mathbb{F}_q$-isomorphism with $f(U) = V$. Denote by $G \in \mathbb{F}_q^{k \times k}$ the matrix associated to $g$ with respect to the canonical basis of $\mathbb{F}_q^k$. Define the rank-metric code $\mathscr{D} = G\mathscr{C} = \{GM \mid M \in \mathscr{C}\}$. Clearly, $\mathscr{D}$ has the same dimension and minimum distance as $\mathscr{C}$. In particular, it is MRD. Observe moreover that $\mathscr{C}(U) = \mathscr{D}(V)$.

Consider the maps

$$\mathscr{D} \xrightarrow{\pi_1} \mathbb{F}_q^{(k-d+1) \times m} \xrightarrow{\pi_2} \mathbb{F}_q^{(k-u) \times m},$$

where $\pi_1$ is the projection on the last $k - d + 1$ coordinates, and $\pi_2$ is the projection on the last $k - u$ coordinates. Since $d(\mathscr{D}) = d$, $\pi_1$ is injective. Since $\mathscr{D}$ is MRD, we have $\log_q(|\mathscr{D}|) = m(k - d + 1)$. Therefore $\pi_1$ is bijective. The map $\pi_2$ is $\mathbb{F}_q$-linear and surjective. Therefore

$$|\pi_2^{-1}(0)| = |\pi_2^{-1}(M)| = q^{m(u-d+1)} \quad \text{for all } M \in \mathbb{F}_q^{(k-u) \times m}.$$

Since $\pi_1$ is bijective and $\pi_2$ is surjective, the map $\pi = \pi_2 \circ \pi_1$ is surjective. Moreover,

$$|\pi^{-1}(0)| = |\pi^{-1}(M)| = q^{m(u-d+1)} \quad \text{for all } M \in \mathbb{F}_q^{(k-u) \times m}.$$

The lemma now follows from the identity $\mathscr{C}(U) = \mathscr{D}(V) = \pi^{-1}(0)$.

We can now show that the dual of a linear MRD code is MRD. The next fundamental result is [13, Theorem 5.5].

**Theorem 7** *Let $\mathscr{C} \subseteq \mathbb{F}_q^{k \times m}$ be a linear MRD code. Then $\mathscr{C}^\perp$ is MRD.*

*Proof* The result is immediate if $\dim(\mathscr{C}) \in \{0, km\}$. Assume $1 \leq \dim(\mathscr{C}) \leq km - 1$, and let $d = d(\mathscr{C})$, $d^\perp = d(\mathscr{C}^\perp)$. Applying Theorem 5 to $\mathscr{C}$ and $\mathscr{C}^\perp$ we obtain

$$\dim(\mathscr{C}) \leq m(k - d + 1), \qquad \dim(\mathscr{C}^\perp) \leq m(k - d^\perp + 1).$$

Therefore $km = \dim(\mathscr{C}) + \dim(\mathscr{C}^\perp) \leq 2mk - m(d + d^\perp) + 2m$, i.e.,

$$d + d^\perp \leq k + 2. \tag{10}$$

Let $U \subseteq \mathbb{F}_q^k$ be any $\mathbb{F}_q$-subspace with $\dim(U) = k - d + 1$. By Proposition 2 we have

$$|\mathscr{C}^\perp(U)| = \frac{|\mathscr{C}^\perp|}{q^{m(d-1)}} |\mathscr{C}(U^\perp)|. \tag{11}$$

Since $\dim(U^{\perp}) = d - 1$, by Lemma 3 we have $|\mathscr{C}(U^{\perp})| = |\mathscr{C}|/q^{m(k-d+1)} = 1$, where the last equality follows from the fact that $\mathscr{C}$ is MRD. Therefore (11) becomes

$$|\mathscr{C}^{\perp}(U)| = \frac{|\mathscr{C}^{\perp}|}{q^{m(d-1)}} = \frac{q^{km}/q^{m(d-1)}}{q^{m(d-1)}} = 1.$$

Since $U$ is arbitrary with $\dim(U) = k - d + 1$, this shows $d^{\perp} \geq k - d + 2$. Using (10) we conclude $d^{\perp} = k - d + 2$. The theorem now follows from

$$\dim(\mathscr{C}^{\perp}) = km - \dim(\mathscr{C}) = km - m(k - d + 1) = m(k - d^{\perp} + 1).$$

The proof of Theorem 7 also shows the following useful characterization of linear MRD codes in terms of their minimum distance and dual minimum distance.

**Proposition 3** *Let $\mathscr{C} \subseteq \mathbb{F}_q^{k \times m}$ be a linear code with $1 \leq \dim(\mathscr{C}) \leq km - 1$. The following are equivalent.*

1. *$\mathscr{C}$ is MRD,*
2. *$\mathscr{C}^{\perp}$ is MRD,*
3. *$d(\mathscr{C}) + d(\mathscr{C}^{\perp}) = k + 2$.*

In the remainder of the section we concentrate on the weight and distance distributions of (possibly non-linear) MRD codes. We start with a result on the weight distribution of MRD codes containing the zero vector (see [14, Theorem 7.46]).

**Theorem 8** *Let $\mathscr{C}$ be an MRD code with $|\mathscr{C}| \geq 2$ and $0 \in \mathscr{C}$. Let $d = d(\mathscr{C})$. Then $W_0(\mathscr{C}) = 1$, $W_i(\mathscr{C}) = 0$ for $1 \leq i \leq d - 1$, and*

$$W_i(\mathscr{C}) = \sum_{u=0}^{d-1} (-1)^{i-u} q^{\binom{i-u}{2}} \begin{bmatrix} k \\ i \end{bmatrix}_q \begin{bmatrix} i \\ u \end{bmatrix}_q + \sum_{u=d}^{i} (-1)^{i-u} q^{\binom{i-u}{2} + m(u-d+1)} \begin{bmatrix} k \\ i \end{bmatrix}_q \begin{bmatrix} i \\ u \end{bmatrix}_q$$

*for $d \leq i \leq k$.*

*Proof* Since $0 \in \mathscr{C}$, we have $W_0(\mathscr{C}) = 1$ and $W_i(\mathscr{C}) = 0$ for $1 \leq i \leq d - 1$. For all subspaces $V \subseteq \mathbb{F}_q^k$ define

$$f(V) = |\{M \in \mathscr{C} \mid \text{colsp}(M) = V\}|, \qquad g(V) = \sum_{U \subseteq V} f(U) = |\mathscr{C}(V)|.$$

Fix $0 \leq i \leq k$ and a vector space $V \subseteq \mathbb{F}_q^k$ of dimension $i$. By Lemma 2 we have

$$f(V) = \sum_{u=0}^{i} (-1)^{i-u} q^{\binom{i-u}{2}} \sum_{\substack{U \subseteq V \\ \dim(U) = u}} g(U).$$

Using Lemma 3 and the fact that $\mathscr{C}$ is MRD with $0 \in \mathscr{C}$ we obtain

$$g(U) = \begin{cases} 1 & \text{if } 0 \le \dim(U) \le d-1, \\ q^{m(u-d+1)} & \text{if } d \le \dim(U) \le k. \end{cases}$$

Therefore

$$f(V) = \sum_{u=0}^{d-1} (-1)^{i-u} q^{\binom{i-u}{2}} \begin{bmatrix} i \\ u \end{bmatrix}_q + \sum_{u=d}^{i} (-1)^{i-u} q^{\binom{i-u}{2}+m(u-d+1)} \begin{bmatrix} i \\ u \end{bmatrix}_q.$$

The result now follows from the identity

$$W_i(\mathscr{C}) = \sum_{\substack{V \subseteq \mathbb{F}_q^k \\ \dim(V)=i}} f(V).$$

Different formulas for the weight distribution of linear MRD codes were obtained in [4] employing elementary methods.

Theorem 8 implies the following [3, Theorem 5.6], which states that the distance distribution of any MRD code is determined by its parameters.

**Corollary 2** *Let $\mathscr{C} \subseteq \mathbb{F}_q^{k \times m}$ be an MRD code with $|\mathscr{C}| \ge 2$ and minimum distance $d$. We have $D_0(\mathscr{C}) = 1$, $D_i(\mathscr{C}) = 0$ for $1 \le i \le d-1$, and*

$$D_i(\mathscr{C}) = \sum_{u=0}^{d-1} (-1)^{i-u} q^{\binom{i-u}{2}} \begin{bmatrix} k \\ i \end{bmatrix}_q \begin{bmatrix} i \\ u \end{bmatrix}_q + \sum_{u=d}^{i} (-1)^{i-u} q^{\binom{i-u}{2}+m(u-d+1)} \begin{bmatrix} k \\ i \end{bmatrix}_q \begin{bmatrix} i \\ u \end{bmatrix}_q,$$

*for $d \le i \le k$.*

*Proof* Fix an $i$ with $d \le i \le k$. For $N \in \mathscr{C}$ define $\mathscr{C} - N = \{M - N \mid M \in \mathscr{C}\}$. By definition of distance distribution we have

$$|\mathscr{C}| \cdot D_i(\mathscr{C}) = |\{(M,N) \in \mathscr{C}^2 \mid \text{rk}(M-N) = i\}| = \sum_{N \in \mathscr{C}} W_i(\mathscr{C} - N).$$

For all $N \in \mathscr{C}$ the code $\mathscr{C} - N$ is MRD. Moreover, $0 \in \mathscr{C} - N$. The result now easily follows from Theorem 8.

Corollary 2 shows in particular that the weight distribution of a linear MRD code is determined by $k$, $m$ and $d(\mathscr{C})$. Recall from Proposition 3 that an MRD code $\mathscr{C} \subseteq \mathbb{F}_q^{k \times m}$ is characterized by the property $d(\mathscr{C}) + d(\mathscr{C}^\perp) = k+2$. We now prove that the weight distribution of a linear code $\mathscr{C}$ with $d(\mathscr{C}) + d(\mathscr{C}^\perp) = k+1$ is determined by $k$, $m$ and $\dim(\mathscr{C})$. The following result is [2, Corollary 28].

**Corollary 3** *Let $\mathscr{C} \subseteq \mathbb{F}_q^{k \times m}$ be a linear rank-metric code with $1 \le \dim(\mathscr{C}) \le km - 1$ and $d(\mathscr{C}) + d(\mathscr{C}^\perp) = k+1$. Then*

$$\dim(\mathscr{C}) \not\equiv 0 \mod m \qquad \text{and} \qquad d(\mathscr{C}) = k - \lceil \dim(\mathscr{C})/m \rceil + 1.$$

*Moreover, for all $d \leq i \leq k$ we have*

$$W_i(\mathscr{C}) = \begin{bmatrix} k \\ i \end{bmatrix}_q \sum_{u=0}^{i-d(\mathscr{C})} (-1)^u q^{\binom{u}{2}} \begin{bmatrix} i \\ u \end{bmatrix}_q \left( q^{\dim(\mathscr{C}) - m(k+u-i)} - 1 \right).$$

*Proof* Assume by contradiction that $\dim(\mathscr{C}) = \alpha m$ for some $\alpha$. Applying Theorem 5 to $\mathscr{C}$ and $\mathscr{C}^\perp$ we obtain

$$d(\mathscr{C}) \leq k - \alpha + 1, \qquad d(\mathscr{C}^\perp) \leq \alpha + 1. \tag{12}$$

By Proposition 3, the two inequalities in (12) are either both equalities, or both strict inequalities. Since $d(\mathscr{C}) + d(\mathscr{C}^\perp) = k + 1$ by assumption, they must be both strict inequalities. Therefore

$$d(\mathscr{C}) \leq k - \alpha, \qquad d(\mathscr{C}^\perp) \leq \alpha,$$

hence $d(\mathscr{C}) + d(\mathscr{C}^\perp) \leq k$, a contradiction. This shows that $\dim(\mathscr{C}) \not\equiv 0 \mod m$.

Now write $\dim(\mathscr{C}) = \alpha m + \beta$ with $1 \leq \beta \leq m - 1$. Applying again Theorem 5 to $\mathscr{C}$ and $\mathscr{C}^\perp$ one finds

$$d(\mathscr{C}) \leq k - \left\lceil \frac{\alpha m + \beta}{m} \right\rceil + 1 = k - \alpha, \qquad d(\mathscr{C}^\perp) \leq k - \left\lceil \frac{km - \alpha m - \beta}{m} \right\rceil = \alpha + 1.$$

Since $d(\mathscr{C}) + d(\mathscr{C}^\perp) = k + 1$, we must have

$$d(\mathscr{C}) = k - \left\lceil \frac{\alpha m + \beta}{m} \right\rceil + 1 = k - \left\lceil \frac{\dim(\mathscr{C})}{m} \right\rceil + 1,$$

as claimed. The last part of the statement follows from Theorem 4.

## 4 Rank-Metric Anticodes

This section is devoted to rank-metric anticodes, i.e., rank-metric codes in which the distance between any two matrices is bounded from above by a given integer $\delta$.

In Theorem 9 we give a bound for the cardinality of a (possibly non-linear) anticode, using a code-anticode-type bound. We also characterize optimal anticodes in terms of MRD codes. Then we show that the dual of an optimal linear anticode is an optimal linear anticode. The main results of this section appear in [12, 14].

**Definition 9** Let $0 \leq \delta \leq k$ be an integer. A **(rank-metric) $\delta$-anticode** is a non-empty subset $\mathscr{A} \subseteq \mathbb{F}_q^{k \times m}$ such that $d(M, N) \leq \delta$ for all $M, N \in \mathscr{A}$. We say that $\mathscr{A}$ is **linear** if it is an $\mathbb{F}_q$-linear subspace of $\mathbb{F}_q^{k \times m}$.

*Example 3* Any $\mathscr{A} \subseteq \mathbb{F}_q^{k \times m}$ with $|\mathscr{A}| = 1$ is a 0-anticode. The ambient space $\mathbb{F}_q^{k \times m}$ is a $k$-anticode. The vector space of $k \times m$ matrices over $\mathbb{F}_q$ whose last $k - \delta$ rows are zero is a linear $\delta$-anticode of dimension $m\delta$.

In the sequel we work with a fixed integer $0 \leq \delta \leq k$. Moreover, for $\mathscr{A}, \mathscr{C} \subseteq \mathbb{F}_q^{k \times m}$ we set $\mathscr{A} + \mathscr{C} = \{M + N \mid M \in \mathscr{A}, N \in \mathscr{C}\}$.

**Theorem 9** *Let $\mathscr{A} \subseteq \mathbb{F}_q^{k \times m}$ be a $\delta$-anticode. Then $|\mathscr{A}| \leq q^{m\delta}$. Moreover, if $\delta \leq k - 1$ then the following are equivalent.*

1. $|\mathscr{A}| = q^{m\delta}$.
2. $\mathscr{A} + \mathscr{C} = \mathbb{F}_q^{k \times m}$ *for some MRD code $\mathscr{C}$ with $d(\mathscr{C}) = \delta + 1$.*
3. $\mathscr{A} + \mathscr{C} = \mathbb{F}_q^{k \times m}$ *for all MRD codes $\mathscr{C}$ with $d(\mathscr{C}) = \delta + 1$.*

*Proof* Let $\mathscr{C} \subseteq \mathbb{F}_q^{k \times m}$ be any MRD code with $d(\mathscr{C}) = \delta + 1$. Such a code exists by Theorem 6. For all $M \in \mathscr{A}$ let $[M] = M + \mathscr{C} = \{M + N \mid N \in \mathscr{C}\}$. Then $[M] \cap [M'] = \emptyset$ for all $M, M' \in \mathscr{A}$ with $M \neq M'$. Moreover, by definition of MRD code we have $|[M]| = |\mathscr{C}| = q^{m(k-\delta)}$ for all $M \in \mathscr{A}$, hence

$$|\mathbb{F}_q^{k \times m}| \geq \left| \bigcup_{M \in \mathscr{A}} [M] \right| = \sum_{M \in \mathscr{A}} |[M]| = |\mathscr{A}| \cdot |\mathscr{C}| = |\mathscr{A}| \cdot q^{m(k-\delta)}.$$

Therefore $|\mathscr{A}| \leq q^{m\delta}$, and equality holds if and only if

$$\mathbb{F}_q^{k \times m} = \bigcup_{M \in \mathscr{A}} [M] = \mathscr{A} + \mathscr{C}.$$

A similar argument shows that properties 1, 2 and 3 are equivalent.

**Definition 10** We say that a $\delta$-anticode $\mathscr{A}$ is (cardinality)-**optimal** if it attains the bound of Theorem 9.

*Remark 3* Example 3 shows the existence of optimal linear $\delta$-anticodes for all choices of $\delta$.

In the remainder of the section we prove that the dual of an optimal linear $\delta$-anticode is an optimal $(k - \delta)$-anticode. The result may be regarded as the analogue of Theorem 7 in the context of rank-metric anticodes. We start with a preliminary result on the weight distribution of MRD codes.

**Lemma 4** *Let $\mathscr{C} \subseteq \mathbb{F}_q^{k \times m}$ be an MRD code with $0 \in \mathscr{C}$, $|\mathscr{C}| \geq 2$ and $d(\mathscr{C}) = d$. Then $W_{d+\ell}(\mathscr{C}) > 0$ for all $0 \leq \ell \leq k - d$.*

*Proof* By Theorem 8, we shall prove the lemma for a given MRD code $\mathscr{C} \subseteq \mathbb{F}_q^{k \times m}$ of our choice with $|\mathscr{C}| \geq 2$, minimum distance $d$, and $0 \in \mathscr{C}$. We will first produce a convenient MRD code with the prescribed properties.

Let $C \subseteq \mathbb{F}_{q^m}^k$ be the vector rank-metric code constructed in the proof of Theorem 6, with evaluation set $E = \{\beta_1, ..., \beta_k\}$ and evaluation map $\mathrm{ev}_E$. Let $\Gamma$ be any basis of $\mathbb{F}_{q^m}$ over $\mathbb{F}_q$. By Proposition 1, $\mathscr{C} = \Gamma(C) \subseteq \mathbb{F}_q^{k \times m}$ is a linear code with $\dim(\mathscr{C}) = m(k - d + 1)$ and the same weight distribution as $C$. In particular, $\mathscr{C}$ is a non-zero linear MRD code with minimum distance $d$.

Now we prove the lemma for the MRD code $\mathscr{C}$ constructed above. Fix $\ell$ with $0 \leq \ell \leq k - d$. Define $t = k - d - \ell$, and let $U \subseteq \mathbb{F}_{q^m}$ be the $\mathbb{F}_q$-subspace generated by $\{\beta_1, ..., \beta_t\}$. If $t = 0$ we set $U$ to be the zero space. By [10], Theorem 3.52,

$$p_U = \prod_{\gamma \in U} (x - \gamma)$$

is a linearized polynomial over $\mathbb{F}_{q^m}$ of degree $t = k - d - \ell \leq k - d$, i.e., by definition, $p_U \in \mathrm{Lin}_q(n, k - d)$. Therefore by Proposition 1 it suffices to prove that $\mathrm{ev}_E(p_U) = (p_U(\beta_1), ..., p_U(\beta_k))$ has rank $d + \ell = k - t$. Clearly, $V(p_U) = U$. In particular we have $\mathrm{ev}_E(p_U) = (0, ..., 0, p_U(\beta_{t+1}), ..., p_U(\beta_k))$. We will show that $p_U(\beta_{t+1}), ..., p_U(\beta_k)$ are linearly independent over $\mathbb{F}_q$. Assume that there exist $a_{t+1}, ..., a_k \in \mathbb{F}_q$ such that $\sum_{i=t+1}^k a_i p_U(\beta_i) = 0$. Then we have $p_U\left(\sum_{i=t+1}^k a_i \beta_i\right) = 0$, i.e., $\sum_{i=t+1}^k a_i \beta_i \in V(p_U) = U$. It follows that there exist $a_1, ..., a_t \in \mathbb{F}_q$ such that $\sum_{i=1}^t a_i \beta_i = \sum_{i=t+1}^k a_i \beta_i$, i.e., $\sum_{i=1}^t a_i \beta_i - \sum_{i=t+1}^k a_i \beta_i = 0$. Since $\beta_1, ..., \beta_k$ are independent over $\mathbb{F}_q$, we have $a_i = 0$ for all $i = 1, ..., k$. In particular $a_i = 0$ for $i = t + 1, ..., k$. Hence $p_U(\beta_{t+1}), ..., p_U(\beta_k)$ are linearly independent over $\mathbb{F}_q$, as claimed.

The following proposition characterizes optimal linear anticodes in terms of their intersection with linear MRD codes.

**Proposition 4** *Assume $0 \leq \delta \leq k - 1$, and let $\mathscr{A} \subseteq \mathbb{F}_q^{k \times m}$ be a linear code with $\dim(\mathscr{C}) = m\delta$. The following are equivalent.*

1. *$\mathscr{A}$ is an optimal $\delta$-anticode.*
2. *$\mathscr{A} \cap \mathscr{C} = \{0\}$ for all non-zero MRD linear codes $\mathscr{C} \subseteq \mathbb{F}_q^{k \times m}$ with $d(\mathscr{C}) = \delta + 1$.*

*Proof* By Theorem 9, it suffices to show that if $\mathscr{A} \cap \mathscr{C} = \{0\}$ for all non-zero MRD linear codes $\mathscr{C} \subseteq \mathbb{F}_q^{k \times m}$ with $d(\mathscr{C}) = \delta + 1$, then $\mathscr{A}$ is a $\delta$-anticode.

By contradiction, assume that $\mathscr{A}$ is not a $\delta$-anticode. Since $\mathscr{A}$ is linear, by definition of $\delta$-anticode there exists $N \in \mathscr{A}$ with $\mathrm{rk}(N) \geq \delta + 1$. Let $\mathscr{D}$ be a non-zero linear MRD code with $d(\mathscr{D}) = \delta + 1$ (see Theorem 6 for the existence of such a code). By Lemma 4 there exists $M \in \mathscr{D}$ with $\mathrm{rk}(M) = \mathrm{rk}(N)$. There exist invertible matrices $A$ and $B$ of size $k \times k$ and $m \times m$, resp., such that $N = AMB$. Define $\mathscr{C} = A\mathscr{D}B = \{APB \mid P \in \mathscr{D}\}$. Then $\mathscr{C} \subseteq \mathbb{F}_q^{k \times m}$ is a non-zero linear MRD code with $d(\mathscr{C}) = \delta + 1$ and such that $N \in \mathscr{A} \cap \mathscr{C}$. Since $\mathrm{rk}(N) \geq \delta + 1 \geq 1$, $N$ is not the zero matrix. Therefore $\mathscr{A} \cap \mathscr{C} \neq \{0\}$, a contradiction.

We conclude the section showing that the dual of an optimal linear anticode is an optimal linear anticode.

**Theorem 10** *Let $\mathscr{A} \subseteq \mathbb{F}_q^{k \times m}$ be an optimal linear $\delta$-anticode. Then $\mathscr{A}^\perp$ is an optimal linear $(k - \delta)$-anticode.*

*Proof* Let $\mathscr{A} \subseteq \mathbb{F}_q^{k \times m}$ be an optimal linear $\delta$-anticode. If $\delta = k$ then the result is trivial. From now on we assume $0 \leq \delta \leq k - 1$. By Definition 10 we have $\dim(\mathscr{A}) = m\delta$, hence $\dim(\mathscr{A}^\perp) = m(k - \delta)$. Therefore by Proposition 4 it suffices to show that $\mathscr{A}^\perp \cap \mathscr{C} = \{0\}$ for all non-zero linear MRD codes $\mathscr{C} \subseteq \mathbb{F}_q^{k \times m}$ with $d(\mathscr{C}) = k - \delta + 1$. Let $\mathscr{C}$ be such a code. Then

$$\dim(\mathscr{C}) = m(k - (k - \delta + 1) + 1) = m\delta < mk.$$

Combining Theorem 7 and Proposition 3 one shows that $\mathscr{C}^\perp$ is a linear MRD code with $d(\mathscr{C}^\perp) = k - (k - \delta + 1) + 2 = \delta + 1$. By Proposition 4 we have $\mathscr{A} \cap \mathscr{C}^\perp = \{0\}$. Since $\dim(\mathscr{A}) + \dim(\mathscr{C}^\perp) = m\delta + m(k - (\delta + 1) + 1) = mk$, we have $\mathscr{A} \oplus \mathscr{C}^\perp = \mathbb{F}_q^{k \times m}$. Therefore $\{0\} = (\mathbb{F}_q^{k \times m})^\perp = (\mathscr{A} \oplus \mathscr{C}^\perp)^\perp = \mathscr{A}^\perp \cap \mathscr{C}$. This shows the theorem.

# References

1. R. Ahlswede, N. Cai, S.-Y.R. Li, R.W. Yeung, Network information flow. IEEE Trans. Inf. Theory **46**(4), 1204–1216 (2000)
2. J. De la Cruz, E. Gorla, H.H. López, A. Ravagnani, Rank distribution of Delsarte codes. Des. Codes Cryptogr. (to appear)
3. P. Delsarte, Bilinear forms over a finite field, with applications to coding theory. J. Comb. Theory A **25**(3), 226–241 (1978)
4. J.G. Dumas, R. Gow, G. McGuire, J. Sheekey, Subspaces of matrices with special rank properties. Linear Algebr. Appl. **433**, 191–202 (2010)
5. E. Gabidulin, Theory of codes with maximum rank distance. Probl. Inf. Transm. **1**(2), 1–12 (1985)
6. M. Gadouleau, Z. Yan, MacWilliams Identities for Codes with the Rank Metric. EURASIP J. Wirel. Commun. Networ. (2008)
7. E. Gorla, A. Ravagnani, Subspace codes from Ferrers diagrams. J. Algebr. Appl. **16**, 7 (2017)
8. R. Kötter, F.R. Kschischang, Coding for errors and erasures in random network coding. IEEE Trans. Inf. Theory **54**(8), 3579–3591 (2008)
9. S.-Y.R. Li, R.W. Yeung, N. Cai, Linear network coding. IEEE Trans. Inf. Theory **49**(2), 371–381 (2003)
10. R. Lidl, H. Niederreiter, *Finite Fields*. Addison-Wesley Publishing Company (1983)
11. F.J. MacWilliams, A theorem on the distribution of weights in a systematic code. Bell Syst. Tech. J. **42**(1), 79–94 (1963)
12. A. Ravagnani, Rank-metric codes and their duality theory. Des. Codes Cryptogr. **80**(1), 197–216 (2016)
13. A. Ravagnani, Duality of codes supported on regular lattices, with an application to enumerative combinatorics. Submitted. Online preprint: https://arxiv.org/abs/1510.02383
14. A. Ravagnani, Properties and Constructions of Codes with the Rank and the Subspace Metric. PhD thesis, Université de Neuchâtel, 2016
15. J. Sheekey, A new family of MRD codes. Adv. Math. Commun. **10**, 3 (2016)
16. D. Silva, F.R. Kschishang, On metrics for error correction in network coding. IEEE Trans. Inf. Theory **55**(12), 5479–5490 (2009)

17. D. Silva, F.R. Kschischang, On metrics for error correction in network coding. IEEE Trans. Inf. Theory **55**(12), 5479–5490 (2009)
18. D. Silva, E.S. Rogers, F.R. Kschishang, R. Koetter, A Rank-Metric Approach to Error Control in Random Network Coding. IEEE Trans. Inf. Theory **54**(9), 3951–3967 (2008)
19. P. Stanley, *Enumerative Combinatorics*, vol. 1, Cambridge Stud. Adv. Math., vol. 49. Cambridge University Press (2012)

# Constructions of Constant Dimension Codes

**Anna-Lena Horlemann-Trautmann and Joachim Rosenthal**

**Abstract** Constant dimension codes are subsets of the finite Grassmann variety. The subspace distance is a natural metric on the Grassmannian. It is desirable to have constructions of codes with large cardinality and efficient decoding algorithm for all parameters. This article provides a survey on constructions of constant dimension codes with a prescribed minimum distance. The article starts with a review of geometric properties of the Grassmann variety. We emphasize the classical Plücker embedding which shows that the Grassmannian describing all $k$-dimensional subspaces of a vector space can be naturally embedded in a projective space and hence has the structure of a projective variety itself. On the construction side we first concentrate on the construction of equidistant codes, which include spreads and partial spreads. Then we review constructions based on matrix codes equipped with the rank metric, in particular those matrices whose non-zero entries constitute certain Ferrers diagrams. A large part of the survey is then concerned with orbit codes, i.e., subsets of the Grassmannian that can be expressed as the orbit under an action of an algebraic group. Finally we review several constructions of constant dimension codes starting with another constant dimension code having good parameters. We conclude by giving references to other results related to constant dimension codes and references to results on constructions of mixed dimension codes.

A.-L. Horlemann-Trautmann (✉)
Faculty of Mathematics and Statistics, University of St. Gallen,
Bodanstrasse 6, 9000 St. Gallen, Switzerland
e-mail: anna-lena.horlemann@unisg.ch

J. Rosenthal
Mathematics Institute, University of Zurich, Winterthurerstr 190,
8057 Zurich, Switzerland
e-mail: rosen@math.uzh.ch

© Springer International Publishing AG 2018
M. Greferath et al. (eds.), *Network Coding and Subspace Designs*,
Signals and Communication Technology,
https://doi.org/10.1007/978-3-319-70293-3_2

# 1  Introduction

In this article we give an overview of general constructions of constant dimension codes, also called Grassmannian codes.

The framework for these codes and their usage for error-correction in random (or non-coherent) network coding was first given in [42]. There a *subspace code* was defined as a set of subspaces of $\mathbb{F}_q^n$, $\mathbb{F}_q$ the finite field of $q$ elements, equipped with the *subspace distance*, defined as

$$d_S(\mathcal{U}, \mathcal{V}) := \dim(\mathcal{U} + \mathcal{V}) - \dim(\mathcal{U} \cap \mathcal{V}) \tag{1}$$

for $\mathcal{U}, \mathcal{V} \subseteq \mathbb{F}_q^n$. For simplicity, one often restricts to *constant dimension codes*, where all codewords have the same fixed dimension $k$. Thus, a constant dimension code is a subset $\mathcal{C}$ of the Grassmann variety $\mathrm{Grass}(k, \mathbb{F}_q^n)$, which we will also denote by $\mathcal{G}_q(k, n)$. Because of this constant dimension codes are also called *Grassmannian codes* in the literature.

As in the classical literature on coding theory one defines the minimum distance $\mathrm{dist}(\mathcal{C})$ of a code $\mathcal{C}$ as the minimal distance of two different code elements; in this case the minimal subspace distance between two subspaces. The goal is once more the construction of subspace codes with many code elements and prescribed minimum distance. For the most part of this paper we will focus on constructions of constant dimension codes; we give a short overview of mixed dimension code constructions in the end.

The paper is structured in the following way. In Sect. 2 we will collect some background material on the geometry of the finite Grassmann variety, which serves as the ambient space for constant dimension codes.

In Sect. 3 we will cover spread codes, partial spread codes and equidistant codes. Spread and partial spread codes are characterized by the property that they have the maximal possible minimum distance $2k$. Spread codes can be defined if $k$ divides $n$; they have maximal cardinality among all codes with this minimum distance. A generalization of (partial) spreads are equidistant codes, where any two codewords must have the same fixed distance, but not necessarily distance $2k$.

One of the most successful ways to construct constant dimension codes with large minimum distance is through a lifting process starting with some rank-metric code of a given minimum distance. In Sect. 4 we will survey some of these constructions.

The general linear group $\mathrm{GL}_n$ acts naturally and transitively on the Grassmannian. In Sect. 5 we will give an overview on constant dimension codes which are also orbits of some subgroup $G$ of $\mathrm{GL}_n$ acting on the Grassmannian. Like classical linear codes the class of orbit codes have a rich mathematical structure.

As it is the case for the construction of classical linear codes inside $\mathbb{F}_q^n$ there are also several known techniques to come up with constructions of constant dimension codes from known constant dimension codes of smaller parameters. These techniques are surveyed in Sect. 6.

The focus of our survey is on algebraic constructions of constant dimension codes. Constructions from combinatorial designs will be covered in chapter "q-analogs of Designs: Subspace Designs" of this book. We also do not treat codes which were derived by computer searches, but we give some references in this regard in the final remarks of this paper.

## 2 Remarks on the Geometry of the Finite Grassmann Variety

Assume $k < n$ are two integers. The set of all $k$-dimensional subspaces in $\mathbb{F}_q^n$ is called the *Grassmann variety* or *Grassmannian* which we will denote by $\mathcal{G}_q(k, n)$. As the name indicates the Grassmannian has a variety structure. To be more precise the set $\mathcal{G}_q(k, n)$ can be embedded into a projective space using the Plücker embedding:

$$\varphi : \mathcal{G}_q(k, n) \longrightarrow \mathbb{P}(\Lambda^k \mathbb{F}_q^n) = \mathbb{P}^{\binom{n}{k}-1}$$
$$\mathrm{span}(u_1, \ldots, u_k) \longmapsto \mathbb{F}_q(u_1 \wedge \ldots \wedge u_k).$$

The image of the map $\varphi$ is described by a set of quadratic equations called *shuffle relations*. In order to make this more precise we will introduce a basis and coordinate functions. For this assume $\{e_1, \ldots, e_n\}$ is the canonical basis of $\mathbb{F}_q^n$. Then $N^k \mathbb{F}_q^n$ has as a canonical basis

$$\{e_{i_1} \wedge \ldots \wedge e_{i_k} \mid 1 \le i_1 < \ldots < i_k \le n\}.$$

If an element of the Grassmannian $\mathcal{G}_q(k, n)$ is described by a full rank $k \times n$ matrix $U$ then an elementary computation shows that the Plücker coordinates of a subspace $\mathrm{rs}(U)$ inside the projective space $\mathbb{P}^{\binom{n}{k}-1}$ are exactly the $k \times k$ full size minors of the $k \times n$ matrix $U$.

In the sequel denote by $U[i_1, \ldots, i_k]$ the submatrix of $U$ given by the columns $i_1, \ldots, i_k$. The numbers $x_{i_1, \ldots, i_k} := \det(U[i_1, \ldots, i_k])$ are then called the *Plücker coordinates* of the subspace $\mathrm{rs}(U)$. In terms of Plücker coordinates the Plücker embedding is simply described through

$$\mathrm{rs}(U) \longmapsto (x_{1,\ldots,k} : x_{1,\ldots,k-1,k+1} : \ldots : x_{n-k+1,\ldots,n}).$$

A vector inside the projective space $\mathbb{P}^{\binom{n}{k}-1}$ is in the image of the Plücker map $\varphi$ exactly when the coordinates satisfy the shuffle relations as formulated in the following proposition:

**Proposition 1** ([40, 47]) *Consider* $x := [x_{1,\ldots,k} : \cdots : x_{n-k+1,\ldots,n}] \in \mathbb{P}^{\binom{n}{k}-1}$. *Then there exists a subspace* $\mathcal{U} \in \mathcal{G}_q(k, n)$ *such that* $\varphi(\mathcal{U}) = x$ *if and only if*

$$\sum_{j \in \{i_1,\dots,i_{k+1}\}} \mathrm{sgn}(\sigma_j) x_{i_1,\dots,i_{k+1}\backslash j} x_{j,i_{k+2},\dots,i_{2k}} = 0 \tag{2}$$

$\forall (i_1, \dots, i_{k+1}) \in \binom{[n]}{k+1}$, $(i_{k+2}, \dots, i_{2k}) \in \binom{[n]}{k-1}$, where $\mathrm{sgn}(\sigma_j)$ denotes the sign of the permutation such that

$$\sigma_{i_\ell}(i_1, \dots, i_{k+1}) = (i_\ell, i_1, \dots, i_{\ell-1}, i_{\ell+1}, \dots, i_{k+1}).$$

Note that the shuffle relations (2) form a set of quadratic equations in terms of the Plücker coordinates, and these equations describe exactly the image of the Plücker map $\varphi$. It follows that the Grassmannian $\mathcal{G}_q(k, n)$ has the structure of a projective variety.

If one considers all subspaces rs($U$) whose first Plücker coordinate $x_{1,\dots,k}$ is nonzero one sees that this is a Zariski open set isomorphic to $\mathbb{F}_q^{k(n-k)}$, in particular the dimension of $\mathcal{G}_q(k, n)$ is equal to $k(n-k)$.

The above dimension calculation is true over any base field $\mathbb{F}$. Over a finite field $\mathcal{G}_q(k, n)$ is a finite set and it is well-known that the cardinality of the Grassmannian is given by the Gaussian binomial, i.e.,

$$|\mathcal{G}_q(k, n)| = \binom{n}{j}_q = \prod_{i=0}^{k-1} \frac{q^n - q^i}{q^k - q^i}.$$

For the purpose of coding theory it is important to understand the set of all subspaces $V \in \mathcal{G}_q(k, n)$ whose subspace distance (1) to a given subspace $U$ is at most $t$. In other words we are interested in the geometric structure of the ball of radius $t$ around a subspace $U$:

$$\mathcal{B}_t := \{V \in \mathcal{G}_q(k, n) \mid d_S(U, V) \le t\}.$$

The subsets $\mathcal{B}_t$ have classically been studied in the algebraic geometry literature and as a matter of fact $\mathcal{B}_t$ is a special case of a so called *Schubert variety*.

In order to make this notion more precise one considers a so called *flag* $\mathcal{F}$, this is a sequence of nested linear subspaces

$$\mathcal{F}: \ \{0\} \subset V_1 \subset V_2 \subset \dots \subset V_n = \mathbb{F}_q^n,$$

where one requires that $\dim V_j = j$ for $j = 1, \dots, n$. Denote by $\nu = (\nu_1, \dots, \nu_k)$ an ordered index set satisfying

$$1 \le \nu_1 < \dots < \nu_k \le n.$$

For a given flag $\mathcal{F}$ one defines the Schubert variety

$$S(\nu; \mathcal{F}) := \{W \in \mathcal{G}_q(k, n) \mid \dim(W \cap V_{\nu_i}) \ge i \ \text{ for } i = 1, \dots, k\}. \tag{3}$$

The Schubert varieties are sub-varieties of the Grassmannian $\mathcal{G}_q(k, n)$. The defining equations of the Schubert variety $S(\nu; \mathcal{F})$ in terms of Plücker coordinates are on one side the defining shuffle relations (2) describing the Grassmannian $\mathcal{G}_q(k, n)$. In addition a set of linear equations (see [40]) have to be satisfied. These equations simplify if one wants to describe the ball $\mathcal{B}_t$ only and the interested reader will find more details in [48].

Using the natural inner product inside the vector space $\mathbb{F}_q^n$ one has a natural duality between a $k$-dimensional subspace $\mathcal{U} \in \mathcal{G}_q(k, n)$ and the $(n - k)$-dimensional subspace $\mathcal{U}^\perp \in \mathcal{G}_q(n - k, n)$. This duality induces a duality between the Grassmann variety $\mathcal{G}_q(k, n)$ and the Grassmannian $\mathcal{G}_q(n - k, n)$, in particular both these Grassmannians are isomorphic. In case $\mathcal{C} \subseteq \mathcal{G}_q(k, n)$ is a subspace code it was pointed out in [42, Sect. 3.A] that the dual code $\mathcal{C}^\perp \subseteq \mathcal{G}_q(n - k, n)$ has the same cardinality and minimum subspace distance as $\mathcal{C}$. For the purpose of studying codes it is therefore enough to restrict the study to codes in $\mathcal{G}_q(k, n)$ with $n \geq 2k$.

Grassmannians $\mathcal{G}_q(k, n)$ are also *homogeneous spaces*. For this note that the general linear group $\mathrm{GL}_n$ acts naturally and transitively on $\mathcal{G}_q(k, n)$ through 'right multiplication':

$$\mathcal{G}_q(k, n) \times \mathrm{GL}_n \to \mathcal{G}_q(k, n)$$
$$(\mathrm{rs}(U), A) \mapsto \mathrm{rs}(UA).$$

As a consequence the tangent space at every point of $\mathcal{G}_q(k, n)$ is isomorphic and it follows that $\mathcal{G}_q(k, n)$ is also a smooth variety. If one restricts above $\mathrm{GL}_n$-action to some subgroup $G \subseteq \mathrm{GL}_n$ then one obtains a partition of $\mathcal{G}_q(k, n)$ into $G$-orbits. In terms of coding theory a particular $G$-orbit defines a so called *orbit code*, which we will say more about in Sect. 5.

# 3 Spread Codes, Partial Spread Codes and Equidistant Codes

The maximal possible distance two elements $\mathcal{U}, \mathcal{V} \in \mathcal{C} \subseteq \mathcal{G}_q(k, n)$ of a constant dimension code can have is $d_S(\mathcal{U}, \mathcal{V}) = 2k$ and this happens exactly when

$$\mathcal{U} \cap \mathcal{V} = \{0\}.$$

As a consequence the maximal possible minimum distance a constant dimension code $\mathcal{C} \subseteq \mathcal{G}_q(k, n)$ can have is $2k$.

The natural question hence arises what the maximal cardinality of non-intersecting $k$-dimensional subspaces in $\mathbb{F}_q^n$ can be?

When $n$ is a multiple of $k$, i.e., $n = rk$ for some natural number $r$, then it is a well known result in finite geometry [34] that one has a collection of constant dimensional subspaces in $\mathbb{F}_q^n$ having the property that any two subspaces intersect only in a trivial manner and in addition every nonzero vector $v \in \mathbb{F}_q^n$ lies in one and only one subspace of this collection.

**Definition 2** Let $n = rk$. Then a set $\mathcal{S} \subseteq \mathcal{G}_q(k, n)$ is called a *spread* if all elements of $\mathcal{S}$ intersect only trivially and they cover the whole space $\mathbb{F}_q^n$.

One immediately computes the cardinality of a spread as

$$\frac{q^n - 1}{q^k - 1} = q^{k(r-1)} + q^{k(r-2)} + \cdots + q^k + 1.$$

i.e., a spread in $\mathcal{G}_q(k, n)$ is a constant dimension code with minimum distance $2k$ and cardinality $(q^n - 1)/(q^k - 1)$. In [44] constant dimension codes which also have the structure of a spread were called *spread codes*. A concrete construction of spread codes can be obtained in the following way:

Let $P$ be the $k \times k$ companion matrix of a monic and irreducible polynomial $p(x) \in \mathbb{F}_q[x]$ of degree $k$. It is then well known (see e.g. [43]) that the $\mathbb{F}_q$-algebra $\mathbb{F}_q[P] \subset \mathbb{F}_q^{k \times k}$ is isomorphic to the finite field $\mathbb{F}_{q^k}$. Denote by $0_k, I_k \in \mathbb{F}_q^{k \times k}$ the zero and the identity matrix respectively. Then one has the following theorem which was proven in [44]:

**Theorem 3** *Let $n = rk$. The collection of subspaces*

$$\mathcal{S} := \bigcup_{i=1}^{r} \left\{ \mathrm{rs} \left[ 0_k \ \cdots \ 0_k \ I_k \ A_{i+1} \ \cdots \ A_r \right] \mid A_{i+1}, \ldots, A_r \in \mathbb{F}_q[P] \right\} \subset \mathcal{G}_q(k, n)$$

*is a spread code with minimum distance $2k$ and cardinality $(q^n - 1)/(q^k - 1)$.*

Spreads are similar to classical MDS codes in the sense that they have maximal possible minimum distance and at the same time maximal cardinality.

In [27] the authors studied the situation when $k$ does not divide $n$. For these cases spreads cannot exist but the construction of Theorem 3 can be done 'up to the last block' and this then leads to the notion of a *partial spread code*. The following definition was inspired by work in finite geometry [7]:

**Definition 4** ([27, 28]) A *partial spread code* in $\mathcal{G}_q(k, n)$ is a subspace code $\mathcal{C} \subset \mathcal{G}_q(k, n)$ having minimum distance $\mathrm{dist}(\mathcal{C}) = 2k$.

In [27] a concrete construction of a partial spread code is provided whose cardinality is maximal among all subspace codes of minimum distance $2k$ inside $\mathcal{G}_q(k, n)$.

For more information on partial spreads the reader is referred to chapters "Geometrical Aspects Of Subspace Codes" and "Partial Spreads And Vector Space Partitions" of this book.

Spread codes and partial spread codes are special cases of *equidistant subspace codes* – these are codes where any two distinct code words have the same distance:

**Definition 5** A code $\mathcal{C} \subseteq \mathcal{G}_q(k, n)$ is an *equidistant code* if for any $\mathcal{U}, \mathcal{V} \in \mathcal{C}$ $d_S(\mathcal{U}, \mathcal{V}) = d$ for some fixed $d$.

Note that for (partial) spread codes any two elements have distance $d = 2k$.

A special class of equidistant codes is the one of *sunflowers*, where all codewords intersect in the same subspace of $\mathbb{F}_q^n$. When the ambient vector space $\mathbb{F}_q^n$ is large it was shown in [15] that an equidistant code of maximal cardinality is a sunflower. The paper [28] provides an almost complete classification in case the ground field is sufficiently large. The authors show that for most parameters an equidistant code of maximal cardinality with minimum subspace distance less than $2k$ is either a sunflower or its dual is a sunflower.

Any partial spread gives rise to a sunflower in a larger ambient space, which gives rise to the following result [28, Theorem 34]:

**Theorem 6** *Let $c \leq k - 1$ be an integer. Write $n - c = h(k - c) + r$, with $0 \leq r \leq k - c - 1, h \geq 2$. Then there exist an equidistant code in $\mathcal{G}_q(k, n)$ with minimum distance $2k - 2c$ and cardinality $(q^{n-c} - q^r)/(q^{k-c} - 1) - q^r + 1$.*

The following result about equidistant codes follows from [30, Theorem 1]. The corresponding codes are constructed from osculating subspaces of Veronese varieties.

**Theorem 7** *Let $\ell, s, t$ be integers such that $t/2 \leq \ell \leq t$, and define $k := \binom{\ell+s}{s}, n := \binom{t+s}{s}$. Then there exist an equidistant code in $\mathcal{G}_q(k, n)$ with minimum distance $2k - 2\binom{2\ell-t+s}{s}$ and cardinality $(q^{s+1} - 1)/(q - 1)$.*

Furthermore, one can construct non-sunflower equidistant codes with minimum distance $2k - 2$ with the help of the Plücker embedding:

**Theorem 8** *[15, Theorem 15] For every $k \geq 2$, there exists an equidistant code in $\mathcal{G}_q(k, \binom{k+1}{2})$ with minimum distance $2k - 2$ and cardinality $(q^{k+1} - 1)/(q - 1)$.*

More examples of equidistant codes which are different from sunflowers can be found in [4].

# 4 Constructions Based on (Ferrers Diagram) Rank-Metric Codes

Some of the first constructions for subspace codes were based on rank-metric codes. These codes are sets of matrices of a fixed size over $\mathbb{F}_q$, equipped with the rank metric, which is defined as

$$d_R(A, B) := \text{rk}(A - B), \quad A, B \in \mathbb{F}_q^{m \times n}.$$

The usefulness of rank-metric codes for the construction of subspace codes is due to the following fact, which follows from [52, Proposition 4]:

**Lemma 9** *Let $C \subseteq \mathbb{F}_q^{k \times (n-k)}$ be a code with minimum rank distance $d$. Then the lifted code*

$$\mathrm{lift}(C) := \{\mathrm{rs}[I_k \mid A] \mid A \in C\}$$

*is a constant dimension code in $\mathcal{G}_q(k, n)$ with minimum subspace distance 2d and cardinality $|C|$.*

Note that geometrically the set of subspaces of the form $\mathrm{rs}[I_k \mid A]$ inside the Grassmannian $\mathcal{G}_q(k, n)$ represents a subset isomorphic to the affine space $\mathbb{F}_q^{k \times (n-k)}$. Geometers also call this subset the 'thick open cell' of the Grassmannian $\mathcal{G}_q(k, n)$.

Constructions of optimal rank-metric codes are known for any set of parameters (see e.g. [21]), the corresponding codes are called *maximum rank distance (MRD) codes*. An MRD code $C \subseteq \mathbb{F}_q^{m \times n}$ with minimum rank distance $d$ has cardinality $q^{\max(m,n)(\min(m,n)-d+1)}$. Therefore, a lifted MRD code $C \subseteq \mathcal{G}_q(k, n)$ with $k \leq n/2$ and minimum subspace distance 2d has cardinality $q^{(n-k)(n-d+1)}$.

For more information on partial spreads the reader is referred to chapters "Geometrical Aspects Of Subspace Codes" and "Partial Spreads And Vector Space Partitions" of this book.

In the following we need the notion of an *identifying vector* $v(\mathcal{U})$ of a subspace $\mathcal{U} \in \mathcal{G}_q(k, n)$, which is defined as the vector of length $n$ that has a one in the coordinates where the reduced echelon form of $\mathcal{U}$ has a pivot, and zeros elsewhere. It was shown in [17, Lemma 2] that for $\mathcal{U}, \mathcal{V} \in \mathcal{G}_q(k, n)$ one has

$$d_S(\mathcal{U}, \mathcal{V}) \geq d_H(v(\mathcal{U}), v(\mathcal{V})).$$

With this we can generalize the lifted MRD construction, as follows.

**Theorem 10** [53, Theorem 2.9] *Let $C_j$ be some MRD code with minimum rank distance d in $\mathbb{F}_q^{k \times (n-k-jd)}$ for $j = 0, \ldots, \lfloor \frac{n-k}{d} \rfloor$. Define*

$$C_j := \left\{ \mathrm{rs}\left[ 0_{k \times jd} \ I_{k \times k} \ A \right] \mid A \in C_j \right\} \quad \text{and} \quad C := \bigcup_{j=0}^{\lfloor \frac{n-k}{d} \rfloor} C_j.$$

*Then $C \subseteq \mathcal{G}_q(k, n)$ has minimum distance 2d and cardinality*

$$N = \sum_{i=0}^{\lfloor \frac{n-2k}{d} \rfloor} q^{(k-d+1)(n-k-di)} + \sum_{i=\lfloor \frac{n-2k}{d} \rfloor + 1}^{\lfloor \frac{n-k}{d} \rfloor} \lceil q^{k(n-k+1-d(i+1))} \rceil.$$

Note that similar statements to the previous theorem can also be found in [20, 22].

One can generalize the lifting idea to general reduced row echelon forms of matrices, where the unit column vectors are not necessarily in the first $k$ columns. This idea was introduced in [16]. First, let us briefly provide some definitions needed for this construction. A *Ferrers diagram* $\mathcal{F}$ is a collection of dots such that the rows have a decreasing and the columns have an increasing number of dots (from top to bottom and from left to right, respectively). We denote the matrix representation of a vector space $\mathcal{U} \in \mathcal{G}_q(k, n)$ in reduced row echelon form by $\text{RE}(\mathcal{U}) \in \mathbb{F}_q^{k \times n}$.

**Definition 11** The *Ferrers diagram form* of a subspace $\mathcal{U} \in \mathcal{G}_q(k, n)$, denoted by $\mathcal{F}(\mathcal{U})$, is obtained from $\text{RE}(\mathcal{U})$ by first removing the zeros to the left of the leading coefficient from each row of $\text{RE}(\mathcal{U})$, and then removing the columns which contain the leading ones. All the remaining entries are shifted to the right. The *Ferrers diagram* of $\mathcal{U}$, denoted by $\mathcal{F}_{\mathcal{U}}$, is obtained from $\mathcal{F}(\mathcal{U})$ by replacing the entries of $\mathcal{F}(\mathcal{U})$ with dots.

**Definition 12** Let $\mathcal{F}$ be a Ferrers diagram with $k$ rows and $n$ columns. A corresponding *Ferrers diagram code* is a matrix code $\mathcal{F} \subseteq \mathbb{F}_q^{k \times n}$, such that all matrix entries not in $\mathcal{F}$ are zero in all codewords. The minimum rank distance of a Ferrers diagram code is defined as usual. Similarly, the lifting of a Ferrers diagram code is defined analogously to the lifting of a rectangular rank-metric code, with a corresponding reduced row echelon form.

We can now state the multi-level construction:

**Theorem 13** *Let $\mathbb{C} \subseteq \mathbb{F}_2^n$ be a binary block code of constant weight $k$ and minimum Hamming distance $2d$. Use each codeword $v_i \in \mathbb{C}$ as the identifying vector of a reduced row echelon form and construct the corresponding lifted Ferrers diagram code $\mathcal{C}_i$ with minimum rank distance $d$. Then the constant dimension code $\mathcal{C} = \bigcup_{i=1}^{|\mathbb{C}|} \mathcal{C}_i \subseteq \mathcal{G}_q(k, n)$ has minimum subspace distance $2d$.*

The size of these codes depends mainly on the size of the corresponding Ferrers diagram codes, for which we do not have a general construction. In [17, Theorem 1] an upper bound on the cardinality of linear Ferrers diagram codes with a prescribed minimum rank distance is given. Moreover, code constructions for special types of Ferrers diagrams attaining this bound are given. The authors conjecture that the bound is always attainable, for any set of parameters. More constructions of codes for certain Ferrers diagrams were derived in [14, 51].

It is an open question, which choice of the identifying vectors is optimal for the multi-level construction. It was suggested in [17] that lexicographic binary constant weight codes are the best choice. However, a counterexample for this statement is given in [29, Example 57]. Furthermore, related results to lexicographic codes and corresponding constant dimension codes found by computer search are presented in [50].

As a next step, it was shown in [56] that one can construct larger codes than with the multi-level construction, if one allows *pending dots* in the construction. We will explain this pending dots construction in the following.

**Definition 14** Let $\mathcal{F}$ be a Ferrers diagram. A dot of $\mathcal{F}$ is called a *pending dot* if the maximal size of a Ferrers diagram code for $\mathcal{F}$ is the same as the maximal size of a Ferrers diagram code for $\mathcal{F}$ without that dot.

We can now state the pending dots construction:

**Theorem 15** *Construct a constant weight-k code $\mathbb{C} \subseteq \mathbb{F}_2^n$ of identifying vectors as follows:*

1. *Let $v_1 = (11...10...0)$ be the first identifying vector in $\mathbb{C}$.*
2. *Choose the next identifying vector such that $d_H(v_i, v_1) \geq 2d$ and fix the set of pending dots (if there are any) of the corresponding Ferrers diagram as $\mu_1$.*
3. *For the next identifying vector choose the first 1 in the same positions as before and use the next lexicode element of distance $\geq 2d - 2$ from the other elements with the same pending dots and $\geq 2d$ from any other element of $\mathbb{C}$. Fix the pending dots as a tuple $\mu_i$ different from the tuples already used for echelon-Ferrers forms, where the Hamming distance of the identifying vectors is $2d - 2$.*
4. *Repeat step 3 until no possibilities for a new skeleton code word with the fixed 1 are left.*
5. *In the skeleton code choose the next vector in lexicographic order that has distance $\geq 2d$ from all other skeleton code words and repeat steps 2,3,4 and 5.*

*Then construct a lifted Ferrers diagram code for all identifying vectors, with the given assignments of the pending dots in the Ferrers diagram. The union of all these codes is a constant dimension code $\mathcal{C} \subseteq \mathcal{G}_q(k, n)$ with minimum subspace distance $2d$.*

One of the disadvantages of the multi-level and the pending dots construction is, that there is no closed formula for the code cardinalities in general. Therefore, in [18, 51] the authors used the pending dots construction (or a slight generalization called the *pending blocks construction*) to derive codes with a closed cardinality formula for variable length $n$. To do so they gave a general construction of the constant weight code for the identifying vectors and for the corresponding Ferrers diagram codes. The results are given in the following theorems.

**Theorem 16** *[18, Theorem 16] Let $k = 3$ and $q, n$ be such that*

$$q^2 + q + 1 \geq \begin{cases} n - 3 & n \text{ even} \\ n - 4 & n \text{ odd} \end{cases}.$$

*Then the code $\mathcal{C} \subseteq \mathcal{G}_q(3, n)$ obtained from Construction I in [18] with minimum subspace distance 4 has cardinality $q^{2(n-3)} + \binom{n-3}{2}_q$.*

A similar result for small $q$ has also been given in [18, Theorem 17].

**Theorem 17** *[51, Theorem 19] Let $k \geq 4$, $s := \sum_{i=3}^{k} i = \frac{k^2+k-6}{2}$, $n \geq s+2+k = \frac{k^2+3k-2}{2}$ and $q^2 + q + 1 \geq \ell$, where $\ell := n - s = n - \frac{k^2+k-6}{2}$ for odd $n - s$ (or $\ell := n - s - 1 = n - \frac{k^2+k-4}{2}$ for even $n - s$). Then the code $\mathcal{C} \subseteq \mathcal{G}_q(k, n)$ obtained from Construction A in [51] has minimum subspace distance $2(k-1)$ and cardinality*

$$q^{2(n-k)} + q^{2(n-(k+(k-1)))} + \ldots + q^{2(n-\frac{k^2+k-6}{2})} + \binom{n - \frac{k^2+k-6}{2}}{2}_q.$$

A similar result for small $q$ has also been given in [51, Corollary 22].

**Theorem 18** [51, Corollary 27] *Let $n \geq 2k + 2$. Then the code $C \subseteq G_q(k, n)$ obtained from Construction B (recursively applied) in [51] has minimum subspace distance 4 and cardinality at least*

$$\sum_{i=1}^{\lfloor \frac{n-2}{k} \rfloor - 1} \left( q^{(k-1)(n-ik)} + \frac{(q^{2(k-2)} - 1)(q^{2(n-ik-1)} - 1)}{(q^4 - 1)^2} q^{(k-3)(n-ik-2)+4} \right).$$

Furthermore, two constructions for codes in $G_q(4, n)$ and $G_q(5, n)$ with minimum subspace distance 4 with closed cardinality formula were presented in [51, Sect. IV.B].

## 5 Orbit Codes

In this section we present several constructions for orbit codes. We will first present theory about single orbits in $G_q(k, n)$, whereas in the second subsection we will review constructions of unions of orbits with a prescribed minimum subspace distance.

Throughout the section we will make use of the isomorphism $\mathbb{F}_{q^n} \cong \mathbb{F}_q^n$. By abuse of notation we will change between the two representations with the same notation.

### 5.1 Single Orbits

The general linear group of order $n$ with entries from $\mathbb{F}_q$, denoted by $GL_n(q)$, defines an action on $G_q(k, n)$ as follows:

$$G_q(k, n) \times GL_n(q) \longrightarrow G_q(k, n)$$
$$\mathcal{U} \times A \longmapsto rs(RE(\mathcal{U})A)$$

An *orbit code* $C \subseteq G_q(k, n)$, in general, is defined as the orbit of a subgroup $G \leq GL_n(q)$ with some initial point $\mathcal{U} \in G_q(k, n)$, i.e.,

$$C = \mathcal{U}G := \{\mathcal{U}A \mid A \in G\}.$$

These codes can be seen as the analog of linear codes in the classical block coding case, since linear block codes are orbit codes with respect to additive subgroups of $\mathbb{F}_q^n$ (see e.g. [54, 55]). One of the consequences of this fact is, that the minimum distance of an orbit code can be determined by comparing all elements of the orbit with the initial point (instead of all possible pairs of codewords), as shown in [54, Theorem 17]:

**Lemma 19** *Let $\mathcal{U} \in \mathcal{G}_q(k, n)$ be an initial point, $G \leq \mathrm{GL}_n(q)$ a subgroup and let $\mathcal{C} = \mathcal{U}G$ be an orbit code. Then*

$$d_S(\mathcal{C}) = \min\{d_S(\mathcal{U}, \mathcal{U}A) \mid A \in G, A \notin \mathrm{Stab}_G(\mathcal{U})\},$$

*where $\mathrm{Stab}_G(\mathcal{U}) := \{A \in G \mid \mathcal{U}A = \mathcal{U}\}$.*

Of particular interest are *cyclic orbit codes*, where the defining subgroup $G \leq \mathrm{GL}_n(q)$ is cyclic, i.e., there is a matrix $P \in \mathrm{GL}_n(q)$ such that $G = \langle P \rangle$ and therefore

$$\mathcal{C} = \mathcal{U}G = \{\mathcal{U}P^i \mid i = 0, \ldots, \mathrm{ord}(P) - 1\}.$$

Understood best is the case where $P$ is the companion matrix of a monic irreducible polynomial $p(x) \in \mathbb{F}_q[x]$ of degree $n$. These codes are also called *irreducible cyclic orbit codes* [54]. If the irreducible polynomial is primitive, we call the corresponding codes *primitive cyclic orbit codes*.

It is well-known that one can construct spread codes as primitive cyclic orbit codes, see e.g. [49, Theorem 3]:

**Theorem 20** *Let $k \mid n$ and $p(x) \in \mathbb{F}_q[x]$ be monic, primitive and of degree $n$, and let $\alpha \in \mathbb{F}_{q^n}$ be a root of $p(x)$. Denote by $P \in \mathrm{GL}_n(q)$ the companion matrix of $p(x)$. Then*

$$\mathcal{U} := \left\{\alpha^{i\frac{q^n-1}{q^k-1}} \mid i = 0, \ldots, q^k - 2\right\} \cup \{0\}$$

*is an element of $\mathcal{G}_q(k, n)$ and $\mathcal{C} = \mathcal{U}\langle P \rangle$ is a spread in $\mathcal{G}_q(k, n)$.*

Note that, in Theorem 20, $\mathcal{U}$ is the subfield $\mathbb{F}_{q^k}$ of $\mathbb{F}_{q^n}$. Moreover, one does not need the companion matrix $P$ in the description of the orbit, instead one can also write $\mathcal{C} = \{\alpha^i \mathbb{F}_{q^k} \mid i = 0, \ldots, (q^n - 1)/(q^k - 1) - 1\}$. The previous construction has been generalized in [6, Section 3.B] as follows:

**Theorem 21** *Let $d \in \mathbb{N}$ divide $k$ and $n$. Then*

$$\mathcal{C} = \left\{\sum_{i=1}^{k/d} \alpha_i \mathbb{F}_{q^d} \mid \alpha_1, \ldots, \alpha_{k/d} \in \mathbb{F}_{q^n} \text{ are linearly independent over } \mathbb{F}_{q^d}\right\}$$

*is a code in $\mathcal{G}_q(k, n)$ with minimum subspace distance $2d$ and cardinality $\binom{n/d}{k/d}_{q^d}$.*

Furthermore, it is conjectured that for all parameters, except some border cases, an irreducible cyclic orbit code of minimum subspace distance $2(k - 1)$ exists:

**Conjecture 22** *For any $k, q$ and $n \geq 6$ there exists an irreducible orbit code $\mathcal{C} \subseteq \mathcal{G}_q(k, n)$ with minimum subspace distance $2(k - 1)$ and cardinality $(q^n - 1)/(q - 1)$.*

This conjecture was stated for general $n$ in [12, 54], but the case $n = 8, k = 4$ has been disproven in [25]. On the other hand, in [6, 45] it was shown that for any $q, k$ there exist infinitely many $n$ where the conjecture holds. The other parameter sets remain as a conjecture.

In general, one can determine the minimum subspace distance of a cyclic orbit code by just looking at the initial point $\mathcal{U} \in \mathcal{G}_q(k, n)$ of the orbit. This was done in [12, 41] for Singer subgroup orbits and, in more general, for any cyclic group in [54, Sect. 4]. This method was later on refined in [25, Theorem 4.11], where the authors defined the *best friend* of a subspace $\mathcal{U} \in \mathcal{G}_q(k, n)$ as the largest subfield $\mathbb{F}_{q^r}$ of $\mathbb{F}_{q^n}$ over which $\mathcal{U}$ is a vector space. With the help of the best friend one gets the following estimates on cardinality and distance of the corresponding code (see [25, Corollary 3.13, Lemma 4.1]):

**Theorem 23** *Let $\mathcal{U} \in \mathcal{G}_q(k, n)$ and $P \in \mathrm{GL}_n(q)$ be a companion matrix of a monic primitive polynomial of degree $n$. Consider the orbit code $\mathcal{C} = \mathcal{U}\langle P \rangle$. If $\mathbb{F}_{q^r}$ is the best friend of $\mathcal{U}$, then*

$$|\mathcal{C}| = \frac{q^n - 1}{q^r - 1}$$

*and*

$$2r \le d_S(\mathcal{C}) \le 2k.$$

Based on the previously mentioned methods to determine the minimum subspace distance of a cyclic orbit code, the paper [24] establishes a connection between the choice of initial point of a cyclic orbit code and the construction of cyclic difference sets of appropriate parameters. Moreover, in [2] some results related to irreducible cyclic orbit codes are given.

## 5.2 Unions of Orbits

In [20] the notion of a *cyclic subspace code* was introduced. In our terminology, this is a union of primitive cyclic orbit codes, although in its general form it could be a union of primitive cyclic orbit codes of different dimensions. In [6] it was shown how *subspace polynomials* can be used to construct good cyclic subspace codes.

**Definition 24** Let $\mathcal{U} \subseteq \mathbb{F}_{q^n}$ be a $\mathbb{F}_q$-subspace. The *subspace polynomial of $\mathcal{U}$* is defined as the monic (linearized) polynomial $P_\mathcal{U}(x)$ of least degree, such that $P_\mathcal{U}(u) = 0$ for any $u \in \mathcal{U}$.

Note that a subspace polynomial of some $\mathbb{F}_q$-subspace of $\mathbb{F}_{q^n}$ is always a linearized polynomial, i.e., it is of the form $\sum_{i=0}^{m} a_i x^{q^i}$ with $a_0, \ldots, a_m \in \mathbb{F}_{q^n}$. The following result, which is a reformulation of [6, Corollary 2], relates the shapes of two subspace polynomials to the subspace distance of the corresponding subspaces:

**Lemma 25** *Let* $\mathcal{U}, \mathcal{V} \in \mathcal{G}_q(k, n)$, *with* $P_{\mathcal{U}}(x) = x^{q^k} + \sum_{i=0}^{\ell} a_i x^{q^i}$ *where* $a_\ell \neq 0$, *and* $P_{\mathcal{V}}(x) = x^{q^k} + \sum_{i=0}^{m} b_i x^{q^i}$ *where* $b_m \neq 0$. *Then*

$$d_S(\mathcal{U}, \mathcal{V}) \geq 2 \min(k - \ell, k - m).$$

Another useful fact is that one can describe the subspace polynomials of an orbit with the help of the subspace polynomial of the initial point [6, Lemma 4]:

$$P_{\alpha\mathcal{U}}(x) = \alpha^{q^k} P_{\mathcal{U}}(\alpha^{-1} x).$$

With the above tools one can derive the following code constructions.

**Theorem 26** [6, Theorem 4] *Let $n$ be a prime and $\gamma \in \mathbb{F}_{q^n}$ be primitive. If $\mathbb{F}_{q^N}$ is the splitting field of $P_{\mathcal{U}}(x) := x^{q^k} + \gamma^q x^q + \gamma x$ and $\mathcal{U} \in \mathcal{G}_q(k, N)$ is the corresponding subspace (i.e., the kernel of $P_{\mathcal{U}}(x)$), then*

$$\mathcal{C} := \bigcup_{i=0}^{n-1} \{\alpha \mathcal{U}^{q^i} \mid \alpha \in \mathbb{F}_{q^N}^*\}$$

*is a cyclic subspace code with minimum subspace distance $2k - 2$ and size $n(q^N - 1)/(q - 1)$.*

This construction was generalized in [45, Theorem 3] to more general trinomials that can be used as subspace polynomials, which allows even larger code sizes. Furthermore, a generalization to subspace polynomials with $s + 2$ summands (for $s > 1$), such that the arising subspace code has minimum subspace distance $2k - 2s$, is given in [45, Corollary 4].

For more information on rank-metric and MRD codes the reader is referred to chapters "Codes Endowed With The Rank Metric" and "Constructions of Cyclic Subspace Codes and Maximum Rank Distance Codes" of this book.

Besides the previous theoretical results, there are many related results using smart computer searches to construct large codes, in particular cyclic subspace codes.

In [41] unions of irreducible cyclic orbit codes are found by solving linear inequalities under Diophantine restrictions. This Diophantine system is only feasible because the authors impose a prescribed automorphism group (namely the one of irreducible cyclic orbit codes) on the solutions. The results of that paper are for length $6 \leq n \leq 14$, dimension $k = 3$, and minimum subspace distance 4. Some of these codes are still the best codes known for the given parameter sets.

In [20, Section III] optimal cyclic codes of length 8 and 9 are presented, in the sense that there is no larger cyclic subspace code of the same length and distance.

These two computational methods prescribe $\mathbb{F}_{q^n}$ as the automorphism group in order to make the computer search feasible. Despite this restriction on unions of primitive cyclic orbit codes, the codes found come quite close to known bounds. This shows that cyclic orbit codes (or cyclic subspace codes) form a powerful class of constant dimension codes.

## 6 New from Old Codes

There are several results about how to construct new constant dimension codes from known constant dimension codes of smaller parameters.

A fairly simple, but very effective construction is given in Construction D of [51].

**Theorem 27** [51, Theorem 37] *Let* $C' \in \mathcal{G}_q(k, n)$ *be a code with minimum subspace distance* $2d$, *let* $\Delta \geq k$, *and let* $C_R \subseteq \mathbb{F}_q^{k \times \Delta}$ *be an MRD code of minimum rank distance* $d$. *Define the code*

$$C := \{\text{rs}[\ \text{RE}(\mathcal{U}) \quad A\ ] \mid \mathcal{U} \in C', A \in C_R\}.$$

*Then* $C$ *is a constant dimension code in* $\mathcal{G}_q(k, n + \Delta)$ *with minimum subspace distance* $2d$ *and cardinality* $|C'|q^{\Delta(k-d+1)}$.

As noted in [51, Corollary 39], one can easily extend this construction by adding constant dimension codes with prescribed zero columns in the beginning, analogously to Theorem 10. This idea is the basis of the linkage construction of [26]:

**Theorem 28** [26, Theorem 2.3] *For* $i = 1, 2$ *let* $\hat{C}_i \subseteq \mathcal{G}_q(k, n_i)$ *with minimum subspace distance* $d_i$. *Furthermore, let* $C_R \subseteq \mathbb{F}^{k \times n_2}$ *be a linear rank-metric code with minimum rank distance* $d_R$. *Then the code* $C := C_1 \cup C_2$, *where*

$$C_1 := \{\text{rs}[\ \text{RE}(\mathcal{U}) \quad A\ ] \mid \mathcal{U} \in \hat{C}_1, A \in C_R\},$$

$$C_2 := \{\text{rs}[\ 0_{k \times n_1} \quad \text{RE}(\mathcal{U})\ ] \mid \mathcal{U} \in \hat{C}_2\},$$

*is a constant dimension code in* $\mathcal{G}_q(k, n_1 + n_2)$, *with minimum subspace distance* $\min\{d_1, d_2, 2d_R\}$ *and cardinality* $|\hat{C}_2| + |\hat{C}_1||C_R|$.

Note that in the original statement of this theorem, the matrix representations of the subspaces $\mathcal{U}$ do not necessarily have to be the reduced row echelon forms. Different basis matrices will give different codewords in $C_3$, but the cardinality and minimum distance of the code remains the same. Furthermore note that this construction can be used to construct good partial spreads [26, Sect. 3].

The coset construction from [33] can be reformulated as follows, for which we need the following map: Let $\mathcal{U} \in \mathcal{G}_q(k, n)$ and $F \in \mathbb{F}_q^{m \times (n-k)}$, then $\varphi_{\mathcal{U}}(F)$ is a matrix $M \in \mathbb{F}_q^{m \times n}$, such that $M$ has a zero column wherever $\text{RE}(\mathcal{U})$ has a pivot and the other columns of $M$ are the columns of $F$.

**Theorem 29** *Let* $\mathcal{A}, \mathcal{B}$ *be disjoint unions of sets* $\mathcal{A}_i \subseteq \mathcal{G}_q(k', n')$, $\mathcal{B}_i \subseteq \mathcal{G}_q(k - k', n - n')$, *respectively. Moreover, let* $C_F \subseteq \mathbb{F}_q^{k' \times (n-n'-k+k')}$ *be an MRD code of minimum rank distance* $d$. *Let* $C$ *be a subset of the set*

$$\bigcup_i \left\{ \text{rs} \begin{pmatrix} \text{RE}(\mathcal{U}) & \varphi_{\mathcal{U}}(F) \\ 0 & \text{RE}(\mathcal{V}) \end{pmatrix} \mid \mathcal{U} \in \mathcal{A}_i, \mathcal{V} \in \mathcal{B}_i, F \in C_F \right\}$$

*such that $d_S(U, U') + d_S(V, V') \geq 2d$ for any two elements of the above form in $C$. Then $C$ is a constant dimension code in $\mathcal{G}_q(k, n)$ with minimum subspace distance $2d$.*

Note that the sets $\mathcal{A}_i$, $\mathcal{B}_i$ can be thought of as known constant dimension codes. A very effective family of unions of subsets, that was used in [33] in the coset construction are parallelisms, i.e., partitions of $\mathcal{G}_q(k, n)$ into spreads.

## 7   Final Remarks

Besides the constructions presented in detail in this survey one can find more results on subspace codes. E.g., a family of constant dimension codes called *linear subspace codes* was defined in [8]. More results on these codes were derived in [5, 46]. Furthermore, one can derive constant dimension codes from Riemann-Roch spaces, as done in [3, 31]. Many results on constant dimension codes with certain fixed parameters, often based on smart computer search, can be found in the literature, see e.g. in [1, 9–12, 35, 36] and references therein.

Moreover, results on non-constant dimension subspace codes, also called mixed-dimension subspace codes, can be found in [17, 23, 37–39, 51]. Note that in the non-constant dimension case there are two different metrics one can consider - the subspace and the injection metric. The before mentioned results include results for both metrics.

The best known subspace codes for given parameters can be looked up in the online database [32] at

$$\text{http://subspacecodes.uni-bayreuth.de.}$$

Finally, some open problems on various aspects of subspace codes can be found in the overview articles [13, 19].

**Acknowledgements** The authors were partially supported by COST – *European Cooperation in Science and Technology*. The second author was also supported by Swiss National Science Foundation grant no. 169510.

## References

1. J. Ai, T. Honold, H. Liu, The expurgation-augmentation method for constructing good plane subspace codes (2016). arXiv:1601.01502 [math.CO]
2. F. Bardestani, A. Iranmanesh, Cyclic orbit codes with the normalizer of a singer subgroup. J. Sci. Islam. Repub. Iran **26**(1), 49–55 (2015)
3. D. Bartoli, M. Bonini, M. Giulietti, Constant dimension codes from Riemann-Roch spaces (2015). arXiv:1508.01727
4. D. Bartoli, F. Pavese, A note on equidistant subspace codes. Discret. Appl. Math. **198**, 291–296 (2016)

5. P. Basu, N. Kashyap. On linear subspace codes closed under intersection. In *2015 Twenty First National Conference on Communications (NCC)*, pages 1–6, February 2015
6. E. Ben-Sasson, T. Etzion, A. Gabizon, N. Raviv, Subspace polynomials and cyclic subspace codes. IEEE Trans. Inf. Theory **62**(3), 1157–1165 (2016)
7. A. Beutelspacher, Blocking sets and partial spreads in finite projective spaces. Geom. Dedic. **9**(4), 425–449 (1980)
8. M. Braun, T. Etzion, A. Vardy, Linearity and complements in projective space. Linear Algebr. Appl. **438**(1), 57–70 (2013)
9. M. Braun, P. Östergard, A. Wassermann. New lower bounds for binary constant-dimension subspace codes. Exp. Math. 1–5 (to appear)
10. A. Cossidente, F. Pavese. Subspace codes in PG(2n-1,q). *Combinatorica* (2016), p. 1–23
11. A. Cossidente, F. Pavese, Veronese subspace codes. Des. Codes Cryptogr. **81**(3), 445–457 (2016)
12. A. Elsenhans, A. Kohnert, A. Wassermann. Construction of codes for network coding. In *Proceedings of the 19th International Symposium on Mathematical Theory of Networks and Systems – MTNS* (Budapest, Hungary, 2010), pp. 1811–1814
13. T. Etzion. Problems on q-analogs in coding theory (2013). arXiv:1305.6126
14. T. Etzion, E. Gorla, A. Ravagnani, A. Wachter-Zeh, Optimal Ferrers diagram rank-metric codes. IEEE Trans. Inf. Theory **62**(4), 1616–1630 (2016)
15. T. Etzion, N. Raviv, Equidistant codes in the Grassmannian. Discret. Appl. Math. **186**, 87–97 (2015)
16. T. Etzion, N. Silberstein. Construction of error-correcting codes for random network coding. In *IEEE 25th Convention of Electrical and Electronics Engineers in Israel, 2008. (IEEEI 2008)* (2008), pp. 070–074
17. T. Etzion, N. Silberstein, Error-correcting codes in projective spaces via rank-metric codes and Ferrers diagrams. IEEE Trans. Inf. Theory **55**(7), 2909–2919 (2009)
18. T. Etzion, N. Silberstein, Codes and designs related to lifted MRD codes. IEEE Trans. Inf. Theory **59**(2), 1004–1017 (2013)
19. T. Etzion, L. Storme, Galois geometries and coding theory. Des. Codes Cryptogr. **78**(1), 311–350 (2016)
20. T. Etzion, A. Vardy, Error-correcting codes in projective space. IEEE Trans. Inf. Theory **57**(2), 1165–1173 (2011)
21. E.M. Gabidulin, Theory of codes with maximum rank distance. Problemy Peredachi Informatsii **21**(1), 3–16 (1985)
22. E.M. Gabidulin, N.I. Pilipchuk. Multicomponent network coding. In *Proceedings of the Seventh International Workshop on Coding and Cryptography (WCC) 2011*, (Paris, France, 2011), pp. 443–452
23. A. Ghatak. Subspace codes for random networks based on Plücker coordinates and Schubert cells (2013). arXiv:1301.6362 [cs.IT]
24. A. Ghatak. Construction of Singer subgroup orbit codes based on cyclic difference sets. In *2014 Twentieth National Conference on Communications (NCC)* (2014), pp. 1–4
25. H. Gluesing-Luerssen, K. Morrison, C. Troha, Cyclic orbit codes and stabilizer subfields. Adv. Math. Commun. **9**(2), 177–197 (2015)
26. H. Gluesing-Luerssen, C. Troha, Construction of subspace codes through linkage. Adv. Math. Commun. **10**(3), 525–540 (2016)
27. E. Gorla, A. Ravagnani, Partial spreads in random network coding. Finite Fields Appl. **26**, 104–115 (2014)
28. E. Gorla, A. Ravagnani, Equidistant subspace codes. Linear Algebr. Appl. **490**, 48–65 (2016)
29. E. Gorla, A. Ravagnani. Subspace codes from Ferrers diagrams. J. Algebr. Appl. 1750131 (2016) (online)
30. J.P. Hansen, in *Osculating spaces of varieties and linear network codes*. Lecture Notes in Computer Science, vol 8080, 83–88 (2013)
31. J.P. Hansen, Riemann-Roch spaces and linear network codes. Comput. Sci. **10**(1), 1–11 (2015)

32. D. Heinlein, M. Kiermaier, S. Kurz, A. Wassermann. Tables of subspace codes (2016). arXiv:1601.02864
33. D. Heinlein, S. Kurz. Coset construction for subspace codes (2015). arXiv:1512.07634 [math.CO]
34. J.W.P. Hirschfeld, *Projective Geometries over Finite Fields*, 2nd edn. (Oxford Mathematical Monographs. The Clarendon Press, Oxford University Press, New York, 1998)
35. T. Honold, M. Kiermaier. On putative q-analogues of the Fano plane and related combinatorial structures, in *Dynamical Systems, Number Theory and Applications: A Festschrift in Honor of Armin Leutbecher's 80th Birthday*. (World Scientific, 2016), pp. 141–175
36. T. Honold, M. Kiermaier, S. Kurz. Optimal binary subspace codes of length 6, constant dimension 3 and minimum subspace distance 4. In *Topics in finite fields*, vol. 632 of Contemporary Mathematics (American Mathematical Society, Providence, RI, 2015), pp. 157–176
37. T. Honold, M. Kiermaier, S. Kurz, Constructions and bounds for mixed-dimension subspace codes. Adv. Math. Commun. **10**(3), 649–682 (2016)
38. A. Khaleghi. *Projective Space Codes for the Injection Metric*. (Masters thesis, University of Toronto, 2009)
39. A. Khaleghi, D. Silva, F.R. Kschischang. Subspace codes. In *IMA International Conference on Cryptography and Coding* (2009), pp. 1–21
40. S.L. Kleiman, D. Laksov, Schubert calculus. Am. Math. Mon. **79**, 1061–1082 (1972)
41. A. Kohnert, S. Kurz. Construction of large constant dimension codes with a prescribed minimum distance. in *MMICS* eds. by J. Calmet, W. Geiselmann, J. Müller-Quade. Lecture Notes in Computer Science, vol. 5393 (Springer, Berlin, 2008), pp. 31–42
42. R. Kötter, F.R. Kschischang, Coding for errors and erasures in random network coding. IEEE Transac. Inf. Theory **54**(8), 3579–3591 (2008)
43. R. Lidl, H. Niederreiter, *Introduction to Finite Fields and their Applications* (Cambridge University Press, Cambridge, London, 1986)
44. F. Manganiello, E. Gorla, J. Rosenthal. Spread codes and spread decoding in network coding. In *Proceedings of the 2008 IEEE International Symposium on Information Theory* (Toronto, Canada, 2008), pp. 851–855
45. K. Otal, F. Özbudak, Cyclic subspace codes via subspace polynomials. Des. Codes Cryptogr. 1–14 (2016)
46. S. Pai, S. Rajan. On the bounds of certain maximal linear codes in a projective space. In *IEEE International Symposium on Information Theory (ISIT)* (2015), pp. 591–595
47. C. Procesi. *A Primer Of Invariant Theory*. (Brandeis lecture notes, Brandeis University, 1982). Notes by G. Boffi
48. J. Rosenthal, N. Silberstein, A.-L. Trautmann, On the geometry of balls in the Grassmannian and list decoding of lifted Gabidulin codes. Des. Codes Cryptogr. **73**(2), 393–416 (2014)
49. J. Rosenthal, A.-L. Trautmann, A complete characterization of irreducible cyclic orbit codes and their Plücker embedding. Des. Codes Cryptogr. **66**(1–3), 275–289 (2013)
50. N. Silberstein, T. Etzion, Large constant dimension codes and lexicodes. Adv. Math. Commun. **5**(2), 177–189 (2011)
51. N. Silberstein, A.-L. Trautmann, Subspace codes based on graph matchings, Ferrers diagrams, and pending blocks. IEEE Trans. Inf. Theory **61**(7), 3937–3953 (2015)
52. D. Silva, F. Kschischang, F. Kötter, A rank-metric approach to error control in random network coding. IEEE Trans. Inf. Theory **54**(9), 3951–3967 (2008)
53. A.-L. Trautmann. Constructions, Decoding and Automorphisms of Subspace Codes. PhD thesis, University of Zurich, 2013
54. A.-L. Trautmann, F. Manganiello, M. Braun, J. Rosenthal, Cyclic orbit codes. IEEE Trans. Inform. Theory **59**(11), 7386–7404 (2013)
55. A.-L. Trautmann, F. Manganiello, J. Rosenthal, Orbit codes - a new concept in the area of network coding, in *2010 IEEE Information Theory Workshop (ITW)* (Dublin, Ireland, 2010), pp. 1–4
56. A.-L. Trautmann, J. Rosenthal. New improvements on the echelon-Ferrers construction. In *Proceedings of the 19th International Symposium on Mathematical Theory of Networks and Systems – MTNS* (Budapest, Hungary, 2010), pp. 405–408

# Constructions of Cyclic Subspace Codes and Maximum Rank Distance Codes

Kamil Otal and Ferruh Özbudak

**Abstract** This chapter is a survey of the recent results on the constructions of cyclic subspace codes and maximum rank distance codes. Linearized polynomials are the main tools used to introduce both constructions in this chapter. In the construction of cyclic subspace codes, codewords are considered as the root spaces of some subspace polynomials (which are a particular type of linearized polynomials). In this set up, some algebraic manipulations on the coefficients and degrees of such polynomials are applied to provide a systematic construction of cyclic subspace codes. In constructions of maximum rank distance codes, linearized polynomials are used as codewords again, but in a different way. Codewords of rank metric codes are considered as the linear maps that these polynomials represent. All known constructions of maximum rank distance codes in the literature are summarized using this linearized polynomial representation. Connections among the constructions and further explanations are also provided.

## 1 Introduction

Subspace codes have an increasing interest in the last decade thank to their applications in random network coding. Similar to subspace codes, the theory of rank metric codes have gained an increasing attraction due to their application to various areas including subspace codes. Linearized polynomials are quite useful mathematical tools to represent and solve some current problems within these areas.

In this chapter, the main goal is to provide all known advances in two recent problems: constructions of cyclic subspace codes and maximum rank distance codes. In particular, Sect. 1 is devoted to the brief introductions of linearized polynomials, cyclic subspace codes and rank metric codes. Further details and an explicit construc-

K. Otal (✉) · F. Özbudak
Middle East Technical University, Ankara, Turkey
e-mail: kotal@metu.edu.tr

F. Özbudak
e-mail: ozbudak@metu.edu.tr

© Springer International Publishing AG 2018
M. Greferath et al. (eds.), *Network Coding and Subspace Designs*,
Signals and Communication Technology,
https://doi.org/10.1007/978-3-319-70293-3_3

tion of cyclic subspace codes are introduced in Sect. 2, whereas the constructions of maximum rank distance codes in the literature are summarized in Sect. 3.

## 1.1 Linearized Polynomials

We recall the definition and some basic properties of linearized polynomials before to explain their applications mentioned above. Let $\mathbb{F}_q$ be a finite field of $q$ elements, $\mathbb{F}_{q^N}$ be its $N$-th degree field extension and $\overline{\mathbb{F}}_q$ be its algebraic closure. A polynomial $f(x) \in \mathbb{F}_{q^N}[x]$ of the form

$$f(x) = \sum_{i=0}^{l} \alpha_i x^{q^i}, \tag{1}$$

for some non-negative integer $l$, is called a $q$-polynomial (or, a linearized polynomial) over $\mathbb{F}_{q^N}$. In this representation, $l$ is called the $q$-degree of $f$ if $\alpha_l \neq 0$. Some useful properties of linearized polynomials of form (1) are listed below.

- $f(c\alpha + \beta) = cf(\alpha) + f(\beta)$ for all $c \in \mathbb{F}_q$ and $\alpha, \beta \in \overline{\mathbb{F}}_q$.
- The multiplicity of each root of $f$ in $\overline{\mathbb{F}}_q$ is the same and equal to $q^r$, where $r$ is the smallest integer satisfying $\alpha_r \neq 0$.
- The set of roots of $f$ in an extension of $\mathbb{F}_{q^N}$ constitutes a vector space over $\mathbb{F}_q$. In particular, the set of roots of $f$ in $\mathbb{F}_{q^N}$ is a subspace of $\mathbb{F}_{q^N}$ over $\mathbb{F}_q$. This set is called the kernel of $f$ and denoted by $\ker(f)$. The rank of $f$ is given by $N - \dim(\ker(f))$ and denoted by $\mathrm{rank}(f)$.

These properties can be proved directly by using the definition. Further information about linearized polynomials can be found, for example, in [19, Sect. 3.4].

   A monic $q$-polynomial is called a subspace polynomial if it splits completely over $\mathbb{F}_{q^N}$ (i.e. all roots are in $\mathbb{F}_{q^N}$) and has no multiple roots (i.e. $\alpha_0 \neq 0$). Therefore, there is a one to one correspondence between an $l$-dimensional subspace $U$ of $\mathbb{F}_{q^N}$ and a subspace polynomial $f$ of $q$-degree $l$ satisfying $f(U) = \{0\}$.

## 1.2 Subspace Codes

The set of all subspaces of $\mathbb{F}_{q^N}$ is called the projective space of (vector) dimension $N$ over $\mathbb{F}_q$ and denoted by $\mathscr{P}_q(N)$. The set of all $k$ dimensional elements of $\mathscr{P}_q(N)$ is called Grassmannian space (or briefly Grassmannian) over $\mathbb{F}_q$ and denoted by $\mathscr{G}_q(N, k)$. The metric $d$ given by

$$d(U, V) := \dim U + \dim V - 2\dim(U \cap V)$$

on $\mathscr{P}_q(N)$ is called the *subspace distance*. A subset $\mathscr{C} \subseteq \mathscr{P}_q(N)$ including at least two elements and equipped with this metric is called a *subspace code*. If moreover $\mathscr{C} \subseteq \mathscr{G}_q(N, k)$, then $\mathscr{C}$ is also called a *constant dimension code* (see chapter "Constructions of Constant Dimension Codes" for various constructions of such codes). We naturally define the *minimum distance* $d(\mathscr{C})$ of a code $\mathscr{C}$ by $d(\mathscr{C}) := \min\{d(U, V) : U, V \in \mathscr{C} \text{ and } U \neq V\}$. A *cyclic shift* of a subspace $U \subseteq \mathbb{F}_{q^N}$ is given by $\alpha U := \{\alpha u : u \in U\}$, where $\alpha \in \mathbb{F}_{q^N}^*$. Observe that a cyclic shift is a subspace of $\mathbb{F}_{q^N}$ over $\mathbb{F}_q$, and the dimension of the cyclic shift is the same as the dimension of $U$. A subspace code $\mathscr{C}$ is called *cyclic* if $\alpha U \in \mathscr{C}$ for all $U \in \mathscr{C}$ and $\alpha \in \mathbb{F}_{q^N}^*$, and *quasi-cyclic* if $\alpha U \in \mathscr{C}$ for all $U \in \mathscr{C}$ and $\alpha \in G$, where $G$ is a multiplicative subgroup of $\mathbb{F}_{q^N}^*$. Quasi-cyclic codes are also known as "(cyclic) orbit codes" (see, for example, [13, 35]).

Subspace codes are the main mathematical tools in random network coding due to their error correction capabilities shown in [17]. Cyclic subspace codes have a particular interest thank to their efficient encoding and decoding algorithms. A systematic construction of cyclic subspace codes including several orbits was given in [1] using subspace polynomials. In [28], this construction was generalized and improved by increasing the code size and the number of possible parameters. In Sect. 2, we aim to give this construction with some computational illustrations.

## 1.3 Rank Metric Codes

Let $\mathbb{F}_q^{m \times n}$ be the set of $m \times n$ matrices over $\mathbb{F}_q$. On $\mathbb{F}_q^{m \times n} \times \mathbb{F}_q^{m \times n}$, the function $d$ defined by

$$d(A, B) := \text{rank}(A - B),$$

which satisfies the usual axioms of a metric, is called the *rank distance* on $\mathbb{F}_q^{m \times n}$.

*Remark 1* Notice that we use $d$ to denote both the subspace distance and the rank distance together, since we think that there is a less possibility to confuse them in this chapter. Observe that they are defined on different ambient spaces, also we examine them in separate sections (Sects. 2 and 3).

A subset $\mathscr{C}$ of $\mathbb{F}_q^{m \times n}$ including at least two elements and equipped with the rank distance is called a *rank metric code*. The *minimum distance* $d(\mathscr{C})$ of a code $\mathscr{C}$ is naturally defined by $d(\mathscr{C}) := \min\{d(A, B) : A, B \in \mathscr{C} \text{ and } A \neq B\}$. Equivalence between any two rank metric codes is determined considering the set of rank distance preserving maps under the light of [36, Theorem 3.4]: Two rank metric codes $\mathscr{C}, \mathscr{C}' \subseteq \mathbb{F}_q^{m \times n}$ are called *equivalent* if there exist $X \in GL(m, \mathbb{F}_q)$, $Y \in GL(n, \mathbb{F}_q)$ and $Z \in \mathbb{F}_q^{m \times n}$ such that

$$\begin{aligned} \mathscr{C}' &= X\mathscr{C}^\sigma Y + Z := \{XC^\sigma Y + Z : C \in \mathscr{C}\} \text{ when } m \neq n, \\ \mathscr{C}' &= X\mathscr{C}^\sigma Y + Z \text{ or } \mathscr{C}' = X(\mathscr{C}^t)^\sigma Y + Z \text{ when } m = n, \end{aligned} \tag{2}$$

for some automorphism $\sigma$ acting on the entries of $C \in \mathscr{C}$, where the superscript $t$ denotes the transposition of matrices. If both $\mathscr{C}$ and $\mathscr{C}'$ are closed under addition, then $Z$ must be the zero matrix. Similarly, if both $\mathscr{C}$ and $\mathscr{C}'$ are linear over $\mathbb{F}_q$, then $\sigma$ can be taken as the identity without loss of generality. This equivalence idea, in different forms and scopes, is used in several studies [3, 20, 22, 23, 25–27, 31].

Rank metric codes have a well-known **Singleton-like bound** given in the following proposition.

**Proposition 1** ([5]) *Assume $m \geq n$ without loss of generality. Let $\mathscr{C} \subseteq \mathbb{F}_q^{m \times n}$ be a rank metric code with the minimum rank distance $d$. Then $|\mathscr{C}| \leq q^{m(n-d+1)}$.*

We may give an elementary proof for this proposition as follows.

*Proof* Let $\mathscr{C}' \subseteq \mathbb{F}_q^{m \times (n-d+1)}$ be the set of matrices obtained by deleting the last $d - 1$ columns of codewords in $\mathscr{C}$. If $A, B \in \mathscr{C}$ are distinct, then their images $A', B' \in \mathscr{C}'$ are also distinct since $d(A, B) = \mathrm{rank}(A - B) > d - 1$, hence $|\mathscr{C}| = |\mathscr{C}'|$. Also the inclusion $\mathscr{C}' \subseteq \mathbb{F}_q^{m \times (n-d+1)}$ implies $|\mathscr{C}'| \leq q^{m(n-d+1)}$. That is, $|\mathscr{C}| \leq q^{m(n-d+1)}$. $\square$

If the Singleton-like bound is met, then the code is called *maximum rank distance (MRD) code*. MRD codes have a rich mathematical structure, inheriting the $q$-analogue dimension of classical coding theory. Besides the theoretical importance, they also have various applications, for example, in random network coding (e.g., [33]), space-time coding (e.g., [21, 34]) and cryptology (e.g., [29, 32]).

Finding new optimal codes, up to an equivalence notion, is one of the basic problems in coding theory. Accordingly, the problem of finding new MRD codes up to the equivalence relation given in (2) has gained interest especially in the last decade. We may summarize the history of the results about this problem as follows.

- Linear MRD codes:

  - 1978, 1985 and 1991: Gabidulin codes in [5, 12, 30].
  - 2005: Generalized Gabidulin codes in [15] (briefly called *GG codes*).
  - 2016: Twisted Gabidulin codes in [31] (briefly called *TG codes*). A particular case was also discovered independently in [25].
  - 2016: Generalized twisted Gabidulin codes in [20, 31] (briefly called *GTG codes*).

- Non-linear additive MRD codes:

  - 2017: Additive generalized twisted Gabidulin codes in [26] (briefly called *AGTG codes*).

- Non-additive (but closed under multiplication by a scalar) MRD codes:

  - 2016: A family for the $m = n = 3$ and $d = 2$ case in [4].
  - 2017: A generalization of [4] to the $d = m - 1$ case in [7].

- Non-additive and not closed under multiplication by a scalar MRD codes:

  - 2017: A family for arbitrary $d$ in [27].

In Sect. 3 we give these constructions explicitly using their linearized polynomial representation. Notice that all families listed above are constructed for $m = n$. From a code $\mathscr{C} \subseteq \mathbb{F}_q^{m \times m}$ of minimum distance $d$, by deleting last $m - n$ rows (or columns), we can obtain another code $\mathscr{C}' \subseteq \mathbb{F}_q^{m \times n}$ of minimum distance larger than or equal to $d - m + n$. The new code $\mathscr{C}'$ is called a *punctured code*. Remark that if a rank metric code $\mathscr{C}$ is MRD then so is its punctured version $\mathscr{C}'$. However, punctured codes are not involved in this chapter, since the $m = n$ case may be considered the more general case in this perspective.

## 2 Construction of Cyclic Subspace Codes

Let $U \in \mathscr{G}_q(N, k)$. We never consider the trivial cases, so assume $1 < k < N$ unless otherwise stated. Here, $U$ is a linear space over $\mathbb{F}_q$, but $U$ can be also a linear space over $\mathbb{F}_{q^t}$ for some $t > 1$ (observe that $t$ always divides both $N$ and $k$ in this case). Such a property affects some important parameters. In case we need to emphasize this property, we briefly say "$\mathbb{F}_q$−linear" and "$\mathbb{F}_{q^t}$−linear" correspondingly.

The $Orb(U) := \{\alpha U : \alpha \in \mathbb{F}_{q^N}^*\}$ set is called the *orbit* of $U$. The cardinality of an orbit can be determined as in the following well-known theorem.

**Theorem 1** *Let $U \in \mathscr{G}_q(N, k)$. Then, $\mathbb{F}_{q^d}$ is the largest field over which $U$ is linear if and only if*

$$|Orb(U)| = \frac{q^N - 1}{q^d - 1}.$$

Similar versions of this theorem are available also in [1, 6, 13]. We may prove this theorem in an elementary way as follows.

*Proof* $(\Rightarrow)$ : Let $\mathbb{F}_{q^d}$ be the largest field over which $U$ is linear. Then we have $|Orb(U)| \leq \frac{q^N - 1}{q^d - 1}$ since $aU = U$ for all $a \in \mathbb{F}_{q^d}$. Assume that the equality does not hold. Then there exists an $\alpha \in \mathbb{F}_{q^N} \setminus \mathbb{F}_{q^d}$ such that $\alpha U = U$, by pigeon hole principle. It implies $U$ is also $\mathbb{F}_{q^d}(\alpha)$−linear, but $\mathbb{F}_{q^d} \subsetneq \mathbb{F}_{q^d}(\alpha)$ since $\alpha \notin \mathbb{F}_{q^d}$. Hence $\mathbb{F}_{q^d}$ is not the largest field over which $U$ is linear, contradiction. Therefore, the equality must hold, i.e. $|Orb(U)| = \frac{q^N - 1}{q^d - 1}$.

$(\Leftarrow)$ : It can be shown by the $(\Rightarrow)$ part. □

If $d = 1$ in Theorem 1, then the orbit is called *full length orbit*. Otherwise, the orbit is called *degenerate orbit*.

*Remark 2* In case we need to study degenerate orbits, equivalently we can study full length orbits over $\mathbb{F}_{q^d}$ and then carry all the data from $\mathscr{G}_{q^d}(N/d, k/d)$ to $\mathscr{G}_q(N, k)$.

## 2.1 A Construction Including Many Full Length Orbits

The following theorem gives a systematic construction of cyclic subspace codes including many full length orbits in $\mathscr{G}_q(N, k)$, when the minimum distance is $2k - 2$.

**Theorem 2** ([28]) *Consider $r$ polynomials*

$$T_i(x) := x^{q^k} + \theta_i x^q + \gamma_i x \in \mathbb{F}_{q^n}[x], \quad 1 \le i \le r$$

*satisfying $\theta_i \ne 0$ and $\gamma_i \ne 0$ for all $1 \le i \le r$, and*

$$\frac{\gamma_i}{\gamma_j} \ne \left( \frac{\gamma_i}{\gamma_j} \left( \frac{\theta_i}{\theta_j} \right)^{-1} \right)^{\frac{q^k-1}{q-1}} \quad when \ i \ne j. \tag{3}$$

*Also let*

- *$N_i$ be the degree of the splitting field of $T_i$ for $1 \le i \le r$,*
- *$U_i \subseteq \mathbb{F}_{q^{N_i}}$ be the kernel of $T_i$ for $1 \le i \le r$,*
- *$N$ be a multiple of $\mathrm{lcm}(N_1, \ldots, N_r)$.*

*Then the code $\mathscr{C} \subseteq \mathscr{G}_q(N, k)$ given by*

$$\mathscr{C} = \bigcup_{i=1}^{r} \{ \alpha U_i : \alpha \in \mathbb{F}_{q^N}^* \}$$

*is a cyclic code of size $r \frac{q^N-1}{q-1}$ and of minimum distance $2k - 2$.*

*Proof* Let $\mathscr{C}_i := \{\alpha U_i : \alpha \in \mathbb{F}_{q^N}^*\}$, for $1 \le i \le r$. Then $\mathscr{C} := \bigcup_{i=1}^{r} \mathscr{C}_i$ is a subset of $\mathscr{G}_q(N, k)$ and it is obviously cyclic. To prove that $|\mathscr{C}| = r \frac{q^N-1}{q-1}$ and $d(\mathscr{C}) = 2k - 2$, it is enough to prove

$$\dim(\alpha U_i \cap \beta U_j) \le 1 \quad \text{when } i \ne j \text{ or } \frac{\alpha}{\beta} \notin \mathbb{F}_q. \tag{4}$$

Let $\theta \in \alpha U_i \cap \beta U_j$, where $1 \le i, j \le r$. Then there exist $u \in U_i$ (i.e. $T_i(u) = 0$) and $v \in U_j$ (i.e. $T_j(v) = 0$) such that $\theta = \alpha u = \beta v$. Eliminating $v$ by taking $v = \frac{\alpha}{\beta}u$ and hence solving the system $T_i(u) = 0$ and $T_j(\frac{\alpha}{\beta}u) = 0$ in terms of $u$, we obtain

$$\left( (\frac{\alpha^{q^k}}{\beta^{q^k}})\theta_i - (\frac{\alpha^q}{\beta^q})\theta_j \right) u^q + \left( (\frac{\alpha^{q^k}}{\beta^{q^k}})\gamma_i - (\frac{\alpha}{\beta})\gamma_j \right) u = 0. \tag{5}$$

The left hand side of equation (5) is a $q$-polynomial in terms of $u$. If we assume that the left hand side is identically zero, then we obtain a contradiction with (3) after eliminating $\alpha$ and $\beta$ in (5). Hence, the polynomial on the left hand side of (5) can not be identically zero. Conclusively, the fundamental theorem of algebra implies (4). □

*Remark 3* Let $T(x) \in \mathbb{F}_{q^N}[x]$ be a subspace polynomial and $U$ be the set of all roots of $T(x)$. We can determine another subspace $\overline{U} \subseteq \mathbb{F}_{q^N}$ associated with $T(x)$ as follows.

$$u \in \overline{U} \Leftrightarrow T(x) = \left( x^q - \frac{1}{u^{q-1}} x \right) \circ Q(x)$$

for some $q-$polynomial $Q(x)$ over $\mathbb{F}_{q^N}$, where $\circ$ denotes the composition of polynomials. This space can be also characterized by

$$u \in \overline{U} \Leftrightarrow u^q \text{ is a root of } \overline{T}(x) := (\alpha_0 x)^{q^k} + \cdots + (\alpha_{k-1}x)^q + x,$$

where

$$T(x) = x^{q^k} + \alpha_{k-1}x^{q^{k-1}} + \cdots + \alpha_0 x.$$

Here, $\overline{U}$ is called the *adjoint space* of $U$ (see [24, Theorem 14, 15 and 16] for the proofs of these facts). Therefore, corresponding to a code $\mathscr{C}$ obtained in Theorem 2, we can construct another code $\overline{\mathscr{C}}$ using the polynomials $\overline{T}_i(x) = \gamma_i^{q^k} x^{q^k} + \theta_i^{q^{k-1}} x^{q^{k-1}} + x$ instead of $T_i(x)$ in Theorem 2. Both $\mathscr{C}$ and $\overline{\mathscr{C}}$ have the same parameters. We call $\overline{\mathscr{C}}$ the *adjoint code* of $\mathscr{C}$.

The following corollary, as a particular case of Theorem 2, shows that we can construct cyclic subspace codes including up to $(q^n - 1)$ orbits.

**Corollary 1** *If $\theta_i = \gamma_i^q$ for all $1 \leq i \leq r$ in Theorem 2, then (3) is satisfied. If moreover $\gamma_i$ and $\gamma_j$ are conjugate as $\gamma_i = \gamma_j^{q^l}$ for some $i$, $j$ and $l$, then $U_i = U_j^{q^l}$ and hence $N_i = N_j$.*

*Proof* The inequality (3) holds clearly when $\theta_i = \gamma_i^q$ for all $1 \leq i \leq r$ in Theorem 2. Now let $\theta_i = \gamma_i^q$ for all $1 \leq i \leq r$, and $\gamma_i = \gamma_j^{q^l}$ for some $i$, $j$ and $l$. Take $u \in U_j^{q^l}$, then $u = v^{q^l}$ for some $v \in U_j$. Hence, $T_j(v) = T_j(u^{q^{-l}}) = 0$ which implies that $T_i(u) = 0$. In that way, we see that $T_i(u) = 0$ and so $U_j^{q^l} \subseteq U_i$. It can be shown similarly that $U_i \subseteq U_j^{q^l}$. Conclusively $U_i = U_j^{q^l}$ and hence $N_i = N_j$. □

*Remark 4* A particular case of Corollary 1, the $r = n$ and $\gamma_i = \gamma^{q^i}$ case for some primitive element $\gamma$ of $\mathbb{F}_{q^n}$ for all $1 \leq i \leq n$, was given in [1].

*Remark 5* What is done in Corollary 1 is to pick some (up to $q^n - 1$) full length orbits satisfying the following.

- The minimum distance in an orbit is $\geq 2k - 2$,
- The minimum distance between any two orbits is $\geq 2k - 2$.

At this point, it is natural to ask for the number of all full length orbits, i.e. the size of the pool we pick orbits from. The next theorem from [6] answers this question. Before stating it, we recall some basic definitions and facts:

- *Möbius function* $\mu$ over the set of positive integers is given by

$$
\mu(n) = \begin{cases} 1 & \text{if } n = 1, \\ (-1)^r & \text{if n is a product of r distinct primes}, \\ 0 & \text{if n is divisible by a square of a prime}. \end{cases}
$$

- *Gaussian coefficient* (or *q-binomial coefficient*) is defined by

$$
\begin{bmatrix} N \\ k \end{bmatrix}_q := \prod_{i=0}^{k-1} \frac{q^N - q^i}{q^k - q^i} = \prod_{i=0}^{k-1} \frac{q^{N-i} - 1}{q^{k-i} - 1}.
$$

The size of $\mathscr{G}_q(N, k)$ is actually $\begin{bmatrix} N \\ k \end{bmatrix}_q$.

**Theorem 3** ([6]) *Let d be a positive integer dividing both N and k. The number of orbits in $\mathscr{G}_q(N, k)$ of size $\frac{q^N - 1}{q^d - 1}$ is*

$$
\frac{q^d - 1}{q^N - 1} \sum_{t:d|t \text{ and } t| \gcd(N,k)} \mu(t/d) \begin{bmatrix} N/t \\ k/t \end{bmatrix}_{q^t}. \tag{6}
$$

The following corollary demonstrates the approximate density of the orbits used in Corollary 1 in the set of all orbits. Similar approximations can be found also in [18].

**Corollary 2** *Let $r \le q^n - 1$ for some n dividing N as in Corollary 1. Also let s be the number of positive integers dividing $\gcd(N, k)$. The ratio of the number of orbits in the code obtained by Corollary 1 to the size of the whole set of full length orbits is approximately*

$$
\frac{r}{sq^{(N-k-1)(k-1)}}.
$$

*Proof* Taking $d = 1$ in Theorem 3, we see that there are $s$ terms in the sum of (6), and the absolute value of each one is less than or equal to $\begin{bmatrix} N \\ k \end{bmatrix}_q \frac{q-1}{q^N-1}$. On the other hand, Corollary 1 gives $r$ orbits. Hence the proportion is

$$
\frac{r}{s \begin{bmatrix} N \\ k \end{bmatrix}_q \frac{q-1}{q^N-1}} \approx \frac{rq^{N-1}}{sq^{kN - \frac{k(k-1)}{2} - \frac{k(k+1)}{2}}} = \frac{r}{sq^{(N-k-1)(k-1)}}.
$$

□

*Remark 6* Corollary 2 says that we pick only a small amount of orbits in Corollary 1 especially when $N$ is large and $k$ is close to $N/2$. Also, the $s$ and $r$ (and so $n$) numbers play an important role.

## 2.2 Generalization to Other Small Minimum Distances

Theorem 2 can be generalized for the codes of other minimum distances as in the following corollary.

**Corollary 3** *Consider r polynomials*

$$T_i(x) := x^{q^k} + \gamma_{s,i} x^{q^s} + \cdots + \gamma_{1,i} x^q + \gamma_{0,i} x \in \mathbb{F}_{q^n}[x], \quad 1 \le i \le r$$

*satisfying*

- $\gamma_{0,i} \ne 0$ *for all* $1 \le i \le r$,
- *there exist* $0 \le l_1 < l_2 \le s$ *such that*

  - $\gcd(l_1, l_2) = 1$,
  - $l_2 - l_1$ *divides* $k - l_2$,
  - $\gamma_{l_2,i} \ne 0$ *and* $\gamma_{l_1,i} \ne 0$ *for all* $1 \le i \le r$,
  - *for all* $1 \le i \le r$, *we have*

  $$\frac{\gamma_{l_1,i}}{\gamma_{l_1,j}} \ne \left( \frac{\gamma_{l_1,i}}{\gamma_{l_1,j}} (\frac{\gamma_{l_2,i}}{\gamma_{l_2,j}})^{-1} \right)^M,$$

  *where*

  $$M = \frac{q^{k-l_1} - 1}{q^{l_2-l_1} - 1}.$$

*Also let*

- $N_i$ *be the degree of the splitting field of* $T_i$, *for* $1 \le i \le r$,
- $U_i \subseteq \mathbb{F}_{q^{N_i}}$ *be the kernel of* $T_i$, *for* $1 \le i \le r$,
- $N$ *be a multiple of* $\mathrm{lcm}(N_1, \ldots, N_r)$.

*Then the code* $\mathscr{C} \subseteq \mathscr{G}_q(N, k)$ *given by*

$$\mathscr{C} = \bigcup_{i=1}^{r} \{\alpha U_i : \alpha \in \mathbb{F}_{q^N}^*\}$$

*is a cyclic code of size* $r \frac{q^N - 1}{q - 1}$ *and of minimum distance* $2k - 2s$.

*Proof* The proof is similar to the proof of Theorem 2. There are a few things to notice about the extra assumptions:

- We want $\frac{q^{k-l_1}-1}{q^{l_2-l_1}-1}$ to be an integer, so we should have "$l_2 - l_1$ divides $k - l_1$" or equivalently "$l_2 - l_1$ divides $k - l_2$".
- We want $U_i$ to be a full length orbit for all $1 \le i \le r$. Notice that $U$ is $\mathbb{F}_{q^t}$-linear if and only if the subspace polynomial of $U$ is a $q^t$-polynomial. Hence we should take $\gcd(k - l_2, l_2 - l_1, l_1) = 1$ to obtain only $q$-polynomials. The "$l_2 -$

$l_1$ divides $k - l_2$" property above implies that $\gcd(k - l_2, l_2 - l_1, l_1) = 1$ if and only if $\gcd(l_1, l_2) = 1$.

$\square$

*Remark 7* In case $s > 1$, we can choose the elements $\gamma_{j,i} \in \mathbb{F}_{q^n}$ freely, for $j \notin \{0, l_1, l_2\}$. So $r$ can be as many as $q^{n(s-1)}(q^n - 1)$ if $l_1 = 0$, and $q^{n(s-2)}(q^n - 1)^2$ otherwise. Case $s = 1, l_1 = 0$ and $l_2 = 1$ is Theorem 2 in particular.

*Remark 8* In case $s > 1$, if there exists an integer $i$ such that $1 < i \le s$ and $i$ divides $k$, then we can insert also degenerate orbits into the code. In that way we can improve Corollary 3 further. We prefer to illustrate this further improvement by only an example below.

*Example 1* Let $k = 6$ and $s = 3$. For some positive integer $n$, consider $r_1$ polynomials

$$x^{q^6} + \gamma_{3,i}x^{q^3} + \gamma_{2,i}x^{q^2} + \gamma_{1,i}x^q + \gamma_{0,i}x \in \mathbb{F}_{q^n}, \quad 1 \le i \le r_1$$

as in Corollary 3 by taking $l_1 = 0$ and $l_2 = 1$. Using them, we can construct a cyclic code $\mathscr{C}_1$ of size up to $q^{2n}(q^n - 1)\frac{q^{N_1}-1}{q-1}$ and of minimum distance $\ge 2k - 2s = 6$ for some suitable $N_1$.

Consider $r_2$ polynomials

$$x^{q^6} + \gamma_{2,i}x^{q^2} + \gamma_{0,i}x \in \mathbb{F}_{q^n}, \quad 1 \le i \le r_2$$

satisfying

- $\gamma_{0,i} \ne 0$ and $\gamma_{2,i} \ne 0$ for all $1 \le i \le r_2$,
- $\frac{\gamma_{0,i}}{\gamma_{0,j}} \ne \left(\frac{\gamma_{0,i}}{\gamma_{0,j}}(\frac{\gamma_{2,i}}{\gamma_{2,j}})^{-1}\right)^{M_2}$ when $i \ne j$,

where $M_2 = \frac{q^k-1}{q^2-1}$. Using them in Theorem 2, we can construct a cyclic code $\mathscr{C}_2$ of size up to $(q^n - 1)\frac{q^{N_2}-1}{q^2-1}$ and of minimum distance $\ge 2k - 2s = 6$ for some suitable $N_2$.

Consider $r_3$ polynomials

$$x^{q^6} + \gamma_{3,i}x^{q^3} + \gamma_{0,i}x \in \mathbb{F}_{q^n}, \quad 1 \le i \le r_3$$

satisfying

- $\gamma_{0,i} \ne 0$ and $\gamma_{3,i} \ne 0$ for all $1 \le i \le r_3$,
- $\frac{\gamma_{0,i}}{\gamma_{0,j}} \ne \left(\frac{\gamma_{0,i}}{\gamma_{0,j}}(\frac{\gamma_{3,i}}{\gamma_{3,j}})^{-1}\right)^{M_3}$ when $i \ne j$,

where $M_3 = \frac{q^k-1}{q^3-1}$. Using them in Theorem 2 again, we can construct a cyclic code $\mathscr{C}_3$ of size up to $(q^n - 1)\frac{q^{N_3}-1}{q^3-1}$ and of minimum distance $\ge 2k - 2s = 6$ for some suitable $N_3$.

Now consider $r_2'$ polynomials

$$x^{q^6} + \gamma_{3,i} x^{q^3} + \gamma_{2,i} x^{q^2} + \gamma_{0,i} x \in \mathbb{F}_{q^n}, \quad 1 \le i \le r_2'$$

satisfying

- $\gamma_{0,i} \ne 0$, $\gamma_{2,i} \ne 0$ and $\gamma_{3,i} \ne 0$ for all $1 \le i \le r_2'$,
- $\frac{\gamma_{2,i}}{\gamma_{2,j}} \ne \left( \frac{\gamma_{2,i}}{\gamma_{2,j}} (\frac{\gamma_{3,i}}{\gamma_{3,j}})^{-1} \right)^{M_2'}$ when $i \ne j$,

where $M_2' = \frac{q^{k-2}-1}{q-1}$. Using them in Corollary 1, we can construct a cyclic code $\mathscr{C}_2'$ of size up to $(q^n - 1)(q^n - 1)\frac{q^{N_2}-1}{q-1}$ and of minimum distance $\ge 2k - 2s = 6$ for some suitable $N_2'$.

Clearly $\mathscr{C}_1$, $\mathscr{C}_2$, $\mathscr{C}_2'$ and $\mathscr{C}_3$ are disjoint (recall the uniqueness of subspace polynomials), and the minimum distance between any two of them is greater than or equal to 6 again. Therefore, we can construct a code $\mathscr{C} := \mathscr{C}_1 \cup \mathscr{C}_2 \cup \mathscr{C}_2' \cup \mathscr{C}_3$ of size up to

$$q^{2n}(q^n - 1)\frac{q^N - 1}{q - 1} + (q^n - 1)(q^n - 1)\frac{q^N - 1}{q - 1} + (q^n - 1)\frac{q^N - 1}{q^2 - 1} + (q^n - 1)\frac{q^N - 1}{q^3 - 1}$$

and of minimum distance 6, where $N$ is a multiple of $\text{lcm}(N_1, N_2, N_2', N_3)$.

## 2.3 Possible Values for the Length Parameter

Notice that parameter $N$ in Theorem 2 is not chosen freely, it is depending on some elements in $\mathbb{F}_{q^n}$. Accordingly, some $N$ values can not be achievable.

In [35] it was conjectured that there exists a cyclic code of size $\frac{q^N-1}{q-1}$ and of minimum distance $2k - 2$ in $\mathscr{G}_q(N, k)$, for all positive integers $N$ and $k$ such that $k \le N/2$. In [13] it was proved by an exhaustive search that there are no cyclic codes with parameters $N = 8$, $k = 4$ and $q = 2$; i.e. the conjecture is not true when $N = 2k$.

In [1] the authors give an approach to determine a set of possible $N$ values. Their proof is given by considering $\mathbb{F}_{q^{N_0}}$ which is the splitting field of $T(x) = x^{q^k} + x^q + x \in \mathbb{F}_q[x]$. Here, the set of infinitely many $N$ values is $N_0\mathbb{N} := \{N_0, 2N_0, \ldots\}$. Hence the proof can be completed using the proof of Theorem 2 taking $n = 1$, $r = 1$ and $\gamma_1 = 1$.

*Remark 9* When $q > 2$, the construction of a set of possible $N$ values given in [1] can be improved so that the set of possible $N$ values is denser. It can be done by taking the polynomials $x^{q^k} + a_1 x^q + a_0 x$ over $\mathbb{F}_q$ for all non-zero $a_1$ and non-zero $a_0$ instead of taking only $x^{q^k} + x^q + x$.

*Proof* Let $T(x) := x^{q^k} + a_1 x^q + a_0 x$ for some non-zero $a_1$ and $a_0$ in $\mathbb{F}_q$. Let $U$ be its kernel in the splitting field $\mathbb{F}_{q^{N_0}}$. It is easy to show that $\dim(\alpha U \cap \beta U) \le$

1 when $\frac{\alpha}{\beta} \notin \mathbb{F}_q$, as in the proof of Theorem 2. Later on, we take another polynomial $T'(x)$ by using another pair $(a'_1, a'_0)$ and obtain $N'_0$. Taking the union $N_0 \mathbb{N} \cup N'_0 \mathbb{N}$, we obtain an opportunity to enlarge the set of possible $N$ values. Then, we take another pair $(a''_0, a''_1)$ and so on.                                                    □

We think that this basic observation is worth emphasizing because of its usefulness that can be seen in the following example.

*Example 2* Let $q = 5$ and $k = 3$. If we consider only $T(x) = x^{q^k} + x^q + x$, then we obtain $N_0 = 62$. That is, $N$ values can be chosen from $62\mathbb{N}$. In addition, by considering $x^{q^k} + x^q + 4x$ we obtain $31\mathbb{N}$. Similarly, if we consider $x^{q^k} + x^q + 2x$, then we obtain $24\mathbb{N}$. By trying other non-zero couples, we get $31\mathbb{N} \cup 24\mathbb{N} \cup 20\mathbb{N}$ as a sample space for $N$.

Now, we recall some tools that we can use to factorize some linearized polynomials in an easier way.

**Definition 1** [19, Definition 3.58] The polynomials

$$l(x) := \sum_{i=0}^{N} \alpha_i x^i \text{ and } L(x) := \sum_{i=0}^{N} \alpha_i x^{q^i}$$

over $\mathbb{F}_{q^N}$ are called $q$−*associates* of each other. More specifically, $l(x)$ is the *conventional* $q$−*associate* of $L(x)$ and $L(x)$ is the *linearized* $q$−*associate* of $l(x)$.

**Lemma 1** ([19, Theorem 3.62]) *Let $L_1(x)$ and $L(x)$ be $q$−polynomials over $\mathbb{F}_q$ with conventional $q$−associates $l_1(x)$ and $l(x)$ respectively. Then $L_1(x)$ divides $L(x)$ if and only if $l_1(x)$ divides $l(x)$.*

**Lemma 2** ([19, Theorem 3.63]) *Let $f(x)$ be irreducible over $\mathbb{F}_q$ and let $F(x)$ be its linearized $q$−associate. Then the degree of every irreducible factor of $F(x)/x$ is equal to the order of $f(x)$.*

*Remark 10* The construction of the set of possible $N$ values can be improved in the following way: Factorize the conventional q-associate $t(x) = x^k + x + 1$ and find the orders of its irreducible factors, then take the least common multiple of these orders and thus determine $N$. In this way, the factorization cost can be decreased significantly.

The proof of Remark 10 is straightforward from Lemmas 1 and 2. In addition, Remarks 9 and 10 can be combined to obtain larger sets in easier ways. We illustrate the efficiency of Remark 10 by the following example.

*Example 3* Let $q = 5$ and $k = 3$ again. Instead of factorizing $T(x) = x^{125} + x^5 + x$, we can easily factorize $t(x) = x^3 + x + 1$, it is actually irreducible since it has no roots in $\mathbb{F}_5$, and its order can be only in $\{e \in \mathbb{N} : e|(5^3 - 1) \text{ and } e \geq k\} = \{4, 31, 62, 124\}$. Notice that $t(x)$ does not divide $x^4 - 1$ and $x^{31} - 1$ but divides $x^{62} - 1$, thus we find $N_0 = 62$ easily.

## 3 Construction of Maximum Rank Distance Codes

Note that chapter "Codes Endowed with the Rank Metric" provides some essential information about the analysis of rank metric codes in vector and matrix representation. In this section we just focus on the various constructions of MRD codes using the linearized polynomial representation.

Let $f(x) \in \mathbb{F}_{q^n}[x]$ be a $q$−polynomial of $q$−degree at most $n - 1$. Let $\{\delta_1, \delta_2, \ldots, \delta_n\}$ and $\{\varepsilon_1, \varepsilon_2, \ldots, \varepsilon_n\}$ be two ordered bases of $\mathbb{F}_{q^n}$ over $\mathbb{F}_q$. Then, for any $\alpha \in \mathbb{F}_{q^n}$ we have

$$
\begin{aligned}
f(\alpha) &= f(c_1\delta_1 + c_2\delta_2 + \cdots + c_n\delta_n) \\
&= c_1 f(\delta_1) + c_2 f(\delta_2) + \cdots + c_n f(\delta_n) \\
&= \begin{bmatrix} f(\delta_1) & f(\delta_2) & \ldots & f(\delta_n) \end{bmatrix} \begin{bmatrix} c_1 & c_2 & \ldots & c_n \end{bmatrix}^t \\
&= \begin{bmatrix} \varepsilon_1 & \varepsilon_2 & \ldots & \varepsilon_n \end{bmatrix} \begin{bmatrix} f(\delta_1)_{\varepsilon_1} & f(\delta_2)_{\varepsilon_1} & \ldots & f(\delta_n)_{\varepsilon_1} \\ f(\delta_1)_{\varepsilon_2} & f(\delta_2)_{\varepsilon_2} & \ldots & f(\delta_n)_{\varepsilon_2} \\ \vdots & \vdots & \ddots & \vdots \\ f(\delta_1)_{\varepsilon_n} & f(\delta_2)_{\varepsilon_n} & \ldots & f(\delta_n)_{\varepsilon_n} \end{bmatrix} \begin{bmatrix} c_1 \\ c_2 \\ \vdots \\ c_n \end{bmatrix}
\end{aligned}
\tag{7}
$$

for some $c_i \in \mathbb{F}_q$, for $1 \leq i \leq n$; where $f(\delta_i)_{\varepsilon_j}$ denotes the coefficient of $\varepsilon_j$ when $f(\delta_i)$ is written as a linear combination of $\varepsilon_1, \ldots, \varepsilon_n$, for all $1 \leq i, j \leq n$. Let $F$ denote the matrix given by $F = [f(\delta_i)_{\varepsilon_j}]_{j,i} \in \mathbb{F}_q^{n \times n}$. Notice that there is a one to one correspondence between $f$ and $F$ with respect to the fixed ordered bases $\{\delta_1, \delta_2, \ldots, \delta_n\}$ and $\{\varepsilon_1, \varepsilon_2, \ldots, \varepsilon_n\}$, also we have rank$(F)$ = rank$(f)$. Moreover, the algebra $\mathbb{F}_q^{n \times n}$ with the matrix addition and the matrix multiplication corresponds to the algebra

$$
\mathscr{L}_n := \{\alpha_0 x + \alpha_1 x^q + \cdots + \alpha_{n-1} x^{q^{n-1}} : \alpha_0, \ldots, \alpha_{n-1} \in \mathbb{F}_{q^n}\}
$$

with the addition and the composition of polynomials modulo $x^{q^n} - x$, respectively.

Let $f(x) = \sum_{i=0}^{n-1} \alpha_i x^{q^i} \in \mathscr{L}_n$ and let $\{\delta_1, \delta_2, \ldots, \delta_n\}$ given above be a normal basis. Also take $\varepsilon_i = \delta_i$ for all $i = 1, \ldots, n$. Then, the correspondence $f \leftrightarrow F$ above implies the correspondence $\widehat{f} \leftrightarrow F^t$, where $\widehat{f}$ is given by $\widehat{f}(x) = \sum_{i=0}^{n-1} \alpha_{n-i}^{q^i} x^{q^i}$ mod $x^{q^n} - x$ and the subscript is for modulo $n$. Here, $\widehat{f}$ is called the *adjoint* of $f$.

Consider the algebra $\mathscr{L}_n$ as the ambient space instead of the algebra $\mathbb{F}_q^{n \times n}$ while studying on rank metric codes. In that way, we can inherit the equivalence notion as follows. Let $\mathscr{C}$ and $\mathscr{C}'$ be non-empty subsets of $\mathscr{L}_n$, then $\mathscr{C}$ and $\mathscr{C}'$ are equivalent if and only if there exist $g_1, g_2, g_3 \in \mathscr{L}_n$ such that $g_1(x)$ and $g_2(x)$ are invertible, and

$$
\begin{aligned}
\mathscr{C}' &= g_1(x) \circ \mathscr{C} \circ g_2(x) + g_3(x) \\
&:= \{g_1(x) \circ f(x) \circ g_2(x) + g_3(x) \quad \text{mod } x^{q^n} - x : f(x) \in \mathscr{C}\}, \text{ or} \\
\mathscr{C}' &= g_1(x) \circ \widehat{\mathscr{C}} \circ g_2(x) + g_3(x) \\
&:= \{g_1(x) \circ \widehat{f}(x) \circ g_2(x) + g_3(x) \quad \text{mod } x^{q^n} - x : f(x) \in \mathscr{C}\},
\end{aligned}
\tag{8}
$$

where the ∘ operation denotes the composition. Notice also that, if $\mathscr{C}$ is closed under addition, then the minimum distance $d(\mathscr{C})$ of $\mathscr{C}$ is indeed the minimum non-zero rank of the elements in $\mathscr{C}$.

In this section, we introduce all MRD codes in the literature. Before that, we give a very useful lemma that efficiently works to construct new MRD codes.

**Lemma 3** ([14]) *Let* $f(x) = \alpha_0 x + \alpha_1 x^q + \cdots + \alpha_l x^{q^l} \in \mathbb{F}_{q^n}[x]$ *be a* $q-$ *polynomial of* $q-$*degree* $l$, *where* $0 < l < n$. *If* $\mathrm{Norm}_{q^n/q}(\alpha_l) \neq (-1)^{nk} \mathrm{Norm}_{q^n/q}(\alpha_0)$, *then* $\dim(\ker(f)) \leq l - 1$.

## 3.1  Linear MRD Codes

The first examples of MRD codes were naturally linear ones. The largest family including all known linear MRD codes can be stated as follows.

**Theorem 4** ([20, 31]) *Let* $n, k, s, h \in \mathbb{Z}^+$ *satisfying* $\gcd(n, s) = 1$ *and* $k < n - 1$. *Let* $\eta \in \mathbb{F}_{q^n}$ *such that* $\mathrm{Norm}_{q^n/q}(\eta) \neq (-1)^{nk}$. *Then the set*

$$\mathscr{H}_{n,k,s}(\eta, h) := \{\alpha_0 x + \alpha_1 x^{q^s} + \cdots + \alpha_{k-1} x^{q^{s(k-1)}} + \eta \alpha_0^{q^h} x^{q^{sk}} : \alpha_0, \alpha_1, \ldots, \alpha_{k-1} \in \mathbb{F}_{q^n}\}$$

*is a linear MRD code of size* $q^{nk}$ *and minimum distance* $n - k + 1$.

The codes constructed in Theorem 4 are called *generalized twisted Gabidulin codes* (or briefly *GTG codes*). The main tool used to prove the bound of minimum rank distance of such codes is Lemma 3. Some subclasses of this family can be listed as follows.

- The $\eta = 0$ and $s = 1$ case: *Gabidulin codes*, firstly given in [5, 12, 30].
- The $\eta = 0$ case: *Generalized Gabidulin codes*, firstly given in [15].
- The $s = 1$ case: *Twisted Gabidulin codes*, firstly given in [31]. A special case was also discovered independently in [25].

Classification of $\mathscr{H}_{n,k,s}(\eta, h)$ under the equivalence given in (8) is dependent on parameters. For example, non-Gabidulin examples of generalized Gabidulin codes do not appear for $1 \leq n \leq 4$. Similarly, the smallest non-Gabidulin example of linear MRD codes for $k = 2$ is $\mathscr{H}_{3,2,1}(2, 1)$. Notice also that there are no examples of linear codes which are not generalized Gabidulin when $q = 2$. We give the following theorem as an example of classification of $\mathscr{H}_{4,2,1}(\eta, 2)$ with respect to the profile of $\eta$.

**Theorem 5** ([25]) *Let* $\eta_1, \eta_2 \in \mathbb{F}_{q^4}$ *with* $\mathrm{Norm}(\eta_1) \neq 1$ *and* $\mathrm{Norm}(\eta_2) \neq 1$, *then* $\mathscr{H}_{4,2,1}(\eta_1, 2)$ *and* $\mathscr{H}_{4,2,1}(\eta_2, 2)$ *are equivalent if and only if*

$$\eta_1^{q^i} = \eta_2 c^{q^2 - 1} \text{ or } \eta_1^{-q^i} = \eta_2 c^{q^2 - 1}$$

*for some* $c \in \mathbb{F}_{q^4}$ *and* $i \in \{0, 1, 2, 3\}$.

*Proof* ($\Rightarrow$) : Assume there are invertible $q$−polynomials $A(x) = a_0 x + a_1 x^q + a_2 x^{q^2} + a_3 x^{q^3}$ and $B(x) = b_0 x + b_1 x^q + b_2 x^{q^2} + b_3 x^{q^3}$ in $\mathbb{F}_{q^4}[x]$ and an invertible map $\mathbb{F}_{q^4} \times \mathbb{F}_{q^4} \to \mathbb{F}_{q^4} \times \mathbb{F}_{q^4}$ given by $(\alpha_0, \alpha_1) \mapsto (\beta_0, \beta_1)$ such that

$$\alpha_0 x + \alpha_1 x^q + \eta_1 \alpha_0^{q^2} x^{q^2} \equiv A(x) \circ (\beta_0 x + \beta_1 x^q + \eta_2 \beta_0^{q^2} x^{q^2}) \circ B(x) \quad \mathrm{mod}\ x^{q^4} - x,$$

or

$$\alpha_0 x + \alpha_1 x^q + \eta_1 \alpha_0^{q^2} x^{q^2} \equiv A(x) \circ (\beta_0 x + \eta_2^{q^2} \beta_0 x^{q^2} + \beta_1^{q^3} x^{q^3}) \circ B(x) \quad \mathrm{mod}\ x^{q^4} - x.$$

for all $\alpha_0, \alpha_1 \in \mathbb{F}_{q^4}$. Then we should have

> (the coefficient of $x^{q^3}$) $\equiv 0 \mod x^{q^4} - x$, and
> (the coefficient of $x^{q^2}$) $\equiv \eta_1$(the coefficient of $x$)$^{q^2} \mod x^{q^4} - x$,

on the right hand side. This property implies a system of 16 equations with 8 unknowns $a_0, a_1, a_2, a_3, b_0, b_1, b_2, b_3$. Some algebraic manipulations on the solution of this system yield the result. Simultaneously, an invertible map $\mathbb{F}_{q^4} \times \mathbb{F}_{q^4} \to \mathbb{F}_{q^4} \times \mathbb{F}_{q^4}$ given by $(\alpha_0, \alpha_1) \mapsto (\beta_0, \beta_1)$ can be constructed.

($\Leftarrow$) : In each case, suitable invertible $A(x)$ and $B(x)$ couples can be determined computationally. Also an invertible map $\mathbb{F}_{q^4} \times \mathbb{F}_{q^4} \to \mathbb{F}_{q^4} \times \mathbb{F}_{q^4}$ given by $(\alpha_0, \alpha_1) \mapsto (\beta_0, \beta_1)$ can be constructed simultaneously. For example, consider case $\eta_1^{q^i} = \eta_2 c^{q^2-1}$ for some $c \in \mathbb{F}_{q^4}$ and $i \in \{0, 1, 2, 3\}$. Then $A(x) = cx^{q^i}$ and $B(x) = x^{q^{4-i}}$, which are clearly invertible, can be used to show the equivalence. Also we have $(\alpha_0, \alpha_1) = (c\beta_0^{q^i}, c\beta_1^{q^i})$. $\qquad\qquad\qquad\qquad\square$

Theorem 5 says that $\mathscr{H}_{4,2,1}(\eta_1, 2)$ and $\mathscr{H}_{4,2,1}(\eta_2, 2)$ are equivalent when $\mathrm{Norm}(\eta_1) = \mathrm{Norm}(\eta_2)$. However, we can have both equivalence and inequivalence when $\mathrm{Norm}(\eta_1) = \mathrm{Norm}(\eta_2)^{-1}$ as we can observe in the following example.

*Example 4* Let $q = 5$ and $\mathbb{F}_{q^4} = \mathbb{F}_q(\gamma)$, where $\gamma$ is a primitive element satisfying $\gamma^4 + \gamma^2 + 2\gamma + 2 = 0$.

1. Let $\eta_1 = \gamma^{39}$ and $\eta_2 = \gamma^{273}$. Then $\mathrm{Norm}(\eta_1) = 3$ and $\mathrm{Norm}(\eta_2) = 2$. Here, $\eta_1^{-1} = \eta_2 c^{q^2-1}$ for $c = \gamma^{26}$. Hence $\mathscr{H}_{4,2,1}(\eta_1, 2)$ and $\mathscr{H}_{4,2,1}(\eta_2, 2)$ are equivalent according to Theorem 5.
2. Let $\eta_1 = \gamma^{39}$ and $\eta_2 = \gamma^{253}$. Then $\mathrm{Norm}(\eta_1) = 3$ and $\mathrm{Norm}(\eta_2) = 2$, that is, $\mathrm{Norm}(\eta_1) = \mathrm{Norm}(\eta_2)^{-1}$. However, $\eta_1^{q^i} \neq \eta_2 c^{q^2-1}$ for all $c \in \mathbb{F}_{q^4}$ and $i \in \mathbb{Z}$. Therefore, $\mathscr{H}_{4,2,1}(\eta_1, 2)$ and $\mathscr{H}_{4,2,1}(\eta_2, 2)$ are not equivalent according to Theorem 5.

### 3.2   Additive MRD Codes

Consider case $q = q_0^u$ for some integer $u > 1$ and a prime power $q_0$. A set of $q$-polynomials, i.e. a rank metric code, can be $\mathbb{F}_{q_0}$-linear but does not have to be $\mathbb{F}_q$-linear. We can observe this situation in the following example.

*Example 5* Consider case $q_0 = 2, u = 2, n = 2$. Let $\mathbb{F}_4 = \mathbb{F}_2(\omega)$ where $\omega^2 + \omega + 1 = 0$, and $\mathbb{F}_{4^2} = \mathbb{F}_4(\zeta)$ where $\zeta^2 + \zeta + \omega = 0$. Then the set

$$\{\alpha x + (\omega\alpha + \omega\alpha^2 + \omega^2\alpha^4 + \omega^2\alpha^8)x^4 \in \mathbb{F}_{4^2}[x] : \alpha \in \mathbb{F}_{4^2}\}$$

is an MRD code with $d = 2$. Taking $\mathbb{F}_4$-bases $(\varepsilon_1, \varepsilon_2) = (1, \zeta)$ and $(\delta_1, \delta_2) = (1, \zeta)$ as in (7), we can construct the matrix version of this code as follows.

$$\mathscr{C} = \mathrm{Span}_{\mathbb{F}_2} \left\{ \begin{bmatrix} 1 & 0 \\ 0 & 1 \end{bmatrix}, \begin{bmatrix} 1 & 1 \\ 1 & 0 \end{bmatrix}, \begin{bmatrix} \omega & 0 \\ 1 & \omega \end{bmatrix}, \begin{bmatrix} \omega & \omega \\ \omega & 1 \end{bmatrix} \right\}$$

as a subset of $\mathbb{F}_4^{2 \times 2}$. Remark that this code is not linear (over $\mathbb{F}_4$), because

$$\omega \begin{bmatrix} 1 & 0 \\ 0 & 1 \end{bmatrix} = \begin{bmatrix} \omega & 0 \\ 0 & \omega \end{bmatrix} \notin \mathscr{C}.$$

At this point, we want to clarify a possible misunderstanding. Note that the field $\mathbb{F}_q = \mathbb{F}_{q_0^u}$ is isomorphic to a suitable subset of $\mathbb{F}_{q_0}^{u \times u}$. This well-known fact is available in [19, Sect. 2.5]. Therefore, it seems that there is a natural and trivial lifting an MRD code $\mathscr{C} \subseteq \mathbb{F}_q^{n \times n}$ with the minimum rank distance $d$ over $\mathbb{F}_q$ to another MRD code $\mathscr{C}' \subseteq \mathbb{F}_{q_0}^{nu \times nu}$ with the minimum rank distance $du$ over $\mathbb{F}_{q_0}$. It is actually true when $d = n$ as we can observe in Example 5 above, also there are some similar observations in [3, 31] considering semifields. However, it is not true when $d < n$, as we can see in the following proposition.

**Proposition 2** *Let $q = q_0^u$ for some prime power $q_0$ and integer $u > 1$. Let $\mathscr{C} \subseteq \mathbb{F}_q^{n \times n}$ be an MRD code with $d(\mathscr{C}) = d$. Let $\mathscr{C}' \subseteq \mathbb{F}_{q_0}^{nu \times nu}$ be the code obtained by writing the entries of matrices from $\mathscr{C}$ in matrix representation. Then $\mathscr{C}'$ is a rank metric code with $d(\mathscr{C}') = du$. Moreover, $\mathscr{C}'$ is an MRD code if and only if $d = n$.*

*Proof* The first part is clear from the isomorphism. The fact about being MRD can be observed from the inequality of sizes $q^{n(n-d+1)} \leq q_0^{nu(nu-du+1)}$, where the equality holds only when $d = n$.                                                                             □

*Remark 11* When we lift a code $\mathscr{C} \subseteq \mathbb{F}_q^{n \times n}$ to another code $\mathscr{C}' \subseteq \mathbb{F}_{q_0}^{nu \times nu}$ using matrix representation, the distance function also changes from $\mathrm{rank}(A - B)$ to $\mathrm{rank}_{\mathbb{F}_{q_0}}(A - B)$ (i.e. the rank of $A - B$ over $\mathbb{F}_{q_0}$). However, keeping it as $\mathrm{rank}(A - B)$ does not change the value of the distance.

Now, we give a family of MRD codes which includes also non-linear but additive MRD codes. Recall that Lemma 3 was used in [31] in order to construct Theorem

4. In case $q$ is not a prime, it is possible to reinterpret this family by using another basic observation, as demonstrated in the following theorem.

**Theorem 6** ([26]) *Let* $n, k, s, u, h \in \mathbb{Z}^+$ *satisfying* $\gcd(n, s) = 1$, $q = q_0^u$ *and* $k < n$. *Let* $\eta \in \mathbb{F}_{q^n}$ *such that* $\mathrm{Norm}_{q^n/q_0}(\eta) \neq (-1)^{nku}$. *Then the set*

$$\mathcal{H}_{n,k,s,q_0}(\eta, h) = \{\alpha_0 x + \alpha_1 x^{q^s} + \cdots + \alpha_{k-1} x^{q^{s(k-1)}}$$
$$+\eta \alpha_0^{q_0^h} x^{q^{sk}} : \alpha_0, \alpha_1, \ldots, \alpha_{k-1} \in \mathbb{F}_{q^n}\}$$

*is an* $\mathbb{F}_{q_0}$*–linear (but does not have to be linear) MRD code of size* $q^{nk}$ *and minimum distance* $n - k + 1$.

*Proof* Code size and $\mathbb{F}_{q_0}$–linearity are clear from the definition. The trick related to Lemma 3 about the norm of $\eta$ can be observed from the transitivity of norm function (see, for example, [19, Theorem 2.29]). □

We call the code family obtained in Theorem 6 as *additive generalized twisted Gabidulin codes*, or briefly *AGTG codes*. Investigation of equivalence among such codes and some further results are available in [26].

*Remark 12* The Venn diagram of families of all known additive MRD codes has been given in Fig. 1. The conditions about the parameters can be summarized as follows.

- When $u$ divides $h$, an AGTG code is a GTG code.
- When $u$ divides $h$ and $s = 1$, an AGTG code is a TG code.
- When $u$ divides $h$ and $\eta = 0$, an AGTG code is a GG code.
- When $u$ divides $h$, $s = 1$ and $\eta = 0$, an AGTG code is a Gabidulin code.

Now, we focus on the equivalence idea for additive codes. Consider $f(x)^{p^i} := x^{p^i} \circ f(x) \mod x^{q^n} - x$, where $p$ is the prime dividing $q$. If we expand $f(x)^{p^i}$ to the corresponding matrix as in (7), then we obtain the matrix $B^{-1} F^{p^i} A$, where $A$ and $B$ are invertible matrices satisfying $[\delta_1^{p^i} \ \ldots \ \delta_n^{p^i}]A = [\delta_1 \ \ldots \ \delta_n]$ and $[\varepsilon_1^{p^i} \ \ldots \ \varepsilon_n^{p^i}]B = [\varepsilon_1 \ \ldots \ \varepsilon_n]$. That is, $f(x)^{p^i}$ does not correspond to $F^{p^i}$ directly. In fact, $f(x)^{p^i}$ is not a $q$–polynomial when $i \not\equiv 0 \mod \log_p(q)$, so it corresponds to another $q$–polynomial which has the matrix form $B^{-1} F^{p^i} A$ when we fix the ordered bases $\{\varepsilon_1, \ldots, \varepsilon_n\}$ and $\{\delta_1, \ldots, \delta_n\}$. Therefore, in polynomial form, it makes sense to write the equivalence between two additive codes $\mathcal{C}$ and $\mathcal{C}'$ as follows: $\mathcal{C} \equiv \mathcal{C}'$ if and only if

$$\mathcal{C}' = g \circ \mathcal{C}^{p^i} \circ h := \{g(x) \circ f(x)^{p^i} \circ h(x) \mod x^{q^n} - x : f(x) \in \mathcal{C}\}, \text{ or}$$
$$\mathcal{C}' = g \circ \widehat{\mathcal{C}}^{p^i} \circ h := \{g(x) \circ \widehat{f}(x)^{p^i} \circ h(x) \mod x^{q^n} - x : f(x) \in \mathcal{C}\}$$

for some invertible $q$–polynomials $g, h \in \mathbb{F}_{q^n}[x]$ and a non-negative integer $i$, even if some elements are not $q$–polynomials.

**Fig. 1** Venn diagram of all
known additive MRD code
families

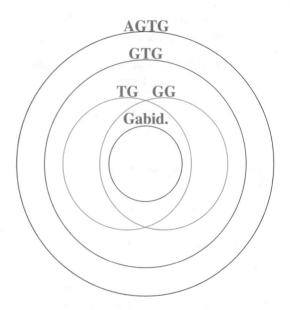

Notice that $\mathscr{H}_{n,k,s,q_0}(\eta, h)$ is not linear when $u \nmid h$, hence an AGTG code is not
equivalent to a GTG code when $u \nmid h$. It is because, a code which is equivalent to a
linear code must be linear, too. When $h$ is a multiple of $u$, $\mathscr{H}_{n,k,s,q_0}(\eta, h)$ is clearly
linear, and equivalent to a GTG code by a new $h := h/u$. This idea can be summed
up and summarized as follows.

**Proposition 3** *Let $\mathscr{C}_1$, $\mathscr{C}_2 \subseteq \mathbb{F}_q^{n \times n}$ be two additive rank metric codes and let $\mathbb{F}_{q_1}$, $\mathbb{F}_{q_2}$
be the largest subfields of $\mathbb{F}_q$ such that $\mathscr{C}_1$ is $\mathbb{F}_{q_1}$–linear and $\mathscr{C}_2$ is $\mathbb{F}_{q_2}$–linear. If
$q_1 \neq q_2$, then $\mathscr{C}_1$ and $\mathscr{C}_2$ are inequivalent.*

In that way, when we compare two AGTG codes, we observe the following.

**Corollary 4** *Let $q = q_1^{u_1} = q_2^{u_2}$. Then $\mathscr{H}_{n,k,s,q_1}(\eta, h)$ and $\mathscr{H}_{n,k,s,q_2}(\eta, h)$ are not
equivalent when $\gcd(h, u_1) \neq \gcd(h, u_2)$.*

## 3.3 A Family of Non-additive MRD Codes

Notice that being non-additive is not enough for a code to be new, since it can be
obtained by adding a fixed non-zero element to each codeword in an additive code.
We call such codes *affine codes*. However, there are also some non-additive codes
which are not affine. In this section, we construct such a family of MRD codes. For
$\lambda \in \mathbb{F}_{q^n} \setminus \{0\}$, let

$$\pi_\lambda := \{(\alpha x) \circ (x + \lambda x^q + \lambda^{1+q} x^{q^2} + \cdots + \lambda^{1+q+\cdots+q^{n-2}} x^{q^{n-1}}) \circ (\beta x) :$$
$$\alpha, \beta \in \mathbb{F}_{q^n} \setminus \{0\}\},$$
$$J_\lambda := \{(\alpha x) \circ (x - \lambda x^{q^{n-1}}) \circ (\beta x) : \alpha, \beta \in \mathbb{F}_{q^n} \setminus \{0\}\},$$
$$A_1 := \{\alpha x : \alpha \in \mathbb{F}_{q^n} \setminus \{0\}\},$$
$$A_2 := \{\alpha x^{q^{n-1}} : \alpha \in \mathbb{F}_{q^n} \setminus \{0\}\}.$$

**Lemma 4** ([7]) *Consider the definitions and notations above, then* $\pi_{\lambda_1} = \pi_{\lambda_2}$ *and* $J_{\lambda_1} = J_{\lambda_2}$ *when* $\mathrm{Norm}(\lambda_1) = \mathrm{Norm}(\lambda_2)$.

Let $\mathrm{Norm}(\lambda) = a$, Lemma 4 allows us to use $\pi_a$ and $J_a$ instead of using the notations $\pi_\lambda$ and $J_\lambda$, respectively.

**Theorem 7** ([7]) *Let* $q > 2$ *and* $n \geq 3$. *For any subset* $I$ *of* $\mathbb{F}_q \setminus \{0, 1\}$, *put* $\Pi_I = \bigcup_{a \in I} \pi_a$, $\Gamma_I = \bigcup_{b \in \mathbb{F}_q \setminus (I \cup \{0\})} J_b$ *and set*

$$\mathscr{A}_{n,I} := \Pi_I \cup \Gamma_I \cup A_1 \cup A_2 \cup \{0\} \subseteq \mathscr{L}_n.$$

*Then* $\mathscr{A}_{n,I}$ *is a non-additive and non-affine MRD code of minimum distance* $n - 1$.

*Remark 13* Notice that $\mathscr{A}_{n,I}$ is closed under multiplication by scalars, see [7] also for the geometric interpretation of Theorem 7.

*Remark 14* Theorem 7 is a generalization of the result in [4] which presents the particular case $\mathscr{A}_{3;I}$.

## 3.4 Another Family of Non-additive MRD Codes

Lemma 3 was used to construct AGTG codes. In this section, as a second application of Lemma 3 to construct new MRD codes, we provide a family of non-additive MRD codes.

**Theorem 8** ([27]) *Let* $I$ *be a subset of* $\mathbb{F}_q$, $k$ *and* $s$ *be positive integers such that* $k < n$ *and* $\gcd(n, s) = 1$. *Also let*

$$\mathscr{C}^{(1)}_{n,k,s,I} := \left\{ \sum_{i=0}^{k-1} \alpha_i x^{q^{si}} \mod x^{q^n} - x : \alpha_0, \ldots, \alpha_{k-1} \in \mathbb{F}_{q^n}, \mathrm{Norm}_{q^n/q}(\alpha_0) \in I \right\},$$

$$\mathscr{C}^{(2)}_{n,k,s,I} := \left\{ \sum_{i=1}^{k} \beta_i x^{q^{si}} \mod x^{q^n} - x : \beta_1, \ldots, \beta_k \in \mathbb{F}_{q^n}, \mathrm{Norm}_{q^n/q}(\beta_k) \notin (-1)^{n(k+1)} I \right\}.$$

*Then* $\mathscr{C}_{n,k,s,I} := \mathscr{C}^{(1)}_{n,k,s,I} \cup \mathscr{C}^{(2)}_{n,k,s,I}$ *is an MRD code with* $d(\mathscr{C}_{n,k,s,I}) = n - k + 1$.

*Proof* Firstly notice that $\mathscr{C}^{(1)}_{n,k,s,I}$ and $\mathscr{C}^{(2)}_{n,k,s,I}$ are disjoint. We can observe it as follows. In case $0 \notin I$, the coefficient of $x$ of each element of $\mathscr{C}^{(1)}_{n,k,s,I}$ is non-zero, whereas the coefficient of $x$ of each element of $\mathscr{C}^{(2)}_{n,k,s,I}$ is zero; therefore, $\mathscr{C}^{(1)}_{n,k,s,I}$ and $\mathscr{C}^{(2)}_{n,k,s,I}$ do not have common elements. In case $0 \in I$, the coefficient of $x^{q^k}$ of each element of $\mathscr{C}^{(2)}_{n,k,s,I}$ is non-zero, whereas the coefficient of $x^{q^k}$ of each element of $\mathscr{C}^{(1)}_{n,k,s,I}$ is zero; therefore, $\mathscr{C}^{(1)}_{n,k,s,I}$ and $\mathscr{C}^{(2)}_{n,k,s,I}$ do not have common elements in this case, too.

In addition, recall that the norm function is a $\frac{q^n-1}{q-1}$ to 1 function on the set of non-zero elements of $\mathbb{F}_{q^n}$, and sends only zero to zero. Therefore, we observe that

$$|\mathscr{C}^{(1)}_{n,k,s,I}| = q^{n(k-1)}|I|\left(\frac{q^n-1}{q-1}\right),$$
$$|\mathscr{C}^{(2)}_{n,k,s,I}| = q^{n(k-1)}\left[1 + (q-1-|I|)\left(\frac{q^n-1}{q-1}\right)\right]$$

if $0 \notin I$, or

$$|\mathscr{C}^{(1)}_{n,k,s,I}| = q^{n(k-1)}\left[1 + (|I|-1)\left(\frac{q^n-1}{q-1}\right)\right],$$
$$|\mathscr{C}^{(2)}_{n,k,s,I}| = q^{n(k-1)}(q-|I|)\left(\frac{q^n-1}{q-1}\right)$$

otherwise. In both cases,

$$|\mathscr{C}^{(1)}_{n,k,s,I}| + |\mathscr{C}^{(2)}_{n,k,s,I}| = q^{nk},$$

which is equal to $|\mathscr{C}_{n,k,s,I}|$ since $\mathscr{C}^{(1)}_{n,k,s,I}$ and $\mathscr{C}^{(2)}_{n,k,s,I}$ are disjoint.

Now, we need to show that $\text{rank}(f - g) \geq n - k + 1$ for all $f, g \in \mathscr{C}_{n,k,s,I}$. Let $f(x) - g(x) = \theta_0 x + \theta_1 x^q + \cdots + \theta_k x^{q^k}$ for arbitrary $f, g \in \mathscr{C}_{n,k,s,I}$. If $f, g \in \mathscr{C}^{(1)}_{n,k,s,I}$ (or $f, g \in \mathscr{C}^{(2)}_{n,k,s,I}$), then the $q$-degree of $f(x) - g(x)$ and the coefficient of $x$ implies that $\dim(\ker(f(x) - g(x))) \leq k - 1$. If $f \in \mathscr{C}^{(1)}_{n,k,s,I}$ and $g \in \mathscr{C}^{(2)}_{n,k,s,I}$ (or vice verse), then we have $\text{Norm}_{q^n/q}(\theta_0) \neq (-1)^{nk}\text{Norm}_{q^n/q}(\theta_k)$ since $\text{Norm}_{q^n/q}(\theta_0) \in I$ whereas $\text{Norm}_{q^n/q}(\theta_k) \in \mathbb{F}_q \setminus (-1)^{nk}I$; i.e. $\dim(\ker(f(x) - g(x))) \leq k - 1$ by Lemma 3.

Combining these four cases, we deduce that $\text{rank}(f(x) - g(x)) \geq n - (k-1) = n - k + 1$ for all $f, g \in \mathscr{C}_{n,k,s,I}$.

Conclusively, we have proved that $|\mathscr{C}_{n,k,s,I}| = q^{nk}$ and $d(\mathscr{C}_{n,k,s,I}) = n - k + 1$, and hence $\mathscr{C}_{n,k,s,I}$ is an MRD code. $\qquad\square$

The code $\mathscr{C}_{n,k,s,I}$ given in Theorem 8 is a generalized Gabidulin code when $q = 2$ or $I = \{0\}$ or $I = \mathbb{F}_q \setminus \{0\}$. However, $\mathscr{C}_{n,k,s,I}$ is not equivalent to any additive MRD codes when $q > 2$, $I \neq \{0\}$ and $I \neq \mathbb{F}_q \setminus \{0\}$.

**Theorem 9** *The code $\mathscr{C}_{n,k,s,I}$ given in Theorem 8 has the following properties.*

1. *If $q = 2$ or $I = \{0\}$ or $I = \mathbb{F}_q \setminus \{0\}$, then $\mathscr{C}_{n,k,s,I}$ is a generalized Gabidulin code.*
2. *If $q > 2$ and $I \neq \{0\}$ and $I \neq \mathbb{F}_q \setminus \{0\}$, then $\mathscr{C}_{n,k,s,I}$ is not an affine code (i.e., not a translated version of an additively closed code).*

When we take two arbitrary $I$ and $J$ subsets of $\mathbb{F}_q$, we can not say anything about the equivalence between $\mathscr{C}_{n,k,s,I}$ and $\mathscr{C}_{n,k,s,J}$ for now. However, we can see that they are equivalent if $J = \mathbb{F}_q \setminus I$, taking the adjoint polynomials into account.

**Corollary 5** *Let $I$ be a subset of $\mathbb{F}_q$, $k$ and $s$ be positive integers such that $k < n$ and $\gcd(n, s) = 1$. Then $\mathscr{C}_{n,k,s,\mathbb{F}_q \setminus I}$ is equivalent to $\mathscr{C}_{n,k,s,I}$.*

In addition, we can give another result about the equivalence between $\mathscr{C}_{n,k,s,I}$ and $\mathscr{C}_{n,k,s,J}$ when both $I$ and $J$ include only one non-zero element.

**Corollary 6** *Let $a$ and $b$ be non-zero elements of $\mathbb{F}_q$, $k$ and $s$ be positive integers such that $k < n$ and $\gcd(n, s) = 1$. Then $\mathscr{C}_{n,k,s,\{a\}}$ is equivalent to $\mathscr{C}_{n,k,s,\{b\}}$.*

The $\mathscr{C}_{n,k,s,I}$ family in Theorem 8 is called *partition codes*.

*Remark 15* Some properties of non-additive code families $\mathscr{A}_{n,I}$ constructed in Theorem 7 and $\mathscr{C}_{n,k,s,I}$ constructed in Theorem 8 in comparison can be listed as follows.

- The construction of $\mathscr{C}_{n,k,s,I}$ works for all parameters, whereas the minimum distance of an $\mathscr{A}_{n,I}$ code is fixed to be $n - 1$.
- An $\mathscr{A}_{n,I}$ code is closed under multiplication by scalars, but a $\mathscr{C}_{n,k,s,I}$ code is not always.

The following example provides partition codes which are not twisted (i.e., not additively closed).

*Example 6* • Let $q = 5$, $n = 3$, $k = 2$, $s = 1$ and $I = \{1, 2, 3\}$. Then,

$$\mathscr{C}_{n,k,s,I}^{(1)} = \{\alpha_0 x + \alpha_1 x^q : \alpha_0, \alpha_1 \in \mathbb{F}_{q^n} \text{ and } \mathrm{Norm}(\alpha_0) \in \{1, 2, 3\}\},$$
$$\mathscr{C}_{n,k,s,I}^{(2)} = \{\beta_1 x^q + \beta_2 x^{q^2} : \beta_1, \beta_2 \in \mathbb{F}_{q^n} \text{ and } \mathrm{Norm}(\beta_2) \in \{0, 1\}\}$$

and the resulting partition code $\mathscr{C}_{n,k,s,I}$ is not additively closed.
- Let $q = 4$, $n = 5$, $k = 1$, $s = 2$ and $I = \{1, u\}$, where $u \in \mathbb{F}_q$ satisfies $u^2 + u + 1 = 0$. Then,

$$\mathscr{C}_{n,k,s,I}^{(1)} = \{\alpha_0 x : \alpha_0 \in \mathbb{F}_{q^n} \text{ and } \mathrm{Norm}(\alpha_0) \in \{1, u\}\},$$
$$\mathscr{C}_{n,k,s,I}^{(2)} = \{\beta_1 x^{q^s} : \beta_1 \in \mathbb{F}_{q^n} \text{ and } \mathrm{Norm}(\beta_1) \in \{0, u + 1\}\}.$$

Similarly, the resulting partition code is not additively closed.

*Example 7* Let $q = 7$, $n = 3$, $k = 2$, $s = 1$ and $I = \{1, 2\}$. Then,

$$\mathscr{C}_{n,k,s,I}^{(1)} = \{\alpha_0 x + \alpha_1 x^q : \alpha_0, \alpha_1 \in \mathbb{F}_{q^n} \text{ and } \mathrm{Norm}(\alpha_0) \in \{1, 2\}\},$$
$$\mathscr{C}_{n,k,s,I}^{(2)} = \{\beta_1 x^q + \beta_2 x^{q^2} : \beta_1, \beta_2 \in \mathbb{F}_{q^n} \text{ and } \mathrm{Norm}(\beta_2) \in \{0, 1, 2, 3, 4\}\}.$$

Remember that the norm function is multiplicative. Therefore, the resulting partition code $\mathscr{C}_{n,k,s,I}$ is not closed under multiplication by a scalar (and also not closed under addition).

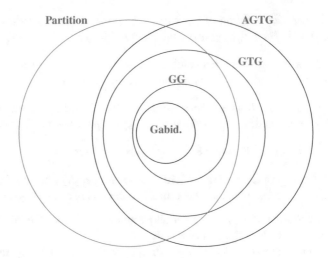

**Fig. 2** Venn diagram of MRD code families

## 3.5 Some Concluding Remarks

We may give the Venn diagram of involvements for all MRD codes except Theorem 7 in Fig. 2.

Singleton-like bound and Lemma 3 provide some connections with some certain optimal sets as we can see in the following corollary.

**Corollary 7** *Let k be a positive integer. The size of the largest possible subset $T \subseteq \mathbb{F}_{q^n} \times \mathbb{F}_{q^n}$ satisfying*

$$\beta_1 - \beta_2 = 0, \text{ or } \alpha_1 - \alpha_2 = 0, \text{ or } \text{Norm}(\beta_1 - \beta_2) \neq (-1)^{nk} \text{Norm}(\alpha_1 - \alpha_2)$$

*for all $(\alpha_1, \beta_1), (\alpha_2, \beta_2) \in T$ is $q^n$.*

*Example 8* Sheekey's example from [31]: Let $h$ be a positive integer and $\eta \in \mathbb{F}_{q^n}$ satisfying Norm $\neq (-1)^{nk}$. Then

$$T_{\eta,h} := \{(\alpha, \eta \alpha^{q^h}) : \alpha \in \mathbb{F}_{q^n}\}$$

is an optimal set given as in Corollary 7.

*Example 9* Example from the construction in Theorem 8: Let $I$ be a subset of $\mathbb{F}_q$. Then

$$T_I := \{(\alpha, 0), (0, \beta) : \alpha, \beta \in \mathbb{F}_{q^n}, \text{Norm}(\alpha) \in I, \text{Norm}(\beta) \notin (-1)^{n(k+1)} I\}$$

is another optimal set given as in Corollary 7.

As observed in Examples 8 and 9, finding new optimal sets as in Corollary 7 gives an opportunity to construct new MRD codes. However, testing the equivalence with the existing codes should be taken into account, too. There are several such sets that can be found by computation easily, but most of their MRD codes were equivalent to some known codes.

**Acknowledgements** The authors thank COST for the support under Action IC 1104. The first author thanks also TÜBİTAK for the support under the program BİDEB 2211.

# References

1. E. Ben-Sasson, T. Etzion, A. Gabizon, N. Raviv, Subspace polynomials and cyclic subspace codes. IEEE Trans. Inf. Theory **62**, 1157–1165 (2016)
2. M. Braun, T. Etzion, P. Ostergard, A. Vardy, A. Wasserman, Existence of q-analogues of Steiner systems. Forum Math. Pi **4**, 14 (2016). https://doi.org/10.1017/fmp.2016.5
3. J. Cruz, M. Kiermaier, A. Wassermann, W. Willems, Algebraic structures of MRD codes. Adv. Math. Commun. **10**, 499–510 (2016)
4. A. Cossidente, G. Marino, F. Pavese, Non-linear maximum rank distance codes. Des. Codes Cryptogr. **79**, 597–609 (2016)
5. P. Delsarte, Bilinear forms over a finite field, with applications to coding theory. J. Comb. Theory A **25**, 226–241 (1978)
6. K. Drudge, On the orbits of Singer groups and their subgroups. Electron. J. Comb. **9** (2002)
7. N. Durante, A. Siciliano, Non-linear maximum rank distance codes in the cyclic model for the field reduction of finite geometries, arXiv: 1704.02110 [cs.IT]
8. T. Etzion, Problems on q-analogs in coding theory, arXiv: 1305.6126 [cs.IT]
9. T. Etzion, E. Gorla, A. Ravagnani, A. Wachter-Zeh, Optimal Ferrers diagram rank-metric codes. IEEE Trans. Inf. Theory **62**, 1616–1630 (2016)
10. T. Etzion, N. Silberstein, Error-correcting codes in projective spaces via rank-metric codes and Ferrers diagrams. IEEE Trans. Inf. Theory **55**, 2909–2919 (2009)
11. T. Etzion, A. Vardy, Error correcting codes in projective space. IEEE Trans. Inf. Theory **57**, 1165–1173 (2011)
12. E.M. Gabidulin, The theory with maximal rank metric distance. Probl. Inform. Trans. **21**, 1–12 (1985)
13. H. Gluesing-Luerssen, K. Morrison, C. Troha, Cyclic orbit codes and stabilizer subfields. Adv. Math. Commun. **25**, 177–197 (2015)
14. R. Gow, R. Quinlan, Galois theory and linear algebra. Linear Algebra Appl. **430**, 1778–1789 (2009)
15. A. Kshevetskiy, E.M. Gabidulin, The new construction of rank codes. In: Proceedings of the IEEE International Symposium on Information Theory, 2105–2108 (2005)
16. A. Kohnert, S. Kurz, Construction of large constant dimension codes with a prescribed minimum distance. Lect. Notes Comput. Sci. **5393**, 31–42 (2008)
17. R. Kötter, F.R. Kschischang, Coding for errors and erasures in random network coding. IEEE Trans. Inf. Theory **54**, 3579–3591 (2008)
18. A. Klein, K. Metsch, L. Storme, Small maximal partial spreads in classical finite polar spaces. Adv. Geom. **10**, 379–402 (2010)
19. R. Lidl, H. Niederreiter, Finite fields. In: Encyclopedia of Mathematics and its Applications, vol. 20. Cambridge University Press, Cambridge (1997)
20. G. Lunardon, R. Trombetti, Y. Zhou, Generalized twisted Gabidulin codes, arXiv: 1507.07855v2 [math.CO]

21. P.J. Lusina, E.M. Gabidulin, M. Bossert, Maximum rank distance codes as space-time codes. IEEE Trans. Inf. Theory **49**, 2757–2760 (2003)
22. K. Marshall, A.-L. Horlemann-Trautmann, New criteria for MRD and Gabidulin codes and some rank-metric code constructions. Adv. Math. Commun. (to appear)
23. K. Morrison, Equivalence of rank-metric and matrix codes and automorphism groups of Gabidulin codes. IEEE Trans. Inf. Theory **60**, 7035–7046 (2014)
24. O. Ore, On a special class of polynomials. Trans. Am. Math. Soc. **35**, 559–584 (1933)
25. K. Otal, F. Özbudak, Explicit constructions of some non-Gabidulin linear MRD codes. Adv. Math. Commun. **10**, 589–600 (2016)
26. K. Otal, F. Özbudak, Additive rank metric codes. IEEE Trans. Inf. Theory **63**, 164–168 (2017)
27. K. Otal, F. Özbudak, Some new non-additive maximum rank distance codes (submitted)
28. K. Otal, F. Özbudak, Cyclic subspace codes via subspace polynomials. Des. Codes Cryptogr. https://doi.org/10.1007/s10623-016-0297-1
29. R. Overbeck, Structural attacks for public key cryptosystems based on Gabidulin codes. J. Cryptol. **21**, 280–301 (2008)
30. R.M. Roth, Maximum-rank array codes and their application to crisscross error correction. IEEE Trans. Inf. Theory **37**, 328–336 (1991)
31. J. Sheekey, A new family of linear maximum rank distance codes. Adv. Math. Commun. **10**, 475–488 (2016)
32. D. Silva, F.R. Kschischang, Universal secure network coding via rank-metric codes. IEEE Trans. Inf. Theory **57**, 1124–1135 (2011)
33. D. Silva, F.R. Kschischang, R. Kötter, A rank-metric approach to error control in random network coding. IEEE Trans. Inf. Theory **54**, 3951–3967 (2008)
34. V. Tarokh, N. Seshadri, A.R. Calderbank, Space-time codes for high data rate wireless communication: performance criterion and code construction. IEEE Trans. Inf. Theory **44**, 744–765 (1998)
35. A.-L. Trautmann, F. Manganiello, M. Braun, J. Rosenthal, Cyclic orbit codes. IEEE Trans. Inf. Theory **59**, 7386–7404 (2013)
36. Z.-X. Wan, Geometry of matrices. In: Memory of Professor L.K. Hua (1910–1985). (World Scientific, Singapore, 1996)

# Generalizing Subspace Codes to Flag Codes Using Group Actions

Dirk Liebhold, Gabriele Nebe and María Ángeles Vázquez-Castro

**Abstract** In this chapter, we first discuss structural properties of subspace codes by applying group actions. We will see that the projective space, the Grassmannian and the Schubert cells can be realized by looking at the orbits of certain linear groups on matrices. The approach naturally leads to a generalization, where we replace subspace codes with flag codes. We describe and analyze a channel model for these new codes that is similar to the operator channel introduced by Koetter and Kschischang, give examples of flag codes and discuss how to compute sphere packing and sphere covering bounds for flag codes.

## 1 Introduction

Network Coding is by now a well-established theoretical concept for transmission of information over a network. The main difference compared to routing is that in a network coded network, nodes cannot only forward network information units but can also perform linear operations on them.

In their seminal work [7], Koetter and Kschischang described such a network coded network as an operator channel for which input and output alphabets are given as subspaces. Their results triggered a wide research into the field of subspace codes, see for example [1, 2, 4, 10, 11].

In this work, we will analyze the structure of subspace codes using group actions. This will suggest flag codes [9, 14] as a natural generalization of subspace codes and

D. Liebhold (✉) · G. Nebe
RWTH Aachen, 52062 Aachen, Germany
e-mail: liebhold@mathb.rwth-aachen.de

G. Nebe
e-mail: nebe@math.rwth-aachen.de

M. Á. Vázquez-Castro
Department of Telecommunications and Systems Engineering, Universitat Autònoma de Barcelona, Cerdanyola del Vallès, 08193 Barcelona, Spain
e-mail: angeles.vazquez@uab.cat

© Springer International Publishing AG 2018
M. Greferath et al. (eds.), *Network Coding and Subspace Designs*,
Signals and Communication Technology,
https://doi.org/10.1007/978-3-319-70293-3_4

67

we will therefore discuss such codes, introducing distance functions, cells, circles and sphere packing bounds.

The chapter is structured as follows. In the first section, we collect some results on subspace codes. We describe the operator channel by Koetter and Kschischang and the Ferrers diagram approach to the construction of subspace codes. Here we also introduce the term of a cell in the set of subspaces. In the second section, we talk about group actions. After briefly introducing the concept and giving some examples, we let certain groups act on matrices and on subspaces, giving us structural results. For example, it will turn out that each cell is an orbit code in itself, that is an orbit under a subgroup of the general linear group.[1] These actions lead to the concept of flags, which we then discuss in the third section. We extend the notion of cells to the set of flags, discuss distance functions and show how the subspace case embeds into this more general theory. In the fourth section, we briefly discuss flag codes and show how to alter the network model to allow for a flag to be transmitted rather than a subspace. We also give sphere packing and sphere covering bounds in a special case and discuss the $\ell - \partial p$ polynomial, which allows us to read off these bounds, although an efficient algorithm for its computation is not yet known.

## 2   Preliminaries

In this section, we collect known results on subspace codes. We will start by briefly describing the network model introduced in [7], and then discuss how to use matrices when working with subspaces, finishing with the Ferrers diagram construction introduced in [1]. We will focus on the basic concepts, which will be reviewed later on with the help of group actions and generalized from subspaces to flags. For a more detailed overview see chapter "Constructions of Constant Dimension Codes".

### 2.1   The Network Model for Subspace Codes

We consider the multicast scenario, having a single source and multiple receivers. A finite field $\mathbb{F}_q$ is fixed and the source injects vectors (packets) of constant length $m$, i.e. elements of $\mathbb{F}_q^m$, into the network. These vectors are the information units that are sent through the network to realize communication between the source and the multiple receivers. We will denote such vectors as information packets. In-network intermediate nodes form linear combinations (e.g. at random) of their received packets and forward the result and each receiver tries to recompute the original message from his received vectors.

Now we assume that the process of communication consists of sending $n$ consecutive packets. This set of packets is called "generation" in the related literature,

---

[1]For more details on orbit codes, see chapter "Constructions of Constant Dimension Codes" by Horlemann-Trautmann and Rosenthal.

with $n$ being the generation size. The channel model introduced in [7] is based on the fact that having a generation of $n$ packets $x_1, x_2, \ldots x_n \in \mathbb{F}_q^m$, what is left invariant when forming linear combinations of these vectors is the corresponding vector space, $U := \langle x_1, x_2, \ldots, x_n \rangle$. Assuming no errors or erasures, a receiver can recover the subspace $U$ – or rather a reduced basis of $U$ – as soon as he has collected $n$ linearly independent vectors. Thus the operator channel given in [7] uses the set $\mathscr{P}(\mathbb{F}_q^m)$ of all subspaces of $\mathbb{F}_q^m$ both as input and output alphabet. This set can be further broken down into the Grassmannians, $\mathrm{Gr}_n(\mathbb{F}_q^m)$, the set of all subspaces of dimension $n$ of $\mathbb{F}_q^m$.

As the codewords are full vector spaces in this setting, a bit-wise error count is not reasonable; one should rather define errors and erasures dimension-wise. Assume that the space $U$ was sent and a receiver observes a space $W$. Then the network can be modeled as an operator channel as follows [7]

$$W = \mathscr{H}_k(U) \oplus E$$

where $\mathscr{H}_k(U) = W \cap U$ is a subspace of dimension $k$ of $U$ and $E$ is some subspace of $W$ satisfying $E \cap U = \{0\}$. Therefore, $\mathscr{H}_k(U)$ is the part of $U$ that was not erased by the network and $E$ is an error space. In this case, we say that the network has introduced $\rho := \dim(U) - k$ erasures and $t := \dim(E)$ errors. Network erasures model the loss of dimension of the original space, which is due to the fact that the underlying protocols of the network erase packets, for example because they are found corrupted. Network errors on the other hand model the injection into the network of exogenous packets that were not contained in the original generation of packets (that generate $U$). The distance function counting errors and erasures is the Grassmannian distance,

$$\rho + t = \dim(U + W) - \dim(U \cap W).$$

As this is a distance function, it also obeys a triangle inequality. This allows to show that given a set of subspaces with minimal distance $d_{min}$, a minimal distance decoder can correct all errors and erasures as long as $2(\rho + t) < d_{min}$ [7, Theorem 2].

## 2.2  Matricial Representation of Subspaces

In this subsection, we discuss the matricial representation of subspaces. We will show how the structural properties of such a representation allow for both efficient computations and for the construction of subspace codes in matricial representation.

Most of the concepts for this subsection were introduced in [11] and refined and used for code construction in [1].

Given a subspace $U \leq \mathbb{F}_q^m$, there are different ways to represent it. First, it is a finite set, so one could write down the elements. However, if $d = \dim(U)$ this would be a set of size $q^d$, which is often too big to handle efficiently. A more efficient way is by computing a basis of $U$. This would be a set of size $d$ which can be computed using Gaussian elimination. However, such a basis is not unique; the space $U$ of dimension $d$ has $\prod_{i=0}^{d-1}(q^d - q^i)$ ordered bases. To achieve uniqueness, we look at matrices in reduced row echelon form.

**Definition 1**  A matrix $X \in \mathbb{F}_q^{d \times m}$ of full rank $d \leq m$ is in reduced row echelon form, if

1. For every $1 \leq i \leq d$ the first non-zero entry in the $i$th row is a 1.
2. If $r_i$ is the column containing the first non-zero entry of row $i$, then we have
   $$r_1 < r_2 < \ldots < r_d.$$
3. The column $r_i$ does only contain the single 1 in row $i$; all other entries are zero.

The binary vector $v(X) \in \mathbb{F}_2^m$ defined by $v(X)_j = 1$ if $j = r_i$ for some $1 \leq i \leq d$ and $v(X)_j = 0$ otherwise, is called the identifying vector of $X$.

The reduced row echelon form is well-known as the form returned by the classical Gaussian elimination. For a quick example consider

$$X = \begin{pmatrix} 1 & 1 & 0 & 1 & 1 & 0 \\ 0 & 0 & 1 & 0 & 1 & 0 \\ 0 & 0 & 0 & 0 & 0 & 1 \end{pmatrix}.$$

This matrix is in reduced row echelon form and the corresponding identifying vector is given by $v(X) = (1, 0, 1, 0, 0, 1)$. By using Gaussian elimination, every matrix can be transformed to a matrix in reduced row echelon form. As this transformation only forms linear combinations of the rows, the row space of $X$ does not change. Hence, by writing an arbitrary basis of $U$ as rows into a matrix and performing Gaussian elimination, we get that every vector space $U$ has a unique ordered basis in reduced row echelon form.

**Definition 2**  Let $U \leq \mathbb{F}_q^m$ be a subspace having dimension $d$. A matrix $X \in \mathbb{F}_q^{d \times m}$ is called the basis matrix of $U$, if $X$ is in reduced row echelon form and the rows of $X$ form a basis of $U$. The identifying vector of $U$ is then defined to be the identifying vector of $X$, $v(U) := v(X)$.

We now have a way to store the information of a subspace uniquely in a matrix. Note that the Hamming weight of $v(U)$ satisfies $w_H(v(U)) = \dim(U)$. This will become important when constructing codes later on.

To store the information of a subspace more compactly, we look at the free positions in $X$. As an example, take the matrix $X$ from above, having as identifying vector $v(X) = (1, 0, 1, 0, 0, 1)$. Every matrix having this identifying vector is of the form

$$\begin{pmatrix} 1 & \bullet & 0 & \bullet & \bullet & 0 \\ 0 & 0 & 1 & \bullet & \bullet & 0 \\ 0 & 0 & 0 & 0 & 0 & 1 \end{pmatrix}$$

where $\bullet$ can be any element of $\mathbb{F}_q$ and we can hence conclude that there are exactly $q^5$ subspaces of $\mathbb{F}_q^m$ having this identifying vector.

**Definition 3** For a given binary vector $w \in \mathbb{F}_2^m$, the set of all subspaces of $\mathbb{F}_q^m$ having this identifying vector is called the Schubert cell of $w$,

$$C(w) := \{U \leq \mathbb{F}_q^m \mid v(U) = w\}.$$

In the case that the base field might vary, we write $C_q(w)$. To store all necessary information for a subspace, we hence only need the identifying vector and the entries for the free positions. The free positions in such a matrix form a Ferrers-diagram, in the above case the diagram

$$\bullet\ \bullet\ \bullet$$
$$\bullet\ \bullet$$

If all free entries are zero, we call the space the standard space with identifying vector $v$, written as $S(v)$.

**Definition 4** Let $n$ be a positive integer and $n = n_1 + n_2 + \ldots + n_k$ be a partition of $n$ into positive integers, ordered such that $n_1 \geq n_2 \geq \ldots \geq n_k$. Then a representation of this partition by patterns of dots such that the $i$th row has the same number of dots as the $i$th term in the partition and in each row the dots are shifted to the right is called a Ferrers diagram.

Note that such diagrams are closely related to Young tableaux. Taking a Ferrers diagram, one can construct a Ferrers diagram rank metric code [1].

**Definition 5** Let $\mathscr{F}$ be a Ferrers diagram with $n$ rows and $k$ columns. A subspace $\mathscr{C} \leq \mathbb{F}_q^{n \times k}$ is called a Ferrers diagram rank metric code with respect to $\mathscr{F}$, if every matrix in $\mathscr{C}$ has zeros in the positions where $\mathscr{F}$ has no dots.

The distance function on such a matrix code is called the rank metric. For two matrices $A$, $B$ it is defined as $d_r(A, B) := \text{rk}(A - B)$. We have seen that a subspace $U$ of $\mathbb{F}_q^m$ can be uniquely described through its identifying vector $v(U)$ and a matrix $A$ with corresponding Ferrers diagram. Given an identifying binary vector $v$ and the corresponding matrix $A$, the subspace $U := \ell(v, A)$ is called the lift of $w$ and $A$. An important relation regarding the distances was shown in [11].

**Lemma 1** *Let $v \in \mathbb{F}_2^n$ and let $A$, $B$ be matrices for the Ferrers diagram induced by $v$. Then $d(\ell(v, A), \ell(v, B)) = 2d_r(A, B)$.*

Thus the rank distance gives us the subspace distance as long as we are in the same cell. To illustrate some ideas, we prove this lemma for the special case of $v = 1^k 0^{m-k} = (1, 1, \ldots, 1, 0, 0, \ldots, 0)$. For given matrices $A$, $B$, the basis matrices for the two subspaces $U := \ell(v, A)$ and $W := \ell(v, B)$ will be $(I_k A)$ and $(I_k B)$ and the sum $U + W$ is the row space of

$$\begin{pmatrix} I_k & B \\ I_k & A \end{pmatrix}.$$

Using Gaussian elimination, we reduce this matrix to

$$\begin{pmatrix} I_k & B \\ 0 & A - B \end{pmatrix}$$

and as $\dim(U + W)$ is the rank of this matrix, we see that

$$\dim(U + W) = k + \mathrm{rk}(A - B).$$

The well-known formula

$$\dim(U + W) = \dim(U) + \dim(W) - \dim(U \cap W)$$

now also allows us to compute $\dim(U \cap W)$, giving $d(U, W) = 2d_r(A, B)$ in the end.

In general, given two vector spaces $U$ and $W$ by a matrix in reduced row echelon form or by an identifying vector and a matrix, computing the representation of $U + W$ can be done efficiently by Gaussian elimination. A representation for $U \cap W$ can be obtained similarly using Zassenhaus' algorithm.

### 2.3 Subspace Codes

Above, we saw that every subspace has an identifying vector and a Ferrers diagram. Here, we will discuss a construction from [1], where a certain set of vectors is prescribed and the problem of finding good Ferrers diagram rank metric codes is tackled for every cell separately.

First, we fix some notations. A subset $\mathscr{C} \subseteq \mathscr{P}(\mathbb{F}_q^m)$ is called a subspace code. The minimal distance of $\mathscr{C}$ is defined as

$$d_{\min}(\mathscr{C}) := \min\{d(U, W) \mid U \neq W \in \mathscr{C}\}.$$

We say that $\mathscr{C}$ is a $(m, \log_q(|\mathscr{C}|), d_{\min})$ subspace code. Furthermore,

- if every space in $\mathscr{C}$ has the same dimension, we say that $\mathscr{C}$ is a constant dimension code.
- if every space in $\mathscr{C}$ has the same identifying vector, we say that $\mathscr{C}$ is a single cell code.

Note that every code is a disjoint union of constant dimension codes and every constant dimension code is a disjoint union of single cell codes. The most common example of a single cell code (see for example [7, Example 1]) has as identifying vector $v = 1^k 0^{n-k}$, the corresponding matrices are often taken from an MRD code. When using a subspace code for transmission, the size of $\mathscr{C}$ gives a bound on how much information can be encoded. However, having a big code does not automatically mean that much information can be sent at once; one would also need efficient encoding. The same holds for the distance. A big minimal distance theoretically allows for good error correction in the sense that a minimal distance decoder can correct all $\rho$ erasures and $t$ errors as long as $2(\rho + t) < d$ (see [7] for more details), but a minimal distance decoder is not efficient in general.

We have already seen that having two subspaces in the same cell, the distance is determined by the rank distance of the corresponding matrices. Having two subspaces in different cells, the identifying vectors give a bound on the distance.

**Lemma 2** *Let* $U, W \leq \mathbb{F}_q^m$ *be subspaces and* $u, w \in \mathbb{F}_2^n$ *their corresponding identifying vectors. Then* $d(U, W) \geq d_H(u, w)$.

Thus, given a minimal distance $d$ and the task to construct a constant dimension subspace code with dimension $k$ achieving this minimal distance with a big size, we first choose a constant weight code $C \subseteq \mathbb{F}_2^n$ with Hamming distance $d$, the so called skeleton code. For every vector $v \in C$, we then construct a Ferrers diagram rank metric code with minimal distance (in the rank metric) at least $d/2$. As the size of a cell is strongly related to how far the ones in the identifying vector are to the left, a skeleton code chosen for example in a lexicographic way leads to good results. Finding big codes with a given Ferrers diagram is harder. An upper bound on the size of such a code was given in [1, Theorem 1] and codes achieving this bound were constructed in many cases (see for instance [4, 10]) and properties of such maximal codes were also discussed (e.g. in [2]). However, the question if this bound is always achievable and the search for a general construction always achieving this bound are still open.

Furthermore, even if these codes are very big, they do not always allow for efficient encoding and decoding. Using results from [1], one can reduce these problems to the question of finding efficient encoding and decoding algorithms for the skeleton code and for every rank metric code used. A big class of important rank metric codes for which efficient algorithms are known (see for example [11]) are Gabidulin codes [3], but not all Ferrers diagram constructions are based on such codes.

## 3 Group Actions

The above matricial representations of subspaces and constructions are related to the structural properties of the input and output alphabet of the network, the set $\mathscr{P}(\mathbb{F}_q^m)$ of all subspaces of $\mathbb{F}_q^m$. We propose to use group actions to identify further structural properties that can be relevant for subspace codes and its possible generalization. A group is a non-empty set of elements $G$ together with a composition rule

$$\circ : G \times G \to G,$$

such that for all $a, b, c \in G$ we have

- Associativity: $a \circ (b \circ c) = (a \circ b) \circ c$,
- a neutral element $1 \in G$, satisfying $1 \circ a = a \circ 1$ for all $a \in G$,
- an inverse element $a^{-1}$, satisfying $a \circ a^{-1} = a^{-1} \circ a = 1$.

Classical examples for groups are the general linear group $GL(V)$ of a vector space $V$, the set of all invertible linear maps from $V$ into itself, and $S_m$, the set of all permutations on $m$ points. A group $G$ is said to act (from the left) on a set $\Omega$ if there is a map $* : G \times \Omega \to \Omega$ satisfying $1 * \omega = \omega$ and $g * (h * \omega) = (g \circ h) * \omega$ for all $\omega \in \Omega$ and all $g, h \in G$. Similarly, one can also define the action on the right. For example the two groups mentioned above come with a natural action: $GL(V)$ acts on $V$ by applying a map to a vector and $S_m$ acts on a set of $m$ points by applying a permutation. Other actions are for example the multiplication in the group itself, setting $* := \circ$ and $\Omega := G$, or the conjugation in the group, where we set $* : G \times G \to G, g * h = ghg^{-1}$.

The acting groups induce structural properties in the sets acted upon, such as a partition of the elements of the set into orbits. For an element $\omega \in \Omega$, the orbit of $\omega$ under $G$ is defined as

$$G * \omega := \{g * \omega \mid g \in G\}.$$

The set of all orbits, denoted by $\mathcal{O}$, is a partition of $\Omega$, being in the same orbit is an equivalence relation on $\Omega$ and an action having only a single orbit is called transitive. In order to understand the structural properties of the set given by the orbits, one looks at invariants.

**Definition 6** Let $G$ be a group that acts on a set $\Omega$, let $\mathcal{O} = \{G * \omega \mid \omega \in \Omega\}$ be the set of orbits and $I$ a set. A map $\varphi : \Omega \to I$ is called an invariant for the group operation if $\varphi$ is constant on the orbits, that is $\varphi(g\omega) = \varphi(\omega)$ for all $g \in G$ and all $\omega \in \Omega$. In this case the invariant induces a map

$$\Phi : \mathcal{O} \to I, G * \omega \mapsto \varphi(\omega)$$

and the invariant is called separating if $\Phi$ is injective.

Note that the set $I$ can often be reduced to make $\Phi$ surjective, in this case a separating invariant classifies the orbits. Every action has a separating invariant by taking $I = \mathcal{O}$ and $\Phi = id$. However, other invariants might be more useful in understanding the structure of orbits or to quickly decide if two given elements lie in the same orbit or not.

As an example, take the set $\Omega$ to be acted on as the projective space $\mathscr{P}(K^m)$ and the group $GL_m(K)$ acting on the right by applying a linear transformation to every vector of a subspace $U \in \mathscr{P}(K^m)$:

$$\mathscr{P}(K^m) \times \mathrm{GL}_m(K) \to \mathscr{P}(K^m), (U, g) \mapsto \{ug \mid u \in U\}.$$

The orbits under this action are the Grassmannians $\mathrm{Gr}_n(K^m)$, the set of subspaces of dimension $n$ of $K^m$, i.e. in this case we have

$$\mathscr{O} \cong \{\mathrm{Gr}_n(K^m) \mid 0 \leq n \leq m\}.$$

Therefore, a separating invariant of this action is the dimension of subspaces.

Now consider $\Omega$ as the matricial representation of subspaces of $\mathbb{F}_q^m$, given as rectangular matrices. In particular, let $\mathbb{F}_q^{n \times m, n}$ denote the set of $n \times m$ matrices of rank $n$. Then the action on the right

$$\mathbb{F}_q^{n \times m, n} \times \mathrm{GL}_m(K) \to \mathbb{F}_q^{n \times m, n}, (M, g) \mapsto Mg$$

is transitive. Note that the rank of a matrix equals the dimension of its row space, and thus we see how the abstract and matricial actions are related. However, there is not a bijection between the elements of an orbit of the abstract action and the elements of the orbit of the matricial action, as different matrices can have the same row space.

Another important action, relating matricial and abstract notion of subspaces, is the action of $\mathrm{GL}_n(\mathbb{F}_q)$ on the left via matrix multiplication,

$$\mathrm{GL}_n(\mathbb{F}_q) \times \mathbb{F}_q^{n \times m, n} \to \mathbb{F}_q^{n \times m, n}, (g, M) \mapsto gM$$

which induces orbits of matrices that can be represented by the same reduced row echelon form (Fig. 1). Thus the map that takes a matrix to its reduced row echelon form is a separating invariant of this action. But also the orbit generated by the rows of the matrix can be used as a separating invariant. Therefore, there exists a bijection between the elements of an orbit of the abstract action (i.e. subspaces) and the orbits of the matricial action on the left (i.e. sets of matrices represented by a unique reduced row echelon form). Note that in the matricial representation, similar to the action on the left, we have an action on the right, giving us as invariants the column echelon form and the column space – looking at rows rather than columns is a convention, not a restriction.

To join all these actions together, we fill a matrix of $\mathbb{F}_q^{n \times m, n}$ with redundant rows, that is linear combinations of the given rows, to get a quadratic $m \times m$ matrix. In this manner, we can construct all $m \times m$ matrices. Looking at the action

$$(\mathrm{GL}_m(\mathbb{F}_q) \times \mathrm{GL}_m(\mathbb{F}_q)) \times \mathbb{F}_q^{m \times m} \to \mathbb{F}_q^{m \times m}, ((g, h), M) \mapsto gMh^{-1},$$

what is left invariant is the rank of the matrices. A set of representatives of the orbits are the rank normal forms

$$\begin{pmatrix} I_n & 0 \\ 0 & 0 \end{pmatrix}$$

for $0 \leq n \leq m$.

$$\mathscr{P}(K^m) \times \mathrm{GL}_m(K) \to \mathscr{P}(K^m)$$

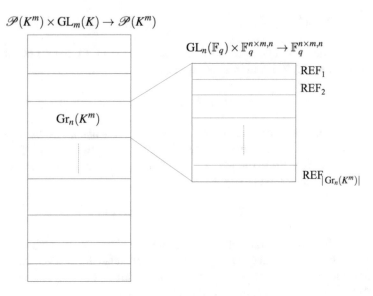

**Fig. 1** Graphical illustration of the orbits of the abstract action $\mathscr{P}(K^m) \times \mathrm{GL}_m(K) \to \mathscr{P}(K^m)$ and the correspondence of the elements of its orbits with the orbits of the matricial action $\mathrm{GL}_n(\mathbb{F}_q) \times \mathbb{F}_q^{n \times m,n} \to \mathbb{F}_q^{n \times m,n}$

We just saw that acting on both sides with the general linear group, only the rank was left invariant. To get more structure, we introduce the groups of upper and lower triangular matrices,

$$B := \{X \in \mathrm{GL}_m(\mathbb{F}_q) \mid X_{ij} = 0 \ \forall \ i < j\} \text{ and } B^T := \{X \in \mathrm{GL}_m(\mathbb{F}_q) \mid X_{ij} = 0 \ \forall \ i > j\}.$$

When acting with $B^T$ on $\mathscr{P}(\mathbb{F}_q^m)$ through

$$\mathscr{P}(\mathbb{F}_q^m) \times B^T \to \mathscr{P}(\mathbb{F}_q^m), (U, b) \mapsto \{ub \mid u \in U\},$$

the orbits turn out to be the Schubert cells.

**Theorem 1** *The orbits of $B^T$ on $\mathscr{P}(\mathbb{F}_q^m)$ are the Schubert cells $C_q(v)$ introduced in Definition 3.*

*Proof* Let $v \in \mathbb{F}_2^m$ be a vector having ones exactly at positions $i_1, i_2, \ldots i_n$, $U = S(v)$ the corresponding standard space and $X \in B^T$ a matrix with rows $x_1, x_2, \ldots, x_m$. Then $(x_{i_1}, x_{i_2}, \ldots, x_{i_n})$ is a basis of $UX$, and as $X$ is an upper triangular matrix, the vector $x_{i_j}$ has its first non-zero entry at position $i_j$ for all $j$. Thus $UX$ lies in the same cell as $U$. On the other hand, every space $W$ in the cell parametrized by $v$ has a basis $(y_1, y_2, \ldots, y_n)$ where the first non zero entry of $y_j$ is at position $i_j$ for all $j$. We can thus define a matrix $Y \in B^T$ having $y_j$ as rows in the desired positions and filled up with rows of the identity matrix. Thus $W = UY$ and hence $UB^T$ is exactly the cell as claimed. □

We thus see that the orbits of $GL_m(\mathbb{F}_q) \times B^T$ on $\mathbb{F}_q^{m \times m}$ are in bijection to the cells, in this case giving us

$$\mathcal{O} \cong \{C(v) \mid v \in \mathbb{F}_2^m\}.$$

As these are indexed by binary vectors $v \in \mathbb{F}_2^m$, let $\mu(v) \in \mathbb{F}_q^{m \times m}$ be a matrix in reduced row echelon form having ones at positions indexed by $v$ and zeros at any other position (meaning that $S(v)$ is the row space of $\mu(v)$). Then our results up to now tell us that for every $M \in \mathbb{F}_q^{m \times m}$, there is a unique $v \in \mathbb{F}_2^m$, such that there exists $g \in GL_m(\mathbb{F}_q)$ and $b \in B^T$ with $M = g\mu(v)b$. This can be written as

$$\mathbb{F}_q^{m \times m} = \biguplus_{v \in \mathbb{F}_2^m} GL_m(\mathbb{F}_q)\mu(v)B^T.$$

Operating with $B$ from the left also gives interesting results when extending from subspaces to flags in the next section.

# 4   Flags

In this section, we introduce flags and generalize the results on subspaces discussed above. We start by defining the basic terms.

**Definition 7**  A flag $\Lambda$ is a chain of distinct, non-trivial, proper subspaces of $\mathbb{F}_q^m$,

$$\Lambda = \{V_1 < V_2 < \ldots < V_k\}.$$

The tuple $(\dim(V_1), \dim(V_2), \ldots, \dim(V_k))$ is called the type of the flag and the flag is said to be a full (or fine) flag if it is of type $(1, 2, 3, \ldots, m - 1)$.

By $\mathscr{F} = \mathscr{F}(m, \mathbb{F}_q)$ we denote the set of all flags and by $\mathscr{F}_f = \mathscr{F}_f(m, \mathbb{F}_q)$ the set of all full flags. The group $G := GL_m(\mathbb{F}_q)$ acts on $\mathscr{F}$ from the right by applying a linear transformation to every single subspace in a flag and the type is a separating invariant of this action. Given a flag $\Lambda$ and a $g \in G$, we write $\Lambda \cdot g$ for this action. In the next two sections, we will focus on full flags, returning to the general case afterwards.

## 4.1   Cells and the Gauss–Jordan Decomposition

Here we try to transfer the concept of cells – indexed by pivot positions – to the set of flags. Note that our cells will not coincide with the Bruhat cells in the theory of flags (see for example [12]) but rather with the cells for subspaces. The Bruhat cells will be called "circles" in this article and will be discussed in the next section.

To talk about pivot positions, we first need to define the matricial representation of a flag. Given a matrix $X \in G$, the rows $x_1, x_2, \ldots, x_m$ of $X$ are linearly independent. Thus the sequence of subspaces

$$\langle x_1 \rangle < \langle x_1, x_2 \rangle < \ldots < \langle x_1, x_2, \ldots, x_{m-1} \rangle$$

forms a full flag, which we will call $\Delta(X)$. Note that we do not need the last row of $X$ to define the flag. If $X = I_m$ is the identity matrix, we write $\Delta_0 := \Delta(I_m)$ and if $\pi \in S_m$ has the corresponding permutation matrix $\Pi \in G$, we write $\Delta_\pi := \Delta(\Pi)$, thus $\Delta_0 = \Delta_{id}$. The set $\mathscr{A} = \{\Delta_\pi \mid \pi \in S_m\}$ is called the standard apartment and plays an important role when looking at cells and circles. Given $X, Y \in G$, we get that $\Delta(X) \cdot Y = \Delta(XY)$, thus explaining why we consider the right action on flags. In fact, we can write $\Delta(X) = \Delta_0 \cdot X$ for all $X \in G$.

When discussing subspaces, we saw that Gaussian elimination does not change the row space of a matrix. However, this is not true for the flag represented by a matrix in general, e.g. it is easy to see that interchanging two rows of a matrix will always change the full flag. The only operations in the Gaussian elimination that leave the flag unchanged are multiplications of rows and adding rows downwards, that is adding row $i$ to row $j$ for $j > i$. These operations can be represented by triangular matrices of the already introduced group $B$, giving us the following result.

**Lemma 3** *Let $X \in G$. Then $Y \in G$ represents the same full flag as $X$, if and only if $Y \in BX = \{bX \mid b \in B\}$.*

We thus see that operating with $B$ on the left, what is left invariant is not only the row space but the whole flag. Furthermore, we can conclude that

$$\Delta_0 \cdot b = \Delta(b) = \Delta(b \cdot I_m) = \Delta_0$$

for all $b \in B$. In fact, $B$ is the stabilizer of $\Delta_0$ under the action of $G$ on $\mathscr{F}$.

For subspaces, we have the reduced row echelon form as a normal form under the operation of $G$. For flags, we call a matrix $X$ a basis matrix of a flag $\Lambda$, if $\Lambda = \Delta(X)$ and $X$ is reduced downwards, that is the first non-zero entry in each row of $X$ is a one and below all these ones, which we once again call pivots, $X$ has only zeros. The two big differences to the classical reduced row echelon form are that we do not enforce zeros above the pivots and that the pivot positions do not have to be ordered. As an example, consider a permutation $\pi \in S_m$ with permutation matrix $\Pi \in G$. Then $\Pi$ is a basis matrix for $\Delta_\pi$.

As the basis matrix is exactly what we get when reducing only downwards, we get that every flag has a unique basis matrix, that is every $B$ orbit on $G$ (via left multiplication) contains exactly one basis matrix. As all our matrices are invertible, we have to have a pivot element in each row and each column. If $p_i$ denotes the position of the pivot element in the $i$th row, we thus have that $(p_1, p_2, \ldots, p_m)$ is a permutation, an element of $S_m$ (in list notation). This yields the definition of a cell in the flag case.

**Definition 8** Let $\Lambda$ be a flag with basis matrix $X$ whose pivot positions yield the permutation $\pi \in S_m$. Then we call $\pi$ the identifying permutation of $\Lambda$, written as $\pi = \mathrm{perm}(\Lambda)$. The cell of a permutation $\pi \in S_m$ is defined as

$$C(\pi) := \{\Lambda \in \mathscr{F}_f \mid \mathrm{perm}(\Lambda) = \pi\}.$$

The biggest cell is $C(id)$, having size $q^k$ with $k = m(m-1)/2$, the smallest cell is $C(w)$ where $w : \{1, 2, \ldots, m\} \to \{1, 2, \ldots, m\}, i \mapsto m - i + 1$; this cell contains only a single element. The general size of a cell will be computed in Sect. 4.3. A representation system for the cells is given by the standard apartment. Just as we did with subspaces, one can show that the cells are exactly the orbits of $B^T \leq G$ under the action of $G$ on flags described above, i.e. $C(\pi) = \Delta_\pi \cdot B^T$. Identifying $\pi \in S_m$ with its permutation matrix, we get the well-known Gauss–Jordan decomposition

$$G = \biguplus_{\pi \in S_m} B\pi B^T.$$

## 4.2 Circles, $S_m$-valued Distance and the Gauss–Bruhat Decomposition

In the last section we saw that the action of $B^T$ on flags yields the Gauss–Jordan decomposition. Different from the subspace case, also the action of $B$ on the right yields interesting results for flags. The following decomposition of $G$ is called the Gauss–Bruhat decomposition and will help us to analyze distance functions on $\mathscr{F}_f$ and $\mathscr{F}$:

$$G = \biguplus_{\pi \in S_m} B\pi B.$$

Using that $B = wB^Tw$ where $w$ is defined as above, this decomposition can also be computed using Gaussian elimination. We now show how the orbits of $B$ on $\mathscr{F}_f$ relate to distance functions.

**Lemma 4** Let $d : \mathscr{F}_f \times \mathscr{F}_f \to \mathbb{Z}_{\geq 0}$ be a distance function that is invariant under the operation of $G$, i.e. $d(\Lambda, \Gamma) = d(\Lambda \cdot g, \Gamma \cdot g)$ for all $\Lambda, \Gamma \in \mathscr{F}_f$ and all $g \in G$. Let further $X, Y \in G$ and $\pi \in S_m$ such that there exist $b_1, b_2 \in B$ with $XY^{-1} = b_1 \Pi b_2$. Then $d(\Delta(X), \Delta(Y))$ depends only on $\pi$.

*Proof* Write $\Delta(X) = \Delta_0 \cdot X$ and $\Delta(Y) = \Delta_0 \cdot Y$. As $d$ is assumed to be invariant under the action of $G$, we have that

$$d(\Delta(X), \Delta(Y)) = d(\Delta_0 \cdot (XY^{-1}), \Delta_0) = d(\Delta(b_1 \Pi b_2), \Delta_0).$$

Now $b_1 \in B$ does not change the flag represented by the matrix $\Pi b_2$ and $b_2 \in B$ satisfies $\Delta_0 \cdot b_2^{-1} = \Delta_0$ and hence we get

$$d(\Delta(b_1 \pi b_2), \Delta_0) = d(\Delta_\pi, \Delta_0 \cdot b_2^{-1}) = d(\Delta_\pi, \Delta_0)$$

and thus the value of $d$ only depends on $\pi$ as claimed.                                       □

The assumption that $d$ should be invariant under the operation allows for more structure to be exploited. In fact, both distance functions that we will discuss in Sect. 4.3 will have this property. Each such distance is thus made up of two parts: A function $f : S_m \to \mathbb{Z}_{\geq 0}$ and the function that we call the $S_m$-valued distance function, defined through

$$d_{S_m}(\Delta(X), \Delta(Y)) := \pi \in S_m$$

if there exist $b_1, b_2 \in B$ such that $XY^{-1} = b_1 \pi b_2$. As in this case we will have $d_{S_m}(\Delta(Y), \Delta(X)) = \pi^{-1}$, the function $f$ should be invariant under taking inverses. Furthermore, to guarantee that $f \circ d_{S_m}$ has all properties of a distance function, it should satisfy $f(\pi) = 0$ if and only if $\pi = id$ and $f(\pi \varphi) \leq f(\pi) + f(\varphi)$ for all $\pi, \varphi \in S_m$.

The length and the depth are two such functions and will be discussed in the next section.

Just as the Gauss–Jordan decomposition gave us the cells, the Gauss–Bruhat decomposition gives us the circles.

**Definition 9** For $\pi \in S_m$, the $\pi$-circle around a flag $\Lambda$ is defined as

$$C_\pi(\Lambda) := \{\Gamma \in \mathscr{F}_f \mid d_{S_m}(\Gamma, \Lambda) = \pi\}.$$

Taking as center $\Lambda = \Delta_0$, the circles around that flag are exactly the $B$ orbits on $\mathscr{F}_f$ and in this case, the standard apartment is once again a set of orbit representatives through $C_\pi(\Delta_0) = \Delta_\pi \cdot B$. As we assumed the original distance $d$ to be invariant under the action of $G$, one should assume the same to be true for $d_{S_m}$. Indeed, taking two matrices $X, Y \in G$ and corresponding flags $\Delta(X), \Delta(Y)$, we get the same $S_m$-valued distance for $\Delta(X) \cdot g = \Delta(Xg)$ and $\Delta(Y) \cdot g = \Delta(Yg)$, due to the fact that $XY^{-1} = (Xg)(Yg)^{-1}$. Using this result and once again the relation $B = wB^Tw$, we can show a relation between cells and circles.

**Lemma 5** *For every $\pi \in S_m$ we have $C(\pi) = C_{\pi w}(\Delta_w)$.*

*Proof* Let $\Lambda \in C(\pi)$. Then there exists $b^T \in B^T$ such that $\Lambda = \Delta_\pi \cdot b^T$. Writing $b^T = wb'w$ for some $b' \in B$, we get $\Lambda = \Delta_{\pi w} \cdot b'w$. Thus $\Lambda \cdot w = \Delta_{\pi w} \cdot b'$ has $S_m$-distance $d_{S_m}(\Lambda_w, \Delta_0) = \pi w$ and thus $d_{S_m}(\Lambda, \Delta_w) = \pi w$, hence giving us $\Lambda \in C_{\pi w}(\Delta_w)$. Retracing these steps, we also get the other inclusion and thus the claimed equality.                                       □

## 4.3 Length and Depth of a Permutation and Corresponding Distance Functions

The two functions on the symmetric group $S_m$ that we introduce in this section are the length and the depth. The length plays an important role when studying $S_m$ as a Coxeter group (see for example [12, 13]), but it also appears in other contexts. The definition of the length is based on the fact that

$$S := \{(1, 2), (2, 3), (3, 4), \ldots, (m - 1, m)\}$$

is a generating set for the group $S_m$, that is every permutation can be constructed by successively interchanging neighboring elements. The length of a permutation $\pi \in S_m$ is then defined as the length of a shortest word in $S^*$ yielding $\pi$, written as $\ell(\pi)$. From this definition we can immediately conclude the desired properties of a function $f$ given at the end of the last section: As every transposition $(i, i + 1)$ is self-inverse, we get $\ell(\pi) = \ell(\pi^{-1})$. We further get $\ell(\pi) = 0$ if and only if $\pi = id$ and given a word $w_1 \in S^*$ for $\pi$ and a word $w_2 \in S^*$ for a permutation $\varphi$, we get that $w_1 w_2$ is a representation of $\pi\varphi$ (although not necessarily a shortest one). Not immediately clear, but also important, is the fact that the permutation $w$, mapping $i$ to $m - i + 1$, is the unique longest element of $S_m$, having length $m(m - 1)/2$.

A well-known way to compute a shortest word for a permutation $\pi$ – given as a list $(\pi(1), \pi(2), \ldots, \pi(m))$ – is the bubble sort algorithm, as it only changes neighboring elements. This fact yields the following formula for the length (see for instance [6] for more details):

$$\ell(\pi) = \sum_{i=1}^{m} |\{1 \leq k \leq i \mid \pi(k) > \pi(i)\}|.$$

This way to compute the length allows us to now introduce the depth, which looks quite similar:

$$\partial p(\pi) := \sum_{i=1}^{m} |\{1 \leq k \leq i \mid \pi(k) > i\}| = \sum_{k=1, k < \pi(k)}^{m} \pi(k) - k = \frac{1}{2} \sum_{k=1}^{m} |\pi(k) - k|.$$

Unlike the length, which has been used for several centuries already, the depth of a permutation received some attention only recently in [5, 8], whereas twice the depth – called the sum of distances – was studied in [6].

Given the similar sum notation of the length and the depth, one might conjecture a relation between these two functions. In [8], such a relation was found, namely for all $\pi \in S_m$ we have

$$\frac{\ell(\pi)}{2} \leq \partial p(\pi) \leq \ell(\pi).$$

We now introduce the corresponding distance functions on flags, starting with the length of a permutation. As in the generating set $S$, only neighboring elements are interchanged in one step, we first define neighboring flags, calling two flags

$$\Lambda = \{V_1 < V_2 < \ldots < V_{m-1}\} \text{ and } \Gamma = \{W_1 < W_2 < \ldots < W_{m-1}\}$$

neighbors, if they differ in exactly one space. Using this definition, we construct a graph on the set $\mathscr{F}_f$, joining two flags by an edge if they are neighbors. A shortest path from a flag $\Lambda$ to a flag $\Gamma$ in this graph is called a gallery between $\Lambda$ and $\Gamma$, the length of such a path is called the gallery distance of $\Lambda$ and $\Gamma$. This distance is the classically used distance when working with (full) flags and it is well-known (see for example [12, 13]) that this distance is invariant under the action of $G$ and that the gallery distance of $\Delta_0$ and $\Delta_\pi$ is exactly $\ell(\pi)$.

If we look at flags as a generalization of subspaces, we would like a distance function that is defined for all flags (not only full ones) such that the distance of two flags of length one $\Lambda = \{U\}$ and $\Gamma = \{W\}$ is exactly the Grassmannian distance $\dim(U + W) - \dim(U \cap W)$. The length function, however, is classically only defined for full flags and finding a generalization to arbitrary flags that is still easy to compute seems to be difficult. We thus take a different approach, generalizing the Grassmannian distance to the set of all flags. This yields the Grassmannian distance on flags, first introduced in [9].

**Definition 10** Let $\Lambda = \{V_1 < V_2 < \ldots < V_k\}$ and $\Gamma = \{W_1 < W_2 < \ldots < W_n\}$ be two flags. Fill the shorter one with $\mathbb{F}_q^m$ such that $n = k$ and then define the Grassmannian distance

$$d_G(\Lambda, \Gamma) := \sum_{i=1}^{k} \dim(V_i + W_i) - \dim(V_i \cap W_i).$$

Following the proof from [7] that the Grassmannian distance for subspaces is a distance, one can show that $d_G$ is a distance function. Note that $d_G$ can be extended to a distance on the set of all sequences (not necessarily ordered as a flag) of subspaces.

As the Grassmannian distance on subspaces is invariant under the operation of $G$, so is the distance $d_G$. The corresponding map on $S_m$ is twice the depth, i.e. $d_G(\Delta_0, \Delta_\pi) = 2 \mathfrak{dp}(\pi)$, the factor two coming from the fact that the Grassmannian distance of two subspaces of the same dimension is always even; hence also a sum over such values is.

Using the length, we can now also compute the size of a cell and a sphere in $\mathscr{F}_f$. The size of a circle – what is called a Bruhat cell in the literature – is well-known (see for example [12, 13]) to be

$$|C_\pi(\Lambda)| = q^{\ell(\pi)}.$$

The size of a cell then follows from Lemma 5 as

$$|C(\pi)| = q^{\ell(\pi w)}.$$

## 4.4 Non-full Flags

In this section, we briefly translate the previous results to the case of non-full flags, taking flags of length one as examples. We already mentioned before that flags of length one are basically subspaces; now we will see that the concepts of cells and distance translate properly.

For full flags, we looked at the stabilizer of the flag $\Delta_0$, which we called $B$. Taking the same approach here, we fix a type $T = (d_1, d_2, \ldots, d_k)$ and for a matrix $X \in G$ we let $\Delta_T(X)$ be the corresponding flag of this type, i.e. we take the full flag $\Delta(X)$ and drop all spaces but the ones that have dimension $d_i$ for some $i$. By dropping these spaces, the stabilizer of the standard flag of type $T$, $\Delta_T(I_m)$, gets bigger and we get a so called standard parabolic group, which is the group of all block triangular matrices, having block sizes $d_{i+1} - d_i$. For example taking the type $T = (n)$ and thus flags of length one, the stabilizer of $\Delta_T(I_m)$ is

$$P_n := \left\{ \begin{pmatrix} A & 0 \\ B & C \end{pmatrix} \mid A \in \mathrm{GL}(n, \mathbb{F}_q),\ C \in \mathrm{GL}(m-n, \mathbb{F}_q),\ B \in \mathbb{F}_q^{(m-n) \times n} \right\}.$$

We can once again define the cell of a permutation for such flags. However, different permutations may yield the same cell. In fact, for a $\pi \in S_m$ we have that the permutation matrix $\Pi$ lies in $P_n$ if and only if $\pi$ maps $\{1, 2 \ldots, n\} \subseteq \{1, 2, \ldots, m\}$ onto itself. Two permutations $\pi, \varphi \in S_m$ will give the same cell if and only if $\Pi \Phi^{-1} \in P_n$. We thus get that the number of cells for flags of type $T = (n)$ is

$$\frac{|S_m|}{|S_n \times S_{m-n}|} = \frac{m!}{n!(m-n)!} = \binom{m}{n},$$

which is exactly the number of weight $n$ vectors in $\mathbb{F}_2^m$. This once again shows that the definition of a cell in the flag case is a generalization of the definition in the subspace case.

## 5 Flag Codes

In this section, we describe how to use flags in a network coded network, define basic properties of a flag code and take a look at sphere packing and sphere covering bounds.

A set of flags $\mathscr{C} \subseteq \mathscr{F}(\mathbb{F}_q^m)$ is called a flag code. Similar to subspaces, we call

$$d_{\min} := \min\{d_G(\Lambda, \Gamma) \mid \Lambda \neq \Gamma \in \mathscr{C}\}$$

the minimum distance of $\mathscr{C}$ and say that $\mathscr{C}$ is a $(m, \log_q(|\mathscr{C}|), d_{\min})$ flag code. Furthermore,

- if every flag in $\mathscr{C}$ has the same type $T$, we say that $\mathscr{C}$ is a constant type code, if $T = (1, 2, \ldots, m-1)$ is the type of full flags, we say that $\mathscr{C}$ is a full flag code.
- if $\mathscr{C}$ is a constant type code and every flag lies in the same cell, we say that $\mathscr{C}$ is a single cell code.

The last ingredient we need to analyze flag codes is the rate of such a code. For that, let $d_{\max}(\mathscr{C})$ be the biggest dimension of a space in a flag in $\mathscr{C}$. Then only the first $m - d_{\max}$ rows of a basis matrix contain information on a flag of $\mathscr{C}$ and hence need to be sent, thus yielding a rate of

$$R(\mathscr{C}) := \frac{\log_q(|\mathscr{C}|)}{m d_{\max}}.$$

If we take for example $\mathscr{C} = C(id)$ to be the biggest cell of full flags, we have a rate of $1/2$. Taking $\mathscr{C} = \mathscr{F}_f$, the set of all full flags, the rate also converges towards $1/2$ if we fix $m$ and increase $q$. On the other hand, fixing $q$ and looking at big $m$ allows for rates bigger than $1/2$.

By introducing additional zeros in the biggest cell, we can get higher distances. Take for example the set

$$D_k = \{X \in B^T \mid X_{ii} = 1 \,\forall i, X_{ij} = 0 \,\forall i < j \leq i + k\}$$

of all strict upper triangular matrices having $k$ secondary diagonals filled with zeros. Then two different matrices of $D_k$ always yield two different full flags and the corresponding flag code has minimal distance $2(k+1)$ and rate

$$\frac{(m-k)(m-k-1)}{2m(m-1)}$$

(see [9] for proofs). Furthermore, the set $D_k$ forms a subgroup of $B^T$, thus allowing for easier computation of distances and structure. Error correction for example is possible efficiently using this code. A similar construction is also possible in other cells, allowing us to further extend this code. To know which cells to choose, we would like a result on the distance between two cells, similar to the one on subspaces.

**Lemma 6** Let $\Lambda \in C(\pi)$ and $\Gamma \in C(\varphi)$ be two full flags. Then

$$d_G(\Lambda, \Delta) \geq 2 \, \mathfrak{dp}(\varphi^{-1}\pi).$$

Note that we use left convention, meaning that $\varphi^{-1}\pi$ is the permutation that maps every $i$ to $\varphi^{-1}(\pi(i))$.

*Proof* Let $\Lambda = \{V_1 < V_2 < \ldots < V_{m-1}\}$ and $\Gamma = \{W_1 < W_2 < \ldots W_{m_1}\}$, and let further $1 \leq i \leq m$ such that $\varphi^{-1}(\pi(i)) = j > i$. Then $V_i$ is the first space in $\Lambda$ having a pivot element at position $\pi(i)$ and $W_j$ is the first space in $\Gamma$ containing a pivot at the same position. Thus we have $V_k \neq W_k$ for all $i \leq k < j$ and as these spaces have the same dimension, they hence have subspace distance at least two. As we sum over all subspace distances to form the Grassmannian distance, we hence get a distance of at least $2(j - i)$ in this case, in total giving us

$$d_G(\Lambda, \Delta) \geq 2 \sum_{i=1, \varphi^{-1}(\pi(i)) > i}^{n} \varphi^{-1}(\pi(i)) - i,$$

which is exactly the definition of twice the depth of $\pi\varphi^{-1}$.                          □

To see that this bound is tight, remember that $\Delta_\pi \in C(\pi)$ and $\Delta_\varphi \in C(\varphi)$ have exactly distance $2\,\partial p(\varphi^{-1}\pi)$.

## 5.1 The Network Model for Flag Codes

In a network coded network, we assume a source to inject packets, that are then linearly combined and forwarded by intermediate nodes. Looking at today's networks (e.g. the Internet), there is, however, something else to exploit: a numbering of packets. Assuming that we want to send a flag $\Lambda = \{V_1 < V_2 < \ldots < V_k\}$ of type $(d_1, d_2, \ldots, d_k)$ through a network. This flag is given to the transmitter in the form of a basis matrix $X$ and the transmitter then injects the rows of $X$ into the network. While doing so, the rows are numbered according to the current space $V_i$ in the flag, that is the first $d_1$ rows get the number one, the next $d_2 - d_1$ rows get number two, etc. An intermediate node now forms and forwards linear combinations in a desired way (e.g. at random) and marks the outgoing vector with the highest number of vectors used in the linear combination. Therefore, a vector with number $i$ always lies in $V_i$ and hence also in all $V_j$ with $j \geq i$. Thus collecting enough vectors with different numbers, a receiver can reconstruct $X$ and hence also $\Lambda$. Just as constant dimension subspace codes simplify this process in the subspace setting, constant type flag codes allow the receiver in this setting to know when he has enough vectors. To assure that every receiver gets enough vectors, a proper choice of type and transmission protocol, one that might take into account the numbering of packets, is necessary.

To introduce errors and erasures, assume that the flag $\Lambda$ as above was sent. The receiver reconstructs the spaces $W_i$ generated by all received vectors with number $j \leq i$ and thus gets a stuttering flag $\Gamma = \{W_1 \leq W_2 \leq \ldots \leq W_k\}$, the term stuttering meaning that some of these spaces might coincide. Similar to the operator channel model in [7], we say that an erasure occurred at some point $i$, if $W_i$ does not contain all of $V_i$. An error occurred, if $W_i$ contains elements that do not lie in $V_i$. Writing $W_i = \mathcal{H}_{r_i}(V_i) \oplus E_i$, where $\mathcal{H}_r(V_i) = W_i \cap V_i$ is a subspace of $V_i$ of dimension $r_i$ and $E_i$ is some subspace of $W_i$ satisfying $E_i \cap V_i = \{0\}$, we thus have $d_i - r_i$ erasures and $\dim(E_i)$ errors in the $i$th space. Then the total number of erasures is $\rho := \sum_{i=1}^{k} d_i - r_i$ and the total number of errors is $t := \sum_{i=1}^{k} \dim(E_i)$. As these definitions and the Grassmannian distance on flags are both generalizations of the constructions for subspaces, we get the following result, analog to [7, Theorem 2].

**Theorem 2** *Using a flag code $\mathscr{C}$ of minimal distance $d_{\min}$ for transmission, a minimal distance decoder can correct all errors and erasures, as long as $2(\rho + t) < d_{\min}$.*

The proof of this generalization uses the triangle inequality and can be found in [9].

## 5.2 Sphere Sizes and the $\ell - \partial\mathsf{p}$ Polynomial

In this section, we compute some sphere packing and sphere covering bounds for full flags and compare them to the bounds on subspaces computed in [7]. To do so, remember that a $\pi$-circle around a full flag has exactly size $q^{\ell(\pi)}$ where $\ell$ denotes the length function on $S_m$. A sphere of radius $2t$ with regard to the Grassmannian distance thus contains

$$\sum_{\pi \in S_m, \partial\mathsf{p}(\pi) \le t} q^{\ell(\pi)}$$

flags. To compute sphere sizes, it would therefore be helpful to be able to efficiently compute the $\ell - \partial\mathsf{p}$ polynomial

$$P_m(x, y) := \sum_{\pi \in S_m} x^{\ell(\pi)} y^{\partial\mathsf{p}(\pi)}.$$

Computing this polynomial once, we can then read off sphere sizes by dropping certain powers of $y$ and then inserting $(x, y) = (q, 1)$. However, an efficient way to compute this polynomial or a closed form is not yet known.

Using computed values of $P_m$, we can give sphere packing and sphere covering bounds for full flag codes and compare them to subspace codes. The following figure shows the bounds for full flag codes in the space $\mathbb{F}_2^8$ and compares them to the bounds on constant dimension subspace codes of dimension 4 computed in [7]. Note that flag codes are able to achieve minimal distances that are not reachable by subspace codes (Fig. 2).

We will now collect some properties of this polynomial. The special value $P_m(x, 1)$ is well-known to be the generating function of the length,

$$P_m(x, 1) = \prod_{i=1}^{m} \frac{x^i - 1}{x - 1}.$$

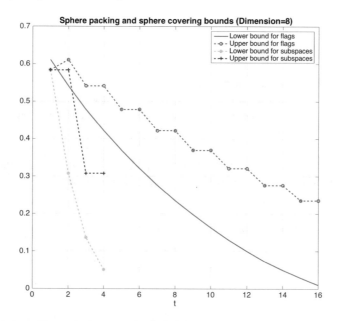

**Fig. 2** Sphere packing and sphere covering bounds

For the generating function of the depth, $P_m(1, y)$, a continued fraction was found recently in [5]. We will here show a nice property, regarding the alternating distribution of the depth.

**Lemma 7** *We have*

$$P_m(-1, y) = \sum_{\pi \in S_m} \text{sign}(\pi) y^{\partial p(\pi)} = (1 - y)^{m-1}.$$

*Proof* We will use the definition

$$\partial p(\pi) = \frac{1}{2} \sum_{k=1}^{m} |\pi(k) - k|.$$

The first equality is clear, as the length is one well-known way to compute the signum of a permutation. We thus show the second equality by induction over $m$. For $m = 1$ there is only the identity, having signum 1 and depth 0. Thus assume that the claim holds for $m - 1$. Let $H \leq S_m$ be the stabilizer of the point $m$. Then $H$ acts on $\{1, 2, \ldots, m - 1\}$ and is isomorphic to $S_{m-1}$. We have

$$S_m = \biguplus_{i=1}^{m} (i, m) H,$$

where $(i, m)$ denotes the transposition interchanging $i$ and $m$ and $(m, m)$ stands for the identity. Now let $i < m - 1$, let $h \in H$ be arbitrary and set $f := (i, m)h$ and $g := (i, m)(i, m - 1)h$. We will now compute twice the depth of $f$ and $g$, seeing that they are the same.

$\underline{2 \, \partial p(f) :}$

Let $j := h^{-1}(i)$. Then $f(j) = m$, $f(m) = i$ and $f(x) = h(x)$ for all other values $x$. Thus we have to add these two terms for $f$ and subtract the original ones for $h$, getting

$$\begin{aligned} 2 \, \partial p(f) &= 2 \, \partial p(h) + |m - j| + |i - m| - |i - j| - |m - m| \\ &= 2 \, \partial p(h) + (m - j) + (m - i) - |i - j| \\ &= 2(\partial p(h) + m - \max(i, j)). \end{aligned}$$

$\underline{2 \, \partial p(g) :}$

Set $j := h^{-1}(i)$ and $k := h^{-1}(m - 1)$. Then $g(j) = m - 1$, $g(k) = m$, $g(m) = i$ and $g(x) = h(x)$ for all other values of $x$. Thus we get

$$\begin{aligned} 2 \, \partial p(g) &= 2 \, \partial p(h) + |m - 1 - j| + |m - k| + |i - m| - |i - j| - |m - 1 - k| \\ &= 2 \, \partial p(h) + (m - 1 - j) + (m - k) + (m - i) - (m - 1 - k) - |i - j| \\ &= 2(\partial p(h) + m - \max(i, j)). \end{aligned}$$

As for $i < m - 1$ we always have that $(i, m - 1) \in H$ is a non-trivial transposition, the elements $f$ and $g$ are different, but in the same class $(i, m)H$. As they have the same depth but opposite signum, the corresponding terms cancel out when taking the sum. We thus only have to look at the cases $i = m - 1$ and $i = m$. For $i = m$ we get the subgroup $H$ itself, thus by induction hypothesis this case contributes $(1 - y)^{m-2}$ to the polynomial. For $i = m - 1$ and $h \in H$, we once again can set $f := (m - 1, m)h$ and $j := h^{-1}(m - 1)$. Using that $m - 1 \geq j$, we then get

$$2 \, \partial p(f) = 2 \, \partial p(h) + (m - j) - (m - 1 - j) + (m - (m - 1)) - (m - m) = 2 \, \partial p(h) + 2$$

and thus $\partial p(f) = \partial p(h) + 1$. This gives us that the set $(m - 1, m)H$ contributes $-y(1 - y)^{m-2}$ to the sum, as the signum of every element of $H$ switches and the depth is increased by one. In total we thus get

$$P_m(-1, y) = -y(1 - y)^{m-2} + (1 - y)^{m-2} = (1 - y)^{m-1}.$$

$\square$

# References

1. T. Etzion, N. Silberstein, Error-correcting codes in projective spaces via rank-metric codes and Ferrers diagrams. IEEE Trans. Inform. Theory **55**(7), 2909–2919 (2009)
2. T. Etzion, E. Gorla, A. Ravagnani, A. Wachter-Zeh, Optimal Ferrers diagram rank-metric codes. IEEE Trans. Inform. Theory **62**(4), 1616–1630 (2016)
3. È.M. Gabidulin, Theory of codes with maximum rank distance. Problemy Peredachi Informatsii **21**(1), 3–16 (1985)
4. E. Gorla, A. Ravagnani, Subspace codes from Ferrers diagrams. J. Algebra Appl. (to appear)
5. M. Guay-Paquet, T. Kyle Petersen, The generating function for total displacement. Electron. J. Combin. **21**(3), Paper 3.37, 13 (2014)
6. D.E. Knuth, *The Art of Computer Programming: Sorting and Searching*, vol. 3 (Addison-Wesley, Reading, 1998)
7. R. Kötter, F.R. Kschischang, Coding for errors and erasures in random network coding. IEEE Trans. Inform. Theory **54**(8), 3579–3591 (2008)
8. T. Kyle Petersen, B. Eileen Tenner, The depth of a permutation. J. Comb. **6**(1–2), 145–178 (2015)
9. D. Liebhold, G. Nebe, A. Vazquez-Castro, Network coding with flags. Des. Codes and Cryptogr. (2017)
10. N. Silberstein, A.-L. Trautmann, Subspace codes based on graph matchings, Ferrers diagrams, and pending blocks. IEEE Trans. Inform. Theory **61**(7), 3937–3953 (2015)
11. D. Silva, F.R. Kschischang, R. Kötter, A rank-metric approach to error control in random network coding. IEEE Trans. Inform. Theory **54**(9), 3951–3967 (2008)
12. T.A. Springer, *Linear Algebraic Groups*, 2nd edn., Modern Birkhäuser Classics (Birkhäuser, Boston, 2009)
13. D.E. Taylor, *The Geometry of the Classical Groups*, vol. 9, Sigma Series in Pure Mathematics (Heldermann Verlag, Berlin, 1992)
14. M.A. Vazquez-Castro, A geometric approach to dynamic network coding, in *2015 IEEE Information Theory Workshop - Fall (ITW)* (2015), pp. 207–211

# Multi-shot Network Coding

**Diego Napp and Filipa Santana**

**Abstract** The seminal paper of Koetter and Kschischang [11] introduced coding concepts for errors and erasures in a random network coding setting and since then has opened a major research area in communication technology. Here, the network is allowed to change very quickly, which is the case in many mobile applications. The problem is suitably modeled via the operator channel, which makes a very clear connection between network coding and classical information theory. However, coding can also be performed over multiple uses of the network, whose internal structure may change at each shot, giving rise to the so-called multi-shot network coding. Although the potential of using multi-shot network coding was already observed in [11], only recently this more involved approach have been investigated. The general idea stems from the fact that creating dependencies among the transmitted codewords of different shots can improve the error-correction capabilities. A very natural way to impose correlation of codewords(subspaces) over time is by means of convolutional codes, originating the notion of rank metric convolutional codes. In this Chapter we review some of the main results and ideas of multi-shot network coding, propose new approaches and point out some avenues for future research.

## 1 Introduction

In many multicast communication networks, like the internet, wireless communication and cloud computing, a transmitter sends packets to several users through a series of intermediate nodes. In 2008, a seminal and award winning paper [11] provided the mathematical foundations for the case where the topology of the net-

D. Napp (✉) · F. Santana
CIDMA - Center for Research and Development in Mathematics and Applications,
Department of Mathematics, University of Aveiro, Campus Universitario de Santiago,
3810 -193 Aveiro, Portugal
e-mail: diego@ua.pt

F. Santana
e-mail: vfssantana@ua.pt

© Springer International Publishing AG 2018
M. Greferath et al. (eds.), *Network Coding and Subspace Designs*,
Signals and Communication Technology,
https://doi.org/10.1007/978-3-319-70293-3_5

work is unknown and the nodes perform a random linear combination of the packets received and forward this random combination to adjacent nodes. If one considers the transmitted packet as columns of a matrix with entries in a finite field $\mathbb{F}_q$, then the linear combinations performed in the nodes are columns operations on this matrix. If no errors occur during the transmission over such a network, the column space of the transmitted matrix remains invariant. In such an scenario the problem of reliable communication is suitably modeled via the operator channel, which makes a very clear connection between network coding and classical information theory. The operator channel can be seen as a standard discrete memoryless channel with input and output alphabets given by the projective space $\mathcal{P}(\mathbb{F}_q^n)$, i.e., the set of all possible vector subspaces of the vector space $\mathbb{F}_q^n$. To achieve a reliable communication over this channel, matrix codes are employed forming the so-called rank metric codes [26]. Rank metric codes such as Gabidulin codes are known to be able to protect packets in such a scenario. We call these codes *one-shot* codes, as they use the (network) channel only once.

However, coding can also be performed over multiple uses of the network as it has been recently shown by several authors, see for instance [2, 13, 16, 20, 28]. The general idea stems from the fact that creating dependencies among the transmitted codewords (subspaces) of different shots can improve the error-correction capabilities of the code. In order to increase the error-correcting capabilities of a block code in one single shot, one necessarily needs to increase the field size or the packet size and this might not be optimal or impossible in many applications and consequently one must create dependencies across multiple nodes to approach the channel capacity. Thus, multi-shot codes constitute an attractive class of codes for such situations.

There are mainly two approaches for building multi-shot codes: one using concatenation of codes and other using rank metric convolutional codes. In [20], a type of concatenated $n$-shot codes ($n \geq 1$) was proposed based on a multilevel code. In [17], a concatenation scheme was presented using a Hamming metric convolutional code as an outer code and a rank metric code as an inner code. A different type of concatenation was introduced in [13] where the authors use codes that layer both Maximum Sum Rank (MSR) codes and Gabidulin in order to achieve the streaming capacity for the Burst Erasure Channel.

Apart from concatenated codes, another very natural way to spread redundancy across codewords is by means of convolutional codes [3, 7, 10, 15, 18]. Adapting this class of codes to the context of networks gave rise to rank metric convolutional codes and interestingly there have been little research on these codes, see [2, 12, 13, 28]. The work in [28] was pioneer in this direction by presenting the first class of rank metric convolutional codes together with a decoding algorithm able to deal with errors, erasures and deviations. However, the results were only valid for unit memory convolutional codes and in [2, 12] (see also the references therein) an interesting and more general class of rank metric convolutional codes was introduced to cope with network streaming applications. For a more general theoretical framework to rank metric convolutional codes, see [16] and for more information on new investigations on rank metric codes properties the reader is also referred to chapters "Codes Endowed with the Rank Metric" and "Constructions of Cyclic Subspace Codes and Maximum Rank Distance Codes".

In this multi-shot setting a new distance, called sum rank distance, was introduced as a generalization of the rank distance used for one-shot network coding. This new distance has proven to be the proper notion in order to deal with delay-free networks, i.e., assuming that the natural delay in the transmission (due, for instance, to the delay of the nodes) is so small that can be disregarded. In this work we show that in order to handle networks with delays, a new metric needs to be introduced: the generalized sum rank distance.

Finally, we note that in the last years others papers have also appeared dealing with convolutional network coding using very different approaches [8, 22]. These codes do not transmit over the operator channel and therefore are not equipped with the rank metric.

In this chapter, we aim to provide a general overview of the area of multi-shot codes for network coding and indicate some open problems and avenues for further research. We will review the approaches and results proposed so far and present some new alternatives such as a new metric that can be considered as an analog of the column distance of Hamming convolutional codes. We note that by no means it is a complete review as important issues such as decoding are not included.

**Notation**: Denote a finite field with $q$ elements by $\mathbb{F}_q$. For a given basis of $\mathbb{F}_{q^M}$ viewed as an $M$ vector space over $\mathbb{F}_q$, any element of $\mathbb{F}_{q^M}$ can be seen as a vector in $\mathbb{F}_q^M$. Analogously, any vector $x$ of length $n$ over $\mathbb{F}_{q^M}$ can be regarded as an element $X$ in $\mathbb{F}_q^{M \times n}$, the set of $M \times n$ matrices over $\mathbb{F}_q$. We commit a harmless abuse of notation and define the rank of a vector $x \in \mathbb{F}_{q^M}^n$ as the rank of $x$ as an $M \times n$ matrix over $\mathbb{F}_q$. If each column of $X$ represents a packet of length $M$ then so it does $x^i$, for $i \in \{1, \ldots n\}$, where $x = (x^1, \ldots, x^n)$. Throughout the paper we denote the vectors in lower-case type whereas a matrix is identified using upper-case type. We use them interchangeably if no confusion arises.

## 2   Preliminaries

In order to state more precisely the results to be presented we introduce in this section the necessary material and notation on standard theory of (one-shot) network coding and convolutional codes.

## 2.1   One-Shot Network Coding

### 2.1.1   The Network Model

Let $v \in \mathbb{F}_{q^M}^n$ (or equivalently $V \in \mathbb{F}_q^{M \times n}$) represents the $n$ packets of length $M$ to be sent through the network at one time instance. We shall follow the approach proposed in [11, 26] and consider the operator channel for one shot given by

$$x = vA + z, \tag{1}$$

where $x \in \mathbb{F}_{q^M}^n$ represents the received packets, $A \in \mathbb{F}_q^{n \times n}$ is the rank deficiency channel matrix and $z \in \mathbb{F}_{q^M}^n$ is the additive error. The adversaries of the matrix channel (1) come as rank deficiency of the channel matrix and as the additive error matrix. The channel matrix $A$ correspond to the overall linear transformations applied by the network over the base field $\mathbb{F}_q$ and it is known by the receiver (as the combinations are carried over in the header bits of the packets). For perfect communications we have that $z = 0$ and rank$(A) = n$, but failing and deactivated links may cause a rank deficient channel matrix. We call $n - \text{rank}(A)$ the rank deficiency of the channel.

### 2.1.2 Rank Metric

A rank metric code $\mathcal{C}$ is defined as any nonempty subset of $\mathbb{F}_q^{M \times n}$. A natural metric for matrix codes is induced by the distance measure $d_{\text{rank}}(V, W) = \text{rank}(V - W)$, where $V, W \in \mathbb{F}_q^{M \times n}$ [11]. In the context of the rank metric, a matrix code is called rank metric code.

The rank distance of a code $\mathcal{C} \subset \mathbb{F}_q^{M \times n}$ is defined as

$$d_{\text{rank}}(\mathcal{C}) = \min_{V, W \in \mathcal{C}, \ V \neq W} d_{\text{rank}}(V, W).$$

Consider linear codes over $\mathbb{F}_{q^M}$ and use $k$ for the dimension of the linear code over $\mathbb{F}_{q^M}$. To simplify presentation we will assume that $M \geq n$. In this case, for linear $(n, k)$ rank metric codes over $\mathbb{F}_{q^M}$ the following analog of the Singleton bound holds:

$$d_{\text{rank}}(\mathcal{C}) \leq n - k + 1.$$

A code that achieves this bound is called Maximum Rank Distance (MRD). Gabidulin codes are a well-known class of MRD codes [6], see also [5, 9].

Although rank metric codes in $\mathbb{F}_q^{M \times n}$ are usually constructed as block codes of length $n$ over the extension field $\mathbb{F}_{q^M}$ [11] a more general construction was considered in [16]. Here, an $(\tilde{n} \times M, \tilde{k})$ rank metric code $\mathcal{C} \subset \mathbb{F}_q^{\tilde{n} \times M}$ of rate $\tilde{k}/\tilde{n}M < 1$ is seen as an image of a monomorphism $\varphi : \mathbb{F}_q^{\tilde{k}} \to \mathbb{F}_q^{\tilde{n} \times M}$ ($\varphi = \psi \circ \gamma$ is a composition of an isomorphism $\psi$ and a monomorphism $\gamma$):

$$\varphi : \mathbb{F}_q^{\tilde{k}} \xrightarrow{\ \gamma\ } \mathbb{F}_q^{\tilde{n}M} \xrightarrow{\ \psi\ } \mathbb{F}_q^{\tilde{n} \times M}$$
$$u \longmapsto v = uG \longmapsto V = \psi(v)$$

where $G \in \mathbb{F}_q^{\tilde{k} \times \tilde{n}M}$ and $[V_{ij}] = v_{i+\tilde{n}j}$ with $0 \leq i < \tilde{n}$ and $0 \leq j < M$.

In [16, Theorem 3.1] the following upper bound on the rank distance of an $(\tilde{n} \times M, \tilde{k})$ linear rank metric code was presented

$$d_{\text{rank}}(\mathcal{C}) \leq \tilde{n} - \left\lfloor \frac{\tilde{k} - 1}{M} \right\rfloor = \tilde{n} - \left\lceil \frac{\tilde{k}}{M} \right\rceil + 1.$$

## 2.2 Convolutional Codes

As opposed to block codes, convolutional encoders take a stream of information bits and converts it into a stream of transmitted bits (by means of shift registers) and therefore they are very suitable for streaming applications. Mathematically speaking, a *convolutional code* $\mathcal{C}$ of rate $k/n$ is an $\mathbb{F}_{q^M}[D]$-submodule of $\mathbb{F}_{q^M}[D]^n$ of rank $k$. A full row rank matrix $G(D) \in \mathbb{F}_{q^M}[D]^{k \times n}$ with the property that

$$\mathcal{C} = \text{im}_{\mathbb{F}_{q^M}[D]} G(D) = \left\{ u(D)G(D) \mid u(D) \in \mathbb{F}_{q^M}^k[D] \right\},$$

is called a *generator matrix*. The *degree* $\delta$ of a convolutional code $\mathcal{C}$ is the maximum of the degrees of the determinants of the $k \times k$ sub-matrices of one, and hence any, generator matrix of $\mathcal{C}$.

A rate $k/n$ code $\mathcal{C}$ with degree $\delta$ is called an $(n, k, \delta)$-convolutional code [15]. An $(n, k)$ block code is an $(n, k, \delta)$ convolutional code with $\delta = 0$.

We say that a generator matrix $G(D)$ is *basic* (see, e.g., [23, 25]) if it has a polynomial right inverse.[1]

Let $\text{wt}(v_i)$ be the number of the nonzero components of a vector $v_i \in \mathbb{F}_{q^M}^n$ and $\text{wt}(v(D))$ the Hamming weight of a polynomial vector $v(D) = \sum_{i \in \mathbb{N}} v_i D^i$ defined as

$$\text{wt}(v(D)) = \sum_{i \in \mathbb{N}} \text{wt}(v_i).$$

An important distance measure for a convolutional code $\mathcal{C}$ is its free distance or *Hamming distance* $d_{\text{H}}(\mathcal{C})$ defined as

$$d_{\text{H}}(\mathcal{C}) = \min \{ \text{wt}(v(D)) \mid v(D) \in \mathcal{C} \text{ and } v(D) \neq 0 \}.$$

Let $v_{[0,j]}(D) = v_0 + v_1 D + \cdots + v_j D^j$ be the $j$th truncation of the codeword $v(D) = \sum_{i \in \mathbb{N}} v_i D^i \in \mathcal{C}$.

---

[1] Using a basic generator matrix it allows to avoid catastrophic situations in which a sequence $u(D)$ with an infinite number of nonzero coefficients can be encoded into a sequence $v(D)$ with a finite number of nonzero coefficients; this case would decode finitely many errors on the received code sequence into infinitely many errors when recovering the original information sequence.

Another important distance measure for a convolutional code $\mathcal{C}$ is the *jth column distance* $d_j^c(\mathcal{C})$, given by the expression

$$d_{\mathrm{H}}^j(\mathcal{C}) = \min \left\{ \mathrm{wt}(v_{[0,j]}(D)) \mid v(D) \in \mathcal{C} \text{ and } v_0 \neq 0 \right\}.$$

This notion is related to the free distance $d_{\mathrm{H}}(\mathcal{C})$ in the following way

$$d_{\mathrm{H}}(\mathcal{C}) = \lim_{j \to \infty} d_{\mathrm{H}}^j(\mathcal{C}).$$

# 3 Multi-shot Network Coding

In this section we explain how to extend the classical theory of (one-shot) network coding to the context of multi-shot network coding. In fact, this is possible as each packet carries a label identifying the shot (or generation) to which it corresponds. Despite the little research in the area, this possibility was already observed in the seminal papers [11, 26].

## 3.1 Encoder and Decoder

The transmitter receives at each time instance $t$ a source packet $u_t \in \mathbb{F}_{q^M}^k$ (constituted by a set of $k$ packets) and a channel packet $v_t \in \mathbb{F}_{q^M}^n$ (constituted by a set of $n$ packets) is constructed using not only $u_t$ but also previous source packets $u_0, \ldots, u_{t-1}$.

A channel packet $v_t$ is sent through the network at each shot (time instance) $t$. The receiver collects the packets $x_t$ as they arrive causally and tries to infer $v_t$ from $(x_0, \ldots, x_t)$.

## 3.2 Channel Model

Following the operator channel in (1) at each shot $t$ the received packets $x_t \in \mathbb{F}_{q^M}^n$ are constituted by corrupted packets $z$ and linear combinations of the packets of $v_t$ and, if there is delay in the transmission, also of combinations of the previous packets $v_0, \ldots, v_{t-1}$. Hence, we have

$$x_{[0,j]} = v_{[0,j]} A_{[0,j]} + z \tag{2}$$

where $x_{[0,j]} = (x_1, x_2, \ldots, x_j)$, $v_{[0,j]} = (v_0, v_1, \ldots v_j) \in \mathbb{F}_{q^M}^{n(j+1)}$, $A_{[0,j]} \in \mathbb{F}_q^{n(j+1) \times n(j+1)}$ is a block upper triangular truncated channel matrix and $z \in \mathbb{F}_{q^M}^{n(j+1)}$ the additive error. So far this channel model has not been proposed nor addressed in the literature in this generality and only the delay-free case has been considered, see [12] and reference therein. In the delay-free case only combinations of packets of $v_t$ arrive at time instance $t$ and not of packets of $v_i$, $i < t$ and therefore in this case the rank deficiency matrix $A_{[0,j]}$ is a block diagonal matrix.

## 3.3 Distances

The sum rank distance is the distance that has been widely considered for multi-shot network coding and can be seen as the analog of the rank distance for one-shot network coding. This distance was first introduced in [20] under the name of *extended rank distance* and is define as follows.

Let $v = (v_0, \ldots, v_t)$ and $w = (w_0, \ldots, w_t)$ be two $t$-tuples of vectors in $\mathbb{F}_{q^M}^n$. The *sum rank distance* (SRD) between them is

$$d_{SR}(v, w) = \sum_{i=0}^{t} \text{rank}(v_i - w_i).$$

Since the rank distance is a metric so it is the SRD and is obviously upper bound by

$$d_{SR}(v, w) \leq (t + 1)(n - k + 1).$$

As we will see below, the SRD is a metric that can be used to fully characterize the error-correcting capabilities of multi-shot codes in the context of delay-free networks. For the general case we propose the following distance,

$$d_{GR}(v, w) = \text{rank}(v - w) = \text{rank}((v_0, \ldots, v_t) - (w_0, \ldots, w_t)).$$

This is a straightforward generalization of the rank distance for one-shot network codes but it is new in the context of multi-shot network codes. We call it *generalized rank distance* (GRD). In the next section we will show that it is the proper metric to deal with network that allows delays. Still, further research needs to be done to understand this less investigated situation.

In the next two sections we present some of the approaches to deal with multi-shot network coding.

## 4 Rank Metric Convolutional Codes

In the context of rank metric convolutional codes two different settings have been proposed. Similarly as in the block code case, rank metric convolutional codes have been typically constructed over the extension field $\mathbb{F}_{q^M}$. However, a more general framework was introduced in [16], where a rank metric convolutional code can be defined for any given rate over the base field $\mathbb{F}_q$.

## 4.1 General Framework

A *rank metric convolutional code* $\mathcal{C} \subset \mathbb{F}_q^{\tilde{n} \times M}$ is defined as the image of an homomorphism $\varphi : \mathbb{F}_q[D]^{\tilde{k}} \to \mathbb{F}_q[D]^{\tilde{n} \times M}$ and written as a composition of a monomorphism $\gamma$ and an isomorphism $\psi$:

$$\varphi : \mathbb{F}_q[D]^{\tilde{k}} \xrightarrow{\gamma} \mathbb{F}_q[D]^{\tilde{n}M} \xrightarrow{\psi} \mathbb{F}_q[D]^{\tilde{n} \times M}$$

$$u(D) \mapsto v(D) = u(D)G(D) \mapsto V(D) \tag{3}$$

where $G(D) \in \mathbb{F}_q^{\tilde{k} \times \tilde{n}M}$ is a full row rank polynomial matrix, called *generator matrix* of $\mathcal{C}$, and $V(D) = [V_{ij}[D]]$, such that $V_{ij}(D) = v_{Mi+j}(D)$.

The *sum rank distance* of a rank metric convolutional code $\mathcal{C}$ is defined as

$$d_{SR}(\mathcal{C}) = \min_{V(D),U(D) \in \mathcal{C}, V(D) \neq U(D)} d_{SR}(V(D), U(D))$$

$$= \min_{0 \neq V(D) \in \mathcal{C}} rank\big(V(D)\big).$$

In the next theorem authors establish the Singleton-like bound for rank metric convolutional codes.

**Theorem 1** ([16, Theorem 4.1]) *Let $\mathcal{C}$ be a $(\tilde{n} \times M, \tilde{k}, \delta)$-rank metric convolutional code. Then the sum rank distance of $\mathcal{C}$ is upper bounded by*

$$d_{\text{sumrank}}(\mathcal{C}) \leq \tilde{n} \left( \left\lfloor \frac{\delta}{\tilde{k}} \right\rfloor + 1 \right) - \left\lceil \frac{\tilde{k}(\lfloor \frac{\delta}{\tilde{k}} \rfloor + 1) - \delta}{M} \right\rceil + 1. \tag{4}$$

## 4.2 Usual Approach

Consider from now on the extension field $\mathbb{F}_{q^M}$ instead of $\mathbb{F}_q$. This leads to the following definitions.

For an $(n, k, \delta)$-convolutional code $\mathcal{C}$ and $v(D) = v_0 + v_1 D + v_2 D^2 + \cdots \in \mathcal{C}$ we define its *free sum rank distance* as

$$d_{\text{SR}}(\mathcal{C}) = \min \left\{ \sum_{i \geq 0} rank(v_i) \mid v(D) \in \mathcal{C} \text{ and } v(D) \neq 0 \right\},$$

and its *column sum rank distance* as

$$d_{SR}^j(\mathcal{C}) = \min \left\{ \sum_{i=0}^{j} \text{rank}(v_i) \mid v(D) \in \mathcal{C} \text{ and } v_0 \neq 0 \right\}.$$

Moreover, based in the GRD we introduce a new distance for convolutional codes, the *generalized column rank distance*, as follows:

$$d_{GR}^j(\mathcal{C}) = \min \left\{ \text{rank}(v_0, \ldots, v_j) \mid v(D) \in \mathcal{C} \text{ and } v_0 \neq 0 \right\}.$$

Adapting (4) to the definitions considered in this paper, this bound reads as follows:

$$d_{SR}(\mathcal{C}) \leq (n - k) \left( \left\lfloor \frac{\delta}{k} \right\rfloor + 1 \right) + \delta + 1. \tag{5}$$

Note that this bound coincides with the so-called generalized Singleton bound of (Hamming) $(n, k, \delta)$-convolutional codes [21, 24]. Thus, the bound (5) could be also derived from the fact that the rank distance is upper bounded by the Hamming distance [6]. The problem of existence and construction of rank metric convolutional codes whose free rank distance achieves the bound (5) remains open.

For the column SRD the following upper bound was presented in [14]:

$$d_{SR}^j(\mathcal{C}) \leq (n - k)(j + 1) + 1,$$

and moreover,

$$\left[ d_{SR}^j(\mathcal{C}) = (n - k)(j + 1) + 1 \right] \Rightarrow \left[ d_{SR}^i(\mathcal{C}) = (n - k)(i + 1) + 1 \text{ for } i = 0, \ldots, j \right].$$

In [28], concrete decoding algorithms for unit memory rank metric convolutional codes were presented using another distance, namely the active rank distance. However, in [14], it was shown that this metric fails to determine the error-correcting capabilities of rank metric convolutional codes with arbitrary memory and the column SRD needs to be considered. In fact, necessary and sufficient conditions were inferred to recover rank deficiencies within a given time interval in delay-free networks when no errors occur (i.e., when $z = 0$ in (2)).

**Theorem 2** ([14, Theorem 2]) *Let $\mathcal{C}$ be a $(n, k, \delta)$ rank metric convolutional code and $v(D) = v_0 + v_1 D + \cdots \in \mathcal{C}$ with $v_0 \neq 0$. Assume a delay-free transmission and let $A_{[0,T]} = diag(A_0, \ldots, A_T)$ represent the block diagonal truncated channel matrix with $A_i \in \mathbb{F}_q^{n \times n}$, i.e., $x_{[0,T]} = v_{[0,T]} A_{[0,T]}$ is the received set of packets with $x_i = v_i A_i$, $i = 0, 1, \ldots, T$. Note that in this case $\text{rank}(A_{[0,T]}) = \sum_{j=0}^{T} \text{rank}(A_j)$. Then, we can recover $v_0$ if*

$$d_{SR}^T(\mathcal{C}) > n(T+1) - \text{rank}(A_{[0,T]}). \tag{6}$$

Theorem 2 illustrates how the column SRD can characterize the rank deficiency correcting capabilities of a rank metric convolutional code in delay-free networks within a time interval. The more column SRD a convolutional codes has the better is its rank deficiencies correcting capabilities.

Constructive constructions of convolutional codes having maximum column SRD profile were also presented in [14] using a superregular matrix derived in [1]. These constructions are not optimal as they require very large finite fields. The problem of deriving optimal constructions remains an interesting open problem for research.

Motivated by streaming applications, rank metric convolutional codes tailor-made to cope with burst of rank deficiency networks where studied in [13]. This can be considered a generalization of the theory of burst erasure convolutional codes, from a Hamming context to a network context. Concrete constructions of optimal rank metric convolutional codes in this setting are not known yet.

The following example illustrates that, in the case the network has delays in the transmission of packets, the previous theorem fails to characterize the rank deficiency correcting capability of $\mathcal{C}$.

*Example 1* Let $G(D) = G_0 + G_1 D \in \mathbb{F}_{2^6}[D]^{2\times 3}$ be a generator matrix of the convolutional code $\mathcal{C} \subset \mathbb{F}_{2^6}[D]^3$, $\alpha$ a primitive element of $\mathbb{F}_{2^6}$ and

$$G_0 = \begin{pmatrix} 1 & 0 & 0 \\ 1 & \alpha & \alpha^2 \end{pmatrix}, \quad G_1 = \begin{pmatrix} 0 & 1 & 0 \\ \alpha^3 & \alpha^4 & \alpha^5 \end{pmatrix}.$$

It is easy to see that $d_{SR}^1(\mathcal{C}) = 2$ and that there exists a $v(D) \in \mathcal{C}$ such that $v_{[0,1]} = (1, 0, 0 \mid 0, 1, 0) \in \mathbb{F}_{2^6}^6$. Theorem 2 says that we can recover $v_0$ if the 1th column SRD of $\mathcal{C}$ is larger than the rank deficiency of a delay-free channel in the window $[0, 1]$, i.e., if $n(T+1) - \text{rank}(A_{[0,T]}) = 6 - \text{rank}(A_{[0,T]}) \leq 1$ or equivalently, if $\text{rank}(A_{[0,1]}) \geq 5$. However, in presence of delays in the network this does not necessarily hold. Take

$$A = \left( \begin{array}{ccc|ccc} 0 & 0 & 0 & 0 & -1 & 0 \\ 0 & 1 & 0 & 0 & 0 & 0 \\ 0 & 0 & 1 & 0 & 0 & 0 \\ \hline 0 & 0 & 0 & 1 & 0 & 0 \\ 0 & 0 & 0 & 0 & 1 & 0 \\ 0 & 0 & 0 & 0 & 0 & 1 \end{array} \right) \in \mathbb{F}_2^{6\times 6}$$

that has rank equal to 5 and yields $v_{[0,1]} A_{[0,1]} = 0$, i.e., $v_{[0,1]}$ is indistinguishable from the zero sequence and therefore cannot be corrected.

The next result can be considered the analog of Theorem 2 for the general case in which the network admits delay in the transmission.

**Theorem 3** *Let $\mathcal{C}$ be a $(n, k, \delta)$ rank metric convolutional code, $v(D) \in \mathcal{C}$ and $A_{[0,T]}$ be the truncated channel matrix. Then, we can recover $v_{[0,T]}$ if*

$$d_{\mathrm{GR}}^T(\mathcal{C}) > n(T+1) - \mathrm{rank}(A_{[0,T]}). \tag{7}$$

*Proof* Let $x_{[0,T]} = v_{[0,T]}A_{[0,T]}$. Due to the linearity of the code it is enough to show that all output channel sequence are distinguishable from the zero sequence, i.e., we need to prove that $v_{[0,T]}A_{[0,T]} = 0$ is impossible if $\mathrm{rank}(A_{[0,T]})$ satisfies (7). It is easy to see that $\mathrm{rank}(v_{[0,T]}) \le n(T+1) - \mathrm{rank}(A_{[0,T]})$. Using this, together with assumption (7), it follows that $\mathrm{rank}(v_{[0,T]}) < d_{\mathrm{GR}}^T(\mathcal{C})$ which is impossible by definition of $d_{\mathrm{GR}}^T(\mathcal{C})$.

## 5 Concatenation Codes

In this section we briefly comment on another alternative to construct multi-shot codes: concatenated codes. A widely used class of concatenated codes in the Hamming context is the one constituted by an inner convolutional code concatenated to a outer block code (typically a Reed-Solomon code). The idea is that the Viterbi decoder will clean up the channel by correcting most of the errors but will occasionally output a burst of errors that will be handled by the Reed-Solomon code. However, in the network coding context little is known about the decoding of rank metric convolutional codes. Next, we present a class of concatenated codes obtained by the concatenation of a Hamming metric outer convolutional code and a rank metric inner block code. The idea is that the rank metric code deals with the possible errors occurring in the network during the transmission at each shot and either delivers the correct symbol or an erasure to the convolutional code. The reason for choosing this non-standard scheme is twofold: Firstly, we want to exploit the fact that convolutional codes perform very efficiently when dealing only with erasures. It was recently shown in [4, 27] that using the flexibility of selecting different sliding windows in convolutional codes allows to recover (using elementary linear algebra) patterns of erasures that cannot be decoded by an MDS block code of the same rate. Secondly, we want to exploit the existing efficient decoding algorithms for rank metric codes. It was shown in [17] that despite the simple way in which these codes add complex dependencies to data streams, this concatenation scheme can exploit the potential of both codes to produce a code that can correct a large number of error patterns.

Let $\mathcal{C}_I$ be a linear $(n_I, k_I)$ rank metric code with (rank) distance $d_{\mathrm{rank}}(\mathcal{C}_I)$ and generator matrix $G_I$. Let $\mathcal{C}_o$ be a $(n_o, k_o, \delta)$ convolutional code over the field $\mathbb{F}_{q^{Mk_I}}$ with (Hamming) distance $d_{\mathrm{H}}(\mathcal{C}_o)$, column distance $d_H^j(\mathcal{C}_o)$ and a basic generator matrix $G_o(D)$. The concatenation scheme is explain as follows.

Let $u(D) = u_0 + u_1 D + u_2 D^2 + \cdots \in \mathbb{F}_{q^{Mk_I}}[D]^{k_o}$ be the information sequence of source packets. Encode it through $G_o(D) \in \mathbb{F}_{q^{Mk_I}}[D]^{k_o \times n_o}$ to obtain

$$v(D) = v_0 + v_1 D + v_2 D^2 + \cdots = u(D)G_o(D) \in \mathcal{C}_o \subset \mathbb{F}_{q^{Mk_I}}[D]^{n_o}.$$

We write

$$v_i = (v_i^0, v_i^1, \ldots, v_i^{n_o-1}), \quad v_i^j \in \mathbb{F}_{q^{Mk_I}}.$$

We identify $v_i^j \in \mathbb{F}_{q^{Mk_I}}$ with a vector $v_i^j \in \mathbb{F}_{q^M}^{k_I}$ (for a given basis of $\mathbb{F}_{q^{Mk_I}}$ over $\mathbb{F}_{q^M}$) and write

$$v_i = (v_i^0, v_i^1, \ldots, v_i^{n_o-1}) \in (\mathbb{F}_{q^M}^{k_I})^{n_o}$$

and therefore

$$v(D) = v_0 + v_1 D + v_2 D^2 + \cdots \in \mathbb{F}_{q^M}^{k_I}[D]^{n_o}.$$

Finally, the codewords $c(D)$ of the concatenated code $\mathcal{C}$ are obtained through the matrix $G_I \in \mathbb{F}_{q^M}^{k_I \times n_I}$ in the following way:

$$c_i^j = v_i^j G_I \in \mathbb{F}_{q^M}^{n_I},$$

$$c_i = (c_i^0, c_i^1, \ldots, c_i^{n_o-1}) \in (\mathbb{F}_{q^M}^{n_I})^{n_o}$$

and

$$c(D) = c_0 + c_1 D + c_2 D^2 + \cdots \in \mathcal{C} \subset \mathbb{F}_{q^M}^{n_I}[D]^{n_o}.$$

Next, we present some of the distance properties of the concatenated code $\mathcal{C}$ as described above.

**Theorem 4** ([17]) *The sum rank distance of the concatenated code $\mathcal{C}$ satisfies*

$$d_{SR}(\mathcal{C}) \geq d_H(\mathcal{C}_o) d_{\mathrm{rank}}(\mathcal{C}_I)$$

*and*

$$d_{SR}^j(\mathcal{C}) \geq d_H^j(\mathcal{C}_o) d_{\mathrm{rank}}(\mathcal{C}_I).$$

*Moreover,*

$$d_{SR}(\mathcal{C}) \leq (n_o n_I - k_o k_I) \left( \left\lfloor \frac{\delta}{k_o} \right\rfloor + 1 \right) + \delta k_I + 1.$$

In [17], decoding algorithms were also presented together with some performance evaluation. Simulation results showed that this class of codes perform very efficiently when transmitting streaming data over a network.

Finally, it is worth mentioning the pioneer work of Nóbrega et al. in [19, 20] where for the first time the general ideas of multi-shot code for network coding were laid out. Moreover, the authors presented an $n$-shot code (for a fixed $n$) by means of a concatenated code using a multilevel construction. Some interesting upper and lower bounds were derived.

**Acknowledgements** The authors are supported by Portuguese funds through the CIDMA - Center for Research and Development in Mathematics and Applications, and the Portuguese Foundation for Science and Technology (FCT-Fundação para a Ciência e a Tecnologia), within project PEst-UID/MAT/04106/2013.

# References

1. P. Almeida, D. Napp, R. Pinto, A new class of superregular matrices and MDP convolutional codes. Linear Algebra Appl. **439**(7), 2145–2157 (2013)
2. A. Badr, A. Khisti, Wai-Tian. Tan, J. Apostolopoulos, Layered constructions for low-delay streaming codes. IEEE Trans. Inform. Theory (2013)
3. J.J. Climent, D. Napp, C. Perea, R. Pinto, Maximum distance separable 2D convolutional codes. IEEE Trans. Inf. Theory **62**(2), 669–680 (2016)
4. J.J. Climent, D. Napp, R. Pinto, R. Simões, Decoding of 2D convolutional codes over the erasure channel. Adv. Math. Commun. **10**(1), 179–193 (2016)
5. Ph Delsarte, Bilinear forms over a finite field, with applications to coding theory. J. Comb. Theory Ser. A **25**(3), 226–241 (1978)
6. É.M. Gabidulin, Theory of codes with maximum rank distance. Prob. Inf. Transm. **21**, 1–12 (1985)
7. H. Gluesing-Luerssen, J. Rosenthal, R. Smarandache, Strongly MDS convolutional codes. IEEE Trans. Inf. Theory **52**(2), 584–598 (2006)
8. W. Guo, X. Shi, N. Cai, M. Medard, Localized dimension growth: a convolutional random network coding approach to managing memory and decoding delay. IEEE Trans. Commun. **61**(9), 3894–3905 (2013)
9. A. Horlemann-Trautmann, K. Marshall, New criteria for MRD and gabidulin codes and some rank-metric code constructions, arXiv: 1507.08641
10. R. Johannesson, KSh Zigangirov, *Fundamentals of Convolutional Coding* (IEEE Press, New York, 1999)
11. R. Kötter, F.R. Kschischang, Coding for errors and erasures in random network coding. IEEE Trans. Inf. Theory **54**(8), 3579–3591 (2008)
12. R. Mahmood, Rank Metric Convolutional Codes with Applications in Network Streaming. Master of applied science (2015)
13. R. Mahmood, A. Badr, A. Khisti, Streaming-codes for multicast over burst erasure channels. IEEE Trans. Inf. Theory **61**(8), 4181–4208 (2015)
14. R. Mahmood, A. Badr, A. Khisti, Convolutional codes with maximum column sum rank for network streaming. IEEE Trans. Inf. Theory **62**(6), 3039–3052 (2016)
15. R.J. McEliece, The algebraic theory of convolutional codes, in *Handbook of Coding Theory*, vol. 1, ed. by V. Pless, W.C. Huffman (Elsevier Science Publishers, Amsterdam, The Netherlands, 1998), pp. 1065–1138
16. D. Napp, R. Pinto, J. Rosenthal, P. Vettori, Rank metric convolutional codes, in *Proceedings of the 22nd International Symposium on Mathematical Theory of Network and Systems (MTNS)* (Minnesota, USA, 2016)
17. D. Napp, R. Pinto, V.R. Sidorenko, Concatenation of convolutional codes and rank metric codes for multi-shot network coding. submitted to Des. Codes Cryptogr
18. D. Napp, R. Pinto, T. Toste, *On MDS convolutional codes over* $\mathbb{Z}_{p^r}$ (Designs, Codes and Cryptography, 2016), pp. 1–14
19. R.W. Nóbrega, B.F. Uchoa-Filho, Multishot codes for network coding: Bounds and a multilevel construction, in *2009 IEEE International Symposium on Information Theory* (Seoul, South Korea, 2009), pp. 428–432
20. R.W. Nóbrega, B.F. Uchoa-Filho, Multishot codes for network coding using rank-metric codes, in *Wireless Network Coding Conference (WiNC)* (IEEE 2010), pp. 1–6

21. F. Pollara, R.J. McEliece, K. Abdel-Ghaffar, Finite-state codes. IEEE Trans. Inf. Theory **34**(5), 1083–1089 (1988)
22. K. Prasad, B.S. Rajan, Network error correction for unit-delay, memory-free networks using convolutional codes. in *2010 IEEE International Conference on Communications (ICC)* (2010), pp. 1 –6
23. J. Rosenthal, Connections between linear systems and convolutional codes, in *Codes, Systems and Graphical Models*, IMA, vol. 123, ed. by B. Marcus, J. Rosenthal (Springer, 2001). pp. 39–66
24. J. Rosenthal, R. Smarandache, Maximum distance separable convolutional codes. Appl. Algebra Eng. Comm. Comput. **10**(1), 15–32 (1999)
25. J. Rosenthal, E.V. York, BCH convolutional codes. IEEE Trans. Autom. Control **45**(6), 1833–1844 (1999)
26. D. Silva, R. Kötter, F.R. Kschischang, A rank-metric approach to error control in random network coding. IEEE Trans. Inf. Theory **54**(9), 3951–3967 (2008)
27. V. Tomas, J. Rosenthal, R. Smarandache, Decoding of convolutional codes over the erasure channel. IEEE Trans. Inf. Theory **58**(1), 90–108 (2012)
28. A. Wachter-Zeh, M. Stinner, V. Sidorenko, Convolutional codes in rank metric with application to random network coding. IEEE Trans. Inf. Theory **61**(6), 3199–3213 (2015)

# Part II
# Finite Geometries and Subspace Designs

# Geometrical Aspects of Subspace Codes

**Antonio Cossidente, Francesco Pavese and Leo Storme**

**Abstract** Subspace codes are codes whose codewords are equal to subspaces of a finite vector space $V(n, q)$. Since the geometry of the subspaces of a finite vector space $V(n, q)$ is equivalent to the geometry of the subspaces of a projective space $PG(n - 1, q)$, problems on subspace codes can be investigated by using geometrical arguments. Here, we illustrate this approach by showing some recent results on subspace codes, obtained via geometrical arguments. We discuss upper bounds on the sizes of subspace codes, by showing the link between the Johnson bound and the size of partial spreads in finite projective spaces. We present geometrical constructions of subspace codes, and we also focus on subspace codes constructed from Maximum Rank Distance (MRD) codes. Here, we also present geometrical links of MRD codes to exterior sets of Segre varieties. Our aim is to motivate researchers on subspace codes to also consider geometrical arguments when investigating problems on subspace codes.

## 1 Introduction

The finite projective space $PG(n - 1, q)$ of dimension $n - 1$ over the finite field $\mathbb{F}_q$ of order $q$ is constructed from the vector space $V(n, q)$ of dimension $n$ over the finite field $\mathbb{F}_q$ of order $q$ in the following way: an $i$-dimensional projective subspace

A. Cossidente (✉)
Dipartimento di Matematica e Informatica, Università della Basilicata,
Contrada Macchia Romana, 85100 Potenza, Italy
e-mail: antonio.cossidente@unibas.it

F. Pavese
Dipartimento di Meccanica, Matematica e Management, Politecnico di Bari, Via Orabona 4,
70125 Bari, Italy
e-mail: francesco.pavese@poliba.it

L. Storme
Department of Mathematics, Ghent University, Krijgslaan 281, 9000 Ghent, Belgium
e-mail: leo.storme@ugent.be

© Springer International Publishing AG 2018                                          107
M. Greferath et al. (eds.), *Network Coding and Subspace Designs*,
Signals and Communication Technology,
https://doi.org/10.1007/978-3-319-70293-3_6

of $PG(n-1, q)$ is defined by an $(i + 1)$-dimensional subspace of the vector space $V(n, q)$.

R. Kötter and F. Kschischang [31] proved that a very good way of transmission in networks is obtained if *subspace codes* are used. Here, the codewords are subspaces of the $n$-dimensional vector space $V(n, q)$ over the finite field $\mathbb{F}_q$ of order $q$. To transmit a codeword, i.e. a $k$-dimensional vector space, through the network, it is sufficient to transmit a basis of this $k$-dimensional vector space. But a $k$-dimensional subspace has different bases. Kötter and Kschischang proved that the transmission can be optimized if the nodes in the network transmit linear combinations of the incoming basis vectors of the $k$-dimensional subspace which represents the codeword.

The geometry of the subspaces of a vector space over a finite field can be investigated in the setting of finite projective spaces, also called *Galois geometries*. The three standard references [24–26] to Galois geometries contain a wealth of information on Galois geometries. In Galois geometries, geometrical methods, and also other methods, such as polynomial methods and links with other domains, can be used to investigate a great variety of problems on substructures of these finite projective spaces. Since a subspace code precisely is a substructure of a finite projective space, this setting has, and still is able, to provide many new interesting results on subspace codes.

This article wishes to highlight this fact. We present some constructions of, and results on, subspace codes, which, presently, are among the best results, and which match results that are found via other techniques. We have concentrated on the constructions and results which can illustrate in an optimal way the ideas which have been used, and which show that Galois geometries and geometrical techniques are of great value for the study of subspace codes. We hope via these examples to motivate many researchers to also investigate problems on subspace codes via geometrical techniques, and to help showing that the study of Galois geometries is not only of purely theoretical importance, but also of practical importance.

We first present the basic ideas on Galois geometries to fix notations and properties. We then present these particular examples of results on subspace codes, obtained via geometrical arguments.

## 2 Preliminaries

Let $\mathbb{F}_q$ be the finite field of order $q$ and let $V(n, q)$ be the vector space of dimension $n$ over $\mathbb{F}_q$. The 1-dimensional subspaces of $V(n, q)$ are called *vector lines*, the 2-dimensional subspaces are called *vector planes*. The number of $k$-dimensional subspaces of $V(n, q)$ is the $q$-binomial coefficient

$$\begin{bmatrix} n \\ k \end{bmatrix}_q = \frac{(q^n - 1)(q^{n-1} - 1) \cdots (q^{n-k+1} - 1)}{(q^k - 1)(q^{k-1} - 1) \cdots (q - 1)}.$$

The number of $k$-dimensional subspaces containing a fixed $t$-dimensional subspace, $t \leq k$, is

$$\begin{bmatrix} n - t \\ k - t \end{bmatrix}_q.$$

The *subspace distance* between subspaces $U$ and $W$ of $V(n, q)$ is defined by

$$d(U, W) = \dim(U + W) - \dim(U \cap W).$$

A *subspace code* $\mathscr{C}$ is a non-empty collection of subspaces of $V(n, q)$. A *codeword* is an element of $\mathscr{C}$. The *minimum distance of* $\mathscr{C}$ is defined by

$$d(\mathscr{C}) := \min\{d(U, W) \mid U, W \in \mathscr{C}, U \neq W\}.$$

A code $\mathscr{C}$ is called an $(n, M, d)$-*subspace code* if $\mathscr{C}$ consists of subspaces of $V(n, q)$, is of cardinality $M$, and has minimum distance $d$. The maximal number of codewords in an $(n, M, d)$-code, defined over $\mathbb{F}_q$, is denoted by $\mathscr{A}_q(n, d)$. A *constant dimension subspace code* is a code with all codewords of the same dimension. If all codewords of an $(n, M, d)$-code $\mathscr{C}$ are $k$-dimensional subspaces of $V(n, q)$, then we say that $\mathscr{C}$ is an $(n, M, d, k)$-*subspace code*. The maximum number of codewords in an $(n, M, d, k)$-code, defined over $\mathbb{F}_q$, is denoted by $\mathscr{A}_q(n, d, k)$.

## 2.1 Codes in Projective Spaces

We will investigate subspace codes in the projective setting.

The *projective space* $\mathrm{PG}(n, q)$ of *projective dimension* $n$ over $\mathbb{F}_q$ is the lattice of subspaces of $V(n + 1, q)$ with respect to set inclusion. In this article, a $k$-*space*, $0 \leq k \leq n$, is a subspace $\mathrm{PG}(k, q)$ of $\mathrm{PG}(n, q)$ of projective dimension $k$. This means that a projective $k$-space is defined by a $(k + 1)$-dimensional subspace of $V(n + 1, q)$, but we interpret the vector lines contained in this $(k + 1)$-dimensional subspace as 0-dimensional objects, called *projective points*. The dimension $k$ of a projective space $\mathrm{PG}(k, q)$ is called the *projective dimension* of this subspace $\mathrm{PG}(k, q)$.

The number of projective points of a $k$-space is $\begin{bmatrix} k+1 \\ 1 \end{bmatrix}_q$. An $i$-dimensional projective space $\mathrm{PG}(i, q), i = 1, 2, 3, n - 1$, is called a *projective line*, a *projective plane*, a *solid* or a *hyperplane* of $\mathrm{PG}(n, q)$, respectively.

Alternatively, $\mathrm{PG}(n, q)$ can be defined as a point-line geometry [15]. Let $\Delta$ be a $k$-space of $\mathrm{PG}(n, q)$. If the subspaces of $\mathrm{PG}(n, q)$ of dimension $k + 1$ and $k + 2$, containing $\Delta$, are considered as new *points* and *lines*, with inclusion as incidence, then they correspond to a projective geometry $\mathrm{PG}(n - k - 1, q)$, called the *quotient geometry of* $\Delta$. We shall denote the quotient geometry of $\Delta$ as $\mathrm{PG}(n - k - 1, q)_\Delta$. An $m$-space of $\mathrm{PG}(n, q)$ that contains $\Delta$ has projective dimension $m - k - 1$ in $\mathrm{PG}(n - k - 1, q)_\Delta$.

To illustrate the close links between subspace codes and finite projective spaces, we illustrate how the study of the Johnson bound immediately relates to the study of a geometrical object in finite projective spaces, called a partial spread, which is already investigated for many years in Galois geometries.

## 3 Johnson Bound and Partial Spreads in Finite Projective Spaces

### 3.1 The Johnson Bound and Partial Spreads

**Definition 1** A $(k-1)$-*spread* of $PG(n-1, q)$ is a partitioning of the point set of $PG(n-1, q)$ in $(k-1)$-spaces.

A *partial* $(k-1)$-*spread* $\mathscr{S}$ of $PG(n-1, q)$ is a set of pairwise disjoint $(k-1)$-spaces of $PG(n-1, q)$.

A $(k-1)$-spread of $PG(n-1, q)$ exists if and only if $n \equiv 0 \pmod{k}$. Then, the size of a $(k-1)$-spread is $\left[{n \atop 1}\right]_q / \left[{k \atop 1}\right]_q$.

Partial spreads are closely related to subspace codes. A $(k-1)$-spread of $PG(n-1, q)$ is a maximal $(n, M, 2k, k)$-code of cardinality

$$M = \mathscr{A}_q(n, 2k, k) = \left[{n \atop 1}\right]_q / \left[{k \atop 1}\right]_q, \qquad \text{when } n \equiv 0 \pmod{k}.$$

A partial $(k-1)$-spread of $PG(n-1, q)$, $n \equiv 0 \pmod{k}$, of size $M$ is also an $(n, M, 2k, k)$-code, not necessarily of the maximal size.

When $k$ does not divide $n$, the following lower bound on $\mathscr{A}_q(n, 2k, k)$ is known.

**Theorem 1** ([6, 20]) *Let* $n \equiv r \pmod{k}$, $0 \leq r \leq k-1$. *Then for all* $q$, *we have*

$$\mathscr{A}_q(n, 2k, k) \geq \frac{q^n - q^k(q^r - 1) - 1}{q^k - 1}.$$

The following upper bound on $\mathscr{A}_q(n, 2k, k)$ is known.

**Theorem 2** ([5, 17]) *Let* $\mathscr{S}$ *be a partial* $t$-*spread in* $PG(d, q)$, *where* $d = k(t + 1) - 1 + r$, $1 \leq r \leq t$.

*Let* $|\mathscr{S}| = q^r \cdot \frac{q^{k(t+1)} - 1}{q^{t+1} - 1} - s$. *Then*

- $s \geq q - 1$.
- $s > \frac{q^r - 1}{2} - \frac{q^{2r-t-1}}{5}$.

Since Theorem 1 states that there exists an example with $s = q^r - 1$, we know that

$$\mathscr{A}_q(n, 2k, k) = \frac{q^n - q^k(q^r - 1) - 1}{q^k - 1},$$

when $r = 1$.

For many years, it was conjectured that

$$\mathscr{A}_q(n, 2k, k) = \frac{q^n - q^k(q^r - 1) - 1}{q^k - 1},$$

so the lower bound of Theorem 1 is sharp.

However, El-Zanati et al. [18] found a counter example: there is a partial 3-spread in the finite vector space $V(8, 2)$, which corresponds to a partial 2-spread in $PG(7, 2)$, of size 34, but there exists no example of size 35. So $\mathscr{A}_2(8, 6, 3) = 34$.

But, very recently, a major breakthrough has been made by E. Năstase and P. Sissokho [37].

**Theorem 3** ([37]) *Let $r \equiv n \pmod{k}$. Then, if*

$$k > \frac{q^r - 1}{q - 1},$$

*the maximal size for a partial $(k - 1)$-spread in $PG(n - 1, q)$ is equal to $\frac{q^n - q^k(q^r - 1) - 1}{q^k - 1}$.*

*Consequently, if $k > \frac{q^r - 1}{q - 1}$, then*

$$\mathscr{A}_q(n, 2k, k) = \frac{q^n - q^k(q^r - 1) - 1}{q^k - 1}.$$

Similar results for $q = 2$ were recently also proven by Kurz [32].

We now make the link between the problem of the largest size of partial spreads in finite projective spaces to the Johnson bound on subspace codes.

In general, in the terminology of projective geometry, an $(n, M, d, k)$-code $\mathscr{C}$ can be viewed as a set of $(k - 1)$-spaces of $PG(n - 1, q)$ of cardinality $M$ with the following property: $t = k - d/2 + 1$ is the smallest integer such that every $(t - 1)$-space of $PG(n - 1, q)$ is contained in at most one $(k - 1)$-space of $\mathscr{C}$. Equivalently, any two distinct codewords of $\mathscr{C}$ intersect in at most a $(t - 2)$-space. Hence, $(k - 1)$-spaces of $\mathscr{C}$ through a $(t - 2)$-space $\Delta$ form a partial $(k - t)$-spread $\mathscr{C}_\Delta$ in the quotient geometry $PG(n - t, q)_\Delta$ of $\Delta$. Note that $\mathscr{C}_\Delta$ is an $(n - k + d/2, M', d, d/2)$-code, called the *quotient code of* $\Delta$. Using this observation, the following upper bound on the size of a constant dimension code was obtained. This upper bound is known under the name *Johnson bound*.

**Theorem 4** (Johnson bound [20]) *The maximal size $\mathscr{A}_q(n, d, k)$ of an $(n, M, d, k)$-code satisfies the upper bound*

$$\mathscr{A}_q(n, d, k) \leq \frac{\left[ {n \atop t-1} \right]_q}{\left[ {k \atop t-1} \right]_q} \mathscr{A}_q(n - k + d/2, d, d/2),$$

*where $t = k - d/2 + 1$ and where $\mathscr{A}_q(n - k + d/2, d, d/2)$ is the maximal size of a partial $(d/2 - 1)$-spread in $\mathrm{PG}(n - k + d/2 - 1, q)$.*

**Corollary 1** *The maximal size $\mathscr{A}_q(n, d, k)$ of an $(n, M, d, k)$-code, with $n - k + d/2 \equiv 0 \pmod{d/2}$, satisfies the upper bound*

$$\mathscr{A}_q(n, d, k) \leq \frac{\left[ {n \atop t} \right]_q}{\left[ {k \atop t} \right]_q}.$$

The main problem is whether there exist constant dimension codes meeting the Johnson bound.

For $d = 2k$, this is the classical partial spreads problem, mentioned above. This shows that a particular problem on subspace codes is completely equivalent to a geometrical problem, already for decades of interest and of importance in Galois geometries. This is one of the classical examples of the close links between the theory of subspace codes and the theory of Galois geometries.

For more detailed information on partial spreads in finite projective spaces, we refer to Chap. "Partial Spreads and Vector Space Partitions" of this Springer special volume.

## 3.2 Subspace Codes from Maximum Rank Distance Codes

Rank distance codes were introduced by Delsarte [14] and are suitable for error correction in the case where the network topology and the underlying network code are known (the coherent case).

The set $\mathscr{M}_{m \times n}(q)$ of $m \times n$ matrices over the finite field $\mathbb{F}_q$ forms a metric space with respect to the *rank distance*, defined by

$$d_r(A, B) = rk(A - B).$$

The maximum size of a code of minimum distance $d$, $1 \leq d \leq \min\{m, n\}$, in $(\mathscr{M}_{m \times n}(q), d_r)$ is $q^{n(m-d+1)}$ for $m \leq n$ and $q^{m(n-d+1)}$ for $m \geq n$. A code $\mathscr{A} \subset \mathscr{M}_{m \times n}(q)$ attaining this bound is said to be an $(m, n, k)_q$ *maximum rank distance code (MRD code)*, where $k = m - d + 1$ for $m \leq n$ and $k = n - d + 1$ for $m \geq n$. A rank distance code $\mathscr{A}$ is called $\mathbb{F}_q$–*linear* if $\mathscr{A}$ is a subspace of $\mathscr{M}_{m \times n}(q)$ considered as a vector space over $\mathbb{F}_q$. We can always assume that $m \leq n$.

In [40], the authors introduced a method, called the *lifting process*, to construct a constant dimension subspace code from a maximum rank distance code. Let $A$ be an $m \times n$ matrix over $\mathbb{F}_q$, and let $I_m$ be the $m \times m$ identity matrix. The rows of the $m \times (n + m)$ matrix $(I_m|A)$ can be viewed as coordinates of points in general position of an $(m - 1)$–dimensional projective space of $PG(n + m - 1, q)$. This subspace is denoted by $L(A)$. Hence, the matrix $A$ can be *lifted* to a subspace $L(A)$. Let $\mathscr{C}$ be an $(m, n, k)_q$ MRD code, and let $A_1$ and $A_2$ be two distinct matrices of $\mathscr{C}$. Since $rk(A_1 - A_2) \geq d$, we have that

$$rk \begin{pmatrix} I_m & A_1 \\ I_m & A_2 \end{pmatrix} = m + rk(A_1 - A_2) \geq m + d.$$

Hence, $L(A_1)$ meets $L(A_2)$ in at most an $(m - d - 1)$–dimensional projective space and $\{L(A) \mid A \in \mathscr{C}\}$ is an $(n + m, q^{nk}, 2d, m)$ subspace code. A constant dimension subspace code such that all of its codewords are lifted codewords of an MRD code is called a *lifted MRD code*.

This lifting process implies the following result.

**Theorem 5** ([40])
$$\mathscr{A}_q(n + m, 2d, m) \geq q^{n(m-d+1)}.$$

Although the size of a lifted MRD code equals the highest power of $q$ in the Johnson bound, it is known that it is not maximal and it can be extended to a larger subspace code.

Several examples of linear $(n, n, k)_q$ MRD codes are known to exist (Gabidulin codes [14, 22], twisted Gabidulin codes [39], generalized twisted Gabidulin codes [35]).

## 3.3 Planes in PG(5, q) Pairwise Intersecting in at Most a Point

Let us consider an other case for the Johnson bound.

Honold, Kiermaier, and Kurz proved that $\mathscr{A}_2(6, 4, 3) = 77$ and there are exactly five codes of size 77 [27].

Geometrically, this means that, in the projective space $PG(5, 2)$, a set of projective planes pairwise intersecting in at most one projective point, has size at most 77, and there are exactly five different examples of such sets of 77 planes.

Honold, Kiermaier, and Kurz managed to extend one of the five examples to an infinite class of $q^6 + 2q^2 + 2q + 1$ planes in $PG(5, q)$, pairwise intersecting in at most one point.

Alternatively, Cossidente and Pavese [12] also constructed an infinite class of $q^6 + 2q^2 + 2q + 1$ planes in $PG(5, q)$, pairwise intersecting in at most one point. We wish to highlight this construction since it relies on many nice geometrical properties and results. This includes bundles of conics in a plane of PG(3, $q$), hyperbolic quadrics in PG(3, $q$), and the Klein correspondence between the lines of PG(3, $q$) and the Klein quadric $Q^+(5, q)$ of PG(5, $q$).

We now present the construction of a set of $q^6 + 2q^2 + 2q + 1$ planes in PG(5, $q$), pairwise intersecting in at most one point, by Cossidente and Pavese. This leads to the result

**Theorem 6**
$$q^6 + 2q^2 + 2q + 1 \le \mathscr{A}_q(6, 4, 3) \le (q^3 + 1)^2.$$

The construction of Cossidente and Pavese uses the *Klein correspondence* between lines of PG(3, $q$) and their Plücker coordinates on the Klein quadric $Q^+(5, q)$ in PG(5, $q$) [24, Sect. 15.4].

We first present the hyperbolic quadric $Q^+(3, q)$ of PG(3, $q$). This is the quadric with standard equation $Q^+(3, q) : X_0 X_2 - X_1 X_3 = 0$.

This quadric contains two particular sets of $q + 1$ lines, which play a symmetrical role, and which are called a *regulus* $\mathscr{R}$ and its *opposite regulus* $\mathscr{R}^\perp$.

Lines of a regulus are pairwise disjoint, while the lines of one regulus intersect the lines of the opposite regulus in one point. Figure 1 presents a drawing of a 3-dimensional hyperbolic quadric, with three lines of every regulus.

We now introduce the basic facts of the Klein correspondence which plays a central role in the construction of Cossidente and Pavese [24].

**Definition 2** Consider a line $\ell$ in PG(3, $q$) and suppose that $y(y_0, y_1, y_2, y_3)$ and $z(z_0, z_1, z_2, z_3)$ are different points of $\ell$. Let $p_{ij} = y_i z_j - y_j z_i$.

**Fig. 1** The hyperbolic quadric $Q^+(3, q)$

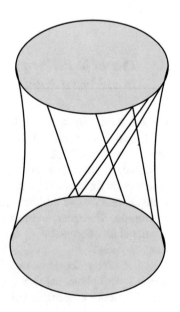

Then the *Plücker coordinates* of the line $\ell$ are the 6-tuple $(p_{01}, p_{02}, p_{03}, p_{23}, p_{31}, p_{12})$.

The Plücker coordinates of a line $\ell$ in $PG(3, q)$ are well-defined. However, not every 6-tuple is Plücker coordinates of a line $\ell$ of $PG(3, q)$.

**Theorem 7** *The 6-tuple* $(p_{01}, p_{02}, p_{03}, p_{23}, p_{31}, p_{12})$ *is Plücker coordinates of a line* $\ell$ *of* $PG(3, q)$ *if and only if*

$$p_{01} p_{23} + p_{02} p_{31} + p_{03} p_{12} = 0. \tag{1}$$

The set of points $\mathscr{Q}$ of $PG(5, q)$, defined by the equation $p_{01} p_{23} + p_{02} p_{31} + p_{03} p_{12} = 0$, is a hyperbolic quadric $Q^+(5, q)$ of $PG(5, q)$, called the *Klein quadric*. This hyperbolic quadric $Q^+(5, q)$ contains points, lines and planes. The planes of the hyperbolic quadric $Q^+(5, q)$ are also called the *generators* of $Q^+(5, q)$. The link with the corresponding sets of lines in $PG(3, q)$ is defined in the following list [24, Sect. 15.4].

- Point of $PG(5, q)$ on the Klein quadric defines a line $\ell$ of $PG(3, q)$.
- Line of $PG(5, q)$ on the Klein quadric defines the set of $q + 1$ lines of $PG(3, q)$ through a fixed point $P$ in a fixed plane $\pi$.
- Planes of $PG(5, q)$ on the Klein quadric define either the set of all lines of $PG(3, q)$ through a fixed point $P$, or the set of all lines of $PG(3, q)$ lying in a fixed plane $\pi$. A *Greek* plane is a plane contained in the Klein quadric, defining the lines of $PG(3, q)$ lying in a fixed plane $\pi$, and a *Latin* plane is a plane contained in the Klein quadric, defining the lines of $PG(3, q)$ through a fixed point $P$.
- Greek planes, respectively Latin planes, on the Klein quadric pairwise intersect in one point, while a Latin and a Greek plane either are skew to each other, or intersect in a line.
- Consider a plane $\pi$ of $PG(5, q)$, intersecting the Klein quadric in a conic $C$. This conic corresponds to a regulus $\mathscr{R}$ of a hyperbolic quadric $Q^+(3, q)$ in $PG(3, q)$. The polar plane $\pi^\perp$ of $\pi$ with respect to the Klein quadric also intersects the Klein quadric in a conic $C^\perp$, corresponding to the opposite regulus $\mathscr{R}^\perp$ of $\mathscr{R}$ of this hyperbolic quadric $Q^+(3, q)$ in $PG(3, q)$.

These two polar planes $\pi$ and $\pi^\perp$ are skew to each other in $PG(5, q)$.

In the preceding description, we recognize already the fact that the set of planes of the Klein quadric can be partitioned into two equivalence classes. Two planes of the Klein quadric are called *equivalent* when they are equal or intersect in a point [26, p. 20]. This definition leads to an equivalence relation on the set of planes of the Klein quadric, having two distinct equivalence classes. They are in the preceding description respectively the class of the Greek planes and the class of the Latin planes. The equivalence classes of the generators of the Klein quadric are also sometimes called the *systems of generators*.

The construction of Cossidente and Pavese of a set of $q^6 + 2q^2 + 2q + 1$ planes of PG$(5, q)$, pairwise intersecting in at most one point, starts by considering specific structures in PG$(3, q)$, and then transferring, via the Klein correspondence, to planes in PG$(5, q)$.

- Consider a fixed plane $\pi$ of PG$(3, q)$.
- A bundle of conics $\mathcal{B}$ in this plane $\pi$ is a set of $q^2 + q + 1$ conics pairwise intersecting in one point.
- Each conic of the bundle $\mathcal{B}$ is contained in $q^3(q - 1)/2$ hyperbolic quadrics $Q^+(3, q)$ of PG$(3, q)$.

This leads to $(q^6 - q^3)/2$ hyperbolic quadrics $Q^+(3, q)$ having $q^6 - q^3$ reguli, and, by applying the Klein correspondence, to a set $\mathcal{L}_1$ of $q^6 - q^3$ planes of PG$(5, q)$, pairwise intersecting in at most one point.

Figure 2 shows a drawing of the plane $\pi$, together with two conics of the bundle of conics $\mathcal{B}$ in this plane $\pi$, and one hyperbolic quadric passing through each one of these two conics.

In the second step of the construction of Cossidente and Pavese, consider this fixed plane $\pi$ in PG$(3, q)$, containing the bundle of conics $\mathcal{B}$. This plane $\pi$ defines a Greek plane $\Pi$ on the Klein quadric.

Add the other $q^3 + q^2 + q$ Greek planes on the Klein quadric to the set $\mathcal{L}_1$. This leads to a larger set $\mathcal{L}_2$ of $q^6 + q^2 + q$ planes of PG$(5, q)$, pairwise intersecting in at most one point.

In the third step, consider again the Greek plane $\Pi$ in PG$(5, q)$, corresponding to the plane $\pi$ in PG$(3, q)$, containing the bundle of conics $\mathcal{B}$.

**Fig. 2** The bundle of conics and the hyperbolic quadrics passing through the conics of the bundle

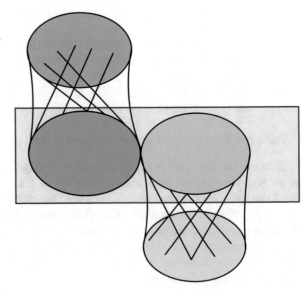

Each line $r$ of $\Pi$ lies in $q - 1$ planes of PG(5, $q$) only intersecting the Klein quadric in the line $r$.

For every line $r$ of $\Pi$, select one such plane, and add it to the set $\mathcal{L}_2$. This gives the desired set $\mathcal{L}$ of $q^6 + 2q^2 + 2q + 1$ planes of PG(5, $q$), pairwise intersecting in at most one point.

## 3.4 Planes in PG(5, $q$) Pairwise Intersecting in at Most a Point (Alternative Construction)

Let $U_i = (0, \ldots, 0, 1, 0, \ldots, 0)$, $i = 0, \ldots, 5$, be the vector with the one in the $(i + 1)$-th position.

Let $\mathcal{C}$ be a linear $(3, 3, 2)_q$ MRD code. Let $\mathcal{L} = \{L(A) \mid A \in \mathcal{C}\}$ be the lifted MRD code obtained by lifting the elements of $\mathcal{C}$. Then $\mathcal{L}$ consists of $q^6$ planes of PG(5, $q$) mutually intersecting in at most a point. In particular, members of $\mathcal{L}$ are disjoint from the special plane $T = \langle U_3, U_4, U_5 \rangle$ and therefore every line covered by an element of $\mathcal{L}$ is disjoint from $T$. Moreover, from [27, Lemma 6], every line of PG(5, $q$) disjoint from $T$ is covered by a member of $\mathcal{L}$ exactly once. We denote by $T'$ the plane $\langle U_0, U_1, U_2 \rangle$. Since the zero matrix belongs to $\mathcal{C}$, we have that $T' \in \mathcal{L}$.

Let $\mathcal{S}$ denote the set of $q^3$ $3 \times 3$ skew–symmetric matrices over $\mathbb{F}_q$. Here, for $q$ even, the diagonal elements of a $3 \times 3$–skew symmetric matrix are zeros. Since there exists a linear $(3, 3, 2)_q$ MRD code containing $\mathcal{S}$, we may assume that $\mathcal{S} \subset \mathcal{C}$, see [10, Proposition 3.1].

Now, we introduce the non–degenerate hyperbolic quadric $\mathcal{Q}$ of PG(5, $q$) having the following equation:

$$X_0 X_3 + X_1 X_4 + X_2 X_5 = 0.$$

The planes $T$ and $T'$ are generators of $\mathcal{Q}$. They belong to different systems of generators of $\mathcal{Q}$. Let $\mathcal{M}, \mathcal{M}'$ be the system of generators of $\mathcal{Q}$ containing $T$, $T'$, respectively.

It can be seen that if $A$ is a $3 \times 3$ skew–symmetric matrix over $\mathbb{F}_q$, then $L(A)$ is a generator of $\mathcal{Q}$ disjoint from $T$.

*Remark 1* Since the number of generators of $\mathcal{Q}$ disjoint from $T$ equals $q^3$, we have that each such a plane is of the form $L(A)$, for some $A \in \mathcal{S}$. Note that, if $A \in \mathcal{S}$, then $L(A)$ belongs to the system $\mathcal{M}'$ of generators containing $T'$.

Let $\mathcal{L}'$ be the set consisting of the $q^3$ planes obtained by lifting the matrices of $\mathcal{S}$. Then, the set $(\mathcal{L} \setminus \mathcal{L}') \cup \mathcal{M}$ consists of $q^6 + q^2 + q + 1$ planes mutually intersecting in at most a point.

Corresponding to a line $r$ of $T$, there corresponds a set $\mathcal{T}_r$ of $q - 1$ planes of PG(5, $q$) meeting $\mathcal{Q}$ exactly in $r$. Varying the line $r$ over the line set of $T$ and choosing one of the planes in $\mathcal{T}_r$, we obtain a set $\mathcal{T}$ of $q^2 + q + 1$ planes mutually intersecting in at most one point.

Finally, we have that the set $(\mathscr{L} \setminus \mathscr{L}') \cup \mathscr{M} \cup \mathscr{T}$ is a set of $q^6 + 2q^2 + 2q + 1$ planes mutually intersecting in at most a point.

## 3.5   Solids in $\mathrm{PG}(7, q)$ Pairwise Intersecting in at Most a Line

Let $\mathscr{C}$ be a linear $(4, 4, 3)_q$ MRD code. Let $\mathscr{L}_1 = \{L(A) \mid A \in \mathscr{C}\}$ be the lifted MRD code obtained by lifting the elements of $\mathscr{C}$. Then $\mathscr{L}_1$ consists of $q^{12}$ solids of $\mathrm{PG}(7, q)$ mutually intersecting in at most a line. In particular, members of $\mathscr{L}_1$ are disjoint from the special solid $T = \langle U_4, U_5, U_6, U_7 \rangle$ and therefore every plane covered by an element of $\mathscr{L}_1$ is disjoint from $T$. Moreover, from [27, Lemma 6], every plane of $\mathrm{PG}(7, q)$ disjoint from $T$ is covered by a member of $\mathscr{L}_1$ exactly once. We denote by $T'$ the solid $\langle U_0, U_1, U_2, U_3 \rangle$. Since the zero matrix belongs to $\mathscr{C}$, we have that $T' \in \mathscr{L}_1$.

Let $\mathscr{C}_r \subset \mathscr{C}$ be the set consisting of all the matrices of $\mathscr{C}$ having rank $r$, with $2 \le r \le 4$. From [22], it is known that a linear $(4, 4, 3)_q$ MRD code contains $(q^4 - 1)(q^2 + 1)(q^2 + q + 1)$ matrices of rank 2, $(q^4 - 1)(q^2 + 1)(q + 1)(q^4 - q^2 - q)$ matrices of rank 3, and $(q^4 - 1)q^3(q^5 - q^4 - q^3 + q + 1)$ matrices of rank 4.

Let $A$ be an element of $\mathscr{C}_2$. As in Sect. 3.2, the rows of the $4 \times 8$ matrix $(A|I_4)$ can be viewed as coordinates of points in general position of a solid, say $L'(A)$, of $\mathrm{PG}(7, q)$. The solid $L'(A)$ is disjoint from $T'$ and meets $T$ in a line. Let $\mathscr{L}_2 = \{L'(A) \mid A \in \mathscr{C}_2\}$ be the set of solids obtained from the elements of $\mathscr{C}_2$. Then we have that $\mathscr{L}_1 \cup \mathscr{L}_2$ is a set of $q^{12} + (q^4 - 1)(q^2 + 1)(q^2 + q + 1)$ solids of $\mathrm{PG}(7, q)$ mutually intersecting in at most a line.

Let $\mathscr{S}$ denote the set of $q^6$ $4 \times 4$ skew–symmetric matrices over $\mathbb{F}_q$. Here, for $q$ even, the diagonal elements of a $4 \times 4$–skew symmetric matrix are zeros. Since there exists a linear $(4, 4, 3)_q$ MRD code containing $\mathscr{S}$, we may assume that $\mathscr{S} \subset \mathscr{C}$, see [10, Proposition 3.1].

Now, we introduce the non-degenerate hyperbolic quadric $\mathscr{Q}_1$ of $\mathrm{PG}(7, q)$ having the following equation:

$$X_0 X_4 + X_1 X_5 + X_2 X_6 + X_3 X_7 = 0.$$

Here again, the solids (generators) of the hyperbolic quadric $\mathscr{Q}_1$ of $\mathrm{PG}(7, q)$ are partitioned into two equivalence classes $\mathscr{M}_1$ and $\mathscr{M}_1'$ [26]. Two generators $\Pi_1$ and $\Pi_2$ are called *equivalent* when they are equal or intersect in a line. This relation is again an equivalence relation, having two equivalence classes $\mathscr{M}_1$ and $\mathscr{M}_1'$ on the set of generators of the hyperbolic quadric $\mathscr{Q}_1$ of $\mathrm{PG}(7, q)$.

The solids $T$ and $T'$ are generators of $\mathscr{Q}_1$. They belong to the same system of generators of $\mathscr{Q}_1$, say $\mathscr{M}_1$. Let $D(X)$ and $I(X)$ denote the set of generators in $\mathscr{M}_1$ disjoint from the solid $X$ or meeting non–trivially $X$, respectively. Then

$$\mathscr{M}_1 = D(T) \cup (D(T') \cap I(T)) \cup (I(T') \cap I(T)),$$

where $D(T)$, $D(T') \cap I(T)$ and $I(T') \cap I(T)$ are trivially intersecting sets.

It can be seen that if $A$ is a $4 \times 4$ skew–symmetric matrix over $\mathbb{F}_q$, then $L(A)$ (resp. $L'(A)$) is a generator of $\mathcal{Q}_1$ disjoint from $T$ (resp. $T'$).

*Remark 2* Since the number of generators of $\mathcal{Q}_1$ disjoint from $T$ equals $q^6$ [30, Lemma 3], we have that each such a solid is of the form $L(A)$, for some $A \in \mathcal{S}$. Note that, if $A \in \mathcal{S}$, then $L(A)$ belongs to $\mathcal{M}_1$.

On the other hand, a solid $L'(A)$ in $D(T')$ is disjoint from $T$ if and only if $A$ is a skew–symmetric matrix of rank 4. Therefore, the number of skew–symmetric matrices of rank 4 is equal to $|D(T) \cap D(T')|$. It follows that

$$|D(T') \cap I(T)| = |D(T')| - |D(T') \cap D(T)| =$$

$$= q^6 - (q-1)q^2(q^3-1) = q^2(q^3+q-1)$$

and

$$|I(T') \cap I(T)| = |\mathcal{M}_1| - 2q^6 + (q-1)q^2(q^3-1) = (q^2+1)(q^2+q+1).$$

Since $\mathcal{S} \subset \mathcal{C}$, we have that $D(T) \subset \mathcal{L}_1$. Moreover, every element $g \in D(T') \cap I(T)$ is of the form $L'(A)$ for some skew–symmetric matrix $A$ having rank 2. Hence, $g \in \mathcal{L}_2$. The set $\mathcal{L}_1 \cup \mathcal{L}_2 \cup (I(T') \cap I(T))$ consists of $q^{12} + (q^4-1)(q^2+1)(q^2+q+1) + (q^2+1)(q^2+q+1)$ solids of $PG(7,q)$ mutually intersecting in at most a line.

Let $\gamma$ be a non–zero element of $\mathbb{F}_q$ such that the polynomial $X^2 - X - \gamma$ is irreducible over $\mathbb{F}_q$. Let $\mathcal{Q}_2$ be the hyperbolic quadric of $PG(7,q)$ having equation

$$X_0X_6 + X_1X_7 + \gamma^{-1}(X_2X_4 + X_3X_5 + X_2X_6 + X_3X_7) = 0.$$

The hyperbolic quadrics $\mathcal{Q}_1$ and $\mathcal{Q}_2$ generate a pencil of hyperbolic quadrics of $PG(7,q)$, say $\mathcal{F}$, containing $q-1$ other distinct quadrics, say $\mathcal{Q}_i$, $3 \le i \le q+1$, none of which is degenerate. Let $\mathcal{X}$ be the base locus of $\mathcal{F}$. Since the hyperbolic quadrics of $\mathcal{F}$ cover all the points of $PG(7,q)$, and any two distinct quadrics in $\mathcal{F}$ intersect precisely in $\mathcal{X}$, we have that $|\mathcal{X}| = (q+1)(q^2+1)^2$. There are $2(q^2+1)$ generators belonging to each hyperbolic quadric of the pencil $\mathcal{F}$ and they all belong to the same system of generators with respect to each of the quadrics $\mathcal{Q}_i$ in $\mathcal{F}$, say $\mathcal{M}_i$. In particular, $T$ and $T'$ belong to each hyperbolic quadric of the pencil $\mathcal{F}$. Let $\mathcal{G}$ be the set of generators meeting both $T$ and $T'$ non–trivially, and belonging to each hyperbolic quadric of the pencil $\mathcal{F}$. We have that $\mathcal{G} \subset \mathcal{M}_i$, for every $1 \le i \le q+1$, and $|\mathcal{G}| = q^2 + 1$.

Let $I_i(X)$ denote the set of solids in $\mathcal{M}_i$ meeting non–trivially $X$, $2 \le i \le q+1$. Then $\mathcal{G} = \bigcap_{i=2}^{q+1}(I_i(T) \cap I_i(T')) \cap (I(T) \cap I(T'))$.

The set $\mathcal{L}_1 \cup \mathcal{L}_2 \cup (\bigcup_{i=2}^{q+1}(I_i(T) \cap I_i(T'))) \cup (I(T) \cap I(T'))$ is a set of $q^{12} + (q^4-1)(q^2+1)(q^2+q+1) + q(q+1)^2(q^2+1) + (q^2+1)$ solids of $PG(7,q)$ mutually intersecting in at most a line.

The set $\mathscr{G}$ consists of $q^2 + 1$ generators belonging to each hyperbolic quadric of the pencil $\mathscr{F}$ such that every element in $\mathscr{G}$ meets both $T$ and $T'$ in a line. The set $\mathscr{D}_T = \{A \cap T \mid A \in \mathscr{G}\}$, $\mathscr{D}_{T'} = \{A \cap T' \mid A \in \mathscr{G}\}$ is a line–spread of $T$, $T'$, respectively. In particular, for a fixed line $\ell \in \mathscr{D}_T$, there exists a unique element in $\mathscr{D}_{T'}$, say $A_\ell$, such that $\langle \ell, A_\ell \rangle$ is in $\mathscr{G}$, and viceversa. Furthermore, if $\ell \in \mathscr{D}_T$ and $B \in \mathscr{D}_{T'} \setminus \{A_\ell\}$, then $\langle \ell, B \rangle$ is a solid meeting a hyperbolic quadric of the pencil $\mathscr{F}$ in a 3-dimensional hyperbolic quadric $\mathscr{Q}^+(3, q)$. Let $\mathscr{D}'$ be the set of solids of the form $\langle \ell, B \rangle$, where $\ell \in \mathscr{D}_T$ and $B \in \mathscr{D}_{T'} \setminus \{A_\ell\}$. Then $\mathscr{D}'$ is disjoint from $\mathscr{G}$ and $|\mathscr{D}'| = q^2(q^2 + 1)$. We have that $\mathscr{L}_1 \cup \mathscr{L}_2 \cup (\bigcup_{i=2}^{q+1}(I_i(T) \cap I_i(T'))) \cup (I(T) \cap I(T')) \cup \mathscr{D}' \cup \{T\}$ is a set of solids mutually intersecting in at most a line, of size

$$q^{12} + (q^4 - 1)(q^2 + 1)(q^2 + q + 1) + (q^3 + 3q^2 + q + 1)(q^2 + 1) + 1.$$

There exists a group $H$ in the orthogonal group $\mathrm{PGO}^+(8, q)$, stabilizing $\mathscr{Q}_1$, fixing both $T$, $T'$, their line–spreads $\mathscr{D}(T)$, $\mathscr{D}(T')$, and permuting in a single orbit the remaining lines of $T$ (respectively $T'$). Let $\perp$ be the orthogonal polarity of $\mathrm{PG}(7, q)$ associated with $\mathscr{Q}_1$. If $r'$ is a line of $T'$, then $r'^\perp$ meets $T$ in a line $r$. If $r'$ belongs to $\mathscr{D}_{T'}$, then $r$ belongs to $\mathscr{D}_T$. Assume that $r'$ does not belong to $\mathscr{D}_{T'}$. Of course, $r'$ meets $q + 1$ lines $l'_1, \ldots, l'_{q+1}$ of $\mathscr{D}_{T'}$ and $r$ meets $q + 1$ lines $l_1, \ldots, l_{q+1}$ of $\mathscr{D}_T$.

The group $H$ contains a subgroup fixing the lines $l'_1, \ldots, l'_{q+1}$ and having $q(q - 1)/2$ orbits of size $q^2 - q$ on the lines of $T$ distinct from $l_1, \ldots, l_{q+1}$. Each one of them, together with $l_1, \ldots, l_{q+1}$, is a line–spread of $T$, one of them being $\mathscr{D}_T$. Let $\mathscr{E}$ be one of the orbits of size $q^2 - q$ disjoint from $\mathscr{D}_T$ and let $Y$ be the solid generated by $r'$ and a line of $\mathscr{E}$. It is possible to prove that $Y^H$ is a set of $q^6 - q^2$ solids mutually intersecting in at most a line.

Finally, we have that the set $\mathscr{L}_1 \cup \mathscr{L}_2 \cup (\bigcup_{i=2}^{q+1}(I_i(T) \cap I_i(T'))) \cup (I(T) \cap I(T')) \cup \mathscr{D}' \cup Y^H \cup \{T\}$ is an $(8, M, 4, 4)_q$–subspace code, where

$$M = q^{12} + q^2(q^2 + 1)^2(q^2 + q + 1) + 1.$$

This leads to the following result.

**Corollary 2**

$$\mathscr{A}_q(8, 4, 4) \geq q^{12} + q^2(q^2 + 1)^2(q^2 + q + 1) + 1.$$

*Remark 3* The previous lower bound was obtained with different techniques in [19], where the authors, among other interesting results, proved that $q^{12} + q^2(q^2 + 1)^2(q^2 + q + 1) + 1$ is also the maximum size of an $(8, M, 4, 4)_q$–subspace code containing a lifted MRD code.

For more information on constant dimension codes, we also refer to Chap. "Constructions of Constant Dimension Codes" of this Springer special volume.

## 4  Optimal Mixed-Dimension Subspace Codes in PG(4, $q$)

Via geometrical arguments, it can be shown that $\mathscr{A}_q(5, 3) = 2(q^3 + 1)$. Here, a $(5, 2(q^3 + 1), 3)$-code consists of subspaces of the vector space $V(5, q)$, equivalently, of subspaces of the projective space PG(4, $q$).

To give an idea which arguments are used to get geometrical insight in which subspaces could be contained in a mixed-dimension (5, $M$, 3)-subspace code, we try to see how the codewords of this code can intersect.

The vector space $V(5, q)$ has four types of subspaces: vector lines, vector planes, subspaces of dimension three, and subspaces of dimension four.

The formula for the subspace distance $d(U, U') = \dim(U + U') - \dim(U \cap U')$ shows that a (5, $M$, 3)-subspace code with minimum distance 3:

1. cannot contain two vector lines, and cannot contain two subspaces of dimension four,
2. two vector planes in the code should only intersect in the zero vector,
3. two subspaces of dimension three should only intersect in a vector line,
4. a vector line in the code cannot be contained in a vector plane or in a 3-dimensional vector space belonging to the code, a vector plane in the code cannot be contained in a 3-dimensional subspace or 4-dimensional subspace belonging to the code, and a 3-dimensional subspace in the code cannot be contained in a 4-dimensional subspace belonging to the code.

We now interpret these conditions in the geometrical setting, so we replace all the codewords in the $(5, 2(q^3 + 1), 3)$-code by their geometrical equivalents in the projective space PG(4, $q$).

Condition (2) implies that two projective lines belonging to the code are skew to each other. Hence, the projective lines belonging to the $(5, 2(q^3 + 1), 3)$-code form a partial line spread of PG(4, $q$).

From Theorem 2, the largest partial line spread of PG(4, $q$) has size $q^3 + 1$. So, if $\mathscr{C}$ is an optimal $(5, 3)_q$ subspace code, then $\mathscr{C}$ contains at most $q^3 + 1$ pairwise skew lines. A dual argument shows that $\mathscr{C}$ contains at most $q^3 + 1$ planes, and these planes pairwise intersect in a projective point.

Hence, if $\mathscr{C}$ consists of projective lines and projective planes, we have that $|\mathscr{C}| \leq 2(q^3 + 1)$ and, if $|\mathscr{C}| = 2(q^3 + 1)$, then $\mathscr{C}$ consists of a set $\mathscr{L}$ of $q^3 + 1$ pairwise skew lines and of a set $\mathscr{P}$ of $q^3 + 1$ planes mutually intersecting in exactly a point, such that no line of $\mathscr{L}$ is contained in a plane of $\mathscr{P}$.

Note that Condition (1) above states that $\mathscr{C}$ contains at most one point and, dually, $\mathscr{C}$ contains at most one solid.

Counting arguments of [11] prove that if $\mathscr{C}$ contains a point, then $\mathscr{C}$ contains at most $q^3$ planes. Dually, if $\mathscr{C}$ contains a solid, then $\mathscr{C}$ contains at most $q^3$ lines.

It follows from these arguments that $\mathscr{A}_q(5, 3) \leq 2(q^3 + 1)$ and there are four possibilities for the code $\mathscr{C}$:

(I)  $\mathscr{C}$ consists of one point, $q^3 + 1$ lines, and $q^3$ planes;
(II) $\mathscr{C}$ consists of $q^3$ lines, $q^3 + 1$ planes, and one solid;

(III) $\mathscr{C}$ consists of one point, $q^3$ lines, $q^3$ planes, and one solid;

(IV) $\mathscr{C}$ consists of $q^3 + 1$ lines and $q^3 + 1$ planes.

We present the construction for $q$ odd, showing that $\mathscr{A}_q(5, 3) = 2(q^3 + 1)$. A similar construction exists for $q$ even, but since it is more technical, we refer to [11] for its description.

Let $q = p^h$, where $p$ is an odd prime. Let $PG(4, q)$ be equipped with homogeneous coordinates $(X_0, X_1, X_2, X_3, X_4)$, let $\pi$ be the projective plane with equations $X_3 = X_4 = 0$ and let $\ell$ be the line of $\pi$ with equations $X_2 = X_3 = X_4 = 0$. Let $\omega$ be a primitive element of $\mathbb{F}_q$ and denote by $\Pi_i$ the solid of $PG(4, q)$ passing through $\pi$ with equation $X_3 = \omega^{i-1} X_4$, if $1 \leq i \leq q - 1$, $X_3 = 0$ if $i = q$, and $X_4 = 0$ if $i = q + 1$.

Let $a, b, c$ be fixed elements of $\mathbb{F}_q$ such that the polynomial $X^3 + aX^2 + bX + c = 0$ is irreducible over $\mathbb{F}_q$ and consider the following matrices

$$M_{r,s,t} = \begin{pmatrix} 1 & 0 & r & r^2 - ar + s & t \\ 0 & 1 & s & 2rs - t & s^2 + bs - cr \\ 0 & 0 & 1 & 2r & 2s \\ 0 & 0 & 0 & 1 & 0 \\ 0 & 0 & 0 & 0 & 1 \end{pmatrix}.$$

Then the group of projective transformations $G = \{x \mapsto M_{r,s,t} \cdot x \mid r, s, t \in \mathbb{F}_q\}$ is a $p$-group of order $q^3$.

This group $G$ has within the set of lines of $PG(4, q)$, exactly $q^3$ orbits of size $q^3$, each consisting of pairwise disjoint lines that are disjoint from $\pi$. Similarly, every plane of $PG(4, q)$, intersecting the plane $\pi$ in exactly one point, not belonging to the line $\ell$, belongs to an orbit of $G$, consisting of $q^3$ planes, pairwise intersecting in one point.

Consider such an orbit under $G$ of $q^3$ planes intersecting the fixed plane $\pi$ in a point not belonging to the line $\ell$. Such a plane contains $q^2$ lines skew to $\pi$. They define only $q^2$ of the $q^3$ orbits of lines that are disjoint from $\pi$. Hence, by taking one of the remaining $q^3 - q^2$ orbits of such lines, a set of $q^3$ planes and $q^3$ lines is obtained forming a subspace code with minimum distance 3.

This $(5, 2q^3, 3)$-code can be extended to a $(5, 2(q^3 + 1), 3)$-code by adding a line $r$ of $\pi$, $r \neq \ell$, and a plane $\xi$ through the line $\ell$, but with $\xi \neq \pi$, to this code. This leads to an optimal $(5, \mathscr{A}_q(5, 3), 3)$-code consisting of $q^3 + 1$ lines and planes.

Figure 3 presents the setting for this mixed dimension code in $PG(4, q)$, $q$ odd. The drawing shows the plane $\pi$, the line $\ell$ with the plane $\xi$ passing through $\ell$, and the line $r$. Two of the $q + 1$ solids through the plane $\pi$ are shown. They are denoted by $\Pi_i$ and $\Pi_j$. The orbit of $q^3$ lines skew to $\pi$ is denoted by the three vertical lines, sharing one point with the solids $\Pi_i$ and $\Pi_j$. Finally, one plane is drawn of the orbit of $q^3$ planes sharing one point with the plane $\pi$, which does not belong to $\ell$, and which are skew to the $q^3$ lines of the selected orbit of lines.

**Fig. 3** The mixed
dimension construction in
$PG(4, q), q$ odd

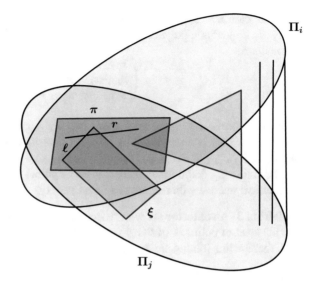

In [11, Remark 2.6], it is shown that minor changes can be made to the construction, to also construct optimal $(5, \mathscr{A}_q(5, 3), 3)$-codes of type (I), (II), and (III).

## 5 Geometrical Links to Non-linear Maximum Rank Distance Codes

In this context, we can make links to the Segre variety of $PG(n^2 - 1, q)$ [13, 26].
The *Segre map*

$$\sigma : PG(n - 1, q) \times PG(n - 1, q) \rightarrow PG(n^2 - 1, q),$$

takes a pair of points $x = (x_0, \ldots, x_{n-1}), y = (y_0, \ldots, y_{n-1})$ of $PG(n - 1, q)$ to their product $(x_0 y_0, x_0 y_1, \ldots, x_{n-1} y_{n-1})$ (the products $x_i y_j$ are taken in lexicographical order). The image of the Segre map is an algebraic variety of $PG(n^2 - 1, q)$, called the *Segre variety*, and is denoted by $\mathscr{S}_{n-1,n-1}$.

When $n = 2$, the Segre variety $\mathscr{S}_{1,1}$ of $PG(3, q)$ is a non–degenerate hyperbolic quadric $\mathscr{Q}^+(3, q)$. In Sect. 3.3, we defined the hyperbolic quadric as the quadric with equation $X_0 X_2 - X_1 X_3 = 0$, but equivalently, this quadric is given as the zero locus of the quadratic polynomial given by the determinant of the matrix

$$\begin{pmatrix} x_0 y_0 & x_0 y_1 \\ x_1 y_0 & x_1 y_1 \end{pmatrix}.$$

In the case $n = 3$, the Segre variety $\mathscr{S}_{2,2}$ of $PG(8, q)$ is defined to be the zero locus of all quadratic polynomials given by the determinants of the $2 \times 2$ matrices of the matrix

$$\begin{pmatrix} x_0 y_0 & x_0 y_1 & x_0 y_2 \\ x_1 y_0 & x_1 y_1 & x_1 y_2 \\ x_2 y_0 & x_2 y_1 & x_2 y_2 \end{pmatrix}.$$

In other terms, in the projective space $PG(\mathscr{M}_{n \times n}(q))$, if $n = 2$, the Segre variety $\mathscr{S}_{1,1}$ of $PG(3, q)$ is represented by all $2 \times 2$ matrices of rank 1 and if $n = 3$, the Segre variety $\mathscr{S}_{2,2}$ of $PG(8, q)$ is represented by all $3 \times 3$ matrices of rank 1.

The following definition considers a set of points that at first sight is of purely geometrical interest with respect to a Segre variety.

**Definition 3** An **exterior set** with respect to a Segre variety $\mathscr{S}_{n-1,n-1}$ of $PG(n^2 - 1, q)$ is a set of points $\mathscr{E}$ of $PG(n^2 - 1, q) \setminus \mathscr{S}_{n-1,n-1}$ of size $(q^{n^2-n} - 1)/(q - 1)$ such that the line joining any two points of $\mathscr{E}$ is disjoint from $\mathscr{S}_{n-1,n-1}$.

But this definition and the previous observations give an immediate link with maximum rank distance codes.

In general, an exterior set $\mathscr{E}$ of $PG(n^2 - 1, q)$ with respect to a Segre variety $\mathscr{S}_{n-1,n-1}$, of size $(q^{n^2-n} - 1)/(q - 1)$, gives rise to a MRD code: this is done by identifying a point of $\mathscr{E}$ and its nonzero scalar multiples together with the zero matrix with members of $\mathscr{M}_{n \times n}(q)$. This is also the key tool of our approach. We formulate this in the next proposition.

**Proposition 1** *An exterior set with respect to $\mathscr{S}_{n-1,n-1}$ gives rise to an $(n, n, n - 1)$ MRD code closed under $\mathbb{F}_q$–multiplication, and viceversa.*

**Corollary 3** *An $(n, n, n - 1)$ $\mathbb{F}_q$-linear Gabidulin code $\mathscr{G}$ is a certain subspace $X$ of $PG(n^2 - 1, q)$ of dimension $n^2 - n - 1$ which is an exterior set with respect to $\mathscr{S}_{n-1,n-1}$.*

The preceding corollary is of particular interest since the maximum dimension of a subspace of $PG(n^2 - 1, q)$ disjoint from $\mathscr{S}_{n-1,n-1}$ is exactly $n^2 - n - 1$ [9].

## 5.1 The Case n = 2

In this subsection, we report the complete classification of linear and non–linear MRD codes that are closed under $\mathbb{F}_q$–multiplication when $n = m = 2$. We do this because this is linked to the solution of a purely geometrical problem, related to the hyperbolic quadric $\mathscr{Q}^+(3, q)$ of the projective space $PG(3, q)$, which is the smallest example of a Segre variety.

A *flock* of the hyperbolic quadric $\mathscr{Q}^+(3, q)$ of the finite projective space $PG(3, q)$ is a partition of $\mathscr{Q}^+(3, q)$ consisting of $q + 1$ irreducible conics. A *linear flock* of $\mathscr{Q}^+(3, q)$ is a flock which consists of $q + 1$ irreducible conics, lying in the $q + 1$

planes through a fixed line $\ell$ skew to the hyperbolic quadric $\mathcal{Q}^+(3, q)$. A *maximal exterior set* (MES) with respect to the hyperbolic quadric $\mathcal{Q}^+(3, q)$ is a set of $q + 1$ points of PG$(3, q)$ such that the line joining any two of them has no point in common with $\mathcal{Q}^+(3, q)$. The polar planes, with respect to the polarity induced by $\mathcal{Q}^+(3, q)$, of the points of a MES, define a flock, and conversely.

In [41], J.A. Thas proved that all flocks of $\mathcal{Q}^+(3, q)$ are linear if $q$ is even, and that $\mathcal{Q}^+(3, q)$ has non–linear flocks (called *Thas flocks*) if $q$ is odd. Furthermore, he showed that, for $q = 3, 7$ and $q \equiv 1 \mod 4$, $\mathcal{Q}^+(3, q)$ has only (up to a projectivity) the linear flock and the Thas flock. For $q = 11, 23, 59$, other flocks of $\mathcal{Q}^+(3, q)$ were discovered, called *exceptional flocks* [1, 3, 29]. Finally, the combined results of Bader and Lunardon [2] and Thas [42] proved that every flock of $\mathcal{Q}^+(3, q)$, $q$ odd, is linear, a Thas flock or one of the exceptional flocks.

The classification theorem is therefore as follows.

**Theorem 8** *Let $\mathcal{E}$ be a MES defined by a flock $F$ of $\mathcal{Q}^+(3, q)$ in the matrix model of* PG$(3, q)$. *Then, either $q$ is even and $\mathcal{E}$ is a line, or $q$ is odd and one of the following possibilities occurs:*

(1) *$\mathcal{E}$ is a line;*
(2) *$\mathcal{E}$ consists of $(q + 1)/2$ points on two lines $\ell$, $\ell^\perp$, where $\perp$ is the polarity of $\mathcal{Q}^+(3, q)$;*
(3) *$\mathcal{E}$ is one of the sporadic examples.*

In our setting, the linear MES corresponds to a $(2, 2, 1)$ $\mathbb{F}_q$-linear MRD-code. In all the other instances (Theorem 8 (2) and (3)), the MES corresponds to a $(2, 2, 1)$ non-linear maximum rank distance code.

## 5.2   The Case $n = 3$

A very useful model of $\mathcal{S}_{2,2}$ arises from the geometry of the Desarguesian projective plane $\pi := \text{PG}(2, q^3)$. Indeed, each point $P$ of PG$(2, q^3)$, when read over $\mathbb{F}_q$, defines a projective plane $X(P)$ of the projective space PG$(8, q)$, and the set $\mathcal{D} = \{X(P) : P \in \text{PG}(2, q^3)\}$ is a *Desarguesian spread* of PG$(8, q)$ [38, Sect. 25]. The incidence structure $\pi := (\mathcal{D}, \mathcal{L})$, whose points are the elements of $\mathcal{D}$ and whose line set $\mathcal{L}$ consists of the 5–dimensional projective subspaces of PG$(8, q)$ joining any two distinct elements of $\mathcal{D}$, is isomorphic to PG$(2, q^3)$. The pair $(\mathcal{D}, \mathcal{L})$ is called the $\mathbb{F}_q$-*linear representation* of PG$(2, q^3)$ (with respect to the Desarguesian spread $\mathcal{D}$).

Let $X_0, X_1, X_2$ denote projective homogeneous coordinates in $\pi \simeq \text{PG}(2, q^3)$ and let $\bar{\pi}$ be a subplane of $\pi$ of order $q$. Let $G$ denote the stabilizer of $\bar{\pi}$ in PGL$(3, q^3)$.

Choose homogeneous coordinates in such a way that $\bar{\pi} := \{(1, x^{q+1}, x^q) : x \in \mathbb{F}_{q^3} \setminus \{0\}, N(x) = 1\}$, where $N(\cdot)$ is the *norm function* from $\mathbb{F}_{q^3}$ over $\mathbb{F}_q$. It turns out that $\bar{\pi}$ is fixed pointwise by the order three semilinear collineation of PG$(2, q^3)$ given by $\phi : (X_0, X_1, X_2) \mapsto (X_2^q, X_0^q, X_1^q)$.

Let $\langle S \rangle$ be a Singer cyclic group of $G$ [28]. We can assume that $S$ is given by

$$\begin{pmatrix} \omega & 0 & 0 \\ 0 & \omega^q & 0 \\ 0 & 0 & \omega^{q^2} \end{pmatrix},$$

where $\omega$ is a primitive element of $\mathbb{F}_{q^3}$.

*Remark 4* The subgroup $\langle S \rangle$ fixes the three points $E_1 = (1, 0, 0)$, $E_2 = (0, 1, 0)$ and $E_3 = (0, 0, 1)$ of $\pi$, and hence the lines $E_i E_j$, $1 \le i < j \le 3$. All the other orbits are subplanes of order $q$ of $\pi$. Note that the line $E_i E_j$ is partitioned into the two points $E_i$ and $E_j$, and into $q - 1$ orbits of $\langle S \rangle$ of size $q^2 + q + 1$. The collineation $\phi$ above normalizes $\langle S \rangle$.

The points of $\bar{\pi}$ correspond to the $q^2 + q + 1$ planes filling the system of a Segre variety $\mathscr{S}_{2,2}$ of PG$(8, q)$ contained in the Desarguesian spread $\mathscr{D}$. Also, the lines of $\pi$, arising from sublines of $\bar{\pi}$, yield a set of $(q^3 - q)(q^2 + q + 1)$ points of $\pi$ that together with the points of $\bar{\pi}$ give rise to the points of the secant variety $\Omega(\mathscr{S}_{2,2})$ of $\mathscr{S}_{2,2}$ [33, 36].

Under the action of the stabilizer $G$ of $\bar{\pi}$ in PGL$(3, q^3)$, the point set of $\pi$ is partitioned into three orbits corresponding to the points of $\bar{\pi}$, points of $\pi \setminus \bar{\pi}$ on extended sublines of $\bar{\pi}$, and the complement. Under the same group, by duality, the line set of $\pi$ is partitioned into three orbits corresponding to sublines of $\bar{\pi}$, lines meeting $\bar{\pi}$ in a point, and lines external to $\bar{\pi}$.

**Proposition 2** *In the linear representation of* PG$(2, q^3)$, *any line of $\pi$ disjoint from $\bar{\pi}$ corresponds to a 5-dimensional projective subspace of* PG$(8, q)$ *disjoint from* $\mathscr{S}_{2,2}$.

Of course, any line of $\pi$ disjoint from $\bar{\pi}$ gives rise to an exterior set with respect to $\mathscr{S}_{2,2}$ and hence, from a coding theoretical point of view, a $(3, 3, 2)$ $\mathbb{F}_q$–linear MRD code.

Now let $q > 2$ and consider the set $\mathscr{X}$ of points of $\pi$ whose coordinates satisfy the equation $X_0 X_1^q - X_2^{q+1} = 0$. The set $\mathscr{X}$ has size $q^3 + 1$ and it is fixed by $\langle S \rangle$. Also, it contains $q - 1$ subplanes of order $q$, one of which is $\bar{\pi}$, and the points $E_1$ and $E_2$. More precisely, the subplanes of order $q$ embedded in $\mathscr{X}$ are the subsets of points of $\pi$ given by

$$\pi_a := \{(1, x^{q+1}, x^q) : x \in \mathbb{F}_{q^3}, \ N(x) = a\},$$

where $a$ is a nonzero element of $\mathbb{F}_q$. In particular, $\pi_1 = \bar{\pi}$. From [16, Proposition 3.1], a line of $\pi$ intersects $\mathscr{X}$ in 0, 1, 2 or $q + 1$ points, and the intersections of size $q + 1$ are actually lines of subplanes of order $q$ of $\pi$ embedded in $\mathscr{X}$. We can assume that the Segre variety corresponding to $\bar{\pi} = \pi_1$ is the only Segre variety of PG$(8, q)$ corresponding to rank one matrices of order three.

We recall the following definition.

**Definition 4** ([4]) Let $\ell_\infty$ be a line of $\pi$ disjoint from the subplane $\bar{\pi}$. The **exterior splash** of $\bar{\pi}$ on $\ell_\infty$ is defined to be the set of $q^2 + q + 1$ points of $\ell_\infty$ that belong to an extended line of $\bar{\pi}$.

The line $E_1 E_2$ is disjoint from all the $q - 1$ subplanes $\pi_a$, $a \in \mathbb{F}_q \setminus \{0\}$, of $\pi$ contained in $\mathcal{X}$. Also, for each subplane $\pi_a$, with $a \in \mathbb{F}_q \setminus \{0\}$, its exterior splash on $E_1 E_2$ is the set of $q^2 + q + 1$ points of $E_1 E_2$ given by

$$Z_a := \{(1, x, 0) : x \in \mathbb{F}_{q^3}, \ N(x) = -a^2\}.$$

Such a set is a so-called $\mathbb{F}_q$-*linear set of pseudoregulus type*. For further details on these linear sets, see [16, 34, 36]. All these subplanes and splashes are of course $\langle S \rangle$-orbits.

Now, let $T$ be the fundamental triangle $E_1 E_2 E_3$ of $\pi$. One can prove that a line of $\pi$ is either a side of $T$, or it contains a vertex of $T$, or it induces a subline of a unique subplane of order $q$ of $\pi$ invariant under $\langle S \rangle$. Consider now the set $K := (\mathcal{X} \setminus \{\pi_1\}) \cup Z_1$. It turns out that $K$ is such that every line defined by any two of its points is disjoint from $\pi_1$. Correspondingly, the set $K'$ corresponding to $K$ in $\mathrm{PG}(8, q)$, $q > 2$, is an exterior set of size $(q^3 + 1)(q^2 + q + 1)$ with respect to the Segre variety $\mathscr{S}_{2,2}$ corresponding to $\pi_1$.

In terms of coding theory, we have the following result.

**Theorem 9** *There exists a* $(3, 3, 2)$ *MRD non-linear code admitting a Singer cyclic group of* $\mathrm{PGL}(3, q)$, $q > 2$, *as an automorphism group.*

*Remark 5* In [21], R. Figueroa presented a new class of non-Desarguesian projective planes of order $q^3$, $q$ a prime power with $q \not\equiv 1 \mod 3$, $q > 2$. C. Hering and H.-J. Schaffer in [23] improved and simplified the construction for all prime powers $q$. From [7, Corollary 3], the set $K$ constructed above represents a line in the Figueroa plane of order $q^3$.

*Remark 6* When $q = 2$, some computer tests performed with MAGMA [8] give that all subsets of $\mathrm{PG}(2, 8)$ yielding exterior sets with respect to a Segre variety $\mathscr{S}_{2,2}$ are precisely the 24 lines disjoint from $\bar{\pi}$. When $q = 2$, no non-linear MRD codes arise from our construction.

We also refer the readers to Chap. "Codes Endowed with the Rank Metric" of this Springer special volume, dedicated to rank metric codes, and to Chap. "Constructions of Cyclic Subspace Codes and Maximum Rank Distance Codes" of this special volume on the construction of cyclic subspace codes and maximum rank distance codes.

# References

1. L. Bader, Some new examples of flocks of $Q^+(3, q)$. Geom. Dedicata **27**, 213–218 (1988)
2. L. Bader, G. Lunardon, On the flocks of $\mathscr{Q}^+(3, q)$. Geom. Dedicata **29**, 177–183 (1989)
3. R.D. Baker, G.L. Ebert, A nonlinear flock in the minkowski plane of order 11. Congr. Numer. **58**, 75–81 (1987)
4. S.G. Barwick, W.-A. Jackson, Exterior splashes and linear sets of rank 3. Discret. Math. **339**, 1613–1623 (2016)
5. A. Beutelspacher, Partial spreads in finite projective spaces and partial designs. Math. Z. **145**, 211–229 (1975)
6. A. Beutelspacher, On $t$-covers in finite projective spaces. J. Geom. **12**, 10–16 (1979)
7. J.M.N. Brown, Some partitions in Figueroa planes. Note Mat. **29**, 33–43 (2009)
8. J. Cannon, C. Playoust, An introduction to MAGMA, University of Sydney, Sydney, Australia (1993)
9. B.N. Cooperstein, External flats to varieties in $PG(M_{n,n}(GF(q)))$. Linear Algebra Appl. **267**, 175–186 (1997)
10. A. Cossidente, F. Pavese, Subspace codes in PG($2n - 1, q$). Combinatorica (to appear). https://doi.org/10.1007/s00493-016-3354-5
11. A. Cossidente, F. Pavese, L. Storme, Optimal subspace codes in PG(4, $q$) (In preparation)
12. A. Cossidente, F. Pavese, On subspace codes. Des. Codes Cryptogr. **78**, 527–531 (2016)
13. A. Cossidente, G. Marino, F. Pavese, Non-linear maximum rank distance codes. Des. Codes Cryptogr. **79**, 597–609 (2016)
14. P. Delsarte, Bilinear forms over a finite field, with applications to coding theory. J. Combin. Theory Ser. A **25**, 226–241 (1978)
15. P. Dembowski, *Finite Geometries* (Springer, Berlin, 1968)
16. G. Donati, N. Durante, Scattered linear sets generated by collineations between pencils of lines. J. Algebr. Combin. **40**, 1121–1134 (2014)
17. J. Eisfeld, L. Storme, (Partial) $t$-spreads and minimal $t$-covers in finite projective spaces. Lecture notes, Universiteit Gent (2000), http://cage.ugent.be/~fdc/courses/GGaGP2.php
18. S. El-Zanati, H. Jordon, G. Seelinger, P. Sissokho, L. Spence, The maximum size of a partial 3-spread in a finite vector space over GF(2). Des. Codes Cryptogr. **54**, 101–107 (2010)
19. T. Etzion, N. Silberstein, Codes and designs related to lifted MRD codes. IEEE Trans. Inform. Theory **59**, 1004–1017 (2013)
20. T. Etzion, A. Vardy, Error-correcting codes in projective space. IEEE Trans. Inform. Theory **57**, 1165–1173 (2011)
21. R. Figueroa, A family of not $(V, l)$-transitive projective planes of order $q^3$, $q \not\equiv 1 \mod 3$ and $q > 2$. Math. Z. **181**, 471–479 (1982)
22. E.M. Gabidulin, Theory of codes with maximum rank distance. Probl. Inform. Trans. **21**, 1–12 (1985)
23. C. Hering, H.-J. Schaffer, On the new projective planes of R. Figueroa, *Combinatorial Theory*, vol. 969, Lecture Notes in Mathematics (Springer, Berlin, 1982), pp. 187–190
24. J.W.P. Hirschfeld, *Finite Projective Spaces of Three Dimensions* (Oxford University Press, Oxford, 1985)
25. J.W.P. Hirschfeld, *Projective Geometries Over Finite Fields*, 2nd edn. (Oxford University Press, Oxford, 1998)
26. J.W.P. Hirschfeld, J.A. Thas, *General Galois Geometries* (Oxford University Press, Oxford, 1991)
27. T. Honold, M. Kiermaier, S. Kurz, Optimal binary subspace codes of length 6, constant dimension 3 and minimum distance 4. Contemp. Math. **632**, 157–176 (2015)
28. B. Huppert, *Endliche Gruppen, I*, Die Grundlehren der Mathematischen Wissenschaften, Band 134 (Springer, Berlin, 1967)
29. N.L. Johnson, Flocks of hyperbolic quadrics and translation planes admitting affine homologies. J. Geom. **34**, 50–73 (1989)

30. A. Klein, K. Metsch, L. Storme, Small maximal partial spreads in classical finite polar spaces. Adv. Geom. **10**, 379–402 (2010)
31. R. Kötter, F. Kschischang, Coding for errors and erasures in random network coding. IEEE Trans. Inform. Theory **54**, 3579–3591 (2008)
32. S. Kurz, Improved upper bound for partial spread. Des. Codes Cryptogr. **85**, 97–106 (2017)
33. M. Lavrauw, G. Van de Voorde, Field reduction and linear sets in finite geometry. Contemp. Math. **632**, 271–293 (2015)
34. M. Lavrauw, C. Zanella, Subgeometries and linear sets on a projective line. Finite Fields Appl. **34**, 95–106 (2015)
35. G. Lunardon, R. Trombetti, Y. Zhou, Generalized twisted Gabidulin codes, arXiv:1507.07855
36. G. Lunardon, G. Marino, O. Polverino, R. Trombetti, Maximum scattered linear sets of pseudoregulus type and the Segre Variety $\mathscr{S}_{n,n}$. J. Algebr. Combin. **39**, 807–831 (2014)
37. E. Nastase, P. Sissokho, The maximum size of a partial spread in a finite projective space. J. Combin. Theory Ser. A **152**, 353–362 (2017)
38. B. Segre, Teoria di Galois, fibrazioni proiettive e geometrie non desarguesiane. Ann. Mat. Pura Appl. **64**, 1–76 (1964)
39. J. Sheekey, A new family of linear maximum rank distance codes. Adv. Math. Commun. **10**, 475–488 (2016)
40. D. Silva, F.R. Kschischang, R. Koetter, A rank-metric approach to error control in random network coding. IEEE Trans. Inform. Theory **54**, 3951–3967 (2008)
41. J.A. Thas, Flocks of non-singular ruled quadrics in PG($3, q$). Atti Accad. Naz. Lincei Rend. Cl. Sci. Fis. Mat. Natur. **59**, 83–85 (1975)
42. J.A. Thas, Flocks, maximal exterior sets and inversive planes, *Finite Geometries and Combinatorial Designs*, vol. 111, Contemporary Mathematics (American Mathematical Society, Providence, 1990), pp. 187–218

# Partial Spreads and Vector Space Partitions

**Thomas Honold, Michael Kiermaier and Sascha Kurz**

**Abstract** Constant-dimension codes with the maximum possible minimum distance have been studied under the name of partial spreads in Finite Geometry for several decades. Not surprisingly, for this subclass typically the sharpest bounds on the maximal code size are known. The seminal works of Beutelspacher and Drake & Freeman on partial spreads date back to 1975 and 1979, respectively. From then until recently, there was almost no progress besides some computer-based constructions and classifications. It turns out that vector space partitions provide the appropriate theoretical framework and can be used to improve the long-standing bounds in quite a few cases. Here, we provide a historic account on partial spreads and an interpretation of the classical results from a modern perspective. To this end, we introduce all required methods from the theory of vector space partitions and Finite Geometry in a tutorial style. We guide the reader to the current frontiers of research in that field, including a detailed description of the recent improvements.

## 1 Introduction

Let $\mathbb{F}_q$ be the finite field with $q$ elements, where $q > 1$ is a prime power. By $\mathbb{F}_q^v$ we denote the standard vector space of dimension $v \geq 1$ over $\mathbb{F}_q$, whose vectors are the $v$-tuples $\mathbf{x} = (x_1, \ldots, x_v)$ with $x_i \in \mathbb{F}_q$. The set of all subspaces of $\mathbb{F}_q^v$, ordered by the incidence relation $\subseteq$, is called $(v - 1)$-*dimensional projective geometry over* $\mathbb{F}_q$ and denoted by $\mathrm{PG}(v - 1, \mathbb{F}_q)$. It forms a finite modular geometric lattice with meet $X \wedge Y = X \cap Y$, join $X \vee Y = X + Y$, and rank function $X \mapsto \dim(X)$. Employing this algebraic notion of dimension instead of the geometric one, we will use the term

T. Honold
Zhejiang University, Hangzhou 310027, China
e-mail: honold@zju.edu.cn

M. Kiermaier · S. Kurz (✉)
University of Bayreuth, 95440 Bayreuth, Germany
e-mail: sascha.kurz@uni-bayreuth.de

M. Kiermaier
e-mail: michael.kiermaier@uni-bayreuth.de

© Springer International Publishing AG 2018
M. Greferath et al. (eds.), *Network Coding and Subspace Designs*,
Signals and Communication Technology,
https://doi.org/10.1007/978-3-319-70293-3_7

*k-subspace* to denote a *k*-dimensional vector subspace of $\mathbb{F}_q^v$.[1] The important geometric interpretation of subspaces will still be visible in the terms *points, lines, planes, solids, hyperplanes* (denoting 1-, 2-, 3-, 4- and $(v-1)$-subspaces, respectively), and in general through our extensive use of geometric language. For more geometrical aspects we refer the reader to Chapter "Geometrical Aspects of Subspace Codes".

In the same way as $\mathbb{F}_q^v$, an arbitrary *v*-dimensional vector space *V* over $\mathbb{F}_q$ gives rise to a projective geometry PG(*V*), and the terminology introduced before (and thereafter) applies to this general case as well. Since a vector space isomorphism $V \cong \mathbb{F}_q^v$ induces a geometric isomorphism ("collineation") $PG(V) \cong PG(\mathbb{F}_q^v) = PG(v-1, \mathbb{F}_q)$, we could in principle avoid the use of non-standard vector spaces, but only at the expense of flexibility—for example, the Singer representation of the point-hyperplane design of $PG(v-1, \mathbb{F}_q)$ is best developed using the field extension $\mathbb{F}_{q^v}/\mathbb{F}_q$ as ambient vector space *V* (and not $\mathbb{F}_q^v$, which would require a discussion of matrix representations of finite fields).

The set of all *k*-subspaces of an $\mathbb{F}_q$-vector space *V* will be denoted by $\begin{bmatrix} V \\ k \end{bmatrix}_q$. The sets $\begin{bmatrix} V \\ k \end{bmatrix}_q$ form finite analogues of the Graßmann varieties studied in Algebraic Geometry. In terms of $v = \dim(V)$, the cardinality of $\begin{bmatrix} V \\ k \end{bmatrix}_q$ is given by the Gaussian binomial coefficient

$$\begin{bmatrix} v \\ k \end{bmatrix}_q := \begin{cases} \frac{(q^v-1)(q^{v-1}-1)\cdots(q^{v-k+1}-1)}{(q^k-1)(q^{k-1}-1)\cdots(q-1)} & \text{if } 0 \leq k \leq v; \\ 0 & \text{otherwise,} \end{cases}$$

which are polynomials of degree $k(v-k)$ in *q* (if they are nonzero) and represent *q*-analogues of the ordinary binomial coefficients in the sense that $\lim_{q \to 1} \begin{bmatrix} v \\ k \end{bmatrix}_q = \binom{v}{k}$. Their most important combinatorial properties are described in [2, Sect. 3.3] and [66, Chap. 24].

Making the connection with the main topic of this book, the geometry $PG(v-1, \mathbb{F}_q)$ serves as input and output alphabet of the so-called *linear operator channel (LOC)*, a clever model for information transmission in coded packet networks subject to noise [43].[2] The relevant metrics on the LOC are given by the *subspace distance* $d_S(X, Y) := \dim(X+Y) - \dim(X \cap Y) = 2 \cdot \dim(X+Y) - \dim(X) - \dim(Y)$, which can also be seen as the graph-theoretic distance in the Hasse diagram of $PG(v-1, \mathbb{F}_q)$, and the *injection distance* $d_I(X, Y) := \max\{\dim(X), \dim(Y)\} - \dim(X \cap Y)$. A set $\mathscr{C}$ of subspaces of $\mathbb{F}_q^v$ is called a *subspace code* and serves as a channel code for the LOC in the same way as classical linear codes over $\mathbb{F}_q$ do for the *q*-ary symmetric channel.[3] The *minimum (subspace) distance* of $\mathscr{C}$ is given by $d = \min\{d_S(X, Y) \mid X, Y \in \mathscr{C}, X \neq Y\}$. If all elements of $\mathscr{C}$ have the same dimen-

---

[1]Using the algebraic dimension has certain advantages—for example, the existence criterion $v = tk$ for spreads (cf. Theorem 1) looks ugly when stated in terms of the geometric dimensions: $v' = t(k'-1) + 1$.

[2]The use of distributed coding at the nodes of a packet-switched network, generally referred to as *Network Coding*, is described in [27, 56, 70] and elsewhere in this volume.

[3]Except that attention is usually restricted to "one-shot subspace codes", i.e. subsets of the alphabet, which makes no sense in the classical case.

sion, we call $\mathscr{C}$ a *constant-dimension code*. For a constant-dimension code $\mathscr{C}$ we have $d_S(X, Y) = 2d_1(X, Y)$ for all $X, Y \in \mathscr{C}$, so that we can restrict attention to the subspace distance. Constant-dimension codes are the most suitable for coding purposes, and the quest for good system performance leads straight to the problem of determining the maximum possible cardinality $A_q(v, d; k)$ of a constant-dimension-$k$ code in $\mathbb{F}_q^v$ with minimum subspace distance $d$. For two codewords $X$ and $Y$ of dimension $k$ the inequality $d_S(X, Y) \geq d$ is equivalent to $\dim(X \cap Y) \leq k - d/2$.[4] Thus, the maximum possible minimum distance of a constant-dimension code with codewords of dimension $k$ is $2k$. This extremal case has been studied under the name "partial spreads" in Finite Geometry for several decades. A *partial k-spread* in $\mathbb{F}_q^v$ is a collection of $k$-subspaces with pairwise trivial, i.e., zero-dimensional intersection. Translating this notion into Projective Geometry and identifying thereby, as usual, subspaces of $\mathbb{F}_q^v$ with their sets of incident points, we have that a partial $k$-spread in $\mathbb{F}_q^v$ is the same as a set of mutually disjoint $k$-subspaces, or $(k-1)$-dimensional flats in the geometric view, of the geometry $\mathrm{PG}(v-1, \mathbb{F}_q)$.[5]

With the history of partial spreads in mind, it comes as no surprise that the sharpest bounds on the maximal code sizes $A_q(v, d; k)$ of constant-dimension codes are typically known for this special subclass. The primary goal of this survey is to collect all available information on the numbers $A_q(v, 2k; k)$ and present this information in an accessible way. Following standard practice in Finite Geometry, we will refer to partial $k$-spreads of size $A_q(v, 2k; k)$ as *maximal partial k-spreads*.[6]

In the case of a perfect packing, i.e., a partition of the point set of $\mathrm{PG}(v-1, \mathbb{F}_q)$, we speak of a *k-spread*. Partitions into subspaces of possibly different dimensions are equivalent to vector space partitions. A *vector space partition* $\mathscr{C}$ of $\mathbb{F}_q^v$ is a collection of nonzero subspaces with the property that every non-zero vector is contained in a unique member of $\mathscr{C}$. If $\mathscr{C}$ contains $m_d$ subspaces of dimension $d$, then $\mathscr{C}$ is said to be of type $k^{m_k} \cdots 1^{m_1}$. Zero frequencies $m_d = 0$ are usually suppressed.[7] So, partial $k$-spreads are just the special case of vector space partitions, in which all members have dimension either $k$ or $1$. For $k \geq 2$ (the case $k = 1$ is trivial) the members of dimension 1 correspond to points not covered by a $k$-subspace of the partial spread and are called *holes* in this context.

Although vector space partitions can be seen as a mixed-dimension analogue of partial spreads, they are not usable as subspace codes of their own.[8] However, it turns out that they provide an appropriate framework to study bounds on the sizes of partial spreads.

---

[4]Note that the distance between codewords of the same dimension, and hence also the minimum distance of a constant-dimension code, is an even integer.

[5]In other words, partial spreads are just packings of the point set of a projective geometry $\mathrm{PG}(v-1, \mathbb{F}_q)$ into subspaces of equal dimension.

[6]The weaker property of *complete* (i.e., inclusion-maximal) partial spreads will not be considered.

[7]Since $\sum_{X \in \mathscr{C}} \dim(X) = \sum_d dm_d = v$, the type of $\mathscr{C}$ can be viewed as an ordinary integer partition of $v$.

[8]The subspace distance $d_S(X, Y)$ depends not only on $\dim(X \cap Y)$ but also on $\dim(X)$ and $\dim(Y)$, which are not constant in this case.

There is a vast amount of related work that we will not cover in this survey: Partial spreads have also been studied for combinatorial designs and in polar spaces; for the latter see, e.g., [5, 21]. In the special case $v = 2k$ spreads can be used to define translation planes and provide a rich source for constructing non-desarguesian projective planes [41, 42, 51]. Also motivated by this geometric point-of-view, partial $k$-spreads in $\mathbb{F}_q^{2k}$ of size close to the maximum size (given by Theorem 1) have been studied extensively. Most of this research has focused on partial spread replacements and complete partial spreads, while we consider only partial spreads of maximum cardinality and hence do not touch the case $v = 2k$ (except for Theorem 1). The classification of all (maximal) partial spreads up to isomorphism, see e.g. [55], is also not treated here. Further, there is a steady stream of literature that characterizes the existing types of vector space partitions in $\mathbb{F}_2^v$ for small dimensions $v$. Here, we touch only briefly on some results that are independent of the ambient space dimension $v$ and refer to [30] otherwise.

The remaining part of this chapter is structured as follows. In Sect. 2 we review some, mostly classical, bounds and constructions for partial spreads. After introducing the concept of $q^r$-divisible sets and codes in Sect. 3, we are able to obtain improved upper bounds for partial spreads in Theorems 9 and 10. Constructions for $q^r$-divisible sets are presented in Sect. 4, some non-existence results for $q^r$-divisible sets are presented in Sect. 5, and we close this survey with a collection of open research problems in Sect. 6.

## 2  Bounds and Constructions for Partial Spreads

Counting points in $\mathbb{F}_q^v$ and $\mathbb{F}_q^k$ gives the obvious upper bound $A_q(v, 2k; k) \leq \left[\begin{smallmatrix} v \\ 1 \end{smallmatrix}\right]_q / \left[\begin{smallmatrix} k \\ 1 \end{smallmatrix}\right]_q = (q^v - 1) / (q^k - 1)$ for the size of a partial $k$-spread in $\mathbb{F}_q^v$. Equality corresponds to the case of spreads, for which a handy existence criterion is known from the work of Segre in 1964.[9]

**Theorem 1** ([62, Sect. VI], [17, p. 29]) $\mathbb{F}_q^v$ contains a $k$-spread if and only if $k$ is a divisor of $v$.

Since $\frac{q^v - 1}{q^k - 1}$ is an integer if and only if $k$ divides $v$ (an elementary number theory exercise), only the constructive part needs to be shown. To this end we write $v = kt$ for a suitable integer $t$, take the ambient space $V$ as the restriction ("field reduction") of $(\mathbb{F}_{q^k})^t$ to $\mathbb{F}_q$, which clearly has dimension $v$, and define the $k$-spread $\mathscr{S}$ in $V/\mathbb{F}_q$ as the set of 1-subspaces of $V/\mathbb{F}_{q^k}$. That $\mathscr{S}$ is indeed a $k$-spread, is easily verified: Each member of $\mathscr{S}$ has dimension $k$ over $\mathbb{F}_q$; the members form a vector space partition of $V$ (this property does not depend on the particular field of scalars); and the size $\left[\begin{smallmatrix} t \\ 1 \end{smallmatrix}\right]_{q^k} = \frac{q^v - 1}{q^k - 1}$ of $\mathscr{S}$ is as required.[10]

---

[9]Segre in turn built to some extent on work of André, who had earlier considered the special case $v = 2k$ in his seminal paper on translation planes [1].

[10]Alternatively, the member of $\mathscr{S}$ containing a nonzero vector $\mathbf{x}$ is the $k$-subspace $\mathbb{F}_{q^k}\mathbf{x}$ of $V/\mathbb{F}_q$.

*Example 1* We consider the parameters $q = 3$, $v = 4$, and $k = 2$. Using canonical representatives in $\mathbb{F}_9 \simeq \mathbb{F}_3[x]/(x^2 + 1)$, the $\begin{bmatrix} 2 \\ 1 \end{bmatrix}_9 = 10$ points in $\mathbb{F}_9^2$ are generated by

$$\begin{pmatrix} 0 \\ 1 \end{pmatrix}, \begin{pmatrix} 1 \\ 0 \end{pmatrix}, \begin{pmatrix} 1 \\ 1 \end{pmatrix}, \begin{pmatrix} 1 \\ 2 \end{pmatrix}, \begin{pmatrix} 1 \\ x \end{pmatrix}, \begin{pmatrix} 1 \\ x+1 \end{pmatrix}, \begin{pmatrix} 1 \\ x+2 \end{pmatrix}, \begin{pmatrix} 1 \\ 2x \end{pmatrix}, \begin{pmatrix} 1 \\ 2x+1 \end{pmatrix}, \begin{pmatrix} 1 \\ 2x+2 \end{pmatrix}.$$

The particular point $P = \mathbb{F}_9\left(\begin{smallmatrix} 1 \\ x+1 \end{smallmatrix}\right) = \left\{\left(\begin{smallmatrix} 0 \\ 0 \end{smallmatrix}\right), \left(\begin{smallmatrix} 1 \\ x+1 \end{smallmatrix}\right), \left(\begin{smallmatrix} 2 \\ 2x+2 \end{smallmatrix}\right), \left(\begin{smallmatrix} x \\ x+2 \end{smallmatrix}\right), \left(\begin{smallmatrix} x+1 \\ 2x \end{smallmatrix}\right),\right.$ $\left(\begin{smallmatrix} x+2 \\ 1 \end{smallmatrix}\right), \left(\begin{smallmatrix} 2x \\ 2x+1 \end{smallmatrix}\right), \left(\begin{smallmatrix} 2x+1 \\ x \end{smallmatrix}\right), \left.\left(\begin{smallmatrix} 2x+2 \\ 2 \end{smallmatrix}\right)\right\}$ on the projective line $\mathrm{PG}(1, \mathbb{F}_9)$ defines a 2-subspace of $\mathbb{F}_9^2/\mathbb{F}_3 \cong \mathbb{F}_3^4$, whose 4 associated points are $\mathbb{F}_3\mathbf{x}$, $\mathbf{x} \in P$, $\mathbf{x} \neq \left(\begin{smallmatrix} 0 \\ 0 \end{smallmatrix}\right)$; and similarly for the other points of $\mathrm{PG}(1, \mathbb{F}_9)$. These ten 2-subspaces form the 2-spread $\mathscr{S}$.

Using any $\mathbb{F}_3$-isomorphism $\mathbb{F}_9^2/\mathbb{F}_3 \cong \mathbb{F}_3^4$, we can translate $\mathscr{S}$ into a 2-spread $\mathscr{S}'$ of the standard vector space $\mathbb{F}_3^4$. Taking, for example, coordinates with respect to the basis $(1, x)$ of $\mathbb{F}_9/\mathbb{F}_3$ and extending to $\mathbb{F}_9^2$ in the obvious way translates $P$ into the 2-subspace of $\mathbb{F}_3^4$ with vectors

$$\begin{pmatrix} 0 \\ 0 \\ 0 \\ 0 \end{pmatrix}, \begin{pmatrix} 1 \\ 0 \\ 1 \\ 1 \end{pmatrix}, \begin{pmatrix} 2 \\ 0 \\ 2 \\ 2 \end{pmatrix}, \begin{pmatrix} 0 \\ 1 \\ 2 \\ 1 \end{pmatrix}, \begin{pmatrix} 1 \\ 1 \\ 0 \\ 2 \end{pmatrix}, \begin{pmatrix} 2 \\ 1 \\ 1 \\ 0 \end{pmatrix}, \begin{pmatrix} 0 \\ 2 \\ 1 \\ 2 \end{pmatrix}, \begin{pmatrix} 1 \\ 2 \\ 0 \\ 1 \end{pmatrix}, \begin{pmatrix} 2 \\ 2 \\ 2 \\ 0 \end{pmatrix}.$$

The other members of $\mathscr{S}'$ are obtained in the same way.

We remark that $k$-spreads are a special case of subspace designs, see chapters "$q$-Analogs of Designs: Subspace Designs" and "Computational Methods in Subspace Designs".

From now on we assume that $k$ does not divide $v$ and write $v = tk + r$ with $1 \le r \le k - 1$. Since the cases $t \in \{0, 1\}$ are trivial ($\mathrm{A}_q(r, 2k; k) = 0$ and $\mathrm{A}_q(k + r, 2k; k) = 1$), we also assume $t \ge 2$. The stated upper bound then takes the form

$$\mathrm{A}_q(v, 2k; k) \le \left\lfloor \frac{q^v - 1}{q^k - 1} \right\rfloor = \frac{q^{tk+r} - q^r}{q^k - 1} + \left\lfloor \frac{q^r - 1}{q^k - 1} \right\rfloor = \sum_{s=0}^{t-1} q^{sk+r} = q^r \begin{bmatrix} t \\ 1 \end{bmatrix}_{q^k}. \tag{2.1}$$

We also see from this computation that the number of holes of a partial $k$-spread is at least $\frac{q^v-1}{q-1} \bmod \frac{q^k-1}{q-1} = \frac{q^r-1}{q-1}$. However, as we will see later, this bound can be improved further.

In accordance with (2.1) we make the following definition (similar to that in [7]): The number $\sigma$ defined by $\mathrm{A}_q(v; 2k; k) = \sum_{s=0}^{t-1} q^{sk+r} - \sigma$ is called the *deficiency* of the maximal partial $k$-spreads in $\mathbb{F}_q^v$.[11] From (2.1) we have $\sigma \ge 0$. In terms of the deficiency, the minimum possible number of holes is $\sigma \cdot \frac{q^k-1}{q-1} + \frac{q^r-1}{q-1}$.

Our next goal is to derive a good lower bound for $\mathrm{A}_q(v, 2k; k)$ (equivalently, a lower bound for the corresponding deficiency) by constructing a large partial $k$-spread

---

[11]This makes sense also for $r = 0$: Spreads are assigned deficiency $\sigma = 0$.

in $\mathbb{F}_q^v$. For this we will employ a special case of the *echelon-Ferrers construction* for general subspace codes [24], which involves only standard maximum rank distance codes of full row rank. More details on the echelon-Ferrers construction can be found in Chapter "Constructions of Constant Dimension Codes". For maximum rank distance codes see Chapters "Codes Endowed with the Rank Metric" and "Constructions of Cyclic Subspace Codes and Maximum Rank Distance Codes".

To this end, recall that every $k$-subspace $X$ of $\mathbb{F}_q^v$ is the row space of a unique "generating" matrix $\mathbf{A} \in \mathbb{F}_q^{k \times v}$ in reduced row-echelon form, which can be obtained by applying the Gaussian elimination algorithm to an arbitrary generating matrix of $X$. This matrix $\mathbf{A}$ is called *canonical matrix* of $X$, and is uniquely specified by its $k$ pivot columns $1 \le j_1 < j_2 < \cdots < j_k \le v$ (forming a $k \times k$ identity submatrix of $\mathbf{A}$) and the complementary submatrix $\mathbf{B} \in \mathbb{F}_q^{k \times (v-k)}$, which has zero entries in positions $(i, j)$ with $j \le j_i - i$ but otherwise can be arbitrary. The $k$-set $\{j_1, \ldots, j_k\}$ will be named *pivot set* of $X$. The positions of the unrestricted entries in $\mathbf{B}$ form the Ferrers diagram of an integer partition, as shown in the following for the cases $v = 8, k = 3$, $(i_1, i_2, i_3) = (1, 2, 3), (2, 4, 7)$.

| matrix shape | Ferrers diagram | integer partition |
|---|---|---|
| $\begin{pmatrix} 1 & 0 & 0 & * & * & * & * & * \\ 0 & 1 & 0 & * & * & * & * & * \\ 0 & 0 & 1 & * & * & * & * & * \end{pmatrix}$ | | $15 = 5 + 5 + 5$ |
| $\begin{pmatrix} 0 & 1 & * & 0 & * & * & 0 & * \\ 0 & 0 & 0 & 1 & * & * & 0 & * \\ 0 & 0 & 0 & 0 & 0 & 0 & 1 & * \end{pmatrix}$ | | $8 = 4 + 3 + 1$ |

$$(2.2)$$

The following lemma is a special case of [24, Lemma 2].

**Lemma 1** *If subspaces $X, Y$ of $\mathbb{F}_q^v$ have disjoint pivot sets, they are itself disjoint (i.e., $X \cap Y = \{\mathbf{0}\}$).*

*Proof* A nonzero vector in $X$ must have its pivot (first nonzero position) in the pivot set of $X$, and similarly for $Y$. The result follows. □

Now we focus on the special case in which the pivot set is $\{1, \ldots, k\}$, i.e., the canonical matrix has the "systematic" form $\mathbf{A} = (\mathbf{I}_k | \mathbf{B})$. For matrices $\mathbf{A}, \mathbf{B} \in \mathbb{F}_q^{k \times v}$ the *rank distance* is defined as $d_R(\mathbf{A}, \mathbf{B}) := \mathrm{rk}(\mathbf{A} - \mathbf{B})$. Codes based on the rank distance are discussed in detail in Chapter "Codes Endowed with the Rank Metric". The subspace distance of two $k$-subspaces with pivot set $\{1, \ldots, k\}$ can be computed from the rank distance of the corresponding canonical matrices:

**Lemma 2** ([64, Proposition 4]) *Let $X, X'$ be $k$-subspaces of $\mathbb{F}_q^v$ with canonical matrices $(\mathbf{I}_k | \mathbf{B})$ and $(\mathbf{I}_k | \mathbf{B}')$, respectively. Then $d_S(X, X') = 2 \cdot d_R(\mathbf{B}, \mathbf{B}')$.* [12]

---

[12]More generally, this formula holds if $X$ and $X'$ have the same pivot set and $\mathbf{B}, \mathbf{B}' \in \mathbb{F}_q^{k \times (v-k)}$ denote the corresponding complementary submatrices in their canonical matrices; see e.g. [63, Corollary 3].

*Proof* The matrix $\begin{pmatrix} \mathbf{I}_k & \mathbf{B} \\ \mathbf{I}_k & \mathbf{B}' \end{pmatrix}$ generates $X + X'$ and reduces via Gaussian elimina-

tion to $\begin{pmatrix} \mathbf{I}_k & \mathbf{B} \\ \mathbf{0} & \mathbf{B}' - \mathbf{B} \end{pmatrix}$. Hence $\dim(X + X') = k + \mathrm{rk}(\mathbf{B}' - \mathbf{B}) = k + d_R(\mathbf{B}, \mathbf{B}')$ and $d_S(X, X') = 2\dim(X + X') - 2k = 2\,d_R(\mathbf{B}, \mathbf{B}')$. $\qquad\square$

The so-called *lifting construction* [64, Sect. IV.A] associates with a matrix code $\mathscr{B} \subseteq \mathbb{F}_q^{k \times (v-k)}$ the constant-dimension code $\mathscr{C}$ in $\mathbb{F}_q^v$ whose codewords are the $k$-spaces generated by $(\mathbf{I}_k | \mathbf{B})$, $\mathbf{B} \in \mathscr{B}$. By Lemma 2, the code $\mathscr{C}$ is isometric to $\mathscr{B}$ with scale factor 2. In particular, $\mathscr{C}$ is a partial $k$-spread if and only if $\mathscr{B}$ has minimum rank distance $d_R(\mathscr{B}) = k$.

**Lemma 3** *There exists a partial $k$-spread $\mathscr{S}$ of size $q^{v-k}$ in $\mathbb{F}_q^v$ whose codewords cover precisely the points outside the $(v - k)$-subspace $S = \{\mathbf{x} \in \mathbb{F}_q^v; x_1 = x_2 = \cdots = x_k = 0\}$.*

*Proof* Write $n = v - k$ and consider a matrix representation $M : \mathbb{F}_{q^n} \to \mathbb{F}_q^{n \times n}$, obtained by expressing the multiplication maps $\mu_\alpha : \mathbb{F}_{q^n} \to \mathbb{F}_{q^n}, x \mapsto \alpha x$ (which are linear over $\mathbb{F}_q$) in terms of a fixed basis of $\mathbb{F}_{q^n}/\mathbb{F}_q$. Then $M(\alpha + \beta) = M(\alpha) + M(\beta)$, $M(\alpha\beta) = M(\alpha)M(\beta)$, $M(1) = \mathbf{I}_n$, and hence all matrices in $M(\mathbb{F}_{q^n})$ are invertible and have mutual rank distance $n$.[13]

Now let $\mathscr{B} \subseteq \mathbb{F}_q^{k \times n}$ be the matrix code obtained from $M(\mathbb{F}_{q^n})$ by deleting the last $n - k$ rows, say, of every matrix. Then $\#\mathscr{B} = q^n$ and $d_R(\mathscr{B}) = k$. Hence by applying the lifting construction to $\mathscr{B}$ we obtain a partial $k$-spread $\mathscr{S}$ in $\mathbb{F}_q^v$ of size $q^n = q^{v-k}$ (Lemma 2).

The codewords in $\mathscr{S}$ cover only points outside $S$ (compare the proof of Lemma 1). It remains to show that every such point is covered. This can be done by a counting argument or in the following more direct fashion: Let $\mathbf{a} \in \mathbb{F}_q^k \setminus \{\mathbf{0}\}, \mathbf{b} \in \mathbb{F}_q^{v-k}$ be arbitrary vectors and consider the equation $\mathbf{aX} = \mathbf{b}$ for $\mathbf{X} \in \mathscr{B}$. Since $\mathrm{rk}(\mathbf{X} - \mathbf{X}') = k$ for $\mathbf{X} \neq \mathbf{X}'$, the $q^{v-k}$ elements $\mathbf{aX}$, $\mathbf{X} \in \mathscr{B}$, are distinct and hence account for all elements in $\mathbb{F}_q^{v-k}$. Thus the equation has a solution $\mathbf{B} \in \mathscr{B}$, and the point $P = \mathbb{F}_q(\mathbf{a}|\mathbf{b}) = \mathbb{F}_q\mathbf{a}(\mathbf{I}_k|\mathbf{B})$ is covered by the codeword in $\mathscr{S}$ with canonical matrix $(\mathbf{I}_k|\mathbf{B})$. $\qquad\square$

Now we are ready for the promised construction of large partial $k$-spreads.

**Theorem 2** ([7]) *Let $v, k$ be positive integers satisfying $v = tk + r, t \geq 2$ and $1 \leq r \leq k - 1$. There exists a partial $k$-spread $\mathscr{S}$ in $\mathbb{F}_q^v$ of size*

$$\#\mathscr{S} = 1 + \sum_{s=1}^{t-1} q^{v-sk} = 1 + \sum_{s=1}^{t-1} q^{sk+r},$$

*and hence we have* $A_q(v, 2k; k) \geq 1 + \sum_{s=1}^{t-1} q^{sk+r}.$

---

[13] In ring-theoretic terms, the matrices in $M(\mathbb{F}_{q^n})$ form a maximal subfield of the ring of $n \times n$ matrices over $\mathbb{F}_q$.

The corresponding bound for the deficiency is $\sigma \le q^r - 1$. It depends only on $k$ and the residue $r = v \bmod k$.

It had been conjectured in [21, Sect. 2.2] that $\sigma = q^r - 1$ in general, but this conjecture was later disproved in [22] by exhibiting a maximal partial plane spread of size 34 in $\mathbb{F}_2^8$, which has deficiency $2^2 - 2$.

*Proof* The proof is by induction on $t$, using Lemma 2 and applying the inductive hypothesis to $S \cong \mathbb{F}_q^{v-k}$. The case $v = k + r$, in which $A_q(v, 2k; k) = 1$, serves as the anchor of the induction. ☐

The partial spread $\mathscr{S}$ exhibited in the proof of Theorem 2 consists of $t - 1$ "layers" $\mathscr{S}_1, \ldots, \mathscr{S}_{t-1}$ of decreasing sizes $\#\mathscr{S}_s = q^{v-sk}$, whose codewords are obtained from matrix representations of $\mathbb{F}_{q^{v-sk}}$ and have their pivots in positions $(s - 1)k + 1$, $(s - 1)k + 2$, $\ldots$, $sk$ (hence vanish on the first $(s - 1)k$ coordinates). The union $\bigcup_{s=1}^{t-1} \mathscr{S}_s$ leaves exactly the points of a $(k + r)$-subspace $S$ of $\mathbb{F}_q^v$ (the span of the last $k + r$ standard unit vectors) uncovered. Finally, one further $k$-subspace $S_0$ of $S$ is selected to form $\mathscr{S} = \bigcup_{s=1}^{t-1} \mathscr{S}_s \cup \{S_0\}$.[14]

*Example 2* We consider the particular case $q = 2$, $v = 5$, $k = 3$, in which $\#\mathscr{S} = 2^3 + 1 = 9$. In this case there is only one layer $\mathscr{S}_1$, which can be obtained from a matrix representation of $\mathbb{F}_8$ as follows: Representing $\mathbb{F}_8$ as $\mathbb{F}_2(\alpha)$ with $\alpha^3 + \alpha + 1 = 0$, we first express the powers $\alpha^j$, $0 \le j \le 6$ in terms of the basis $1, \alpha, \alpha^2$ of $\mathbb{F}_8/\mathbb{F}_2$, as in the following matrix:

$$
\mathbf{M} = \begin{array}{c|ccccccc|cc}
 & \alpha^0 & \alpha^1 & \alpha^2 & \alpha^3 & \alpha^4 & \alpha^5 & \alpha^6 & \alpha^0 & \alpha^1 \\
\hline
\alpha^0 & 1 & 0 & 0 & 1 & 0 & 1 & 1 & 1 & 0 \\
\alpha^1 & 0 & 1 & 0 & 1 & 1 & 1 & 0 & 0 & 1 \\
\alpha^2 & 0 & 0 & 1 & 0 & 1 & 1 & 1 & 0 & 0
\end{array}
\tag{2.3}
$$

The seven consecutive $3 \times 3$ submatrices of this matrix, which has been extended to the right in order to mimic the cyclic wrap-around, form a matrix field isomorphic to $\mathbb{F}_8$ (with $\mathbf{0} \in \mathbb{F}_2^{3 \times 3}$ added). Similarly, the code $\mathscr{B} \subset \mathbb{F}_2^{2 \times 3}$ is obtained by extracting the first seven consecutive $3 \times 2$ submatrices, adding $\mathbf{0} \in \mathbb{F}_2^{3 \times 2}$, and transposing; cf. the proof of Lemma 3. Prepending the $2 \times 2$ identity matrix then gives the canonical matrices of the 8 codewords of $\mathscr{S}_1$:

$$
\begin{pmatrix} 1 & 0 & 0 & 0 & 0 \\ 0 & 1 & 0 & 0 & 0 \end{pmatrix}, \begin{pmatrix} 1 & 0 & 1 & 0 & 0 \\ 0 & 1 & 0 & 1 & 0 \end{pmatrix}, \begin{pmatrix} 1 & 0 & 0 & 1 & 0 \\ 0 & 1 & 0 & 0 & 1 \end{pmatrix}, \begin{pmatrix} 1 & 0 & 0 & 0 & 1 \\ 0 & 1 & 1 & 1 & 0 \end{pmatrix},
$$

$$
\begin{pmatrix} 1 & 0 & 1 & 1 & 0 \\ 0 & 1 & 0 & 1 & 1 \end{pmatrix}, \begin{pmatrix} 1 & 0 & 0 & 1 & 1 \\ 0 & 1 & 1 & 1 & 1 \end{pmatrix}, \begin{pmatrix} 1 & 0 & 1 & 1 & 1 \\ 0 & 1 & 1 & 0 & 1 \end{pmatrix}, \begin{pmatrix} 1 & 0 & 1 & 0 & 1 \\ 0 & 1 & 1 & 0 & 0 \end{pmatrix}.
$$

Finally, from the seven lines in the plane $S = \{\mathbf{x} \in \mathbb{F}_2^5; x_1 = x_2 = 0\}$ a 9th codeword $L_0$ (moving line) is selected to form $\mathscr{S} = \mathscr{S}_1 \cup \{L_0\}$.

---

[14] The space $S_0$ has been named *moving subspace* of $\mathscr{S}$, since it can be freely "moved" within $S$ without affecting the partial spread property of $\mathscr{S}$.

The partial line spread $\mathscr{S}$ is in fact maximal, as we will see in a moment, and represents one of the 4 isomorphism types of maximal partial line spreads in $PG(\mathbb{F}_2^5) = PG(4, \mathbb{F}_2)$.[15]

Now we will reduce the upper bound (2.1) by a summand of $q - 1$, which is sufficient to settle the case $r = 1$ and hence determine the numbers $A_q(tk + 1, 2k; k)$. The key ingredient will be the observation that a partial $k$-spread induces in every hyperplane a vector space partition, whose members have dimension $k$, $k - 1$, or 1. Before turning to the general case, which is a little technical, we continue the preceding example and illustrate the method for partial line spreads in $\mathbb{F}_2^5$.

The geometry $PG(4, \mathbb{F}_2)$ has 31 points, each line containing 3 points, and thus it is conceivable that a partial line spread $\mathscr{S}$ of size 10 exists in $PG(4, \mathbb{F}_2)$. But in fact it does not. To prove this, consider a hyperplane (solid) $H$ in $PG(4, \mathbb{F}_2)$. If $H$ contains $\alpha$ lines of $\mathscr{S}$, it meets the remaining $\#\mathscr{S} - \alpha$ lines in a point, giving the constraint $\alpha \cdot 3 + (\#\mathscr{S} - \alpha) \cdot 1 \leq 15$, the total number of points in $H$. This is equivalent to $\#\mathscr{S} \leq 15 - 2\alpha$. In order to complete the proof, we need to show that there exists a hyperplane containing at least 3 lines of $\mathscr{S}$. This can be done by an averaging argument. On average, a hyperplane contains

$$\sum_H \frac{\#\{L \in \mathscr{S}; L \subset H\}}{2^5 - 1} = \sum_{L \in \mathscr{S}} \frac{\#\{H; H \supset L\}}{2^5 - 1} = \frac{\#\mathscr{S}(2^3 - 1)}{2^5 - 1} = \frac{7}{31} \cdot \#\mathscr{S}$$

(2.4)

lines of $\mathscr{S}$. If $\#\mathscr{S} \geq 9$, this number is $> 2$, implying the desired conclusion $\sigma \geq 1$.[16]

The general case is the subject of the following

**Theorem 3** ([21, Theorem 2.7(a)]) *The deficiency of a maximal $k$-spread in $\mathbb{F}_q^v$, where $k$ does not divide $v$, is at least $q - 1$.*

*Proof* Reasoning as in the preceding example gives $\alpha \cdot \frac{q^k - 1}{q - 1} + (\#\mathscr{S} - \alpha)\frac{q^{k-1} - 1}{q - 1} \leq \frac{q^{tk+r-1} - 1}{q - 1}$ and hence the bound

$$\#\mathscr{S} \leq \frac{q^{tk+r-1} - 1 - \alpha(q^k - q^{k-1})}{q^{k-1} - 1}$$

(2.5)

for any partial $k$-spread $\mathscr{S}$ having a hyperplane incident with $\alpha$ members of $\mathscr{S}$.

Now suppose $\#\mathscr{S} = 1 + \sum_{s=1}^{t-1} q^{sk+r}$, the same size as the partial $k$-spread in Theorem 2. In this case the average number of codewords contained in a hyperplane is

---

[15]The full classification, including also partial line spreads of smaller size, can be found in [26]. To our best knowledge there is only one further nontrivial parameter case, where a classification of maximal (proper) partial spreads is known, viz. the case of plane spreads in $PG(6, \mathbb{F}_2)$, settled in [38].

[16]In this particular case one may also argue as follows: If $\#\mathscr{S} = 10$ then there is only one hole and the hyperplane constraint becomes $3\alpha + (10 - \alpha) + h = 15$, where $h \in \{0, 1\}$. This forces $\alpha = 2$ and $h = 1$, i.e., every hyperplane should contain the hole. This is absurd, of course.

$$\frac{q^{(t-1)k+r} - 1}{q^{tk+r} - 1}\left(1 + \sum_{s=1}^{t-1} q^{sk+r}\right) = \frac{1}{q^{tk+r} - 1}\left(\sum_{s=t}^{2t-2} q^{sk+2r} - \sum_{s=1}^{t-2} q^{sk+r} - 1\right)$$

$$= \frac{1}{q^{tk+r} - 1}\left(\sum_{s=t}^{2t-2} q^{sk+2r} - \sum_{s=0}^{t-2} q^{sk+r} + q^r - 1\right)$$

$$= \sum_{s=0}^{t-2} q^{sk+r} + \frac{q^r - 1}{q^{tk+r} - 1}.$$

It follows that $\mathscr{S}$, and likewise all partial $k$-spreads of deficiency $\leq q^r - 1$, have a hyperplane containing at least $1 + \sum_{s=0}^{t-2} q^{sk+r}$ codewords. Substituting this number into (2.5) gives

$$\#\mathscr{S} \leq \frac{1}{q^{k-1} - 1}\left(q^{tk+r-1} - 1 - \left(1 + \sum_{s=0}^{t-2} q^{sk+r}\right)(q^k - q^{k-1})\right)$$

$$= \frac{1}{q^{k-1} - 1}\left(q^{tk+r-1} - 1 - q^k + q^{k-1} - \sum_{s=1}^{t-1} q^{sk+r} + \sum_{s=0}^{t-2} q^{sk+r+k-1}\right)$$

$$= 1 + \sum_{s=1}^{t-2} q^{sk+r} + \frac{q^{tk+r-1} - q^k - q^{(t-1)k+r} + q^{r+k-1}}{q^{k-1} - 1}$$

$$= 1 + \sum_{s=1}^{t-1} q^{sk+r} + \frac{q^{r+k-1} - q^k}{q^{k-1} - 1}$$

$$= 1 + \sum_{s=1}^{t-1} q^{sk+r} + q^r - q + \frac{q^r - q}{q^{k-1} - 1}$$

$$= \sum_{s=0}^{t-1} q^{sk+r} - (q - 1) + \frac{q^r - q}{q^{k-1} - 1},$$

valid now for any partial $k$-spread $\mathscr{S}$ in $\mathbb{F}_q^v$. Since the last summand is $< 1$, we obtain the desired conclusion $\sigma \geq q - 1$.                                                               $\square$

Theorem 3 has the following immediate corollary, established by Beutelspacher in 1975, which settles the case $r = 1$ completely.

**Corollary 1** ([Theorem 4.1]; **see also** [7, 36] **for the special case** $q = 2$)  *For integers $k \geq 2$ and $v = tk + 1$ with $t \geq 1$ we have $A_q(v, 2k; k) = 1 + \sum_{s=1}^{t-1} q^{sk+1}$,[17] with corresponding deficiency $\sigma = q - 1$.[18]*

In particular, maximal partial line spreads in $\mathbb{F}_q^v$, $v$ odd (the case where no line spreads exist), have size $q^{v-2} + q^{v-4} + \cdots + q^3 + 1$, deficiency $q - 1$, and $q^2$ holes.

---

[17]This can also written as $A_q(v, 2k; k) = q^1 \cdot \frac{q^{v-1}-1}{q^k-1} - q + 1 = \frac{q^v - q^{k+1} + q^k - 1}{q^k - 1}$.

[18]The corresponding number of holes is $q^k$.

In his original proof of the corollary Beutelspacher considered the set of holes $N$ and the average number of holes per hyperplane, which is less than the total number of holes divided by $q$. An important insight was the relation $\#N \equiv \#(H \cap N)$ (mod $q^{k-1}$) for each hyperplane $H$, i.e., the number of holes per hyperplane satisfies a certain modulo constraint. We will see this concept in full generality in Sect. 3. In terms of integer linear programming, the upper bound is obtained by an integer rounding cut. The construction in [7, Theorem 4.2] recursively uses arbitrary $k'$-spreads, so that it is more general than the one of Theorem 2.

For a long time the best known upper bound on $A_q(v, 2k; k)$, i.e., the best known lower bound on $\sigma$, was the one obtained by Drake and Freeman in 1979:

**Theorem 4** (Corollary 8 in [20]) *The deficiency of a maximal partial $k$-spread in $\mathbb{F}_q^v$ is at least* $\lfloor \theta \rfloor + 1 = \lceil \theta \rceil$,[19] *where* $2\theta = \sqrt{1 + 4q^k(q^k - q^r)} - (2q^k - 2q^r + 1)$.

The authors concluded from the existence of a partial spread the existence of a (group-constructible) $(s, r, \mu)$-net and applied [11, Theorem 1B]—a necessary existence criterion formulated for orthogonal arrays of strength 2 by Bose and Bush in 1952. The underlying proof technique can be further traced back to [60] and is strongly related to the classical second-order Bonferroni Inequality [10, 25]; see also [39, Sect. 2.5] for an application to bounds for subspace codes.

Given Theorem 1 and Corollary 1, the first open binary case is $A_2(8, 6; 3)$. The construction from Theorem 2 gives a partial spread of cardinality 33, while Theorem 4 implies an upper bound of 34. As already mentioned, in 2010 El-Zanati et al. [22] found a sporadic partial plane spread in $\mathbb{F}_2^8$ of cardinality 34 by a computer search. Together with the following easy lemma, this completely answers the situation for partial plane spreads in $\mathbb{F}_2^v$; see Corollary 2 below.

**Lemma 4** *For fixed $q$, $k$ and $r$ the deficiency $\sigma$ is a non-increasing function of $v = kt + r$.*

*Proof* Let $\mathscr{S}$ be a maximal partial $k$-spread in $\mathbb{F}_q^{tk+r}$ and $\sigma$ its deficiency, so that $A_q(tk + r, 2k; k) = \sum_{s=0}^{t-1} q^{sk+r} - \sigma$. We can embed $\mathscr{S}$ into $\mathbb{F}_q^{(t+1)k+r}$ by prepending $k$ zeros to each codeword. Then Lemma 3 can be applied and yields a partial $k$-spread $\mathscr{S}'$ in $\mathbb{F}_q^{(t+1)k+r}$ of size $q^{tk+r}$, whose codewords are disjoint from those in $\mathscr{S}$. This implies $A_q((t + 1)k + r, 2k; k) \geq \#\mathscr{S} \cup \mathscr{S}' = \sum_{s=0}^{t} q^{sk+r} - \sigma$, and hence the deficiency $\sigma'$ of a maximal partial $k$-spread in $\mathbb{F}_q^{(t+1)k+r}$ satisfies $\sigma' \leq \sigma$.     □

So, any improvement of the best known lower bound for a single parameter case gives rise to an infinite series of improved lower bounds. Unfortunately, so far, the sporadic construction in [22] is the only known example being strictly superior to the general construction of Theorem 2.

---

[19]Assuming $1 + 4q^k(q^k - q^r) = 1 + 4q^{k+r}(q^{k-r} - 1) = (2z - 1)^2 = 1 + 4z(z - 1)$ for some integer $z > 1$ implies $q^{k+r} \mid z$ or $q^{k+r} \mid z - 1$, so that $z \geq q^{k+r}$, which is impossible for $(k, r) \neq (1, 0)$. Thus, $2\theta \notin \mathbb{Z}$, so that $\theta \notin \mathbb{Z}$ and $\lfloor \theta \rfloor + 1 = \lceil \theta \rceil$.

**Corollary 2** *For each integer* $m \geq 2$ *we have* $A_2(3m, 6; 3) = \frac{2^{3m}-1}{7}$, $A_2(3m + 1, 6; 3) = \frac{2^{3m+1}-9}{7}$, *and* $A_2(3m + 2, 6; 3) = \frac{2^{3m+2}-18}{7}$. *The corresponding deficiencies are* 0, 1 *and* 2, *respectively.*

Very recently, the case $q = r = 2$ was completely settled. For $k = 3$ the answer is given in the preceding corollary, and for $k \geq 4$ by the following

**Theorem 5** ([46, Theorem 5]) *For integers* $k \geq 4$ *and* $v = tk + 2$ *with* $t \geq 1$ *we have* $\sigma = 3$ *and* $A_2(kt + 2, 2k; k) = 1 + \sum_{s=1}^{t-1} 2^{tk+2} = \frac{2^{kt+2}-3\cdot 2^k-1}{2^k-1}$. [20]

The technique used to prove this theorem is very similar to the one presented in the proof of Theorem 3.

**Corollary 3** *We have* $A_2(4m, 8; 4) = \frac{2^{4m}-1}{15}$. $A_2(4m + 1, 8; 4) = \frac{2^{4m+1}-17}{15}$, $A_2(4m + 2, 8; 4) = \frac{2^{4m+2}-49}{15}$, *and* $\frac{2^{4m+3}-113}{15} \leq A_2(4m + 3, 8; 4) \leq \frac{2^{4m+3}-53}{15}$ *for all* $m \geq 2$. *The corresponding deficiencies are* 0, 1, 3 *and* $3 \leq \sigma \leq 7$, *respectively*

As a consequence, the first unknown binary case is now $129 \leq A_2(11, 8; 4) \leq 133$.[21] For $r = 2$ and $q = 3$ the upper bound of Theorem 4 has been decreased by 1:

**Lemma 5** (cf.[46, Lemma 4]) *For integers* $t \geq 2$ *and* $k \geq 4$, *we have* $\sigma \geq 5$ *and* $A_3(kt + 2, 2k; k) \leq \frac{3^{kt+2}-3^2}{3^k-1} - 5$.

Again, the proof technique is very similar to that used in the proof of Theorem 3.

Theorem 2 is asymptotically optimal for $k \gg r = v \bmod k$, as recently shown by Năstase and Sissokho:

**Theorem 6** ([59, Theorem 5]) *If* $k > \begin{bmatrix} r \\ 1 \end{bmatrix}_q$ *then* $\sigma = q^r - 1$ *and* $A_q(v, 2k; k) = 1 + \sum_{s=1}^{t-1} q^{sk+r}$.[22]

Choosing $q = r = 2$, this result covers Theorem 5. The same authors have refined their analysis, additionally using Theorem 14 from the theory of vector space partitions, to obtain improved upper bounds for some of the cases $k \leq \begin{bmatrix} r \\ 1 \end{bmatrix}_q$, see [58, Theorem 6 and 7]. Using the theory of $q^r$-divisible codes, presented in the next section, we extend their results further in Corollary 7 and Theorem 10.

## 3 $q^r$-divisible Sets and Codes

The currently most effective approach to good upper bounds for partial spreads follows the original idea of Beutelspacher and considers the set of holes as a stand-alone object. As it appears in the proof of Beutelspacher, the number of holes in a

---

[20]Thus in all these cases $\sigma = 2^2 - 1$ and the partial spreads of Theorem 2 are maximal. This notably differs from the case $k = 3$.

[21]The upper bound can be sharpened to 132, as we will see later.

[22]This corresponds again to the upper bound $\sigma = q^r - 1$.

hyperplane satisfies a certain modulo constraint. In this section we consider sets of points in $PG(v - 1, \mathbb{F}_q)$ having the property that modulo some integer $\Delta > 1$ the number of points in each hyperplane is the same. Such point sets are equivalent to $\Delta$-divisible codes [68, 69] with projectively distinct coordinate functionals (so-called *projective* codes), and this additional restriction forces $\Delta$ to be a power of the same prime as $q$. Writing $q = p^e$, $p$ prime, and $\Delta = p^f$, we have $\Delta = q^r$ with $r = f/e \in \frac{1}{e}\mathbb{Z}$.

We will derive several important properties of these $q^r$-*divisible sets* and *codes* and in particular observe that the set of holes of a partial spread is exactly of this type. Without the notion of $q^r$-divisible sets and the reference to the linear programming method, almost all results of this section are contained in [47, 48]. A more extensive introduction to the topic, including constructions and relations to other combinatorial objects, is currently in preparation [31].

In what follows, we denote the point set of $PG(v - 1, \mathbb{F}_q)$ by $\mathscr{P}$ and call for subsets $\mathscr{C} \subseteq \mathscr{P}$ and subspaces $X$ of $\mathbb{F}_q^v$ the integer $\#(\mathscr{C} \cap X) = \#\{P \in \mathscr{C}; P \subseteq X\}$ the *multiplicity* of $X$ with respect to $\mathscr{C}$.

**Definition 1** Let $\Delta > 1$ be an integer. A set $\mathscr{C}$ of points in $PG(v - 1, \mathbb{F}_q)$ is called *weakly $\Delta$-divisible* if there exists $u \in \mathbb{Z}$ with $\#(\mathscr{C} \cap H) \equiv u \pmod{\Delta}$ for each hyperplane $H$ of $PG(v - 1, \mathbb{F}_q)$. If $u \equiv \#\mathscr{C} \pmod{\Delta}$, we call $\mathscr{C}$ *(strongly) $\Delta$-divisible*.

Trivial cases are $\mathscr{C} = \emptyset$ (strongly $\Delta$-divisible for any $\Delta$) and $\mathscr{C} = \mathscr{P}$ (weakly $\Delta$-divisible for any $\Delta$, with largest strong divisor $\Delta = q^{v-1}$).[23]

It is well-known (see, e.g., [18, 65, Proposition 1]) that the relation $C \to \mathscr{C}$, associating with a full-length linear $[n, v]$ code $C$ over $\mathbb{F}_q$ the $n$-multiset $\mathscr{C}$ of points in $PG(v - 1, \mathbb{F}_q)$ defined by the columns of any generator matrix, induces a one-to-one correspondence between classes of (semi-)linearly equivalent spanning multisets and classes of (semi-)monomially equivalent full-length linear codes. Point sets correspond in this way to projective linear codes, which are also characterized by the condition $d(C^\perp) \geq 3$. The importance of the correspondence lies in the fact that it relates coding-theoretic properties of $C$ to geometric or combinatorial properties of $\mathscr{C}$. An example is the formula

$$w(\mathbf{a}G) = n - \#\{1 \leq j \leq n; \mathbf{a} \cdot \mathbf{g}_j = 0\} = n - \#(\mathscr{C} \cap \mathbf{a}^\perp), \tag{3.1}$$

where w denotes the Hamming weight, $\mathbf{G} = (\mathbf{g}_1 | \ldots | \mathbf{g}_n) \in \mathbb{F}_q^{v \times n}$ a generating matrix of $C$, $\mathbf{a} \cdot \mathbf{b} = a_1 b_1 + \cdots + a_v b_v$, and $\mathbf{a}^\perp$ is the hyperplane in $PG(v - 1, \mathbb{F}_q)$ with equation $a_1 x_1 + \cdots + a_v x_v = 0$.

A linear code $C$ is said to be $\Delta$-*divisible* ($\Delta \in \mathbb{Z}_{>1}$) if all nonzero codeword weights are multiples of $\Delta$. A lower bound on the cardinality of $\Delta$-divisible point sets in $PG(v - 1, \mathbb{F}_q)$ has been given in [3]. Following the Gleason-Pierce-Ward Theorem on the divisibility of self-dual codes (see, e.g., [40, Chap. 9.1]), a rich theory of divisible codes has been developed over time, mostly by H. N. Ward; cf.

---

[23]This is also true for $v = 1$, where $\mathscr{C} = \emptyset$, $\mathscr{P}$ exhausts all possibilities.

his survey [69]. One of Ward's results implies that nontrivial weakly $\Delta$-divisible point sets in $PG(v - 1, \mathbb{F}_q)$ are strongly $\Delta$ divisible and exist only in the case $\Delta = p^f$. The proof uses the so-called *standard equations* for the hyperplane spectrum of $\mathscr{C}$, which we state in the following lemma. The standard equations are equivalent to the first three MacWilliams identities for the weight enumerators of $C$ and $C^\perp$ (stated as Equation (3.6) below), specialized to the case of projective linear codes. The geometric formulation, however, seems more in line with the rest of the paper.

**Lemma 6** *Let $\mathscr{C}$ be a set of points in $PG(v - 1, \mathbb{F}_q)$ with $\#\mathscr{C} = n$, and let $a_i$ hyperplanes of $PG(v - 1, \mathbb{F}_q)$ contain exactly $i$ points of $\mathscr{C}$ $(0 \leq i \leq n)$. Then we have*

$$\sum_{i=0}^{n} a_i = \begin{bmatrix} v \\ 1 \end{bmatrix}_q, \tag{3.2}$$

$$\sum_{i=1}^{n} i a_i = n \cdot \begin{bmatrix} v - 1 \\ 1 \end{bmatrix}_q, \tag{3.3}$$

$$\sum_{i=2}^{n} \binom{i}{2} a_i = \binom{n}{2} \cdot \begin{bmatrix} v - 2 \\ 1 \end{bmatrix}_q. \tag{3.4}$$

*Proof* Double-count[24] incidences of the tuples $(H)$, $(P_1, H)$, and $(\{P_1, P_2\}, H)$, where $H$ is a hyperplane and $P_1 \neq P_2$ are points contained in $H$.  $\square$

In the proof of Theorem 7 we will need that (3.2) and (3.3) remain true for any multiset $\mathscr{C}$ of points in $PG(v - 1, \mathbb{F}_q)$, provided points are counted with their multiplicities in $\mathscr{C}$ and the cardinality $\#\mathscr{C}$ is defined in the obvious way. We will also need the following concept of a *quotient multiset*. Let $\mathscr{C}$ be a set of points in $PG(v - 1, \mathbb{F}_q)$ and $X$ a subspace of $\mathbb{F}_q^v$. Define the multiset $\mathscr{C}/X$ of points in the quotient geometry $PG(\mathbb{F}_q^v/X)$ by assigning to a point $Y/X$ of $PG(\mathbb{F}_q^v/X)$ (i.e., $Y$ satisfies $\dim(Y/X) = 1$) the difference $\#(\mathscr{C} \cap Y) - \#(\mathscr{C} \cap X) = \#(\mathscr{C} \cap Y \setminus X)$ as multiplicity.[25] With this definition it is obvious that $\#(\mathscr{C}/X) = \#\mathscr{C} - \#(\mathscr{C} \cap X)$. In particular, if $\mathscr{C}$ is an $n$-set and $X = P$ is a point then $\#(\mathscr{C}/P) = n - 1$ or $n$, according to whether $P \in \mathscr{C}$ or $P \notin \mathscr{C}$, respectively.[26]

---

[24]The general (multiset) version of (3.4) has an additional summand of $q^{v-2} \cdot \sum_{P \in \mathscr{P}} \binom{\mathscr{C}(P)}{2}$ on the right-hand side, accounting for the fact that "pairs of equal points" are contained in $\begin{bmatrix} v-1 \\ 1 \end{bmatrix}_q$ hyperplanes.

[25]This definition can be extended to multisets $\mathscr{C}$ by defining the multiplicity of $Y/X$ in $\mathscr{C}/X$ as the sum of the multiplicities in $\mathscr{C}$ of all points in $Y \setminus X$.

[26]If $C \leftrightarrow \mathscr{C}$ then the multisets $\mathscr{C}/P$, $P \in \mathscr{P}$, are associated to the $(v - 1)$-dimensional subcodes $D \subset C$, and the $n$ points $P \in \mathscr{C}$ correspond to the $n$ subcodes $D$ of effective length $n - 1$ ("$D$ is $C$ shortened at $P$"). This correspondence between points and subcodes extends to a correlation between $PG(v - 1, \mathbb{F}_q)$ and $PG(C/\mathbb{F}_q)$, which includes the familiar correspondence between hyperplanes and codewords as a special case; see [18, 65] for details.

**Theorem 7** *Let $\mathscr{C} \neq \emptyset$, $\mathscr{P}$ be a weakly $\Delta$-divisible point set in $\mathrm{PG}(v-1, \mathbb{F}_q)$, $n =$ #$\mathscr{C}$, and C any linear $[n, v]$-code over $\mathbb{F}_q$ associated with $\mathscr{C}$ as described above.[27] Then*

*(i)  $\mathscr{C}$ is strongly $\Delta$-divisible;*
*(ii)  C is $\Delta$-divisible;*
*(iii)  $\Delta$ is a divisor of $q^{v-2}$.*

*Proof* (i) and (ii) are equivalent in view of (3.1).

First we prove (i). Let $u$ be as in Definition 1. Choose a point $P \notin \mathscr{C}$ and let $\mathscr{C}' = \mathscr{C}/P$. Then $\mathscr{C}'$ is an $n$-multiset of points in $\mathrm{PG}(\mathbb{F}_q^v/P) \cong \mathrm{PG}(v-2, \mathbb{F}_q)$ with #$(\mathscr{C}' \cap H') \equiv u \pmod{\Delta}$ for each hyperplane $H'$ of $\mathrm{PG}(\mathbb{F}_q^v/P)$. The hyperplane spectrum $(a_i')$ of $\mathscr{C}'$ satisfies (3.3) with $v$ replaced by $v-1$. Multiplying the identity for $\mathscr{C}'$ by $q$ and subtracting from it the identity for $\mathscr{C}$ gives

$$\sum_{i \geq 0}(u + i\Delta)(a_{u+i\Delta} - qa_{u+i\Delta}') = n\left(\frac{q^{v-1}-1}{q-1} - \frac{q^{v-1}-q}{q-1}\right) = n.$$

Reading this equation modulo $\Delta$ and using (3.2) further gives

$$n \equiv u\sum_{i \geq 0}(a_{u+i\Delta} - qa_{u+i\Delta}') = u\left(\frac{q^v-1}{q-1} - \frac{q^v-q}{q-1}\right) = u \pmod{\Delta},$$

as desired. Thus (i) and (ii) hold.

For the proof of (iii) we use a point $Q \in \mathscr{C}$ and its associated quotient multiset $\mathscr{C}'' = \mathscr{C}/Q$, which satisfies #$\mathscr{C}'' = n - 1$ and #$(\mathscr{C}'' \cap H'') \equiv u - 1 \pmod{\Delta}$ for each hyperplane $H''$ of $\mathrm{PG}(\mathbb{F}_q^v/Q)$. Subtracting (3.3) for $\mathscr{C}'$ and $\mathscr{C}''$ gives

$$\sum_{i \geq 0}(u + i\Delta)a_{u+i\Delta}' - \sum_{i \geq 0}(u - 1 + i\Delta)a_{u-1+i\Delta}'' = \frac{q^{v-2}-1}{q-1}.$$

Again reading the equation modulo $\Delta$ and using (3.2) gives

$$\frac{q^{v-2}-1}{q-1} \equiv u\sum_{i \geq 0}a_{u+i\Delta}' - (u-1)\sum_{i \geq 0}a_{u-1+i\Delta}'' = \frac{q^{v-1}-1}{q-1} \pmod{\Delta}$$

or $q^{v-2} \equiv 0 \pmod{\Delta}$, as asserted.  □

Let us remark that Part (ii) of Theorem 7 also follows from [68, Theorem 3], which asserts that a not necessarily projective code C satisfying the assumption of the theorem must be either $\Delta$-divisible or the juxtaposition of a $\Delta$-divisible code and a $v$-dimensional linear constant weight code. Since the latter is necessarily a repetition

---

[27]It is not required that $\mathscr{C}$ is spanning; if it is not then (iii) sharpens to "$\Delta$ is a divisor of $q^{\dim(\mathscr{C})-2}$".

of simplex codes, this case does not occur for projective codes. Our proofs of (i), (iii) use the very same ideas as in [68], translated into the geometric framework.[28]

Part (iii) of Theorem 7 says that exactly $\Delta$-divisible point sets in $\mathrm{PG}(v - 1, \mathbb{F}_q)$ exist only if $\Delta = q^r$ with $r \in \frac{1}{e}\mathbb{Z}$ and $r \leq v - 1$[29]; the whole point set $\mathcal{P}$ has $\Delta = q^{v-1}$, and $v - 2 < r < v - 1$ does not occur. Conversely, it is not difficult to see that every divisor $\Delta > 1$ of $q^{v-2}$ is the largest divisor of some point set in $\mathrm{PG}(v - 1, \mathbb{F}_q)$.[30]

In the proof of Theorem 7 we have used that the (weak) divisibility properties of $\mathcal{C}$ and its quotient multisets $\mathcal{C}/X$ are the same. Now we consider the restrictions $\mathcal{C} \cap X$, which correspond to residual codes of the associated code $C$.

**Lemma 7** *Suppose that $\mathcal{C}$ is a $q^r$-divisible set of points in $\mathrm{PG}(v - 1, \mathbb{F}_q)$ and $X$ a $(v - j)$-subspace of $\mathbb{F}_q^v$ with $1 \leq j < r$. Then the restriction $\mathcal{C} \cap X$ is $q^{r-j}$-divisible.*

*Proof* By induction, it suffices to consider the case $j = 1$, i.e., $X = H$ is a hyperplane in $\mathrm{PG}(v - 1, \mathbb{F}_q)$.

The hyperplanes of $\mathrm{PG}(H)$ are the $(v - 2)$-subspaces of $\mathbb{F}_q^v$ contained in $H$. Hence the assertion is equivalent to $\#(\mathcal{C} \cap U) \equiv \#\mathcal{C} = u \pmod{q^{r-1}}$ for every $(v - 2)$-subspace $U \subset \mathbb{F}_q^v$. By assumption we have $\#(\mathcal{C} \cap H_i) \equiv u \pmod{q^r}$ for the $q + 1$ hyperplanes $H_1, \ldots, H_{q+1}$ lying above $U$. This gives

$$(q + 1)u \equiv \sum_{i=1}^{q+1} \#(\mathcal{C} \cap H_i) = q \cdot \#(\mathcal{C} \cap U) + \#\mathcal{C} \equiv q \cdot \#(\mathcal{C} \cap U) + u \pmod{q^r}$$

and hence $u \equiv \#(\mathcal{C} \cap U) \pmod{q^{r-1}}$, as claimed. $\qquad\qquad\square$

As mentioned at the beginning of this section, the set of holes of a partial spread provides an example of a $q^r$-divisible set. The precise statement is given in the following theorem, which is formulated for general vector space partitions.

**Theorem 8** (i) *Let $\mathcal{C}$ be a vector space partition of $\mathbb{F}_q^v$ of type $t^{m_t} \cdots s^{m_s} 1^{m_1}$ with $m_s > 0$ (i.e., $\mathcal{C}$ has a member of dimension $> 1$ and $s$ chosen as the smallest such dimension). Then the points (i.e., 1-subspaces) in $\mathcal{C}$ form a $q^{s-1}$-divisible set.*

(ii) *The holes of a partial $k$-spread in $\mathbb{F}_q^v$ form a $q^{k-1}$-divisible set.*

*Proof* It is immediate that (i) implies (ii). For the proof of (i) let $H$ be a hyperplane of $\mathrm{PG}(v - 1, \mathbb{F}_q)$. The points in $\mathcal{P} \setminus H$ are partitioned into the affine subspaces $X \setminus H$ for those $X \in \mathcal{C}$ satisfying $X \nsubseteq H$. If such a $t$-subspace $X$ is not a point, we have $t \geq s$ and hence $\#(X \setminus H) = q^{t-1} \equiv 0 \pmod{q^{s-1}}$. Moreover, we also have $\#(\mathcal{P} \setminus H) = q^{v-1} \equiv 0 \pmod{q^{s-1}}$. It follows that the number of points in $\mathcal{C}$ that are not contained in $H$ is divisible by $q^{s-1}$ as well, completing the proof. $\qquad\square$

---

[28] Readers may have noticed that, curiously, the 3rd standard equation (which characterizes projective codes) was not used at all in the proof.

[29] By this we mean that $\Delta$ is the largest divisor in the sense of Definition 1 or Theorem 7.

[30] If $t = \lfloor r \rfloor$ and $r' \in \{0, 1, \ldots, e - 1\}$ is defined by $r = t + r'/e$, then the union of $p^{r'}$ parallel affine subspaces of dimension $t + 1$ has this property.

Theorem 8 explains our motivation for studying $q^r$-divisible points sets in $PG(v - 1, \mathbb{F}_q)$. Before delving deeper into this topic, we pause for a few example applications to partial spreads, which may help advertising our approach.

First we consider the problem of improving the upper bound (2.1) for the size of a partial $k$-spread in $PG(v - 1, \mathbb{F}_q)$. The bound is equivalent to $\#\mathscr{C} \geq \frac{q^r-1}{q-1}$ for the corresponding hole sets, which are $q^{k-1}$-divisible by Theorem 8(ii). But the smallest nontrivial $q^{k-1}$-divisible point sets in $PG(v - 1, \mathbb{F}_q)$ are the $k$-subspaces of $\mathbb{F}_q^v$, since these are associated to the constant-weight-$q^{k-1}$ simplex code.[31] Thus $\#\mathscr{C} \geq \frac{q^k-1}{q-1} > \frac{q^r-1}{q-1}$, and equality in (2.1) is not possible. Together with Theorem 2 this already gives the numbers $A_2(tk + 1, 2k; k)$.

The preceding argument gives $A_2(8, 6; 3) \leq 35$, and as a second application, we now exclude the existence of a partial plane spread of size 35 in $\mathbb{F}_2^8$. As already mentioned, this also follows from the Drake-Freeman bound (Theorem 4) and forms an important ingredient in the determination of the numbers $A_2(v, 6; 3)$. The hole set $\mathscr{C}$ of such a partial plane spread has size $2^8 - 1 - 35 \cdot 7 = 10$ and is 4-divisible, i.e., it meets every hyperplane in 2 or 6 points.

We claim that $\dim\langle\mathscr{C}\rangle = 4$. The inequality $\dim\langle\mathscr{C}\rangle \geq 4$ is immediate from $\#\mathscr{C} = 10$. The reverse inequality follows from the fact that the linear code $C$ associated with $\mathscr{C}$ is doubly-even, hence self-orthogonal, but cannot be self-dual.[32]

Given that $\dim\langle\mathscr{C}\rangle = 4$, the existence of $\mathscr{C}$ is readily excluded using the standard equations:

$$\begin{aligned}
a_2 + \quad a_6 &= 15, \\
2a_2 + \quad 6a_6 &= 10 \cdot 7, \\
\binom{2}{2}a_2 + \binom{6}{2}a_6 &= \binom{10}{2} \cdot 3.
\end{aligned} \qquad (3.5)$$

The unique solution of the first two equations is $a_2 = 5$, $a_6 = 10$ (corresponding to the 2-fold repetition of the $[5, 4, 2]$ even-weight code), but it does not satisfy the third equation (since this code is not projective).[33]

As already mentioned, the standard equations in Lemma 6 have a natural generalization in the language of linear codes. To this end let $\mathscr{C}$ be a point set of size $\#\mathscr{C} = n$ in $PG(k - 1, \mathbb{F}_q)$, which is spanning,[34] and $C$ the corresponding projective linear $[n, k]$ code over $\mathbb{F}_q$. The hyperplane spectrum $(a_i)_{0 \leq i \leq n}$ of $\mathscr{C}$ and the weight distribution $(A_i)_{0 \leq i \leq n}$ of $C$ are related by $A_i = (q - 1)a_{n-i}$ for $1 \leq i \leq n$ (supplemented by $A_0 = 1$, $a_n = 0$) and hence provide the same information about $\mathscr{C}$. The famous *MacWilliams Identities*, [53]

---

[31]This follows, e.g., by applying the Griesmer bound to the associated linear code, which has minimum distance $\geq q^{k-1}$ and dimension $\geq k$.

[32]Just recall that the length of any doubly-even self-dual binary code must be a multiple of 8.

[33]Adding $20 = 5 \cdot 4$, which accounts for the 5 pairs of equal points in the code, to the right-hand side "corrects" the third equation.

[34]This assumption is necessary for the relation $A_i = (q - 1)a_{n-i}$ to hold.

$$\sum_{j=0}^{n-i} \binom{n-j}{i} A_j = q^{k-i} \cdot \sum_{j=0}^{i} \binom{n-j}{n-i} A_j^{\perp} \quad \text{for } 0 \le i \le n, \qquad (3.6)$$

relate the weight distributions $(A_i)$, $(A_i^{\perp})$ of the (primal) code $C$ and the dual code $C^{\perp} = \{\mathbf{y} \in \mathbb{F}_q^n; x_1 y_1 + \cdots + x_n y_n = 0 \text{ for all } \mathbf{x} \in C\}$. They can be solved for $A_i$ (or $A_i^{\perp}$), resulting in linear relations whose coefficients are values of *Krawtchouk polynomials*; see, e.g., [40, Chap 7.2] for details. In our case we have $A_1^{\perp} = A_2^{\perp} = 0$, since $C^{\perp}$ has minimum distance $d^{\perp} \ge 3$, and the first three equations in (3.6) are equivalent to the equations in Lemma 6.

Of course the $A_i$ and the $A_i^{\perp}$ in (3.6) have to be non-negative integers. Omitting the integrality condition yields the so-called *linear programming method*, see e.g. [40, Sect. 2.6], where the $A_i$ and $A_i^{\perp}$ are variables satisfying the mentioned constraints.[35] Given some further constraints on the weights of the code and/or the dual code, one may check whether the corresponding polyhedron contains non-negative rational solutions. In general, this is a very powerful approach and was used to compute bounds for codes with a given minimum distance; see [15, 52]. Here we consider a subset of the MacWilliams identities and use analytical arguments.[36]

By considering the average number of points per hyperplane, we can guarantee the existence of a hyperplane containing a relatively small number of points of $\mathscr{C}$. If this number is nonzero, Lemma 7 allows us to lower-bound $\#\mathscr{C}$ by induction.[37]

**Lemma 8** *Suppose that $\mathscr{C} \ne \emptyset$ in $\mathrm{PG}(v - 1, \mathbb{F}_q)$ is $q^r$-divisible with $\#\mathscr{C} = a \cdot q^{r+1} + b$ for some $a, b \in \mathbb{Z}$ and $y \in \mathbb{N}_0$ with $y \equiv (q - 1)b \pmod{q^{r+1}}$. Then there exists a hyperplane $H$ such that $\#(\mathscr{C} \cap H) \le (a - 1) \cdot q^r + \frac{b+y}{q} \in \mathbb{Z}$ and $\#(\mathscr{C} \cap H) \equiv b \pmod{q^r}$.*

*Proof* Set $n = \#\mathscr{C}$ and choose a hyperplane $H$ such that $n' := \#(\mathscr{C} \cap H)$ is minimal. Then, by considering the average number of points per hyperplane, we have

$$n' \le \frac{1}{\begin{bmatrix} v \\ 1 \end{bmatrix}_q} \cdot \sum_{\text{hyperplane } H'} \#(\mathscr{C} \cap H') = n \cdot \begin{bmatrix} v-1 \\ 1 \end{bmatrix}_q \Big/ \begin{bmatrix} v \\ 1 \end{bmatrix}_q < \frac{n}{q} = a \cdot q^r + \frac{b}{q} \le a \cdot q^r + \frac{b+y}{q}.$$

Since $n' \equiv b \equiv \frac{b+y}{q} \pmod{q^r}$, this implies $n' \le (a - 1)q^r + \frac{b+y}{q}$. $\qquad\square$

Note that the stated upper bound does not depend on the specific choice of $a$ and $b$, i.e., there is no need to take a non-negative or *small* $b$. Choosing $y$ as small as

---

[35] Typically, the $A_i^{\perp}$ are removed from the formulation using the explicit formulas based on the Krawtchouk polynomials, which may of course also be done automatically in the preprocessing step of a customary linear programming solver.

[36] The use of a special polynomial, like we will do, is well known in the context of the linear programming method, see e.g. [9, Sect. 18.1].

[37] This result is not *new* at all. In [7] Beutelspacher used such an average argument in his upper bound on the size of partial spreads. Recently, Năstase and Sissokho used it in [59, Lemma 9]. In coding theory it is well known in the context of the Griesmer bound. One may also interpret it as an easy implication of the first two MacWilliams identities, see Lemma 20 and Corollary 9.

possible clearly gives the sharpest bound.[38] If $b \geq 0$, which one can always achieve by suitably decreasing $a$, it is always possible to choose $y = (q-1)b$. However, for $q = 3$, $r = 2$, and $\#\mathscr{C} = 1 \cdot 3^3 + 16 = 43$, i.e., $a = 1$ and $b = 16$, Lemma 8 with $y = (2-1)16 = 16$ provides the existence of a hyperplane $H$ with $n' = \#(\mathscr{C} \cap H) \leq 0 \cdot 3^2 + 16 = 16$. Using $y = 7$ gives $n' \leq 7$ and $n' \equiv 7 \pmod{3^2}$, so that $n' = 7$. Applying the argument again yields a subspace of co-dimension 2 containing exactly one hole. Indeed, Equation (3.4) is needed additionally in order to exclude the possibility of $n' = 7$, so that $A_3(8, 6; 3) \leq 248$, i.e., $\sigma \geq 4$, cf. Theorem 4 stating the same bound.

**Corollary 4** *Suppose that $\mathscr{C} \neq \emptyset$ in $\mathrm{PG}(v - 1, \mathbb{F}_q)$ is $q^r$-divisible with $\#\mathscr{C} = a \cdot q^{r+1} + b$ for some $a, b, y \in \mathbb{Z}$ with $y \equiv (q-1)b \pmod{q^{r+1}}$ and $y \geq 1$. Further, let $g \in \mathbb{Q}$ is the largest number with $q^g \mid y$, and $j \in \mathbb{Z}$ satisfies $1 \leq j < r + 1 - \max\{0, g-1\}$. Then there exists a $(v - j)$-subspace $U$ such that $\#(\mathscr{C} \cap U) \leq (a - j) \cdot q^{r+1-j} + \frac{b + [^{j}_{1}]_q \cdot y}{q^j}$ and $\#(\mathscr{C} \cap U) \equiv b \pmod{q^{r+1-j}}$.*

*Proof* In order to apply induction on $j$, using Lemmas 7 and 8, we need to ensure $n' > 0$ in all but the last step. The latter holds due to $pq^g \nmid b$.[39] $\qquad\square$

Choosing the same value of $y$ in every step, in general is not the optimal way to iteratively apply Lemma 8, even if $y$ is chosen optimal for the first step. To this end, consider a $3^3$-divisible set $\mathscr{C} \in \mathrm{PG}(v - 1, \mathbb{F}_3)$ with $\#\mathscr{C} = 31 \cdot 3^4 + 49 = 2560$, which indeed exists as the disjoint union of 64 solids is an example. Here $y = 17$ with $y \equiv (3-1) \cdot 49 \pmod{3^4}$ is the optimal choice in Lemma 8, so that Corollary 4 guarantees the existence of a subspace $U$ with co-dimension 3, $\#(\mathscr{C} \cap U) \leq (31 - 3) \cdot 3^{4-3} + \frac{49 + [^{3}_{1}]_3 \cdot 17}{3^3} = 94$ and $\#(\mathscr{C} \cap U) \equiv 49 \equiv 1 \pmod{3^1}$. However, applying Corollary 4 with $j = 2$ and $y = 17$ guarantees the existence of a subspace $U'$ with co-dimension 2, $\#(\mathscr{C} \cap U') \leq (31 - 2) \cdot 3^{4-2} + \frac{49 + [^{2}_{1}]_3 \cdot 17}{3^2} = 29 \cdot 3^2 + 13 = 30 \cdot 3^2 + 4 = 274$, and $\#(\mathscr{C} \cap U') \equiv 49 \equiv 4 \pmod 9$. Since $\mathscr{C} \cap U'$ is $3^1$-divisible and $8 \equiv 2 \cdot 4 \pmod{3^2}$, we can apply Lemma 8 with $y = 8$ and deduce the existence of a hyperplane $H$ of $U'$ with $\#(\mathscr{C} \cap (U' \cap H)) \leq 29 \cdot 3^1 + \frac{4+8}{3} = 91$ and $\#(\mathscr{C} \cap (U' \cap H)) \equiv 4 \equiv 1 \pmod{3^1}$, while $U' \cap H$ has co-dimension 3.

In the context of partial spreads or, more generally, vector space partitions another parametrization using the number of non-hole elements of the vector space partition turns out to be very useful in order to state a suitable formula for $y$. In what follows we will say that a vector space partition $\mathscr{P}$ of $\mathbb{F}_q^v$ has *hole-type* $(t, s, m_1)$ if $\mathscr{P}$ has $m_1$ holes (1-subspaces), $2 \leq s \leq t < v$, and $s \leq \dim(X) \leq t$ for all non-holes in $\mathscr{P}$. Additionally, we assume that there is at least one non-hole.

**Corollary 5** *Let $\mathscr{P}$ be a vector space partition of $\mathbb{F}_q^v$ of hole-type $(t, s, m_1), l, x \in \mathbb{N}_0$ with $\sum_{i=s}^{t} m_i = lq^s + x$, and $b, c \in \mathbb{Z}$ with $m_1 = bq^s + c \geq 1$. If $x \geq 2$ and $g$ is the*

---

[38] Another parametrization for $y$ is given by $y = qb' - b$, where $b' \in \mathbb{Z}$ with $b' \geq \frac{b}{q}$ and $b' \equiv b \pmod{q^{r+1}}$, so that $y \in \mathbb{N}_0$. Due to $b' = \frac{b+y}{q}$, $y$ is minimal if and only if $b'$ is minimal.

[39] The proof shows that the second assertion of the Corollary is true for all $(v - j)$-subspaces $U$.

*largest integer such that $q^g$ divides $x - 1$, then for each $0 \leq j \leq s - \max\{1, g\}$ there exists a $(v - j)$-dimensional subspace $U$ containing $\widehat{m}_1$ holes with $\widehat{m}_1 \equiv \widehat{c}$ (mod $q^{s-j}$) and $\widehat{m}_1 \leq (b - j) \cdot q^{s-j} + \widehat{c}$, where $\widehat{c} = \frac{c + \left[\begin{smallmatrix} j \\ 1 \end{smallmatrix}\right]_q \cdot (x-1)}{q^j} \in \mathbb{Z}$.*

*Proof* We have $\left[\begin{smallmatrix} v \\ 1 \end{smallmatrix}\right]_q = m_1 + \sum_{i=s}^{t} m_i \left[\begin{smallmatrix} i \\ 1 \end{smallmatrix}\right]_q$. Multiplication by $q - 1$ and reduction modulo $q^s$ yields $-1 \equiv (q - 1)c - x \pmod{q^s}$, allowing us to apply Corollary 4 with $x = y - 1$. Observe that the parameters $g$ from Corollaries 4 and 5 differ by at most $1 - 1/e$ if $q = p^e$. □

So far, we can guarantee that some subspace contains not *too many* holes, since the average number of holes per subspace would be too large otherwise. The modulo-constraints captured in the definition of a $q^r$-divisible set enable iterative rounding, thereby sharpening the bounds. First we consider the special case of partial spreads, and then we will derive some non-existence results for vector space partitions with *few* holes.

**Lemma 9** *Let $\mathscr{P}$ be a vector space partition of type $k^{m_k} 1^{m_1}$ of $\mathbb{F}_q^v$ with $m_k = lq^k + x$, where $l = \frac{q^{v-k} - q^r}{q^k - 1}$, $x \geq 2$, $k = \left[\begin{smallmatrix} r \\ 1 \end{smallmatrix}\right]_q + 1 - z + u > r$, $q^g \mid x - 1$, $q^{g+1} \nmid x - 1$, and $g, u, z, r, x \in \mathbb{N}_0$. For $\max\{1, g\} \leq y \leq k$ there exists a $(v - k + y)$-subspace $U$ with $L \leq (z + y - 1 - u)q^y + w$ holes, where $w = -(x - 1)\left[\begin{smallmatrix} y \\ 1 \end{smallmatrix}\right]_q$ and $L \equiv w \pmod{q^y}$.*

*Proof* Due to $m_1 = \left[\begin{smallmatrix} v \\ 1 \end{smallmatrix}\right]_q - m_k \cdot \left[\begin{smallmatrix} k \\ 1 \end{smallmatrix}\right]_q = \left[\begin{smallmatrix} r \\ 1 \end{smallmatrix}\right]_q q^k - \left[\begin{smallmatrix} k \\ 1 \end{smallmatrix}\right]_q (x - 1)$, we have $m_1 = bq^k + c$ for $b = \left[\begin{smallmatrix} r \\ 1 \end{smallmatrix}\right]_q$ and $c = -\left[\begin{smallmatrix} k \\ 1 \end{smallmatrix}\right]_q (x - 1)$, where $q^{g'} \mid x - 1$ if and only if $q^{g'} \mid c$. Setting $s = t = k$ and $j = k - y$, we observe $0 \leq j \leq k - \max\{1, g\}$, since $\max\{1, g\} \leq y \leq k$. With this, we apply Corollary 5 and obtain an $(n - k + y)$-subspace $U$ with

$$L \leq (b - j) \cdot q^{k-j} + \frac{c + \left[\begin{smallmatrix} j \\ 1 \end{smallmatrix}\right]_q \cdot (x-1)}{q^j} = (z + y - 1 - u) \cdot q^y - (x - 1) \cdot \frac{\left[\begin{smallmatrix} k \\ 1 \end{smallmatrix}\right]_q - \left[\begin{smallmatrix} k-y \\ 1 \end{smallmatrix}\right]_q}{q^{k-y}}$$

$$= (z + y - 1 - u)q^y - (x - 1)\left[\begin{smallmatrix} y \\ 1 \end{smallmatrix}\right]_q = (z + y - 1 - u)q^y + w$$

holes, so that $L \leq (z + y - 1 - u)q^y + w$ and $L \equiv w \pmod{q^y}$. □

The parameter $l$ is chosen in such a way that $m_k = lq^k + x$ matches the cardinality of the partial $k$-spread given by the construction in Theorem 2 for $x = 1$. Thus the assumption $x \geq 2$ is no real restriction. Actually, the chosen parametrization using $x$ in Corollary 5 makes it very transparent why the construction of Theorem 2 is *asymptotically optimal*—as stated in Theorem 6. If the dimension $k$ of the elements of the partial spread is large enough, a sufficient number of rounding steps can be performed while the rounding process is stopped at $x = 1$ for the other direction. For *small $k$* we will not reach the lower bound of the construction of Theorem 2, so that there remains some room for better constructions.

**Lemma 10** *Let* $\Delta = q^{s-1}$, $m \in \mathbb{Z}$, *and* $\mathscr{P}$ *be a vector space partition of* $\mathbb{F}_q^v$ *of hole-type* $(t, s, c)$. *Then,* $\tau_q(c, \Delta, m) \cdot \frac{q^{v-2}}{\Delta^2} - m(m-1) \geq 0$ *and* $\tau_q(c, \Delta, m) \geq 0$, *where* $\tau_q(c, \Delta, m) = m(m-1)\Delta^2 q^2 - c(2m-1)(q-1)\Delta q + c(q-1)\big(c(q-1)+1\big)$. *If* $c > 0$, *then* $\tau_q(c, \Delta, m) = 0$ *if and only if* $m = 1$ *and* $c = \begin{bmatrix} s \\ 1 \end{bmatrix}_q$.

*Proof* Adding $(c - m\Delta)\big(c - (m-1)\Delta\big)$ times the first, $-\big(2c - (2m-1)\Delta - 1\big)$ times the second and twice the third standard equation from Lemma 6, and dividing the result by $\Delta^2/(q-1)$ gives $(q-1) \cdot \sum_{h=0}^{\lfloor c/\Delta \rfloor}(m-h)(m-h-1)a_{c-h\Delta} = \tau_q(c, \Delta, m) \cdot \frac{q^{n-2}}{\Delta^2} - m(m-1)$, due to the $q^{s-1}$-divisibility. We observe $a_i \geq 0$ and $(m-h)(m-h-1) \geq 0$ for all $m, h \in \mathbb{Z}$. If $m \notin \{0, 1\}$, then $\tau_q(c, \Delta, m) > 0$. Solving $\tau_q(c, \Delta, 0) = 0$ yields $c \in \left\{0, -\frac{q^s+1}{q-1}\right\}$ and solving $\tau_q(c, \Delta, 1) = 0$ yields $c \in \left\{0, \begin{bmatrix} s \\ 1 \end{bmatrix}_q\right\}$.

$\square$

We remark that, in the case of $\tau_q(c, \Delta, m) = 0$, their are either no holes at all or the holes form an $s$-subspace. [11, Theorem 1.B] is quite similar to Lemma 10 and its implications. The multipliers used in the proof can be directly read off from the inverse matrix of

$$\mathbf{A} = \begin{pmatrix} 1 & 1 & 1 \\ a & b & c \\ a^2 - a & b^2 - b & c^2 - c \end{pmatrix},$$

which is given by

$$\mathbf{A}^{-1} = \frac{1}{(c-a)(c-b)(b-a)} \begin{pmatrix} bc(c-b) & -(c+b-1)(c-b) & (c-b) \\ -ac(c-a) & (c+a-1)(c-a) & -(c-a) \\ ab(b-a) & -(b+a-1)(b-a) & (b-a) \end{pmatrix}$$

for distinct numbers $a, b, c$. With this, Lemma 10 can be derived in a conceptual way. Consider the linear programming method with just the first three MacWilliams identities. For parameters excluded by Lemma 10 this small linear program is infeasible, which can be seen at a certain basis solution, i.e., a choice of linear inequalities that are satisfied with equality. Solving for these equations, i.e., a change of basis, corresponds to a non-negative linear combination of the inequality system.[40] In the parametric case we have to choose the basis solution also depending on the parameters. Actually, we have implemented a degree of freedom in Lemma 10 using the parameter $m$. Here, the basis consists of two neighboring non-zero $a_i$-entries, parametrized by $m$, and an arbitrary $a_i$, which plays no role when the resulting equation is solved for all remaining $a_i$-terms. In this way we end up with an equation of

---

[40]If we relax $\geq$ 0-inequalities by adding some auxiliary variable on the left hand side and the minimization of this variable, we can remove the infeasibility, so that we apply the duality theorem of linear programming. Then, the mentioned multipliers for the inequalities are given as the solution values of the dual problem.

the form $\sum_{h=0}^{\lfloor c/\Delta \rfloor}(m-h)(m-h-1)a_{c-h\Delta} = \beta$, where the $a_i$ and their coefficients are non-negative. The use of the underlying quadratic polynomial is well known and frequently applied in the literature; see the remarks after Theorem 4.

**Lemma 11** *For integers $v > k \geq s \geq 2$ and $1 \leq i \leq s - 1$, there exists no vector space partition $\mathcal{P}$ of $\mathbb{F}_q^v$ of hole-type $(k, s, c)$, where $c = i \cdot q^s - \begin{bmatrix} s \\ 1 \end{bmatrix}_q + s - 1$.[41]*

*Proof* Since we have $c < 0$ for $i \leq 0$, we can assume $i \geq 1$ in the following. Let, to the contrary, $\mathcal{P}$ be such a vector space partition and apply Lemma 10 with $m = i(q-1)$ onto $\mathcal{P}$. We compute $\tau_q(c, q^{s-1}, m) = (m-1-a)q^s + a(a+1)$ using $c(q-1) = q^s(m-1) + a$, where $a := 1 + (s-1)(q-1)$. Setting $i = s-1-y$, we have $0 \leq y \leq s-2$ and $\tau_q(c, q^{s-1}, m) = -q^s(y(q-1)+2) + (s-1)^2q^2 - q(s-1)(2s-5) + (s-2)(s-3)$. If $q = 2$, then $y \geq 0$ and $s \geq 2$ yields

$$\tau_2(c, 2^{s-1}, m) = -2^s(y+2) + s^2 + s \leq (s^2 - s - 2^s) + (2s - 2^s) < 0.$$

If $s = 2$, then we have $y = 0$ and $\tau_q(c, q^{s-1}, m) = -q^2 + q < 0$. If $q, s \geq 3$, then we have $q(2s-5) \geq s-3$, so that $\tau_q(c, q^{s-1}, m) \leq -2q^s + (s-1)^2q^2 \leq -2 \cdot 3^{s-2}q^2 + (s-1)^2q^2$ due to $y \geq 0$ and $q \geq 3$. Since $2 \cdot 3^{s-2} > (s-1)^2$ for $s \geq 3$, we have $\tau_q(c, q^{s-1}, m) < 0$ in all cases. Thus, Lemma 10 yields a contradiction, since $q^{n-2s} > 0$ and $m(m-1) \geq 0$ for every integer $m$. $\qquad\square$

Now we are ready to present the first improved (compared to Theorem 4) upper bound for partial spreads, which also covers Theorem 6 setting $z = 0$.

**Theorem 9** *For integers $r \geq 1$, $t \geq 2$, $u \geq 0$, and $0 \leq z \leq \begin{bmatrix} r \\ 1 \end{bmatrix}_q/2$ with $k = \begin{bmatrix} r \\ 1 \end{bmatrix}_q + 1 - z + u > r$ we have $A_q(v, 2k; k) \leq lq^k + 1 + z(q-1)$, where $l = \frac{q^{v-k}-q^r}{q^k-1}$ and $v = kt + r$.*

*Proof* Apply Lemma 9 with $x = 2 + z(q-1) \geq 2$ in order to deduce the existence of a $(v-k+y)$-subspace $U$ with $L \leq (z+y-1-u)q^y - (x-1)\begin{bmatrix} y \\ 1 \end{bmatrix}_q$ holes, where $L \equiv -(x-1)\begin{bmatrix} y \\ 1 \end{bmatrix}_q \pmod{q^y}$. Now, we set $y = z + 1$. Observe that $q^g \mid x - 1$ implies $g \leq z < y$ and we additionally have $1 \leq y = z+1 \leq \begin{bmatrix} r \\ 1 \end{bmatrix}_q + 1 - z \leq t$. If $z = 0$, then $y = 1$, $x = 2$, and $L \leq -uq - 1 < 0$. For $z \geq 1$, we apply Lemma 11 to the subspace $U$ with $s = y$, $c = (z+y-1-u)q^y - (x-1)\begin{bmatrix} y \\ 1 \end{bmatrix}_q - jq^y = (y - 1 - j - u)q^y - \begin{bmatrix} y \\ 1 \end{bmatrix}_q + y - 1$ for some $j \in \mathbb{N}_0$, and $i = y - 1 - j - u \in \mathbb{Z}$. Thus, $A_q(n, 2k; k) \leq lq^k + x - 1$. $\qquad\square$

The case $z = 0$ covers Theorem 6. The non-negativity of the number of holes in a certain carefully chosen subspace is sufficient to prove this fact. The case $z = 1$ was announced in [59, Lemma 10] and proven in [58]. Since the known constructions

---

[41] For more general non-existence results of vector space partitions see e.g. [29, Theorem 1] and the related literature. Actually, we do not need the assumption of an underlying vector space partition of the mentioned type. The result is generally true for $q^{s-1}$-divisible codes, since the parameter $x$ is just a nice technical short-cut to ease the notation.

for partial $k$-spreads give $A_q(kt + r, 2k; k) \geq lq^k + 1$, see e.g. [7] or Theorem 2, Theorem 9 is tight for $k \geq \begin{bmatrix} r \\ 1 \end{bmatrix}_q + 1$ and $A_2(8, 6; 3) = 34$.

So far Lemma 10 was just applied in the case of Lemma 11 excluding the existence of some very special vector space partitions. Next, we look at a subspace and consider the number of holes, i.e., we apply Lemma 9 giving us the freedom to choose the dimension of the subspace. Then Lemma 10, stating that a certain quadratic polynomial is non-negative, can be applied. By minimizing this function in terms of the free parameter $m$, we obtain the following result.

**Theorem 10** *For integers $r \geq 1$, $t \geq 2$, $y \geq \max\{r, 2\}$, $z \geq 0$ with $\lambda = q^y$, $y \leq k$, $k = \begin{bmatrix} r \\ 1 \end{bmatrix}_q + 1 - z > r$, $v = kt + r$, and $l = \frac{q^{v-k} - q^r}{q^k - 1}$, we have $A_q(v, 2k; k) \leq lq^k + \lceil \lambda - \frac{1}{2} - \frac{1}{2}\sqrt{1 + 4\lambda(\lambda - (z + y - 1)(q - 1) - 1)} \rceil$.*

*Proof* From Lemma 9 we conclude $L \leq (z + y - 1)q^y - (x - 1)\begin{bmatrix} y \\ 1 \end{bmatrix}_q$ and $L \equiv -(x - 1)\begin{bmatrix} y \\ 1 \end{bmatrix}_q \pmod{q^y}$ for the number of holes of a certain $(v-k+y)$-subspace $U$. Using the notation of Lemma 9, $\mathscr{P} \cap U := \{P \cap U \mid P \in \mathscr{P}\}$ is of hole-type $(k, y, L)$ if $y \geq 2$. Next, we will show that $\tau_q(c, \Delta, m) \leq 0$, where $\Delta = q^{y-1}$ and $c = iq^y - (x - 1)\begin{bmatrix} y \\ 1 \end{bmatrix}_q$ with $1 \leq i \leq z + y - 1$, for suitable integers $x$ and $m$. Note that, in order to apply Lemma 9, we have to satisfy $x \geq 2$ and $y \geq g$ for all integers $g$ with $q^g \mid x - 1$. Applying Lemma 10 then gives the desired contradiction, so that $A_q(n, 2k; k) \leq lq^k + x - 1$.

We choose[42] $m = i(q - 1) - (x - 1) + 1$, so that $\tau_q(c, \Delta, m) = x^2 - (2\lambda + 1)x + \lambda(i(q - 1) + 2)$. Solving $\tau_q(c, \Delta, m) = 0$ for $x$ gives $x_0 = \lambda + \frac{1}{2} \pm \frac{1}{2}\theta(i)$, where $\theta(i) = \sqrt{1 - 4i\lambda(q - 1) + 4\lambda(\lambda - 1)}$. We have $\tau_q(c, \Delta, m) \leq 0$ for $|2x - 2\lambda - 1| \leq \theta(i)$. We need to find an integer $x \geq 2$ such that this inequality is satisfied for all $1 \leq i \leq z + y - 1$. The strongest restriction is attained for $i = z + y - 1$. Since $z + y - 1 \leq \begin{bmatrix} r \\ 1 \end{bmatrix}_q$ and $\lambda = q^y \geq q^r$, we have $\theta(i) \geq \theta(z + y - 1) \geq 1$, so that $\tau_q(c, \Delta, m) \leq 0$ for $x = \lceil \lambda + \frac{1}{2} - \frac{1}{2}\theta(z + y - 1) \rceil$. (Observe $x \leq \lambda + \frac{1}{2} + \frac{1}{2}\theta(z + y - 1)$ due to $\theta(z + y - 1) \geq 1$.) Since $x \leq \lambda + 1$, we have $x - 1 \leq \lambda = q^y$, so that $q^g \mid x - 1$ implies $g \leq y$ provided $x \geq 2$. The latter is true due to $\theta(z + y - 1) \leq \sqrt{1 - 4\lambda(q - 1) + 4\lambda(\lambda - 1)} \leq \sqrt{1 + 4\lambda(\lambda - 2)} < 2(\lambda - 1)$, which implies $x \geq \lceil \frac{3}{2} \rceil = 2$.

So far we have constructed a suitable $m \in \mathbb{Z}$ such that $\tau_q(c, \Delta, m) \leq 0$ for $x = \lceil \lambda + \frac{1}{2} - \frac{1}{2}\theta(z + y - 1) \rceil$. If $\tau_q(c, \Delta, m) < 0$, then Lemma 10 gives a contradiction, so that we assume $\tau_q(c, \Delta, m) = 0$ in the following. If $i < z + y - 1$ we have $\tau_q(c, \Delta, m) < 0$ due to $\theta(i) > \theta(z + y - 1)$, so that we assume $i = z + y - 1$. Thus, $\theta(z + y - 1) \in \mathbb{N}_0$. However, we can write $\theta(z + y - 1)^2 = 1 + 4\lambda(\lambda - (z + y - 1)(q - 1) - 1) = (2w - 1)^2 = 1 + 4w(w - 1)$ for some integer $w$. If $w \notin \{0, 1\}$, then $\gcd(w, w - 1) = 1$, so that either $\lambda = q^y \mid w$ or $\lambda = q^y \mid w - 1$. Thus, in any case, $w \geq q^y$, which is impossible since $(z + y - 1)(q - 1) \geq 1$.

---

[42] Solving $\frac{\partial \tau_q(c, \Delta, m)}{\partial m} = 0$, i.e., minimizing $\tau_q(c, \Delta, m)$, yields $m = i(q - 1) - (x - 1) + \frac{1}{2} + \frac{x-1}{q^y}$. For $y \geq r$ we can assume $x - 1 < q^y$ due to Theorem 2, so that up-rounding yields the optimum integer choice. For $y < r$ the interval $[\lambda + \frac{1}{2} - \frac{1}{2}\theta(i), \lambda + \frac{1}{2} + \frac{1}{2}\theta(i)]$ may contain no integer.

Finally, $w \in \{0, 1\}$ implies $w(w - 1) = 0$, so that $\lambda - (z + y - 1)(q - 1) - 1 = 0$. Thus, $z + y - 1 = \begin{bmatrix} y \\ 1 \end{bmatrix}_q \geq \begin{bmatrix} r \\ 1 \end{bmatrix}_q$ since $y \geq r$. The assumptions $y \leq k$ and $k = \begin{bmatrix} r \\ 1 \end{bmatrix}_q + 1 - z$ imply $z + y - 1 = \begin{bmatrix} r \\ 1 \end{bmatrix}_q$ and $y = r$. This gives $k = r$, which is excluded. □

An example where Theorem 10 is strictly superior to the results of [58, Theorems 6,7] is given by $A_3(15, 12; 6) \leq 19695$.[43] Setting $y = k$, we obtain Theorem 4. Compared to [11, 20], the new ingredients essentially are $q^r$-divisible sets and Corollary 5, which allows us to choose $y < k$. Theorem 4, e.g., gives $A_2(15, 12; 6) \leq 516$, $A_2(17, 14; 7) \leq 1028$, and $A_9(18, 16; 8) \leq 3486784442$, while Theorem 10 gives $A_2(15, 12; 6) \leq 515$, $A_2(17, 14; 7) \leq 1026$, and $A_9(18, 16; 8) \leq 3486784420$. For $2 \leq q \leq 9, 5 \leq k \leq 19$, there are 66 improvements in total, i.e., almost 19%, and the maximum gap is 22. Next, we provide an estimation of the bound of Theorem 4.

**Lemma 12** *For integers $1 \leq r < k$ and $q \geq 2$ we have*

$$2q^k - q^r - \frac{q^{2r-k}}{\underline{b}} < \sqrt{1 + 4q^k(q^k - q^r)} \leq 2q^k - q^r - \frac{q^{2r-k}}{\overline{b}},$$

*where $\underline{b} = \frac{3+2\sqrt{2}}{2} > 2.91$ and $\overline{b} = \frac{16}{3} < 5.34$.*

*Proof* Due to $q \geq 2$ and $k \geq r + 1 \geq 2$, we have

$$1 + 4q^k(q^k - q^r) > 4q^k(q^k - q^r) - q^{2r} \cdot \overbrace{\left(\frac{4}{\underline{b}} - 1 - \frac{2}{\underline{b}q} - \frac{1}{\underline{b}^2 q^2}\right)}^{\geq 0}$$

$$\geq 4q^k(q^k - q^r) - \frac{4}{\underline{b}}q^{2r} + q^{2r} + \frac{2}{\underline{b}}q^{3r-k} + \frac{1}{\underline{b}^2}q^{4r-2k} = \left(2q^k - q^r - \frac{q^{2r-k}}{\underline{b}}\right)^2.$$

Similarly, $1 + 4q^k(q^k - q^r) \leq 4q^k(q^k - q^r) - q^{2r} \cdot \left(\frac{4}{\overline{b}} - 1\right) \leq \left(2q^k - q^r - \frac{q^{2r-k}}{\overline{b}}\right)^2.$

□

**Corollary 6** *For integers $1 \leq r < k$ and $t \geq 2$ we have $A_q(kt + r, 2k; k) < lq^k + \frac{q^r}{2} + \frac{1}{2} + \frac{q^{2r-k}}{3+2\sqrt{2}}$, where $l = \frac{q^{(t-1)k+r} - q^r}{q^k - 1}$. If $k \geq 2r$, then $A_q(kt + r, 2k; k) < lq^k + 1 + \frac{q^r}{2}$.*

**Corollary 7** *For integers $r \geq 1, t \geq 2$, and $u, z \geq 0$ with $k = \begin{bmatrix} r \\ 1 \end{bmatrix}_q + 1 - z + u > r$ we have $A_q(v, 2k; k) \leq lq^k + 1 + z(q - 1)$, where $l = \frac{q^{v-k} - q^r}{q^k - 1}$ and $v = kt + r$.*

*Proof* Using Corollary 6, we can remove the upper bound $z \leq \begin{bmatrix} r \\ 1 \end{bmatrix}_q / 2$ from Theorem 9. If $z > \begin{bmatrix} r \\ 1 \end{bmatrix}_q / 2$, then $z \geq \begin{bmatrix} r \\ 1 \end{bmatrix}_q / 2 + 1/2$, so that $A_q(v, 2k; k) < lq^k + 1 + \frac{q^r}{2} \leq lq^k + 1 + \frac{q^r-1}{2} + \frac{q-1}{2} \leq lq^k + 1 + z(q - 1)$ for $k \geq 2r$. Thus, we can assume

---

[43]For $2 \leq q \leq 9, 1 \leq v, k \leq 100$ the bounds of [58, Theorem 6,7] are covered by Theorem 10 and Corollary 7. In many cases the bounds coincide.

$r + 1 \leq k \leq 2r - 1$ and $r \geq 2$. With this, we have $z \geq \begin{bmatrix} r \\ 1 \end{bmatrix}_q - 2(r - 1)$ and $lq^k + 1 + z(q - 1) \geq lq^k + q^r - 2(q - 1)(r - 1)$. It remains to show $lq^k + q^r - 2(q - 1)(r - 1) \geq lq^k + \frac{q^r}{2} + \frac{1}{2} + \frac{q^{2r-k}}{3+2\sqrt{2}} \geq lq^k + \frac{q^r}{2} + \frac{1}{2} + \frac{q^{r-1}}{3+2\sqrt{2}}$, i.e., $q^r \geq 1 + \frac{2q^{r-1}}{3+2\sqrt{2}} + 4(q - 1)(r - 1)$. The latter inequality is valid for all pairs $(r, q)$ except $(2, 2)$, $(2, 3)$, and $(3, 2)$. In those cases it can be verified directly that $lq^k + 1 + z(q - 1)$ is not strictly less than the upper bound of Theorem 4. Indeed, both bounds coincide. □

We remark that the first part of Corollary 6 can be written as $\sigma \geq \frac{q^r-1}{2} - \frac{q^{2r-k}}{3+2\sqrt{2}}$. Unfortunately, Theorem 10 is not capable to obtain $\sigma \geq \lfloor (q^r - 1)/2 \rfloor$. For $A_2(17, 12; 6)$, i.e., $q = 2$ and $r = 5$, it gives $\sigma \geq 13$ while $\lfloor (q^r - 1)/2 \rfloor = 15$. In Lemma 23 we give a cubic analog to Lemma 10, which yields $\sigma \geq 14$ for these parameters.

## 4 Constructions for $q^r$-divisible Sets

First note that we can embed every $\Delta$-divisible point set $\mathscr{C}$ in $\mathrm{PG}(v - 1, \mathbb{F}_q)$ into ambient spaces with dimension larger than $v$ and, conversely, replace $\mathbb{F}_q^v$ by the span $\langle \mathscr{C} \rangle$ without destroying the $\Delta$-divisibility. Since in this sense $v$ is not determined by $\mathscr{C}$, we will refer to $\mathscr{C}$ as a $\Delta$-*divisible point set over* $\mathbb{F}_q$. In the sequel we develop a few basic constructions of $q^r$-divisible sets. For the statement of the first lemma recall our convention that subspaces of $\mathbb{F}_q^v$ are identified with subsets of the point set $\mathscr{P}$ of $\mathrm{PG}(v - 1, \mathbb{F}_q)$.

**Lemma 13** *Every $k$-subspace $\mathscr{C}$ of $\mathrm{PG}(v - 1, \mathbb{F}_q)$ with $k \geq 2$ is $q^{k-1}$-divisible.*

*Proof* By the preceding remark we may assume $k = v$ and hence $\mathscr{C} = \mathscr{P}$. In this case the result is clear, since $\#\mathscr{P} - \#H = q^{v-1}$ for each hyperplane $H$. □

In fact a $k$-subspace of $\mathbb{F}_q^v$ is associated to the $k$-dimensional simplex code over $\mathbb{F}_q$ and Lemma 13 is well-known.

For a point set $\mathscr{C}$ in $\mathrm{PG}(v - 1, \mathbb{F}_q)$ we denote by $\chi_\mathscr{C}$ its characteristic function, i.e., $\chi_\mathscr{C} : \mathscr{P} \to \{0, 1\} \subset \mathbb{Z}$ with $\chi_\mathscr{C}(P) = 1$ if and only if $P \in \mathscr{C}$.

**Lemma 14** *Let $\mathscr{C}_i$ be $\Delta_i$-divisible point sets in $\mathrm{PG}(v - 1, \mathbb{F}_q)$ and $a_i \in \mathbb{Z}$ for $1 \leq i \leq m$. If $\mathscr{C} \subseteq \mathscr{P}$ satisfies $\chi_\mathscr{C} = \sum_{i=1}^m a_i \chi_{\mathscr{C}_i}$ then $\mathscr{C}$ is $\gcd(a_1\Delta_1, \ldots, a_m\Delta_m)$-divisible.*

*Proof* We have $\#\mathscr{C} = \sum_{i=1}^m a_i \cdot \#\mathscr{C}_i$ and $\#(\mathscr{C} \cap H) = \sum_{i=1}^m a_i \cdot \#(\mathscr{C}_i \cap H)$ for each hyperplane $H$. Since $\#(\mathscr{C}_i \cap H) \equiv \#\mathscr{C}_i \pmod{\Delta_i}$, the result follows. □

Lemma 14 shows in particular that the union of mutually disjoint $q^r$-divisible sets is again $q^r$-divisible. Another (well-known) corollary is the following, which expresses the divisibility properties of the MacDonald codes.[44]

---

[44] The generalization to more than one "removed" subspace is also quite obvious and expresses the divisibility properties of optimal linear codes of type BV in the projective case [4, 35, 50].

**Corollary 8** *Let $X \subsetneq Y$ be subspaces of $\mathbb{F}_q^v$ and $\mathscr{C} = Y \setminus X$. If $\dim(X) = s$ then $\mathscr{C}$ is $q^{s-1}$-divisible.*

In particular affine $k$-subspaces of $\mathbb{F}_q^v$ are $q^{k-1}$ divisible.

**Lemma 15** *Let $\mathscr{C}_1 \in \mathscr{P}_1$, $\mathscr{C}_2 \in \mathscr{P}_2$ be $q^r$-divisible point sets in $\mathrm{PG}(v_1 - 1, \mathbb{F}_q)$, respectively, $\mathrm{PG}(v_2 - 1, \mathbb{F}_q)$. Then there exists a $q^r$-divisible set $\mathscr{C}$ in $\mathrm{PG}(v_1 + v_2 - 1, \mathbb{F}_q)$ with $\#\mathscr{C} = \#\mathscr{C}_1 + \#\mathscr{C}_2$.*

*Proof* Embed the point sets $\mathscr{C}_1, \mathscr{C}_2$ in the obvious way into $\mathrm{PG}(\mathbb{F}_q^{v_1} \times \mathbb{F}_q^{v_2}) \cong \mathrm{PG}(v_1 + v_2 - 1, \mathbb{F}_q)$, and take $\mathscr{C}$ as their union. □

Let us note that the embedding dimension $v$ in Lemma 15 is usually not the smallest possible, and the isomorphism type of $\mathscr{C}$ is usually not determined by $\mathscr{C}_1$ and $\mathscr{C}_2$.[45]

In analogy to the *Frobenius Coin Problem*, cf. [8, 13, 28], we define $\mathrm{F}(q, r)$ as the smallest positive integer such that a $q^r$-divisible set over $\mathbb{F}_q$ (i.e., with some ambient space $\mathbb{F}_q^v$) with cardinality $n$ exists for all integers $n > \mathrm{F}(q, r)$. Using Lemma 13, Corollary 8, and Lemma 15, we conclude that $\mathrm{F}(q, r) \leq \begin{bmatrix} r+1 \\ 1 \end{bmatrix}_q \cdot q^{r+1} - \begin{bmatrix} r+1 \\ 1 \end{bmatrix}_q - q^{r+1}$, the largest integer not representable as $a_1 \begin{bmatrix} r+1 \\ 1 \end{bmatrix}_q + a_2 q^{r+1}$ with $a_1, a_2 \in \mathbb{Z}_{\geq 0}$.[46] The bound may also be stated as $\mathrm{F}(q, r) \leq \sum_{i=r+2}^{2r+1} q^i - \sum_{i=0}^{r} q^i$.

As the disjoint union of $q^r$-divisible sets is again $q^r$-divisible, one obtains a wealth of constructions. Consequently, the $q^r$-divisible point sets not arising in this way are of particular interest. They are called *indecomposable*.

The next construction uses the concept of a "sunflower" of subspaces, which forms the $q$-analogue of the $\Delta$-systems, or sunflowers, considered in extremal set theory [23].[47]

**Definition 2** Let $X$ be a subspace of $\mathbb{F}_q^v$ and $t \geq 2$ an integer. A *$t$-sunflower* in $\mathbb{F}_q^v$ with *center* $X$ is a set $\{Y_1, \ldots, Y_t\}$ of subspaces of $\mathbb{F}_q^v$ satisfying $Y_i \neq X$ and $Y_i \cap Y_j = X$ for $i \neq j$. The point sets $Y_i \setminus X_i$ are called *petals* of the sunflower.

**Lemma 16** *(i) The union of the petals of a $q$-sunflower in $\mathbb{F}_q^v$ with $r$-dimensional center forms a $q^r$-divisible point set.*

*(ii) The union of the petals and the center of a $q + 1$-sunflower in $\mathbb{F}_q^v$ with $r$-dimensional center forms a $q^r$-divisible point set.*

*Proof* (i) Let $\mathscr{F} = \{Y_1, \ldots, Y_q\}$ and $\mathscr{C} = \bigcup_{i=1}^{q}(Y_i \setminus X_i) = (\bigcup \mathscr{F}) \setminus X$. We have $\chi_{\mathscr{C}} = \sum_{i=1}^{q} \chi_{Y_i} - q \chi_X$. Since $\dim(Y_i) \geq r + 1$, $Y_i$ is $q^r$-divisible, and so is $q \chi_X$. Hence, by Lemma 14, $\mathscr{C}$ is $q^r$-divisible as well.

(ii) follows from (i) by adding one further space $Y_{q+1}$ to $\mathscr{F}$. □

---

[45] If not both $\mathscr{C}_1$ and $\mathscr{C}_2$ are subspaces, then disjoint embeddings into a geometry $\mathrm{PG}(v - 1, \mathbb{F}_q)$ with $v < \dim\langle\mathscr{C}_1\rangle + \dim\langle\mathscr{C}_2\rangle$ exist as well.

[46] Note that $\gcd\left(\begin{bmatrix} r+1 \\ 1 \end{bmatrix}_q, q^{r+1}\right) = 1$ and recall the solution of the ordinary Frobenius Coin Problem.

[47] Our sunflowers need not have constant dimension, however.

**Lemma 17** *Let $r \geq 1$ be an integer and $1 \leq i \leq q^r + 1$. There exists a $q^r$-divisible set $\mathscr{C}_i$ over $\mathbb{F}_q$ with $\#\mathscr{C}_i = \begin{bmatrix} 2r \\ 1 \end{bmatrix}_q + i \cdot \left( q^{r+1} - q^r - \begin{bmatrix} r \\ 1 \end{bmatrix}_q \right)$.*

*Proof* Let $Y = \mathbb{F}_q^{2r}$ and $X_1, \ldots, X_{q^r+1}$ an $r$-spread in $Y$. After embedding $Y$ in a space $\mathbb{F}_q^v$ of sufficiently large dimension $v$, it is possible to choose $q$-sunflowers $\mathscr{F}_1, \ldots, \mathscr{F}_{q^r+1}$ in $\mathbb{F}_q^v$ with the following properties: $Y \in \mathscr{F}_i$ for all $i$; $\dim(Z) = r + 1$ for $Z \in \mathscr{F}_i \setminus \{Y\}$; $\mathscr{F}_i$ has center $X_i$; petals in different sunflowers $\mathscr{F}_i$ and $\mathscr{F}_j$ are either equal (to $Y$) or disjoint. Having made such a choice, we set $\mathscr{C}_i = \left( \bigcup \mathscr{F}_1 \cup \cdots \cup \bigcup \mathscr{F}_i \right) \setminus (X_1 \cup \cdots \cup X_i)$ for $1 \leq i \leq q^r + 1$. Then $\chi_{\mathscr{C}_i} = \sum_{Z \in \mathscr{F}_1 \cup \cdots \cup \mathscr{F}_i} \chi_Z - q\chi_{X_1} - \cdots - q\chi_{X_i}$ is $q^r$-divisible (again by Lemma 14), and $\#\mathscr{C}_i$ is as asserted. $\qquad\square$

Replacing $S$ by some arbitrary $q^1$-divisible set, we similarly obtain:

**Lemma 18** *Let $\mathscr{C}'$ be a $q^1$-divisible set of cardinality $n$, then there exist $q^1$-divisible sets of cardinality $n + i \cdot \left( q^2 - q - 1 \right)$ for all $0 \leq i \leq n$.*

Our last construction in this section uses the concept of a cone.

**Definition 3** Let $X$, $Y$ be complementary subspaces of $\mathbb{F}_q^v$ with $\dim(X) = s$, $\dim(Y) = t$ (hence $v = s + t$) and $\mathscr{B}$ a set of points in $\mathrm{PG}(Y)$. The *cone* with *vertex* $X$ and *base* $\mathscr{B}$ is the point set $\mathscr{C} = \bigcup_{P \in \mathscr{B}} (X + P)$.

**Lemma 19** *Let $\mathscr{B}$ be a $q^r$-divisible point set in $\mathrm{PG}(v - 1, \mathbb{F}_q)$ with $\#\mathscr{B} = m$ and $s \geq 1$ an integer.*

*(i) If $m \equiv 0 \pmod{q^{r+1}}$ then there exists a $q^{r+s}$-divisible point set $\mathscr{C}$ in $\mathrm{PG}(v + s - 1, \mathbb{F}_q)$ of cardinality $\#\mathscr{C} = mq^s$.*

*(ii) If $m(q - 1) \equiv -1 \pmod{q^{r+1}}$ then there exists a $q^{r+s}$-divisible point set $\mathscr{C}$ in $\mathrm{PG}(v + s - 1, \mathbb{F}_q)$ of cardinality $\#\mathscr{C} = \begin{bmatrix} s \\ 1 \end{bmatrix}_q + mq^s$.*

*Proof* Embed $\mathbb{F}_q^v$ into $\mathbb{F}_q^{v+s}$ as $Y$ and consider a cone $\mathscr{K}$ in $\mathrm{PG}(v + s - 1, \mathbb{F}_q)$ with base $\mathscr{B}$ and $s$-dimensional vertex $X$. The hyperplanes $H \supseteq X$ satisfy $\#(\mathscr{K} \setminus H) = q^s \cdot \#(\mathscr{K} \setminus H) \equiv 0 \pmod{q^{r+s}}$. The hyperplanes $H \not\supseteq X$ intersect $X + P$, $P \in \mathscr{B}$, in an $s$-subspace $\neq X$, hence contain $\begin{bmatrix} s \\ s-1 \end{bmatrix}_q = $ points in $X$ and $q^{s-1}$ points in $\mathscr{K} \setminus X$. It follows that $\#(\mathscr{K} \setminus H) = q^{s-1} + m(q^s - q^{s-1}) = (1 + m(q - 1))q^{s-1}$. Thus in Case (ii) we can take $\mathscr{C} = \mathscr{K}$ and in Case (i) we can take $\mathscr{C} = \mathscr{K} \setminus X$.[48] $\qquad\square$

The preceding constructions can be combined in certain nontrivial ways to yield further constructions of $q^r$-divisible point sets. We will return to this topic in Sect. 6.1.

Nonetheless, we are only scratching the surface of a vast subject. Projective two-weight codes with weights $w_1, w_2$ satisfying $w_2 > w_1 + 1$ are $q^r$-divisible by Delsarte's Theorem [16, Corollary 2]. This yields many further examples of $q^r$-divisible point sets; see [14] and the online-table at http://moodle.tec.hkr.se/~chen/research/2-weight-codes. Codes meeting the Griesmer bound whose minimum distance is a

---

[48]Note that $m(q - 1) \equiv 0 \pmod{q^{r+1}}$ is equivalent to $m \equiv 0 \pmod{q^{r+1}}$.

multiple of $q$ are $q^r$-divisible [69, Proposition 13].[49] Optimal codes of lengths strictly above the Griesmer bound tend to have similar divisibility properties; see, e.g., *Best Known Linear Codes* in `Magma`.

## 5 More Non-existence Results for $q^r$-divisible Sets

For a point set $\mathscr{C}$ in $\mathrm{PG}(v-1, \mathbb{F}_q)$ let $\mathscr{T}(\mathscr{C}) := \{0 \le i \le c \mid a_i > 0\}$, where $a_i$ denotes the number of hyperplanes with $\#(\mathscr{C} \cap H) = i$.

**Lemma 20** *For integers $u \in \mathbb{Z}$, $m \ge 0$ and $\Delta \ge 1$ let $\mathscr{C}$ in $\mathrm{PG}(v-1, \mathbb{F}_q)$ be $\Delta$-divisible of cardinality $n = u + m\Delta \ge 0$. Then, we have $(q-1) \cdot \sum_{h \in \mathbb{Z}, h \le m} h a_{u+h\Delta}$ $= (u + m\Delta - uq) \cdot \frac{q^{v-1}}{\Delta} - m$, where we set $a_{u+h\Delta} = 0$ if $u + h\Delta < 0$.*

*Proof* Rewriting the equations from Lemma 6 yields $(q-1) \cdot \sum_{h \in \mathbb{Z}, h \le m} a_{u+h\Delta} = q \cdot q^{v-1} - 1$ and $(q-1) \cdot \sum_{h \in \mathbb{Z}, h \le m} (u + h\Delta) a_{u+h\Delta} = (u + m\Delta)(q^{v-1} - 1)$. $u$ times the first equation minus the second equation gives $\Delta$ times the stated equation. □

**Corollary 9** *Let $\mathscr{C}$ in $\mathrm{PG}(v-1, \mathbb{F}_q)$ satisfy $n = \#\mathscr{C} = u + m\Delta$ and $\mathscr{T}(\mathscr{C}) \subseteq \{u, u+\Delta, \ldots, u+m\Delta\}$. Then $u < \frac{n}{q}$ or $u = n = 0$.*

While the quadratic inequality of Lemma 10 is based on the first three MacWilliams identities, the linear inequality of Lemma 20 is based on the first two MacWilliams identities. Corollary 9 corresponds to the average argument that we have used in the proof of Lemma 8. Lemma 10 can of course be applied in the general case of $q^r$-divisible sets. First we characterize the part of the parameter space where $\tau_q(c, \Delta, m) \le 0$ and then we analyze the right side of the corresponding interval,

**Lemma 21** *For $m \in \mathbb{Z}$, we have $\tau_q(c, \Delta, m) \le 0$ if and only if $(q-1)c - (m - 1/2)\Delta q + \frac{1}{2} \in$*

$$\left[ -\frac{1}{2} \cdot \sqrt{q^2\Delta^2 - 4qm\Delta + 2q\Delta + 1}, \frac{1}{2} \cdot \sqrt{q^2\Delta^2 - 4qm\Delta + 2q\Delta + 1} \right]. \quad (5.1)$$

*The last interval is non-empty, i.e., the radicand is non-negative if and only if $m \le \lfloor (q\Delta + 2)/4 \rfloor$. We have $\tau_q(u, \Delta, 1) = 0$ if and only if $u = (\Delta q - 1)/(q - 1)$ or $u = 0$.*

*Proof* Solving $\tau_q(c, \Delta, m) = 0$ for $c$ yields the boundaries for $c$ stated in (5.1)). Inside this interval we have $\tau_q(c, \Delta, m) \le 0$. Now, $q^2\Delta^2 - 4qm\Delta + 2q\Delta + 1 \ge 0$ is equivalent to $m \le \frac{q\Delta}{4} + \frac{1}{2} + \frac{1}{4q\Delta}$. Rounding downward the right-hand side, while observing $\frac{1}{4q\Delta} < \frac{1}{4}$, yields $\lfloor (q\Delta + 2)/4 \rfloor$. □

---

[49] In the case $q = p$, and in general for codes of type BV, such codes are even $q^e$-divisible, where $q^e$ is the largest power of $p$ dividing the minimum distance [67, Theorem 1 and Proposition 2].

**Lemma 22** *For* $1 \le m \le \lfloor \sqrt{(q-1)q\Delta} - q + \frac{3}{2} \rfloor$, *we have* $(q-1)n - (m-1/2)$ $\Delta q + \frac{1}{2} \le \frac{1}{2} \cdot \sqrt{q^2 \Delta^2 - 4qm\Delta + 2q\Delta + 1}$, *where* $n = m \cdot \begin{bmatrix} r+1 \\ 1 \end{bmatrix}_q - 1$ *and* $\Delta = q^r$.

*Proof* Plugging in yields $\frac{1}{2} \cdot (q\Delta + 3 - 2m - 2q) \le \frac{1}{2} \sqrt{q^2 \Delta^2 - (4m-2)q\Delta + 1}$, so that squaring and simplifying gives $m \le \sqrt{(q-1)q\Delta + 1/4} - q + \frac{3}{2}$.  □

**Theorem 11** *Let* $\mathscr{C}$ *in* $\mathrm{PG}(v-1, \mathbb{F}_q)$ *be* $q^1$-*divisible with* $2 \le n = \#\mathscr{C} \le q^2$, *then either* $n = q^2$ *or* $q + 1$ *divides* $n$.

*Proof* First we show $n \notin [(m-1)(q+1)+2, m(q+1)-1]$ for $1 \le m \le q - 1$. For $m = 1$ this statement follows from Lemmas 21 and 10. For $m \ge 2$ let $(m-1)(q+1)+2 \le n \le m(q+1)-1$. Due to Lemma 10 it suffices to verify $\tau_q(n, q, m) \le 0$. From $n \ge (m-1)(q+1)+2$ we conclude

$$(q-1)n - (m-1/2)\Delta q + \frac{1}{2} \ge -\frac{1}{2} \cdot \left( q^2 - 4q + 1 + 2m \right) \ge -\frac{1}{2} \cdot \left( q^2 - 2m - 3 \right)$$
$$\ge -\frac{1}{2} \cdot \sqrt{q^4 - 4mq^2 + 2q^2 + 1} = -\frac{1}{2} \cdot \sqrt{q^2 \Delta^2 - 4qm\Delta + 2q\Delta + 1}$$

and from $n \le m(q+1)-1$ we conclude

$$(q-1)n - (m-1/2)\Delta q + \frac{1}{2} \le \frac{1}{2} \cdot \left( q^2 - 2m - 2q + 3 \right) \overset{\star}{\le} \frac{1}{2} \cdot \sqrt{q^2 \Delta^2 - 4qm\Delta + 2q\Delta + 1}.$$

With respect to the estimation $\star$, we remark that $-4q^3 + 8q^2 - 12q + 8 + 4m(m + 2q - 3) \overset{m \le q-1}{\le} -4(q-1)(q^2 - 4q + 6) \overset{q \ge 2}{\le} 0$. Thus, Lemma 21 gives $\tau_q(n, q, m) \le 0$.

Applying Corollary 9 with $u = m$ and $\Delta = q$ yields $n \neq (m-1)(q+1)+1$ for all $1 \le m \le q - 1$.  □

The existence of ovoids shows that the upper bound $n \le q^2$ is sharp in Theorem 11.

**Theorem 12** *For the cardinality* $n$ *of a* $q^r$-*divisible set* $\mathscr{C}$ *over* $\mathbb{F}_q$ *we have*

$$n \notin \left[ (a(q-1)+b) \begin{bmatrix} r+1 \\ 1 \end{bmatrix}_q + a + 1, \; (a(q-1)+b+1) \begin{bmatrix} r+1 \\ 1 \end{bmatrix}_q - 1 \right],$$

*where* $a, b \in \mathbb{N}_0$ *with* $b \le q - 2$, $a \le r - 1$, *and* $r \in \mathbb{N}_{>0}$.

*In other words, if* $n \le rq^{r+1}$, *then* $n$ *can be written as* $a \begin{bmatrix} r+1 \\ 1 \end{bmatrix}_q + bq^{r+1}$ *for some* $a, b \in \mathbb{N}_0$.

*Proof* We prove by induction on $r$, set $\Delta = q^r$, and write $n = (m-1) \begin{bmatrix} r+1 \\ 1 \end{bmatrix}_q + x$, where $a + 1 \le x \le \begin{bmatrix} r+1 \\ 1 \end{bmatrix}_q - 1$ and $m - 1 = a(q-1) + b$ for integers $0 \le b \le q - 2$, $0 \le a \le r - 1$. The induction start $r = 1$ is given by Theorem 11.

Now, assume $r \ge 2$ and conclude that for $0 \le b' \le q - 2$, $0 \le a' \le r - 2$ we have $n' \notin \left[ (a'(q-1)+b') \begin{bmatrix} r \\ 1 \end{bmatrix}_q + a' + 1, \; (a'(q-1)+b'+1) \begin{bmatrix} r \\ 1 \end{bmatrix}_q - 1 \right]$ for the cardinality $n'$ of a $q^{r-1}$-divisible set. If $a \le r - 2$ and $x \le \begin{bmatrix} r \\ 1 \end{bmatrix}_q - 1$, then $b' = b$, $a' = a$

yields $\mathcal{T}(\mathscr{C}) \subseteq \{u, u + \Delta, \ldots, u + (m-2)\Delta\}$ for $u = \Delta + (m-1)\begin{bmatrix} r \\ 1 \end{bmatrix}_q + x$. We compute $(q-1)u = q^{r+1} - q^r + (m-1)q^r - (m-1) + (q-1)x \overset{x \geq a+1}{\geq} (m-2) q^r + q^{r+1} > (m-2)\Delta$, so that we can apply Corollary 9. If $a = r - 1$ and $a + 1 \leq x \leq \begin{bmatrix} r \\ 1 \end{bmatrix}_q - 1$, then $b' = b$, $a' = a - 1$ yields $\mathcal{T}(\mathscr{C}) \subseteq \{u, u + \Delta, \ldots, u + (m-1)\Delta\}$ for $u = (m-1)\begin{bmatrix} r \\ 1 \end{bmatrix}_q + x$. We compute $(q-1)u = (m-1)q^r - (m-1) + x(q-1) > (m-1)\Delta$ using $x \geq a + 1$, so that we can apply Corollary 9. Thus, we can assume $\begin{bmatrix} r \\ 1 \end{bmatrix}_q \leq x \leq \begin{bmatrix} r+1 \\ 1 \end{bmatrix}_q - 1$ in the remaining part. Additionally we have $m \leq r(q-1)$.

We aim to apply Lemma 21. Due to Lemma 22 for the upper bound of the interval it suffices to show $r(q-1) \leq \lfloor \sqrt{(q-1)q\Delta} - q + \frac{3}{2} \rfloor$. For $q = 2$ the inequality is equivalent to $r \leq \lfloor \sqrt{2^{r+1}} - \frac{1}{2} \rfloor$, which is valid for $r \geq 2$. Since the right hand side is larger then $(q-1)(\sqrt{\Delta} - 1)$, it suffices to show $q^{r/2} - 1 \geq r$, which is valid for $q \geq 3$ and $r \geq 2$. For the left hand side of the interval if suffices to show

$$(q-1)n - (m - 1/2)\Delta q + \frac{1}{2} \geq -\frac{1}{2} \cdot \sqrt{(\Delta q)^2 - (4m - 2)\Delta q + 1},$$

which can be simplified to $\Delta q + 2m - 3 - 2(q-1)x \leq \sqrt{(\Delta q)^2 - (4m-2)\Delta q + 1}$ using $n = (m-1)\begin{bmatrix} r+1 \\ 1 \end{bmatrix}_q + x$. Since $(q-1)x \geq q^r - 1$ and $m \leq r(q-1)$ it suffices to show

$$-\Delta^2 + 2rq\Delta - 2r\Delta - \Delta - r + r^2q - r^2 \leq 0. \tag{5.2}$$

For $q = 2$ this inequality is equivalent to $-2^{2r} + r2^{r+1} + r^2 - 2 - 2^r \leq 0$, which is valid for $r \geq 2$. For $r = 2$ Inequality (5.2) is equivalent to $-q^4 + 4q^3 - 4q^2 - q^2 + 4q - 6$, which is valid for $q \in \{2, 3\}$ and $q \geq 4$. For $q \geq 3$ and $r \geq 3$ we have $\Delta \geq 3rq$, so that Inequality (5.2) is satisfied. $\square$

This classification result enables us to decide the existence problem for $q^r$-divisible sets over $\mathbb{F}_q$ of cardinality $n$ in many further cases. We restrict ourselves to the cases $q = 2$, $r \in \{1, 2, 3\}$, and refer to [31] for further results.

**Theorem 13** (i) $2^1$-divisible sets over $\mathbb{F}_2$ of cardinality $n$ exist for all $n \geq 3$ and do not exist for $n \in \{1, 2\}$; in particular, $F(2, 1) = 2$.

(ii) $2^2$-divisible sets over $\mathbb{F}_2$ of cardinality $n$ exist for $n \in \{7, 8\}$ and all $n \geq 14$, and do not exist in all other cases; in particular, $F(2, 2) = 13$.

(iii) $2^3$-divisible sets over $\mathbb{F}_2$ of cardinality $n$ exist for $n \in \{15, 16, 30, 31, 32, 45, 46, 47, 48, 49, 50, 51\}$, for all $n \geq 60$, and possibly for $n = 59$; in all other cases they do not exist; thus $F(2, 3) \in \{58, 59\}$.

*Proof* (i) The non-existence for $n \in \{1, 2\}$ is obvious. Existence for $n \geq 3$ can be shown by taking $\mathscr{C}$ as a projective basis in $PG(n - 2, \mathbb{F}_q)$. The corresponding code $C$ is the binary $[n, n - 1, 2]$ even-weight code.

(ii) The non-existence part follows from Theorem 12. Existence for $n \in \{7, 8\}$ is shown by the $[7, 3, 4]$ simplex code and its dual (the $[8, 4, 4]$ extended Hamming code). These two examples and Lemma 15 in turn yield examples of 4-divisible point sets for $n \in \{14, 15, 16\}$.[50] For $n \in \{17, 18, 19, 20\}$ Lemma 17 provides examples.[51] Together these represent all arithmetic progressions modulo 7, showing existence for $n > 20$.

(iii) Existence of 8-divisible sets for the indicated cases with $n \leq 48$ is shown in the same way as in (ii). Examples for $n = 49$, $n = 50$, and $n = 74$ were found by computer search; we refer to [31] for generator matrices of the corresponding 8-divisible codes. The binary irreducible cyclic $[51, 8]$ code, which is a two-weight code with nonzero weights 24 and 32 (see, e.g., [54]), provides an example for $n = 51$.[52]

For $n \in \{63, \ldots, 72\}$ Lemma 17 provides examples. For $n = 73$, a suitable example is given by the projective $[73, 9]$ two-weight code with non-zero weights 32 and 40 in [44]. Together with the mentioned example for $n = 74$ these represent all arithmetic progressions modulo 15, showing existence for $n > 74$.

The non-existence part follows for $n \leq 48$ from Theorem 12 and for $53 \leq n \leq 58$ from Lemma 10 with $m = 4$. It remains to exclude an 8-divisible point set in $\mathrm{PG}(v - 1, \mathbb{F}_q)$ with $\#\mathscr{C} = 52$. For this we will use a variant of the linear programming method, which treats different ambient space dimensions simultaneously. Since $\mathscr{C}$ is in particular 4-divisible, we conclude from Lemma 7 and Part (ii) that there are no 4- or 12-hyperplanes, i.e., $A_{40} = A_{48} = 0$. Using the parametrization $y = 2^{v-3}$, the first four MacWilliams identities for the associated code $C$ are

$$
\begin{aligned}
1 + \quad A_8 + \quad A_{16} + \quad A_{24} + \quad A_{32} &= 8y, \\
52 + \quad 44A_8 + \quad 36A_{16} + \quad 28A_{24} + \quad 20A_{32} &= 4y \cdot 52, \\
\tbinom{52}{2} + \tbinom{44}{2}A_8 + \tbinom{36}{2}A_{16} + \tbinom{28}{2}A_{24} + \tbinom{20}{2}A_{32} &= 2y \cdot \tbinom{52}{2}, \\
\tbinom{52}{3} + \tbinom{44}{3}A_8 + \tbinom{36}{3}A_{16} + \tbinom{28}{3}A_{24} + \tbinom{20}{3}A_{32} &= y\left(\tbinom{52}{3} + A_3^{\perp}\right).
\end{aligned}
$$

Substituting $x = yA_3^{\perp}$ and solving for $A_8, A_{16}, A_{24}, A_{32}$ yields $A_8 = -4 + \frac{1}{512}x + \frac{7}{64}y$, $A_{16} = 6 - \frac{3}{512}x - \frac{17}{64}y$, $A_{24} = -4 + \frac{3}{512}x + \frac{397}{64}y$, and $A_{32} = 1 - \frac{1}{512}x + \frac{125}{64}y$. Since $A_{16}, x \geq 0$, we have $y \leq \frac{384}{17} < 23$. On the other hand, since $3A_8 + A_{16} \geq 0$, we also have $-6 + \frac{y}{16} \geq 0$, i.e., $y \geq 96$—a contradiction. $\qquad\square$

---

[50]The three examples are realized in dimensions 6, 7 and 8, respectively. Alternative solutions for $n \in \{15, 16\}$, having smaller ambient space dimensions, are the $[15, 4, 8]$ simplex code and the $[16, 5, 8]$ first-order Reed-Muller code.

[51]These examples can be realized in $\mathbb{F}_2^6$ for $n \in \{17, 18\}$ and in $\mathbb{F}_2^7$ for $n \in \{19, 20\}$.

[52]It might look tempting to construct a projective 8-divisible binary code of length 50 by shortening such a code $C$ of length 51. However, this does not work: By Lemma 24, $C$ is the concatenation of an ovoid in $\mathrm{PG}(3, \mathbb{F}_4)$ with the binary $[3, 2]$ simplex code. By construction, the corresponding 8-divisible point set $\mathscr{C}$ is the disjoint union of 17 lines. In particular, each point of $\mathscr{C}$ is contained in a line in $\mathscr{C}$. Consequently, shortening $C$ in any coordinate never gives a projective code.

The non-existence of a $2^3$-divisible set of cardinality $n = 52$ implies the (compared with Theorem 4) tightened upper bound $A_2(11, 8; 4) \le 132^{53}$ and can also be obtained from a more general result, viz., Corollary 10 with $t = 3$. Combining the first four MacWilliams identities we obtain an expression involving a certain cubic polynomial [31]:

**Lemma 23** *Let $\mathscr{C}$ be $\Delta$-divisible over $\mathbb{F}_q$ of cardinality $n > 0$ and $t \in \mathbb{Z}$. Then*
$\sum_{i \ge 1} \Delta^2(i - t)(i - t - 1) \cdot (g_1 \cdot i + g_0) \cdot A_{i\Delta} + qhx = n(q - 1)(n - t\Delta)$
$(n - (t + 1)\Delta)g_2$, *where* $g_1 = \Delta qh$, $g_0 = -n(q - 1)g_2$, $g_2 = h - (2\Delta qt + \Delta q - 2nq + 2n + q - 2)$ *and* $h = \Delta^2 q^2 t^2 + \Delta^2 q^2 t - 2\Delta n q^2 t - \Delta n q^2 + 2\Delta n q t + n^2 q^2 + \Delta n q - 2n^2 q + n^2 + nq - n$.

**Corollary 10** *If there exists $t \in \mathbb{Z}$, using the notation of Lemma 23, with $n/\Delta \notin [t, t+1]$, $h \ge 0$, and $g_2 < 0$, then there is no $\Delta$-divisible set over $\mathbb{F}_q$ of cardinality $n$.*

Applying Corollary 10 with $t = 6$ implies the non-existence of a $2^4$-divisible set $\mathscr{C}$ over $\mathbb{F}_2$ with $\#\mathscr{C} = 200$, so that $A_2(16, 12; 6) \le 1032$, while Theorem 10 gives $A_2(16, 12; 6) \le 1033$. There is no $8^5$-divisible set $\mathscr{C}$ over $\mathbb{F}_8$ with $\#\mathscr{C} = 6 \cdot 8^6 + 3 = 1572867$, which can be seen as follows. Corollary 4 implies the existence of a subspace $U$ of co-dimension 4 such that $\mathscr{C} \cap U$ is $8^1$-divisible, $\#\mathscr{C} \cap U \le 2 \cdot 8^2 + 3 = 131$, and $\#\mathscr{C} \cap U \equiv 3 \pmod{8^2}$. Applying Lemma 10 with $m = 1$ and $m = 8$ excludes the existence of a $8^1$-divisible set with cardinality 3 or 67, respectively. Cardinality 131 is finally excluded by Corollary 10 with $t = 14$. Thus, $A_8(14; 12; 6) \le 16777237$, while Theorem 10 gives $A_8(14; 12; 6) \le 16777238$ and Theorem 4 gives $A_8(14; 12; 6) \le 16777248$. See also [33, 47] for a few more such examples.

Most of the currently best known bounds for $A_q(v, 2k; k)$ can also be directly derived by linear programming; cf. the online tables at http://subspacecodes.uni-bayreuth.de [33]. The following lemma gives a glimpse on coding-theoretic arguments dealing with the MacWilliams equations and its non-negative integer solutions.

**Lemma 24** *Let $C$ be a projective 8-divisible binary code of length $n = 51$. Then $C$ is isomorphic to the concatenation of an ovoid in $\mathrm{PG}(3, \mathbb{F}_4)$ with the binary $[3, 2]$ simplex code. The code $C$ has the parameters $[51, 8]$ and the weight enumerator $1 + 204X^{24} + 51X^{32}$.*

*Proof* With the notation $k = \dim(C)$ and $y = 2^{k-3}$, solving the equation system of the first three MacWilliams equations yields $A_0 = 1$, $A_{16} = -6 - 3A_8 + \frac{3}{16}y$, $A_{24} = 8 + 3A_8 + \frac{49}{8}y$, and $A_{32} = -3 - A_8 + \frac{27}{16}y$. Since $A_{16} \ge 0$, we have $y \ge 32$ and hence $k \ge 8$. Plugging the stated equations into the fourth MacWilliams equation and solving for $A_8$ gives $A_8 = \frac{yA_3^\perp}{512} + \frac{47y}{512} - 4$ and $A_{16} = 6 - \frac{3yA_3^\perp}{512} - \frac{45y}{512}$. Since $A_{16} \ge 0$ and $yA_3^\perp \ge 0$, we have $6 - \frac{45y}{512} \ge 0$, so that $y \le 68 + \frac{4}{15}$ and therefore $k \le 9$.

---

[53] Consequently, for all $t \ge 2$ the upper bound for $A_2(4t + 3, 8; 4)$ is tightened by one; cf. Lemma 4.

For $k = 9$, i.e., $y = 64$, $A_{16} \geq 0$ gives $A_3^{\perp} \leq 1$. $A_3^{\perp} = 0$ leads to $A_8 = \frac{15}{8}$, which is impossible. For $A_3^{\perp} = 1$ the resulting weight enumerator of $C$ is $1 + 2X^8 + 406X^{24} + 103X^{32}$. However, there is no such code, as the sum of the two codewords of weight 8 would be a third non-zero codeword of weight at most 16, which does not exist.

In the case $k = 8$, i.e., $y = 32$, the first MacWilliams equation forces $A_{16} = A_8 = 0$. The resulting weight enumerator of $C$ is given by $1 + 204X^{24} + 51X^{32}$. In particular, $C$ is a projective $[51, 8]$ two-weight code. By [12], this code is unique. The proof is concluded by the observation that an ovoid in $\mathrm{PG}(3, \mathbb{F}_4)$ is a projective quaternary $[17, 4]$ two-weight code, such that the concatenation with the binary $[3, 2]$ simplex code yields a projective binary $[51, 8]$ two-weight code. □

## 6 Open Research Problems

In this closing section we have collected some open research problems within the scope of this article. All of them presumably are accessible using the theoretical framework of $q^r$-divisible sets. Considerably more challenging is the question whether similar methods can be developed for arbitrary constant-dimension codes in place of of partial spreads (or vector space partitions). We only mention the following example: The proof of $A_2(6, 4; 3) = 77$ still depends on extensive computer calculations providing the upper bound $A_2(6, 4; 3) \leq 77$ [37]. The known theoretical upper bound of 81 may be sharpened to 77 (along the lines of [57]), if only the existence of a $(6, 81, 4; 3)_2$ code can be excluded. The 81 planes in such a code would form an exact 9-cover of the line set of $\mathrm{PG}(5, \mathbb{F}_2)$.

### 6.1 Better Constructions for Partial Spreads

The only known cases in which the construction of Theorem 2 has been surpassed are derived from the sporadic example of a partial 3-spread of cardinality 34 in $\mathbb{F}_2^8$ [22], which has 17 holes and can be used to show $A_2(3m + 2, 6; 3) \geq (2^{3m+2} - 18)/7$ by adding $m - 2$ layers of lifted MRD codes; cf. Lemma 4 and Corollary 2. A first step towards the understanding of the sporadic example is the classification of all $2^2$-divisible point sets of cardinality 17 in $\mathrm{PG}(k - 1, \mathbb{F}_2)$. It turns out that there are exactly 3 isomorphism types, one configuration $\mathcal{H}_k$ for each dimension $k \in \{6, 7, 8\}$. Generating matrices for the corresponding doubly-even codes are

$$\begin{pmatrix} 10000110010101110 \\ 01000010111011100 \\ 00100100000011000 \\ 00010111001110100 \\ 00001001100111110 \\ 00000011100111011 \end{pmatrix}, \begin{pmatrix} 10000011110100110 \\ 01000001111111000 \\ 00100010000110000 \\ 00010010000101000 \\ 00001001001000100 \\ 00000010100100010 \\ 00000010101011111 \end{pmatrix}, \begin{pmatrix} 10000000111011110 \\ 01000000010110000 \\ 00100000011100000 \\ 00010000001110000 \\ 00001001100000010 \\ 00000101000001010 \\ 00000011000000110 \\ 00000001111011101 \end{pmatrix}. \quad (6.1)$$

While the classification, so far, is based on computer calculations,[54] one can easily see that there are exactly three solutions of the MacWilliams identities.

**Lemma 25** *Let $\mathscr{C}$ be a $2^2$-divisible point set over $\mathbb{F}_2$ of cardinality* 17. *Then $k = \dim\langle\mathscr{C}\rangle \in \{6, 7, 8\}$, and the solutions of the MacWilliams identities are unique for each $k$: $(k; a_5, a_9, a_{13}; A_3^{\perp}) = (6; 12, 49, 2; 6)$, $(7; 25, 95, 7; 2)$, $(8; 51, 187, 17; 0)$.*

*Proof* The unique solution of the standard equations is given by $a_5 = \frac{13}{16} \cdot 2^{k-2} - 1$, $a_9 = \frac{23}{8} \cdot 2^{k-2} + 3$, and $a_{13} = \frac{5}{16} \cdot 2^{k-2} - 3$. Hence $k \geq 6$, since otherwise $a_{13} < 0$, and $k \leq 8$, since $C$ is self-orthogonal.[55] □

The hole set of the partial 3-spread in [22] corresponds to $\mathscr{H}_7$. A geometric description of this configuration is given in [49, p. 84]. We have computationally checked that indeed all three hole configurations can be realized by a partial 3-spread of cardinality 34 in $\mathbb{F}_2^8$.[56] All three configurations have a non-trivial automorphism group, and hence there is a chance to find a partial 3-spread with nontrivial symmetries and eventually discover an underlying more general construction.[57] So far we can only describe the geometric structure of $\mathscr{H}_6$, $\mathscr{H}_7$, $\mathscr{H}_8$.

The hole configuration $\mathscr{H}_6$ consists of two disjoint planes $E_1$, $E_2$ in $\mathrm{PG}(5, \mathbb{F}_2)$ and a solid $S$ spanned by lines $L_1 \subset E_1$ and $L_2 \subset E_2$. The $17 = 4 + 4 + 9$ points of $\mathscr{H}_6$ are those in $E_1 \setminus L_1$, $E_2 \setminus L_2$, and $S \setminus (L_1 \cup L_2)$. An application of Lemma 17 (the case $q = r = i = 2$) gives that $\mathscr{H}_6$ is 4-divisible. In sunflower terms, $\mathscr{F}_1 = \{S, E_1\}$, $\mathscr{F}_2 = \{S, E_2\}$ with centers $L_1$, $L_2$, respectively.

The hole configuration $\mathscr{H}_7$ can be obtained by modifying a 3-sunflower $\mathscr{F} = \{E, S_1, S_2\}$, whose petals are a plane $E$ and two solids $S_1$, $S_2$ and whose base is a line $L$. By Lemma 16(ii), the point set $E \cup S_1 \cup S_2$, of cardinality $3 + 4 + 12 + 12 = 31$ is 4-divisible. From this set we remove two planes $E_1 \subset S_1$, $E_2 \subset S_2$ intersecting $L$ in different points. This gives $\mathscr{H}_7$.

The hole configuration $\mathscr{H}_8$ can be obtained by modifying the cone construction of Lemma 19. We start with a projective basis $\mathscr{B}$ in $\mathbb{F}_2^4$, i.e., $m = 5$ points with no 4 of them contained in a plane. Such point sets $\mathscr{B}$ are associated to the

---

[54] See http://www.rlmiller.org/de_codes and [19] for the classification of, possibly non-projective, doubly-even codes over $\mathbb{F}_2^v$.

[55] Alternatively, the 4th MacWilliams identity yields $64 - 2^{k-2} = 2^{k-3} \cdot A_3^{\perp}$ and hence $k \leq 8$.

[56] 624 non-isomorphic examples can be found at http://subspacecodes.uni-bayreuth.de [33].

[57] In a forthcoming paper we classify the 2612 non-isomorphic partial 3-spreads of cardinality 34 in $\mathbb{F}_2^8$ that admit an automorphism group of order exactly 8, which is possible for $\mathscr{H}_6$ only, and show that the automorphism groups of all other examples have order at most 4.

$[5, 4, 2]_2$ even-weight code and hence 2-divisible. Since $m \equiv 1 \pmod 4$, the proof of Lemma 19 shows that a generalized cone $\mathcal{K}$ over $\mathcal{B}$ with 1-dimensional vertex $Q$ of multiplicity $-m(q-1) \equiv -m \equiv 3 \pmod 4$ is 4-divisible. Working over $\mathbb{Z}$, we can set $\chi_{\mathcal{K}}(Q) = -1$ as well. Adding any 4-divisible point set $\mathcal{D}$ containing $Q$ (i.e., $\chi_{\mathcal{D}}(Q) = 1$) to $\mathcal{K}$ then produces a 4-divisible multiset set $\mathcal{C}$ with $\#\mathcal{C} = 10 - 1 + \#\mathcal{D}$ and $\chi_{\mathcal{C}}(Q) = 0$. By making the ambient spaces of $\mathcal{K}$ and $\mathcal{C}$ intersect only in $Q$ (which requires embedding the configuration into a larger space $\mathbb{F}_2^v$), we can force $\mathcal{C}$ to be a set. The configuration $\mathcal{H}_8$ is obtained by choosing $\mathcal{D}$ as an affine solid.[58]

From the preceding discussion it is clear that all possible hole types of a partial 3-spread of cardinality 34 in $\mathbb{F}_2^8$ belong to infinite families of $q^r$-divisible sets. Can further $2^2$-divisible sets of small cardinality be extended to an infinite family?

**Construction 1** *For integers $r \geq 1$ and $0 \leq m \leq r$ let $S$ be a $2r$-subspace. By $F_1, \ldots, F_{\begin{bmatrix}m\\1\end{bmatrix}_q}$ we denote the $(2r+1)$-subspaces of $\mathbb{F}_2^{2r+m}$ that contain $S$ and by $L_1, \ldots, L_{\begin{bmatrix}r\\1\end{bmatrix}_q}$ we denote a list of $r$-subspaces of $S$ with pairwise trivial intersection. Let $0 \leq a \leq \begin{bmatrix}m\\1\end{bmatrix}_q$ and $0 \leq b_i \leq q^{r-1} - 1$ for all $1 \leq i \leq a$. For each $1 \leq i \leq a$ we choose $q - 1 + b_i q =: c_i$ different $(r+1)$-subspaces $E_{i,j}$ of $F_i$ with $F_i \cap S = L_i$. With this, we set $\mathcal{C} = \left(S \setminus \cup_{i=1}^a L_i\right) \cup \left(\cup_{i=1}^a \cup_{j=1}^{c_i} \left(E_{i,j} \setminus L_i\right)\right)$ and observe $\dim(\mathcal{C}) \leq 2r + m$, $\#\mathcal{C} = \begin{bmatrix}2r\\1\end{bmatrix}_q + a \cdot \left(q^{r+1} - \begin{bmatrix}r+1\\1\end{bmatrix}_q\right) + b \cdot q^{r+1}$, where $b = \sum_{i=1}^a b_i$, and that $\mathcal{C}$ is $q^r$-divisible.*

*Proof* Apply Lemmas 13, 14 using $\chi_{\mathcal{C}}^v = \chi_S^v + \sum_{i=1}^a \sum_{j=1}^{c_i} \chi_{E_{i,j}}^v - q \sum_{i=1}^a (b_i + 1) \chi_{L_i}^v$. $\square$

We remark that the construction can easily be modified to obtain $q^r$-divisible sets of cardinality $n = \begin{bmatrix}2r\\1\end{bmatrix}_q + a \cdot \left(q^{r+1} - \begin{bmatrix}r+1\\1\end{bmatrix}_q\right) + b \cdot q^{r+1}$ and dimension $v$ for all $2r + m \leq v \leq 2r + a(q-1) + bq$. Choosing $m = r$, $a = q^{r-1}$, and $b = 0$ we obtain a $q^r$-divisible set $\mathcal{C}$ of cardinality $n = q^{2r} + \begin{bmatrix}r-1\\1\end{bmatrix}_q$ and dimension $v = 3r$. For which parameters $r$ and $q$ do partial $(r+1)$-spreads of cardinality $q^{2r+1} + q^{r-1}$, with hole configuration $\mathcal{C}$, in $\mathbb{F}_q^{3r+2}$ exist? So far, such partial spreads are known for $r = 1$, $(q, r) = (2, 2)$ and all further examples would be strictly larger than the ones from Theorem 2.

For the corresponding parameters over the ternary field we currently only know the bounds $244 \leq A_3(8, 6; 3) \leq 248$. A putative plane spread in $\mathbb{F}_3^8$ of size 248 would have a $3^2$-divisible hole configuration $\mathcal{H}$ of cardinality 56. Such a point set is unique up to isomorphism and has dimension $k = 6$. It corresponds to an optimal two-weight code with weight distribution $0^1 36^{616} 45^{112}$. The set $\mathcal{H}$ was first described by R. Hill in [34] and is known as the *Hill cap*. A generator matrix for is

---

[58]Since punctured affine solids are associated to the $[7, 4, 3]_2$ Hamming code, we may also think of $\mathcal{H}_8$ as consisting of the 2-fold repetition of the $[5, 4, 2]$-code and the Hamming code "glued together" in $Q$. In fact the doubly-even $[17, 8, 4]_2$ code associated with $\mathcal{H}_8$ is the code $\overline{I}_{17}^{(3)}$ in [61, p. 234]. The glueing construction is visible in the generator matrix.

$$\begin{pmatrix}
1\,0\,0\,0\,0\,0\,2\,2\,1\,1\,0\,1\,0\,0\,1\,1\,0\,2\,0\,2\,1\,1\,1\,1\,0\,0\,1\,0\,1\,2\,0\,1\,0\,2\,1\,2\,1\,1\,1\,1\,1\,2\,2\,0\,0\,1\,2\,0\,0\,2\,0\,1\,2\,2\,1\,1 \\
0\,1\,0\,0\,0\,0\,1\,1\,1\,0\,1\,2\,1\,0\,1\,0\,1\,1\,2\,1\,1\,2\,0\,0\,1\,0\,2\,1\,1\,2\,2\,2\,1\,1\,1\,2\,1\,0\,0\,0\,0\,2\,1\,2\,0\,2\,2\,2\,0\,0\,2\,2\,2\,0\,1\,0 \\
0\,0\,1\,0\,0\,0\,2\,2\,2\,2\,0\,2\,2\,1\,0\,2\,0\,0\,1\,1\,2\,0\,0\,1\,0\,1\,1\,2\,0\,0\,2\,0\,2\,0\,2\,0\,0\,2\,1\,1\,1\,2\,2\,1\,2\,1\,1\,2\,2\,2\,0\,0\,1\,1\,1\,2 \\
0\,0\,0\,1\,0\,0\,1\,0\,1\,1\,2\,2\,2\,2\,0\,2\,2\,1\,0\,2\,0\,0\,2\,2\,1\,0\,0\,1\,0\,1\,0\,1\,0\,0\,2\,2\,2\,2\,1\,0\,0\,2\,2\,2\,1\,1\,2\,1\,2\,2\,2\,2\,1\,2\,0\,0 \\
0\,0\,0\,0\,1\,0\,2\,0\,1\,2\,1\,0\,2\,2\,1\,1\,2\,1\,1\,2\,0\,0\,1\,0\,2\,1\,1\,0\,2\,2\,1\,1\,1\,2\,1\,0\,0\,0\,0\,2\,1\,2\,0\,2\,2\,2\,0\,2\,1\,2\,2\,0\,1\,0\,0\,1 \\
0\,0\,0\,0\,0\,1\,1\,2\,2\,0\,2\,0\,0\,2\,2\,0\,1\,0\,1\,2\,1\,2\,2\,0\,0\,2\,0\,1\,1\,0\,2\,0\,1\,2\,1\,2\,2\,2\,2\,2\,1\,2\,0\,0\,2\,1\,0\,0\,2\,0\,2\,1\,1\,2\,2\,2
\end{pmatrix}\cdot$$

The automorphism group has order 40320. Given the large automorphism group of $\mathcal{H}$, is it possible to construct a partial plane spread in $\mathbb{F}_3^8$ with size larger than 244?

For $q = 2$, the first open case is $129 \leq A_2(11, 8; 4) \leq 132$. A putative partial 4-spread of size 132 has a $2^3$-divisible hole configuration of cardinality 67. Such exist for all dimensions $9 \leq k \leq 11$; cf. Theorem 13(iii). Can one such $2^3$-divisible set be completed to a partial 4-spread?

Already the smallest open cases pose serious computational challenges. A promising approach is the prescription of automorphisms in order to reduce the computational complexity; see, e.g., Chapter "Computational Methods in Subspace Designs" and [45] for an application of this so-called *Kramer-Mesner method* to constant-dimension codes. Of course, the automorphisms have to stabilize the hole configuration, whose automorphism group is known or can be easily computed in many cases.

## 6.2   Existence Results for $q^r$-divisible Sets

Even for rather small parameters $q$ and $r$ we cannot decide the existence question, see Table 1 and [32].

**Table 1**  Undecided cases for the existence of $q^r$-divisible sets

| $q$ | $r$ | $n$ |
|---|---|---|
| 2 | 3 | 59 |
| 2 | 4 | 130, 131, 163, 164, 165, 185, 215, 216, 232, 233, 244, 245, 246, 247 |
| 3 | 2 | 70, 77, 99, 100, 101, 102, 113, 114, 115, 128 |
| 4 | 2 | 129, 150, 151, 172, 173, 193, 194, 195, 215, 216, 217, 236, 237,238, 239, 251, 258, 259, 261, 272, 279, 280, 282, 283, 293, 301, 305, 313, 314, 322, 326, 333, 334, 335,... |
| 5 | 1 | 40 |
| 7 | 1 | 75, 83, 91, 92, 95, 101, 102, 103, 109, 110, 111, 117, 118, 119, 125, 126, 127, 133, 134, 135, 142, 143, 151, 159, 167 |
| 8 | 1 | 93, 102, 111, 120, 121, 134, 140, 143, 149, 150, 151, 152, 158, 159, 160, 161, 167, 168, 169, 170, 176, 177, 178, 179, 185, 186, 187, 188, 196, 197, 205, 206, 214, 215, 223, 224, 232, 233, 241, 242, 250, 251 |
| 9 | 1 | 123, 133, 143, 153, 154, 175, 179, 185, 189, 195, 196, 199, 206, 207, 208, 209, 216, 217, 218, 219, 226, 227, 228, 229, 236, 237, 238, 239, 247, 248, 249, 257, 258, 259, 267, 268, 269, 277, 278, 279, 288, 289, 298, 299, 308, 309, 318, 319, 329, 339, 349, 359 |

## 6.3 Vector Space Partitions

The most general result on the non-existence of vector space partitions (without conditions on the ambient space dimension $v$) seems to be:

**Theorem 14** (Theorem 1 in [29]) *Let $\mathcal{C}$ be a vector space partition of type $k^z \cdots d_2{}^b d_1{}^a$ of $\mathbb{F}_q^v$, where $a, b > 0$.*

(i) *If $q^{d_2-d_1}$ does not divide $a$ and if $d_2 < 2d_1$, then $a \geq q^{d_1} + 1$;*

(ii) *if $q^{d_2-d_1}$ does not divide $a$ and if $d_2 \geq 2d_1$, then $a > 2q^{d_2-d_1}$ or $d_1$ divides $d_2$ and $a = \left(q^{d_2} - 1\right) / \left(q^{d_1} - 1\right)$;*

(iii) *if $q^{d_2-d_1}$ divides $a$ and $d_2 < 2d_1$, then $a \geq q^{d_2} - q^{d_1} + q^{d_2-d_1}$;*

(iv) *if $q^{d_2-d_1}$ divides $a$ and $d_2 \geq 2d_1$, then $a \geq q^{d_2}$.*

For the special case $d_1 = 1$, Theorems 11 and 12, presented in Sect. 5, provide tighter results. For $d_1 > 1$ we can replace $d_1$-subspaces by the corresponding point sets and apply results for $q^r$-divisible sets. For vector space partitions of type $k^z \cdots 4^b 2^a$ in $\mathbb{F}_2^v$ we obtain $2^3$-divisible sets of cardinality $n = 3a$, so that $a \in \{5, 10, 15, 16\}$ or $a \geq 20$ by Theorem 13(iii). Theorem 14 gives $a = 5$ or $a \geq 9$, and $4 \mid a$ implies $a \geq 16$. However, not all results of Theorem 14 can be obtained that easy. For vector space partitions of type $k^z \cdots 4^b 3^a$ in $\mathbb{F}_2^v$ we obtain $2^3$-divisible sets of cardinality $n = 7a$, giving $a = 7$ or $a \geq 9$. Theorem 14 gives $a \geq 9$, and $2 \mid a$ implies $a \geq 10$. In order to exclude $a = 7$ one has to look at hyperplane intersections in this new setting. So far, we have used $\#(H \cap \mathcal{C}) \equiv \#\mathcal{C} \pmod{q^r}$. The sets $H \cap \mathcal{C}$ have to come as a partition of type $s^d (s-1)^c$, where $c + d = a$. Here the possible values for $c$ are restricted by the cases of $q^{r-1}$-divisible sets admitting the partition type $(s-1)^c$. This further reduces the possible hyperplane types, so that eventually the linear programming method can be applied. In our situation we have $\mathcal{T}(\mathcal{C}) \subseteq \{25, 49\}$, which is excluded by Lemma 20. For the general case we introduce the following notation: A point set $\mathcal{C}$ in $\mathrm{PG}(v - 1, \mathbb{F}_q)$ admits the partition type $s^{m_s} \cdots 1^{m_1}$ if there exists a vector space partition of $\mathbb{F}_q^v$ of type $s^{m_s} \cdots 1^{m_1}$ that covers the points in $\mathcal{C}$ and no other points. In terms of this, we may restate the previous result as "there is no $2^3$-divisible set admitting the partition type $3^7$". However, we are very far from the generality and compactness of Theorem 14. Nevertheless, the sketched approach seems to be a very promising research direction (and a natural extension of the study of $q^r$-divisible sets).

**Acknowledgements** The authors would like to acknowledge the financial support provided by COST – *European Cooperation in Science and Technology*. The first author was also supported by the National Natural Science Foundation of China under Grant 61571006. The third author was supported in part by the grant KU 2430/3-1 – Integer Linear Programming Models for Subspace Codes and Finite Geometry from the German Research Foundation.

# References

1. J. André, Über nicht-desarguessche Ebenen mit transitiver Translationsgruppe. Mathematische Zeitschrift **60**(1), 156–186 (1954)
2. G. Andrews, *The Theory of Partitions*, 2nd edn. (Cambridge University Press, Cambridge, 1998)
3. S. Ball, A. Blokhuis, A. Gács, P. Sziklai, Z. Weiner, On linear codes whose weights and length have a common divisor. Adv. Math. **211**(1), 94–104 (2007)
4. B.I. Belov, V.N. Logachev, V.P. Sandimirov, Construction of a class of linear binary codes achieving the Varshamov-Griesmer bound. Probl. Inf. Transm. **10**, 211–217 (1974)
5. J.D. Beule, A. Klein, K. Metsch, Substructures of finite classical polar spaces. In Beule and Storme [6]
6. J.D. Beule, L. Storme (eds.), *Current Research Topics in Galois Geometry* (Nova Science Publishers, 2011)
7. A. Beutelspacher, Partial spreads in finite projective spaces and partial designs. Mathematische Zeitschrift **145**(3), 211–229 (1975)
8. A. Beutelspacher, Partitions of finite vector spaces: an application of the Frobenius number in geometry. Archiv der Mathematik **31**(1), 202–208 (1978)
9. J. Bierbrauer, *Introduction to Coding Theory* (2005)
10. C. Bonferroni, Libreria internazionale Seeber, *Teoria Statistica Delle Classi e Calcolo Delle Probabilità* (1936)
11. R. Bose, K. Bush, Orthogonal arrays of strength two and three. Ann. Math. Statistics, 508–524 (1952)
12. I. Bouyukliev, V. Fack, W. Willems, J. Winne, Projective two-weight codes with small parameters and their corresponding graphs. Des. Codes Cryptogr. **41**(1), 59–78 (2006)
13. A. Brauer, On a problem of partitions. Am. J. Math. **64**(1), 299–312 (1942)
14. R. Calderbank, W. Kantor, The geometry of two-weight codes. Bull. Lond. Math. Soc. **18**(2), 97–122 (1986)
15. P. Delsarte, Bounds for unrestricted codes, by linear programming. Philips Res. Rep **27**, 272–289 (1972)
16. P. Delsarte, Weights of linear codes and strongly regular normed spaces. Discret. Math. **3**, 47–64 (1972)
17. P. Dembowski, *Finite Geometries: Reprint of the 1968 Edition* (Springer Science and Business Media, 2012)
18. S. Dodunekov, J. Simonis, Codes and projective multisets. Electron. J. Comb. **5**(R37), 1–23 (1998)
19. C. Doran, M. Faux, S. Gates, T. Hübsch, K. Iga, G. Landweber, R. Miller, Codes and supersymmetry in one dimension. Adv. Theor. Math. Phys. **15**(6), 1909–1970 (2011)
20. D. Drake, J. Freeman, Partial $t$-spreads and group constructible $(s, r, \mu)$-nets. J. Geom. **13**(2), 210–216 (1979)
21. J. Eisfeld, L. Storme, $t$-spreads and minimal $t$-covers in finite projective spaces. Lecture Notes, 29 (2000)
22. S. El-Zanati, H. Jordon, G. Seelinger, P. Sissokho, L. Spence, The maximum size of a partial 3-spread in a finite vector space over $GF(2)$. Des. Codes Cryptogr. **54**(2), 101–107 (2010)
23. P. Erdős, R. Rado, Intersection theorems for systems of sets. J. Lond. Math. Soc. (1) **35**(1), 85–90 (1960)
24. T. Etzion, N. Silberstein, Error-correcting codes in projective spaces via rank-metric codes and Ferrers diagrams. IEEE Trans. Inf. Theory **55**(7), 2909–2919 (2009)
25. J. Galambos, Bonferroni inequalities. Ann. Probab. 577–581 (1977)
26. N.A. Gordon, R. Shaw, L.H. Soicher, Classification of partial spreads in **PG**(4, 2) (2004). www.maths.qmul.ac.uk/~leonard/partialspreads/PG42new.pdf
27. X. Guang, Z. Zhang, *Linear Network Error Correction Coding* (Springer, SpringerBriefs in Computer Science, 2014)

28. O. Heden, The Frobenius number and partitions of a finite vector space. Archiv der Mathematik **42**(2), 185–192 (1984)
29. O. Heden, On the length of the tail of a vector space partition. Discret. Math. **309**(21), 6169–6180 (2009)
30. O. Heden, A survey of the different types of vector space partitions. Discret. Math. Algorithms Appl. **4**(1), 14 (2012). nr. 1250001
31. D. Heinlein, T. Honold, M. Kiermaier, S. Kurz, A. Wassermann, On projective $q^r$-divisible codes. In preparation (2016)
32. D. Heinlein, T. Honold, M. Kiermaier, S. Kurz, A. Wassermann, Projective divisible binary codes (2017). Preprint at arXiv: 1703.08291
33. D. Heinlein, M. Kiermaier, S. Kurz, A. Wassermann, Tables of subspace codes (2015). http://subspacecodes.uni-bayreuth.de
34. R. Hill, Caps and codes. Discret. Math. **22**(2), 111–137 (1978)
35. R. Hill, Optimal linear codes, in *Cryptography and Coding II*, ed. by C. Mitchell (Oxford University Press, Oxford, 1992), pp. 75–104
36. S. Hong, A. Patel, A general class of maximal codes for computer applications. IEEE Trans. Comput. **100**(12), 1322–1331 (1972)
37. T. Honold, M. Kiermaier, S. Kurz, Optimal binary subspace codes of length 6, constant dimension 3 and minimum distance 4. Contemp. Math. **632**, 157–176 (2015)
38. T. Honold, M. Kiermaier, S. Kurz, Classification of large partial plane spreads in $PG(6, 2)$ and related combinatorial objects (2016). Preprint at arXiv: 1606.07655
39. T. Honold, M. Kiermaier, S. Kurz, Constructions and bounds for mixed-dimension subspace codes. Adv. Math. Commun. **10**(3), 649–682 (2016)
40. W. Huffman, V. Pless, *Fundamentals of Error-Correcting Codes* (Cambridge University Press, Cambridge, 2003)
41. D.R. Hughes, F.C. Piper, *Graduate Texts in Mathematics*, 6th edn., Graduate Texts in Mathematics (Springer, 1973)
42. N.L. Johnson, V. Jha, M. Biliotti, *Handbook of Finite Translation Planes* (CRC Press, 2007)
43. R. Koetter, F. Kschischang, Coding for errors and erasures in random network coding. IEEE Trans. Inf. Theory **54**(8), 3579–3591 (2008)
44. A. Kohnert, Constructing two-weight codes with prescribed groups of automorphisms. Discret. Appl. Math. **155**(11), 1451–1457 (2007)
45. A. Kohnert, S. Kurz, Construction of large constant dimension codes with a prescribed minimum distance, *Mathematical Methods in Computer Science* (Springer, 2008), pp. 31–42
46. S. Kurz, Improved upper bounds for partial spreads. Designs, Codes and Cryptography (2016). published online on Oct 25 (2016)
47. S. Kurz, Improved upper bounds for partial spreads. Des. Codes. Cryptogr. **85**(1), 97–106 (2017–10). https://doi.org/10.1007/s10623-016-0290-8
48. S. Kurz, Packing vector spaces into vector spaces. Australas. J. Comb. **68**(1), 122–130 (2017)
49. L. Lambert, Random network coding and designs over $\mathbb{F}_q$. Ph.D. thesis, Ghent University, Master Thesis, 2013
50. I.N. Landjev, Linear codes over finite fields and finite projective geometries. Discret. Math. **213**, 211–244 (2000)
51. M. Lavrauw, O. Polverino, Finite semifields. In Beule and Storme [6], chapter 6, pp. 129–157
52. S. Lloyd, Binary block coding. Bell Syst. Tech. J. **36**(2), 517–535 (1957)
53. F.J. MacWilliams, A theorem on the distribution of weights in a systematic code. Bell Syst. Tech. J. **42**(1), 79–94 (1963)
54. F.J. MacWilliams, J. Seery, The weight distributions of some minimal cyclic codes. IEEE Trans. Inf. Theory **27**, 796–806 (1981)
55. Z. Mateva, S. Topalova, Line spreads of **PG**(5, 2). J. Combin. Des. **17**(1), 90–102 (2009)
56. M. Médard, A. Sprintson, *Network Coding: Fundamentals and Applications* (Elsevier, 2012)
57. A. Nakić, L. Storme, On the extendability of particular classes of constant dimension codes. Des. Codes Cryptogr. **79**(3), 407–422 (2016)

58. E. Năstase, P. Sissokho, The maximum size of a partial spread II: upper bounds. Discret. Math. **340**(7), 1481–1487 (2017). https://doi.org/10.1016/j.disc.2017.02.001
59. E.L. Năstase, P.A. Sissokho, The maximum size of a partial spread in a finite projective space. J. Comb. Theor. Ser. A **152**, 353–362 (2017–11). https://doi.org/10.1016/j.jcta.2017.06.012
60. R. Plackett, J. Burman, The design of optimum multifactorial experiments. Biometrika **33**(4), 305–325 (1946)
61. V. Pless, N. Sloane, On the classification and enumeration of self-dual codes. J. Comb. Theory Ser. A **18**(3), 313–335 (1975)
62. B. Segre, Teoria di galois, fibrazioni proiettive e geometrie non desarguesiane. Annali di Matematica Pura ed Applicata **64**(1), 1–76 (1964)
63. N. Silberstein, T. Etzion, Large constant dimension codes and lexicodes. Adv. Math. Commun. **5**(2), 177–189 (2011)
64. D. Silva, F. Kschischang, R. Koetter, A rank-metric approach to error control in random network coding. IEEE Trans. Inf. Theory **54**(9), 3951–3967 (2008)
65. M.A. Tsfasman, S.G. Vlăduţ, Geometric approach to higher weights. IEEE Trans. Inf. Theory **41**, 1564–1588 (1995)
66. J.H. van Lint, R.M. Wilson, *A Course in Combinatorics* (Cambridge University Press, Cambridge, 1992)
67. H. Ward, Divisibility of codes meeting the griesmer bound. J. Comb. Theory Ser. A **83**(1), 79–93 (1998)
68. H. Ward, An introduction to divisible codes. Des. Codes Cryptogr. **17**(1), 73–79 (1999)
69. H. Ward, Divisible codes - a survey. Serdica Math. J. **27**(4), 263p–278p (2001)
70. R.W. Yeung, S.-Y.R. Li, N. Cai, Z. Zhang, Network coding theory. Found. Trends Commun. Inf. Theory **2**(4/5), 241–381 (2006)

# q-Analogs of Designs: Subspace Designs

## Michael Braun, Michael Kiermaier and Alfred Wassermann

**Abstract** For discrete structures which are based on a finite ambient set and its subsets there exists the notion of a "q-analog": For this, the ambient set is replaced by a finite vector space and the subsets are replaced by subspaces. Consequently, cardinalities of subsets become dimensions of subspaces. *Subspace designs* are the q-analogs of combinatorial designs. Introduced in the 1970s, these structures gained a lot of interest recently because of their application to random network coding. In this chapter we give a thorough introduction to the subject starting from the subspace lattice and its automorphisms, the Gaussian binomial coefficient and counting arguments in the subspace lattice. This prepares us to survey the known structural and existence results about subspace designs. Further topics are the derivation of subspace designs with related parameters from known subspace designs, as well as infinite families, intersection numbers, and automorphisms of subspace designs. In addition, q-Steiner systems and so called large sets of subspace designs will be covered. Finally, this survey aims to be a comprehensive source for all presently known subspace designs and large sets of subspace designs with small parameters.

## 1 Introduction

Concepts, theories, and discrete structures based on finite sets and their subsets turn into a combinatorial "q-analog" if they are considered over vector spaces over a finite field $\mathbb{F}_q$ with $q$ elements. In this case, "subsets" of a finite set become "subspaces"

M. Braun (✉)
Faculty of Computer Science, Darmstadt University of Applied Sciences, 64295 Darmstadt, Germany
e-mail: michael.braun@h-da.de

M. Kiermaier · A. Wassermann
Department of Mathematics, University of Bayreuth, 95440 Bayreuth, Germany
e-mail: michael.kiermaier@uni-bayreuth.de

A. Wassermann
e-mail: alfred.wassermann@uni-bayreuth.de

© Springer International Publishing AG 2018  171
M. Greferath et al. (eds.), *Network Coding and Subspace Designs*,
Signals and Communication Technology,
https://doi.org/10.1007/978-3-319-70293-3_8

of a vector space and "cardinalities" of subsets become "dimensions" of subspaces. In fact, combinatorics on finite sets can be considered as the limiting case "$q$ tends to 1" of combinatorics on vector spaces over $\mathbb{F}_q$, see Tits [1] and Sect. 2.2.

A very prominent example is the "$q$-analog of a combinatorial $t$-design"—the major object of this chapter. We start with the well-known definition of a *combinatorial design*.

**Definition 1** Let $V$ be a set of size $v$, $0 \le t \le k \le v$ integers and $\lambda$ a non-negative integer. A pair $\mathcal{D} = (V, \mathcal{B})$, where $\mathcal{B}$ a (multi-)set of $k$-subsets (*blocks*) of $V$, is called a $t - (v, k, \lambda)$ *design* on $V$, if each $t$-subset of $V$ is contained in exactly $\lambda$ blocks.

If $\mathcal{B}$ is a set, i.e. if every $k$-subset appears at most once in $\mathcal{B}$, the design is called *simple*.

For a $t$-$(v, k, \lambda)$ design, the integer $\lambda$ is the *index* of the design and $t$ is called its *strength*. Combinatorial designs have a long history starting in the 19th century with the works of Plücker, Kirkman and Steiner. There are many prominent applications of designs, among them are *design of experiments* and *coding theory*, to name the most prominent ones. See the voluminous books *Design Theory* [2] and *Handbook of Combinatorial Designs* [3] for a comprehensive overview.

Now, we turn to the $q$-analogs of combinatorial designs. Throughout this chapter, $q$ will be a prime power and $V$ a vector space over the finite field $\mathbb{F}_q$ of finite dimension $v$. Subspaces of dimension $k$ will be called $k$-*subspaces*. Then, the $q$-analog of a design is defined as follows.

**Definition 2** Let $0 \le t \le k \le v$ be integers and $\lambda$ a non-negative integer. A pair $\mathcal{D} = (V, \mathcal{B})$, where $\mathcal{B}$ is a collection of $k$-subspaces (*blocks*) of $V$, is called a $t$-$(v, k, \lambda)_q$ *subspace design* on $V$ if each $t$-subspace of $V$ is contained in exactly $\lambda$ blocks.

If $\mathcal{B}$ is a set, i.e. if every $k$-subset appears at most once in $\mathcal{B}$, the design is called *simple*.

We will see that combinatorial designs can be seen as the case $q = 1$ of subspace designs. Consequently, a $t$-$(v, k, \lambda)$ design may also be denoted as a $t$-$(v, k, \lambda)_1$ design, see Sect. 2.2.

In the following, all subspace designs will be assumed to be simple unless explicitly stated otherwise. If we allow multiple occurrence of $k$-subspaces in the set of blocks we speak of *non-simple* (subspace) designs.

A particular class of subspace designs are $t$-$(v, k, 1)_q$ designs which are called *$q$-analogs of Steiner systems* or simply *$q$-Steiner systems*. Analogously to ordinary Steiner systems on sets, we use the notation $S(t, k, v)_q$ in order to indicate a $q$-Steiner system with parameters $t$-$(v, k, 1)_q$. An important subclass is given by the *$q$-Steiner triple systems* $STS(v)_q$, which are Steiner systems with the parameters $S(2, 3, v)_q$. Viewed as subspace codes, $q$-Steiner system are *diameter perfect codes*, see [4]. For more information on subspaces codes the reader is referred to part I, in particular to chapters "Codes Endowed with the Rank Metric"–"Construction of Cyclic Subspace Codes and Maximum Rank Distance Codes", and for geometric aspects to chapters

"Geometrical Aspects of Subspace Codes" and "Partial Spreads and Vector Space Paritions".

Remarkably, many different names for *q*-analogs of designs can be found in the literature including

- *subspace designs,*
- *designs over finite fields,*
- *designs over* $\mathbb{F}_q$,
- *q-designs,*
- *geometric designs,*
- *designs in vector spaces,* and
- *designs in the q-Johnson scheme.*

The name "subspace design" has the advantage of being short and not involving symbols like *q*. Furthermore, it nicely fits the closely related terms "subspace lattice" and "subspace code".

## 2 The Subspace Lattice

A closer look at Definition 1 reveals that the definition of a combinatorial design can entirely be expressed in terms of the subset lattice of the *v*-element set *V*. The idea behind *q*-analogs in combinatorics is to replace the subset lattice by the subspace lattice of a *v*-dimensional $\mathbb{F}_q$-vector space *V*.

An introduction to the theory of lattices can be found in the classical textbooks by Birkhoff [5] and Stanley [6].

We denote the set of all subspaces of *V* by $\mathscr{L}(V)$. The set of all *k*-subspaces of *V* is called the *Graßmannian* and denoted by $\begin{bmatrix} V \\ k \end{bmatrix}_q$ (Fig. 1).

The partially ordered set (poset) $(\mathscr{L}(V), \leq)$ forms a lattice—the *subspace lattice*—with containment "$\leq$" of subspaces as partial order relation, intersection "$\cap$" of subspaces as meet and sum "$+$" of subspaces as join operator. Moreover, the subspace lattice is *graded* with the dimension "dim" of subspaces as rank function.

The *dimension formula*

$$\dim(U) + \dim(U') = \dim(U \cap U') + \dim(U + U') \quad \text{for all } U, U' \in \mathscr{L}(V)$$

implies that $\mathscr{L}(V)$ is a *modular* lattice.

Projective Geometry

From a slightly different point of view, for $v \geq 2$ the lattice $\mathscr{L}(V)$ is also known as the *projective geometry* PG(*V*). As the projective geometric dimension is always one less then the corresponding algebraic dimension, *k*-subspaces are called $(k-1)$-*flats*, and the isomorphism class of the finite geometry PG(*V*) is denoted by PG($v-1, q$). The 1-subspaces of *V* are called *points*, the 2-subspaces *lines*, the 3-subspaces *planes*, the 4-subspaces *solids* and the $(v-1)$-subspaces *hyperplanes*.

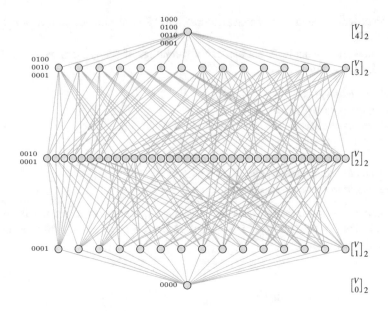

**Fig. 1** Subspace lattice of $\mathbb{F}_2^4$

Coordinates

After a choice of a basis, from now on we may identify $V$ with $\mathbb{F}_q^v$, whose elements will be considered as row vectors. Now, subspaces $U \in \mathcal{L}(V)$ may be represented as the row space of a $\dim(U) \times v$ *generator matrix* $A$. To get a unique matrix, the reduced row echelon form of $A$ may be used.

*Example 1* The following five matrices whose rows form bases of 2-dimensional subspaces define a set of 2-subspaces of the standard vector space $\mathbb{F}_2^4$ yielding a $1\text{-}(4, 2, 1)_2$ design:

$$\left\{ \begin{pmatrix} 0110 \\ 0001 \end{pmatrix}, \begin{pmatrix} 1000 \\ 0010 \end{pmatrix}, \begin{pmatrix} 1100 \\ 0011 \end{pmatrix}, \begin{pmatrix} 1001 \\ 0100 \end{pmatrix}, \begin{pmatrix} 1011 \\ 0101 \end{pmatrix} \right\}.$$

As can be seen in Fig. 2, every 1-subspace is contained in exactly one block of the design.

## 2.1 Automorphisms

Lattice Automorphisms

A bijection $\phi : \mathcal{L} \to \mathcal{L}$ on a lattice $(\mathcal{L}, \leq, \vee, \wedge)$ is an *automorphism* of $\mathcal{L}$ if one of the following equivalent conditions is satisfied for all subspaces $x, y \in \mathcal{L}$:

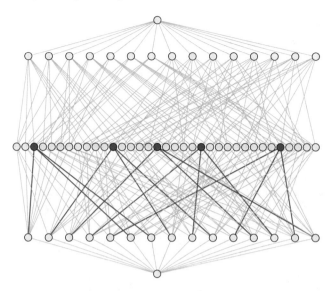

**Fig. 2** A 1-$(4, 2, 1)_2$ design in the subspace lattice of $\mathbb{F}_2^4$

- $x \le y$ if and only if $\alpha(x) \le \alpha(y)$,
- $\alpha(x \vee y) = \alpha(x) \vee \alpha(y)$,
- $\alpha(x \wedge y) = \alpha(x) \wedge \alpha(y)$.

Moreover, $\phi$ is an *antiautomorphism* of $\mathscr{L}$ if one of the following equivalent conditions is satisfied for all subspaces:

- $x \le y$ if and only if $\alpha(y) \le \alpha(x)$,
- $\alpha(x \vee y) = \alpha(x) \wedge \alpha(y)$,
- $\alpha(x \wedge y) = \alpha(x) \vee \alpha(y)$.

The lattice $\mathscr{L}$ is called *self-dual* if and only if there exists an antiautomorphism of $\mathscr{L}$. The set of all automorphisms of $\mathscr{L}$ is a subgroup of the symmetric group on $\mathscr{L}$. It is called the *automorphism group* of $\mathscr{L}$ and denoted by $\mathrm{Aut}(\mathscr{L})$. As the composition of two antiautomorphisms is always an automorphism, we see that the set of all automorphisms and antiautomorphisms forms a group, too, containing $\mathrm{Aut}(\mathscr{L})$ as a normal subgroup of index 1 or 2. The second case occurs if and only if $\# \mathscr{L} \ge 2$ and $\mathscr{L}$ is self-dual.

Group Actions

Let $G$ be a group acting on a set $X$. The action is said to be *transitive* if there is only a single orbit, so for each $x \in X$ we have $\{\alpha(x) : \alpha \in G\} = X$. The *kernel* of the action is defined as

$$G_X = \{\alpha \in G \mid \alpha(x) = x \text{ for all } x \in X\}.$$

The action is called *faithful* if $G_X = \{1\}$. The stabilizers of elements in the same orbit are related by the formula

$$G_{\alpha(x)} = \alpha G_x \alpha^{-1}. \tag{1}$$

Automorphisms of the Subspace Lattice

As usual, the group of all *linear* bijective mappings on $V$ is denoted by $\mathrm{GL}(V)$, and the group of all *semilinear* bijective mappings on $V$ is denoted by $\Gamma\mathrm{L}(V)$. $\mathrm{GL}(V)$ is a normal subgroup of $\Gamma\mathrm{L}(V)$ and $\Gamma\mathrm{L}(V) \cong \mathrm{Aut}(\mathbb{F}_q) \rtimes \mathrm{GL}(V)$.

Now, the groups $\mathrm{GL}(V)$ and $\Gamma\mathrm{L}(V)$ act transitively and faithfully on the elements of $V$. However in general, the induced actions on $\mathscr{L}(V)$ are neither transitive nor faithful. The orbits of the action of these groups on subspaces of $V$ are given by $\begin{bmatrix} V \\ k \end{bmatrix}_q$ with $k \in \{0, \ldots, v\}$. In both cases, the kernel of the action is the set $C = \{\mathbf{v} \mapsto \lambda \mathbf{v} \mid \lambda \in \mathbb{F}_q\}$, which equals the center of $\mathrm{GL}(V)$.[1] So we arrive at the groups $\mathrm{PGL}(V) = \mathrm{GL}(V)/C$ and $\mathrm{P}\Gamma\mathrm{L}(V) = \Gamma\mathrm{L}(V)/C$ which act faithfully on $\mathscr{L}(V)$. Again, $\mathrm{PGL}(V)$ is a normal subgroup of $\mathrm{P}\Gamma\mathrm{L}(V)$ and $\mathrm{P}\Gamma\mathrm{L}(V) \cong \mathrm{Aut}(\mathbb{F}_q) \rtimes \mathrm{PGL}(V)$.

The elements of $\mathrm{PGL}(V)$ will be called the *linear* elements of $\mathrm{P}\Gamma\mathrm{L}(V)$. In the case that $q$ is prime, the group $\mathrm{P}\Gamma\mathrm{L}(V)$ equals $\mathrm{PGL}(V)$, and for the important case $q = 2$, it reduces further to $\mathrm{GL}(V)$. As the mappings in $\mathrm{PGL}(V)$ and $\mathrm{P}\Gamma\mathrm{L}(V)$ are already determined by their images on the point set $\begin{bmatrix} V \\ 1 \end{bmatrix}_q$, we may consider them as permutations of $\begin{bmatrix} V \\ 1 \end{bmatrix}_q$.

The following famous result which is called the "Fundamental Theorem of Projective Geometry" can be found for instance in Artin's book [7].

**Theorem 1** *For $v \geq 3$ the automorphism group of the lattice $(\mathscr{L}(V), \leq)$ is given by the natural action of $\mathrm{P}\Gamma\mathrm{L}(V)$ on $\mathscr{L}(V)$.*

In this way, the group $\mathrm{P}\Gamma\mathrm{L}(V)$ (and its subgroup $\mathrm{PGL}(V)$) provide a notion of (linear) automorphisms and (linear) equivalence on $\mathscr{L}(V)$ and on derived sets like the power set of $\mathscr{L}(V)$.

The group $\mathrm{GL}(V)$ is represented by the group $\mathrm{GL}(v, q)$ of all invertible $v \times v$ matrices over $\mathbb{F}_q$, via assigning the map $\mathbf{x} \mapsto \mathbf{x}A$ to the matrix $A \in \mathrm{GL}(v, q)$. Moreover, $\Gamma\mathrm{L}(V)$ is represented by the set $\Gamma\mathrm{L}(v, q)$ of the pairs $(A, \sigma) \in \mathrm{GL}(v, q) \times \mathrm{Aut}(\mathbb{F}_q)$ via $\mathbf{x} \mapsto \sigma(\mathbf{x})A$, where $\sigma$ is applied simultaneously to all entries of $\mathbf{x}$. By composition of the represented maps (acting from the right), $\Gamma\mathrm{L}(v, q)$ carries the structure of a semidirect product $\mathrm{GL}(v, q) \rtimes \mathrm{Aut}(\mathbb{F}_q)$, where the explicit multiplication law is given by

$$(A, \sigma) \cdot (A', \sigma') = (\sigma'(A)A', \sigma \circ \sigma').$$

The center of $\mathrm{GL}(v, q)$ is the set $\mathbb{F}_q I_v$ of the scalar matrices ($I_v$ denoting the $v \times v$ identity matrix). Therefore, $\mathrm{PGL}(V)$ is represented by $\mathrm{PGL}(v, q) = \mathrm{GL}(v, q)/(\mathbb{F}_q I_v)$ and $\mathrm{P}\Gamma\mathrm{L}(V)$ by $\mathrm{P}\Gamma\mathrm{L}(v, q) = \Gamma\mathrm{L}(V, q)/(\mathbb{F}_q I_v, \mathrm{id}_{\mathbb{F}_q})$.

---

[1] We would like to point out that the center of $\Gamma\mathrm{L}(V)$ is smaller, as it consists only of the elements of $C$ where $\lambda$ is an element of the prime field of $\mathbb{F}_q$.

Duality

We fix some non-singular bilinear form $\beta : V \times V \to \mathbb{F}_q$. For $V = \mathbb{F}_q^v$, the standard inner product $(\mathbf{x}, \mathbf{y}) \mapsto \sum_{i=1}^{v} x_i y_i$ may be taken. The *dual* subspace of $U \in \mathcal{L}(V)$ is defined as

$$U^{\perp} = \{\mathbf{x} \in V \mid \beta(\mathbf{x}, \mathbf{y}) = 0 \text{ for all } \mathbf{y} \in U\}.$$

For all $U \in \mathcal{L}(V)$, $(U^{\perp})^{\perp} = U$. In particular, $U \mapsto U^{\perp}$ is a bijection on $\mathcal{L}(V)$. Moreover, for all $U_1, U_2 \in \mathcal{L}(V)$ we have

$$(U_1 \cap U_2)^{\perp} = U_1^{\perp} + U_2^{\perp} \quad \text{and} \quad (U_1 + U_2)^{\perp} = U_1^{\perp} \cap U_2^{\perp}.$$

Thus, the mapping $U \mapsto U^{\perp}$ is an *antiautomorphism* of $\mathcal{L}(V)$, showing that the subspace lattice is self-dual.

## 2.2   The Gaussian Binomial Coefficient

A *complete chain* in the subspace lattice of $V$ has the form

$$\{\mathbf{0}\} = U_0 < U_1 < U_2 < \ldots < U_v = V$$

with $\dim(U_i) = i$.

To count the number of complete chains, we denote the number of partial chains $U_0 < U_1 < \ldots < U_k$ with $\dim(U_i) = i$ by $a_k$. For $k \geq 1$, $U_k = \langle U_{k-1} \cup \{\mathbf{x}\}\rangle$ for any $\mathbf{x} \in V \setminus U_{k-1}$. As there are $q^v - q^{k-1} = q^{k-1}(q^{v-k+1} - 1)$ vectors $\mathbf{x} \in V \setminus U_k$, and any $q^k - q^{k-1} = q^{k-1}(q - 1)$ vectors $\mathbf{x}$ yield the same space $U_k$, we see that

$$a_k = \frac{q^{v-k+1} - 1}{q - 1} a_{k-1}.$$

Now inductively,

$$a_k = \prod_{i=1}^{k} \frac{q^{v-i+1} - 1}{q - 1} = \prod_{i=v-k+1}^{v} \frac{q^i - 1}{q - 1}$$

and therefore the number of complete chains of $V$ is the *q-factorial*

$$[v]_q! := a_v = \prod_{i=1}^{v} \frac{q^i - 1}{q - 1}.$$

The name *q*-factorial stems from the observation that the complete flags of the subset lattice of a *v*-element set $V$ correspond to the permutations of $V$ and therefore, their number is the ordinary factorial $v!$. This can be carried a step further: Accepting $[v]_q = \frac{q^v - 1}{q - 1}$ as the $q$-analog of the non-negative integer $v$, the $q$-factorial $[v]_q! = \prod_{i=1}^{v} [i]_q$ arises in the very same way from the $q$-numbers as the ordinary factorial from ordinary numbers. For a deeper discussion of $q$-analogs of permutations, see [6].

Let $U_k \in \begin{bmatrix} V \\ k \end{bmatrix}_q$ and let $H = V/U_k = \{v + U_k \mid v \in V\}$ be the corresponding *factor vector space*. It is easy to see that the mapping

$$\{J \in \begin{bmatrix} V \\ j \end{bmatrix}_q \mid U_k \leq J\} \rightarrow \begin{bmatrix} H \\ j - k \end{bmatrix}_q, \, J \mapsto J/U_k = \{v + U_k \mid v \in J\}$$

is a well-defined bijection.

Next, we derive a formula for the number $\begin{bmatrix} v \\ k \end{bmatrix}_q$ of $k$-subspaces of $V$. For this purpose, we count the complete chains $U_0 < \ldots < U_k < \ldots < U_v$ of $V$ in a second way: There are $\begin{bmatrix} v \\ k \end{bmatrix}_q$ possibilities for the choice of $U_k$. For fixed $U_k$, the number of possible head parts $U_0 < \ldots < U_k$ is given by $[k]_q!$. Modding out $U_k$,[2] the possible tail parts $U_k < \ldots < U_v = V$ uniquely correspond to the complete chains of $V/U_k$, whose number is $[v - k]_q!$. So,

$$[v]_q! = \begin{bmatrix} v \\ k \end{bmatrix}_q \cdot [k]_q! \cdot [v - k]_q!,$$

and therefore

$$\# \begin{bmatrix} V \\ k \end{bmatrix}_q = \begin{bmatrix} v \\ k \end{bmatrix}_q = \frac{[v]_q!}{[k]_q! [v - k]_q!}.$$

The expression $\begin{bmatrix} v \\ k \end{bmatrix}_q$ is known as the *Gaussian binomial coefficient* or also as *$q$-binomial coefficient*, as it arises from the $q$-factorial in the usual way. If clear from the context, the index $q$ may be dropped from $\begin{bmatrix} v \\ k \end{bmatrix}_q$ or $\begin{bmatrix} V \\ k \end{bmatrix}_q$ for simplicity. Like for the ordinary binomial coefficient, it is convenient to define $\begin{bmatrix} v \\ k \end{bmatrix}_q = 0$ for integers $k < 0$ or $k > v$.

Many well-known formulas for binomial coefficients have $q$-analogs, and quite often, bijective proofs can be $q$-analogized. As easy examples, we mention

$$\begin{bmatrix} v \\ k \end{bmatrix} = \begin{bmatrix} v \\ v - k \end{bmatrix} \quad \text{and} \quad \begin{bmatrix} v \\ h \end{bmatrix}\begin{bmatrix} v - h \\ k \end{bmatrix} = \begin{bmatrix} v \\ k \end{bmatrix}\begin{bmatrix} v - k \\ h \end{bmatrix},$$

and for $v \geq 1$ the $q$-Pascal triangle identities

---

[2]For a subspace $U$ of $V$, the quotient space is given by $V/U = \{\mathbf{v} + U \mid \mathbf{v} \in V\}$ with the associated canonical projection $\pi : V \rightarrow V/U, \mathbf{v} \mapsto \mathbf{v} + U$. Extending the domain to the set of subspaces of $V$, $\pi$ provides an lattice isomorphism between the lattice interval $[U, V] = \{U' \in \mathscr{L}(V) \mid U \leq U' \leq V\}$ and the subspace lattice $\mathscr{L}(V/U)$.

$$\begin{bmatrix} v \\ k \end{bmatrix}_q = \begin{bmatrix} v-1 \\ k-1 \end{bmatrix}_q + q^k \begin{bmatrix} v-1 \\ k \end{bmatrix}_q = q^{v-k} \begin{bmatrix} v-1 \\ k-1 \end{bmatrix}_q + \begin{bmatrix} v-1 \\ k \end{bmatrix}_q.$$

The expressions $[v]_q = 1 + q + \ldots + q^{v-1}$ and $[v]_q!$ are polynomials in $\mathbb{Z}[q]$ of degrees $v - 1$ and $v(v-1)/2$, respectively. By the formula

$$\begin{bmatrix} v \\ k \end{bmatrix}_q = \prod_{i=1}^{k} \frac{q^{v-i+1} - 1}{q^i - 1},$$

it is checked that $\begin{bmatrix} v \\ k \end{bmatrix}_q$ is a polynomial in $q$, too, and its degree is $\binom{v-1}{k-1}$. The factorization of $X^n - 1$ in the UFD $\mathbb{Z}[X]$ is given by $\prod_d \Phi_d$, where $d$ runs over the positive divisors of $n$ and $\Phi_d \in \mathbb{Z}[X]$ denotes the $d$-th cyclotomic polynomial. Therefore, $q$-numbers and $q$-binomial coefficients factor into a product of pairwise distinct cyclotomic polynomials $\Phi_d$ with $d \geq 2$.

*Remark 1* According to [8], $q$-analog numbers were introduced in [9] and their binomial coefficients in [10]. For a deeper discussion, see [6, 8, 11–13].

## 2.3 Counting Subspaces

The number of $k$-subspaces of $V$ is given by $\begin{bmatrix} v \\ k \end{bmatrix}_q$. Applying this argument to the quotient space $V/U$ for some $U \in \begin{bmatrix} V \\ u \end{bmatrix}_q$, we see that the number of $k$-subspaces $K$ with $U \leq K \leq V$ equals $\begin{bmatrix} v-u \\ k-u \end{bmatrix}_q$.

Every $k$-subspace $K$ of $V$ has a complement, that is a subspace $U \in \mathcal{L}(V)$ with $K \cap U = \{0\}$ and $K + U = V$. By the dimension formula, $\dim(U) = v - k$. An important difference to the subset lattice is that in general, the complement is not unique. More precisely, $K$ has exactly $q^{k(v-k)}$ complements in $V$. For this, one checks that after picking some complement $U$ of $K$ in $\mathcal{L}(V)$, the complements of $K$ in $\mathcal{L}(V)$ bijectively correspond to the homomorphisms $\phi \in \mathrm{Hom}(U, K)$ via $\phi \mapsto \mathrm{im}(\mathrm{id} + \phi)$. Now the claim follows as

$$\dim(\mathrm{Hom}(U, K)) = \dim(U)\dim(K) = (v-k)k.$$

Alternatively, after the choice of a suitable basis the subspace $U$ can be represented as the row space of the $k \times v$-matrix $(0 \mid I_k)$. One checks that the complements of $U$ are given by the row spaces of the $k \times v$ matrices $(A \mid I)$ with $A \in \mathbb{F}_q^{(v-k)v}$, whose number is $q^{k(v-k)}$.

**Lemma 1** *Let $J \leq K \leq V$ be a chain of $\mathbb{F}_q$-vector spaces of finite dimensions $j$, $k$ and $v$. The number of $u$-subspaces $U$ of $V$ with $U \cap K = J$ is*

$$q^{(k-j)(u-j)} \begin{bmatrix} v-k \\ u-j \end{bmatrix}_q.$$

*Proof* By the dimension formula, $\dim(U+K) = u+k-j$. So the possibilities for $U+K$ are given by the $\begin{bmatrix} v-k \\ u-j \end{bmatrix}_q$ intermediate spaces $W$ of $K \le V$ of dimension $u+k-j$. Now modding out $J$, the $u$-subspaces $U$ of $V$ with $U \cap K = J$ and $U+K = W$ correspond to the complements of $K/J$ in $W/J$. Their number is

$$q^{\dim(K/J)(\dim(W/J)-\dim(K/J))} = q^{(k-j)(u-j)}.$$

$\square$

## 2.4 q-Analogs of Combinatorial Structures

As already discussed, we consider the subspace lattice $\mathscr{L}(V)$ of a vector space $V$ of dimension $v$ as the $q$-analog of the subset lattice of a $v$-element set $V$. In Sect. 2.2 we have used this approach to derive $q$-analog numbers, factorials and binomial coefficients, which are polynomials in $\mathbb{Z}[q]$. For $q = 1$, they reduce to ordinary numbers, factorials and binomial coefficients. Hence the subset lattice may be considered as the limit case $q = 1$ of a subspace lattice over $\mathbb{F}_q$. For a deeper discussion see [11], where even the notion of an unary "field" $\mathbb{F}_1$ is used.

Many combinatorial areas, like design theory and coding theory, are based on the subset lattice of a $v$-element ambient set $V$. Given such a class of combinatorial objects, we may define its $q$-analog by replacing the subset lattice by the subspace lattice of some $\mathbb{F}_q$-vector space $V$ of dimension $v$. Thereby, all notions depending on the subset lattice get replaced by the corresponding counterpart in the subspace lattice. Some of these correspondents are shown in Table 2.4. The original class of objects is then considered as the case $q = 1$ of the $q$-analog classes. An important part of the theory of $q$-analogs is the investigation of results in the set-theoretic case for their applicability in the $q$-analog case.

| $q = 1$ | q-analog |
|---|---|
| $v$-element set $V$ | $v$-dim. $\mathbb{F}_q$ vector space $V$ |
| elements of $V$ (*points*) | $1$ − subspaces of $V$ (*points*) |
| subset lattice of $V$ | subspace lattice of $V$ |
| $\binom{V}{k}$ | $\begin{bmatrix} V \\ k \end{bmatrix}_q$ |
| $\binom{v}{k}$ | $\begin{bmatrix} v \\ k \end{bmatrix}_q$ |
| cardinality | dimension |
| $\cap$ | $\cap$ |
| $\cup$ | $+$ |

## 3 Subspace Designs

For all suitable parameters there exist always at least two designs. First, for all $t \in \{0, \ldots, k\}$ and $\mathcal{B} = \emptyset$, $(V, \mathcal{B})$ is a $t$-$(v, k, 0)_q$ subspace design. Second, $(V, \mathcal{B})$ with $\mathcal{B}$ being the full Graßmannian, i.e. $\mathcal{B} = \begin{bmatrix} V \\ k \end{bmatrix}$, is also a design for all $t \leq k$, called the *complete design*.

**Lemma 2** *For* $0 \leq t \leq k \leq v$, $(V, \begin{bmatrix} V \\ k \end{bmatrix}_q)$ *is a* $t$-$(v, k, \begin{bmatrix} v-t \\ k-t \end{bmatrix}_q)_q$ *subspace design.*

*Proof* As we have seen in Sect. 2.3, any $t$-subspace $T$ of $V$ is contained in exactly $\begin{bmatrix} v-t \\ k-t \end{bmatrix}$ $k$-subspaces of $V$. This number is independent of the choice of $T$.  □

Both designs are called *trivial*. Since the full Graßmannian $\begin{bmatrix} V \\ k \end{bmatrix}$ is the design with the largest possible value of $\lambda$ among all simple designs with block dimension $k$, we define

$$\lambda_{\max} = \begin{bmatrix} v-t \\ k-t \end{bmatrix}_q.$$

It is clear that for any $t$-$(v, k, \lambda)_q$ subspace design $\mathscr{D} = (V, \mathcal{B})$, the *supplementary design* $(V, \begin{bmatrix} V \\ k \end{bmatrix} \setminus \mathcal{B})$ is again a subspace design with the parameters $t$-$(v, k, \begin{bmatrix} v-t \\ k-t \end{bmatrix}_q - \lambda)_q$. The empty set and the full Graßmannian are supplementary to each other.

The earliest reference for $q$-analogs of designs we are aware of, is from Ray-Chaudhuri [14], stated in the problem section of the Hypergraph Seminar at Ohio State University 1972. One year later, Cameron [15] presented the same problem at the British Combinatorial conference. The origin of the idea is unclear, since it is stated there that "Several people have observed that the concept of a $t$-design can be generalized [...]". Subspace designs have also been mentioned—independently—in a more general context by Delsarte in [16].

The existence of nontrivial subspace designs with $t \geq 2$ was open for more than a decade, until in [17] subspace designs with the parameters 2-$(v, 3, 7)_2$ and $v \equiv 1, 5 \mod 6$, $v \geq 7$ have been constructed. This construction was generalized to 2-$(v, 3, q^2 + q + 1)_q$ designs for arbitrary $q$ in [18, 19]. In [20] a 3-$(8, 4, 11)_2$ design was given, the first nontrivial subspace design with $t = 3$. Until now, no explicit example with $t \geq 4$ is known.

### 3.1 Divisibility Conditions

The following statement is the $q$-analog of a well-known property of combinatorial designs and was first shown in [21, Lemma 4.1(1)].

**Lemma 3** *Let* $\mathscr{D} = (V, \mathcal{B})$ *be a* $t$-$(v, k, \lambda)_q$ *subspace design. For each* $s \in \{0, \ldots, t\}$, $\mathscr{D}$ *is an* $s$-$(v, k, \lambda_s)_q$ *subspace design with*

$$\lambda_s = \frac{\begin{bmatrix} v-s \\ t-s \end{bmatrix}_q}{\begin{bmatrix} k-s \\ t-s \end{bmatrix}_q} \cdot \lambda = \frac{\begin{bmatrix} v-s \\ k-s \end{bmatrix}_q}{\begin{bmatrix} v-t \\ k-t \end{bmatrix}_q} \cdot \lambda.$$

*In particular, the number of blocks is given by* $\#\mathcal{B} = \lambda_0$.

*Proof* Let $s \in \{0, \ldots, t\}$, $S \in \begin{bmatrix} V \\ t \end{bmatrix}_q$ and $\lambda_s$ the number of blocks of $\mathcal{D}$ containing $S$. We count the set $X$ of all pairs $(T, B) \in \begin{bmatrix} V \\ t \end{bmatrix}_q \times \mathcal{B}$ with $S \leq T \leq B$ in two ways: There are $\begin{bmatrix} v-s \\ t-s \end{bmatrix}_q$ subspaces $T$ containing $S$, and by the design property each $T$ is contained in $\lambda$ blocks $K$. So $\#X = \begin{bmatrix} v-s \\ t-s \end{bmatrix}_q \cdot \lambda$. On the other hand, there are $\lambda_s$ blocks $B \in \mathcal{B}$ containing $S$, each containing $\begin{bmatrix} k-s \\ t-s \end{bmatrix}_q$ subspaces $T \in \begin{bmatrix} V \\ t \end{bmatrix}_q$. This shows $\#X = \lambda_s \cdot \begin{bmatrix} k-s \\ t-s \end{bmatrix}_q$. The proof is concluded by equalizing these two expressions for $\#X$. □

**Definition 3** For $s = t - 1$, the resulting design of Lemma 3 is called the *reduced design* of $\mathcal{D}$.

Furthermore, by Lemma 3, the existence of a $t$-$(v, k, \lambda)_q$ design implies the *integrality conditions*

$$\lambda_s \in \mathbb{Z}, \text{ i.e. } \begin{bmatrix} v-s \\ t-s \end{bmatrix}_q \cdot \lambda \equiv 0 \pmod{\begin{bmatrix} k-s \\ t-s \end{bmatrix}_q}$$

for all $s \in \{0, \ldots, t\}$. Without requiring the actual existence of a corresponding design, any parameter set $t$-$(v, k, \lambda)_q$ fulfilling the integrality conditions will be called *admissible*. Moreover, it will be called *realizable* if a $t$-$(v, k, \lambda)_q$ design does actually exist. Of course, realizability implies admissibility. For $q = 1$, counterexamples are known showing that the contrary is not always true. For $q \geq 2$, this is an open question.

It is easily seen that a 1-$(v, k, \lambda)_q$ design is admissible if and only if $k \mid v$. In fact, for $\lambda = 1$, all those subspace designs are realizable:

**Lemma 4** *There exists a* 1-$(v, k, 1)_q$ *design if and only if $k$ divides $v$.*

*Proof* The direction "$\Rightarrow$" follows from the integrality conditions. For "$\Leftarrow$", let $v$ be a multiple of $k$. Then $\begin{bmatrix} \mathbb{F}_{q^v} \\ 1 \end{bmatrix}_{q^k}$ forms a 1-$(v, k, 1)_q$ design on the $\mathbb{F}_q$-vector space $\mathbb{F}_{q^v}$. □

The above result has been shown in [22] (see also [23] for the case $v = 2k$) in geometric terms: A 1-$(v, k, 1)_q$ design is known as a $(k-1)$-spread in $\mathrm{PG}(v-1, q)$.

The admissibility of a Steiner triple system $\mathrm{STS}_q(v)$ $(v \geq 3)$ depends on the numbers $\lambda_1 = \frac{[v-1]_q}{\Phi_2}$ and $\lambda_0 = \frac{[v-1]_q[v]_q}{\Phi_2 \Phi_3}$ (see Example 4) being integers. The resulting condition is that $\mathrm{STS}_q(v)$ is admissible if and only if $v \equiv 1 \pmod 6$ or $v \equiv 3 \pmod 6$. This condition does not depend on $q$ and coincides with the admissibility in the case $q = 1$ of a combinatorial Steiner triple system $\mathrm{STS}(v)$.

The smallest admissible parameter set of a *q*-analog of a Steiner system with $t \geq 2$ is the Steiner triple system $\mathrm{STS}_q(7)$ or, in other words, a 2-$(7, 3, 1)_q$ design. For $q = 1$, there is a unique design with these parameters, namely the *Fano plane*. Arguably, the most important open problem in the theory of subspace designs is the question for the existence of a 2-$(7, 3, 1)_q$ design, also known as a *q-analog of the Fano plane*. This question was already stated in 1972, in the introduction on subspace designs by Ray-Chaudhuri in [14].

The only *q*-analog of a Steiner triple system known to be realizable is $\mathrm{STS}(13)_2$, see Sect. 6.1.

## 3.2  New Designs from Old Ones

**Lemma 5** (Suzuki [21]) *Let* $\mathscr{D} = (V, \mathscr{B})$ *be a* $t$-$(v, k, \lambda)_q$-*design. For subspaces* $I \leq J \leq V$, *let* $i = \dim(I)$, $j = \dim(V/J)$ *and*

$$\Lambda_{I,J} = \{B \in \mathscr{B} \mid I \leq B \leq J\}.$$

*If* $i + j \leq t$, *the number*

$$\lambda_{i,j} := \#\Lambda_{I,J}$$

*only depends on the dimensions* $i$ *and* $j$, *but not on the exact choice of* $I$ *and* $J$. *The numbers* $\lambda_{i,j}$ *are determined by the recurrence formula*

$$\lambda_{i,0} = \lambda_i \quad and \quad \lambda_{i,j} = \lambda_{i,j-1} - q^{v-k-j+1}\lambda_{i+1,j-1}.$$

*In closed form,*

$$\lambda_{i,j} = \lambda \frac{\left[{v-i-j \atop k-i}\right]_q}{\left[{v-t \atop k-t}\right]_q}.$$

*Proof* Let $i + j \leq t$. We proceed by induction over $j$.

For $j = 0$, we have $J = V$ and $\#\Lambda_{I,J} = \lambda_i$ is independent of the choice of $I$. Now assume $j \geq 1$ and let $\hat{J}$ be a superspace of $J$ with $\dim(\hat{J}/J) = 1$. We count the set $X$ of all pairs $(\hat{I}, B) \in \left[{V \atop i+1}\right]_q \times \mathscr{B}$ with $\hat{I} \leq B \leq \hat{J}$ and $\hat{I} \cap J = I$ in two ways.

By Lemma 1, the number of $\hat{I} \in \left[{\hat{J} \atop i+1}\right]_q$ with $\hat{I} \cap J = I$ is $q^{((v-j)-i)((i+1)-i)} = q^{v-j-i}$. By the induction hypothesis, for each such subspace $\hat{I}$ there are $\lambda_{i+1,j+1}$ blocks $B \in \mathscr{B}$ with $\hat{I} \leq B \leq \hat{J}$. So,

$$\#X = q^{v-j-i}\lambda_{i+1,j-1}.$$

Again by the induction hypothesis, there are $\lambda_{i,j-1}$ blocks $B$ with $I \leq B \leq \hat{J}$. By the dimension formula, those blocks have either

$$B \le J \quad \text{or} \quad \dim(J \cap B) = k - 1.$$

The blocks of the first kind are the blocks in $\Lambda_{I,J}$, which do not have a suitable subspace $\hat{I}$.

For each of the $\lambda_{i,j-1} - \#\Lambda_{I,J}$ blocks $B$ of the second kind, Lemma 1 gives the number of choices for $\hat{I} \in \begin{bmatrix} B \\ i+1 \end{bmatrix}_q$ with $\hat{I} \cap J = \hat{I} \cap (B \cap J)$ as $q^{((k-1)-i)((i+1)-i)} = q^{k-i-1}$. So,

$$\#X = (\lambda_{i,j-1} - \#\Lambda_{I,J})q^{k-i-1}.$$

Equalizing these two expressions for $\#X$ yields

$$\#\Lambda_{I,J} = \lambda_{i,j-1} - q^{v-k-j+1}\lambda_{i+1,j-1},$$

which only depends on $i$ and $j$. The closed formula is readily checked by induction. □

The "refined" $\lambda$'s of the above lemma are a $q$-analog of the values $b_i^j$ in [24]. Lemma 5 has originally been proven as [21, Lemma 4.1] (in the notation $\gamma_i^j$).

*Example 2* If we look at the 1-$(4, 2, 1)_2$ design from Example 1 and take as $I$ the trivial subspace $I = \{0\}$ and as $J$ an arbitrary hyperplane, i.e. $i = 0$ and $j = 1$, then of course all blocks of the design contain $I$. But moreover, from

$$\begin{bmatrix} 4 - 0 - 1 \\ 2 - 0 \end{bmatrix}_2 \Big/ \begin{bmatrix} 4 - 1 \\ 2 - 1 \end{bmatrix}_2 = 1$$

we can conclude that every hyperplane $J$ contains exactly one block of the design.

**Definition 4** Let $\mathscr{D} = (V, \mathscr{B})$ be a $t$-$(v, k, \lambda)_q$ design.

(a) For any point $P \in \begin{bmatrix} V \\ 1 \end{bmatrix}$, we define

$$\text{Der}_P(\mathscr{D}) = (V/P, \{B/P \mid B \in \mathscr{B}, P \le B\}).$$

$\text{Der}_P(\mathscr{D})$ is called the *derived design* of $\mathscr{D}$ with respect to $P$.

(b) For any hyperplane $H \in \begin{bmatrix} V \\ v-1 \end{bmatrix}$, we define

$$\text{Res}_H(\mathscr{D}) = (H, \{B \mid B \in \mathscr{B}, \ B \le H\}).$$

$\text{Res}_H(\mathscr{D})$ is called the *residual design* of $\mathscr{D}$ with respect to $H$.

(c) The *dual design* of $\mathscr{D}$ is defined by

$$\mathscr{D}^{\perp} = (V, \{B^{\perp} \mid B \in \mathscr{B}\}).$$

Different choices for the point $P$ and the hyperplane $H$ may lead to non-isomorphic derived and residual designs. In contrast, up to isomorphism, the definition of the dual design does not depend on the choice of the bilinear form $\beta$.

For the special case of Steiner systems, the derived design shows up in [4, Theorem 2]. The residual design was introduced in [25] and the dual design in [21, Lemma 4.2].

Definition 4 is a $q$-analog of the established notions of the derived, the residual and the complementary design of a combinatorial block design. However, for $q = 1$ the residual design is usually defined as

$$(V \setminus \{P\}, \ \{B \in \mathcal{B} | P \notin B\}) \tag{2}$$

for a point $P$ of $V$. For $q = 1$, (2) is equivalent to Definition 4 via $H = V \setminus \{P\}$. This shows that for $q = 1$, the derived and the residual design with respect to the same point $P$ provide a decomposition of the original $t$-design into two $(t - 1)$-designs. In particular, $\#D = \#\text{Der}(D) + \#\text{Res}(D)$. Unfortunately, this property is not true for $q \geq 2$. One might try to preserve the decomposition property by defining the $q$-analog of the residual design directly via (2). However, this is out of question as for $q \geq 2$, the base set $V \setminus \{B\}$ is not a vector space, and staying with the original ambient space $V$, the resulting set of blocks is not necessarily a design anymore. Definition 4 of the residual design is further backed by all the properties stated below, which are all $q$-analogs of well-known properties of classical combinatorial designs:

**Theorem 2** Let $\mathcal{D} = (V, \mathcal{B})$ be a $t$-$(v, k, \lambda)_q$ design, $P \in \begin{bmatrix} V \\ 1 \end{bmatrix}$ and $H \in \begin{bmatrix} V \\ v-1 \end{bmatrix}$.

(a)  The derived design $\text{Der}_P(\mathcal{D})$ is a design on $V/P$ with the parameters

$$(t - 1)\text{-}(v - 1, k - 1, \lambda)_q.$$

(b)  The residual design $\text{Res}_H(\mathcal{D})$ is a design on $H$ with the parameters

$$(t - 1)\text{-}(v - 1, k, \frac{q^{v-k} - 1}{q^{k-t+1} - 1} \lambda)_q.$$

(c)  The dual design $\mathcal{D}^{\perp}$ is a design on $V$ with the parameters

$$t\text{-}(v, v - k, \frac{\begin{bmatrix} v-t \\ k \end{bmatrix}}{\begin{bmatrix} v-t \\ k-t \end{bmatrix}} \lambda)_q.$$

*Proof* (a) We see that for any $t$-subspace $T$ containing $P$, the blocks $B \in \mathcal{B}$ passing through $T$ uniquely correspond to the blocks $B/P$ of $\text{Der}_U(\mathcal{D})$ passing through $T/P$.
(b) The $\lambda$-value of $\text{Res}_H(\mathcal{D})$ is given by the number $\lambda_{t-1,1}$ of Lemma 5.
(c) For $T \in \begin{bmatrix} V \\ t \end{bmatrix}$, the blocks $B^{\perp}$ of $\mathcal{D}^{\perp}$ passing through $T$ uniquely correspond to the blocks $B \in \mathcal{B}$ contained in $T^{\perp}$. By $\dim(T^{\perp}) = v - t$, the $\lambda$-value of $\mathcal{D}^{\perp}$ is the number $\lambda_{0,t}$ of Lemma 5.                                                                      □

We would like to mention the following straightforward relations:

$$(\mathscr{D}^{\perp})^{\perp} = \mathscr{D},$$
$$\mathrm{Der}_P(\mathscr{D})^{\perp} = \mathrm{Res}_{P^{\perp}}(\mathscr{D}) \text{ and}$$
$$\mathrm{Res}_H(\mathscr{D})^{\perp} = \mathrm{Der}_{H^{\perp}}(\mathscr{D}).$$

The numbers $\lambda_{i,j}$ are nicely visualized in triangle form:

$$
\begin{array}{ccccccccc}
& & & & \lambda_{0,0} & & & & \\
& & & \lambda_{1,0} & & \lambda_{0,1} & & & \\
& & \lambda_{2,0} & & \lambda_{1,1} & & \lambda_{0,2} & & \\
& \cdot\cdot\cdot & & \cdot\cdot\cdot & & \ddots & & \ddots & \\
\lambda_{t,0} & & \lambda_{t-1,1} & & \cdots & & \lambda_{1,t-1} & & \lambda_{0,t}
\end{array}
$$

In this way, the $\lambda_{i,j}$-triangles of various modifications can be read off directly:

- Reduced design: The upper sub-triangle (arising after the deletion of the last row).
- Derived design: The lower left sub-triangle.
- Residual design: The lower right sub-triangle.
- Dual design: The left-right mirror image.

*Example 3* The parameters $3\text{-}(8, 4, 1)_q$ are admissible for any $q$, but known to be realizable only in the ordinary case $q = 1$. Below, the $\lambda_{i,j}$-triangle for these parameters is shown, for $q = 2$ and for general $q$.

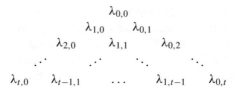

The symmetry of the triangles reflects the property of the parameters being self-dual, meaning that the dual design would have the same parameters. As the $q$-analog of the Fano plane has the derived parameters of $3\text{-}(8, 4, 1)_q$, its triangle can be found as the lower left sub-triangle.

*Example 4* The $\lambda_{i,j}$-triangle for a Steiner triple system $\mathrm{STS}_q(v)$ (a $2\text{-}(v, 3, 1)_q$ design) is:

$$
\begin{array}{ccccc}
& & \frac{[v-1]_q[v]_q}{\Phi_2\Phi_3} & & \\
& \frac{[v-1]_q}{\Phi_2} & & \frac{[v-3]_q[v-1]_q}{\Phi_2\Phi_3} & \\
1 & & \frac{[v-3]_q}{\Phi_2} & & \frac{[v-4]_q[v-3]_q}{\Phi_2\Phi_3}
\end{array}
$$

By construction, for a design its derived and residual designs exist. It may be surprising that also the converse is true, but not necessarily for the designs itself but for the parameter sets. The following statements are the $q$-analogs of results independently found by Tran van Trung [26], van Leijenhorst [27] and Driessen [28]. It has been shown in [25, Lemma 4.1 and Theorem 1].

**Theorem 3** *Let* $t$-$(v, k, \lambda)_q$ *be some parameters.*

(a) *If the derived and the residual parameters are admissible, then the parameters* $t$-$(v, k, \lambda)_q$ *are admissible, too.*
(b) *If the derived and the residual parameters are realizable, then the parameters* $(t - 1)$-$(v, k, \lambda_{t-1})_q$, *i.e. the parameters of the reduced design, are realizable, too.*

Further, there are cases where the derived parameters are realizable and the residual parameters coincide with the parameters of the dual of the derived design. Then, the realizability of derived parameters suffices to get the following corollary.

**Corollary 1** ([25, Corollary 2]) *The realizability of the parameters*

$$t\text{-}(2k - 1, k - 1, \lambda)_q$$

*implies the realizability of the parameters*

$$t\text{-}(2k, k, \lambda \cdot (q^{2k-t} - 1)/(q^{k-t} - 1))_q .$$

## 3.3 Intersection Numbers

Now we discuss intersection numbers, which describe the intersection sizes of the blocks of a design with a fixed subspace $S$ and can be seen as an extension of the numbers $\lambda_{i,j}$ (Lemma 6). For combinatorial designs they have been originally defined in [29] for blocks $S$ and independently as "$i$-Treffer" for general sets $S$ in [30]. We follow [31], where intersection numbers for subspace designs have been introduced.

**Definition 5** Let $\mathscr{D} = (V, \mathscr{B})$ be a $t$-$(v, k, \lambda)_q$ subspace design. For any subspace $S$ of $V$ and $i \in \{0, \ldots, k\}$, we define the $i$-th intersection number of $S$ in $\mathscr{D}$ as

$$\alpha_i(S) = \#\{B \in \mathscr{B} \mid \dim(B \cap S) = i\} .$$

If the set $S$ is clear from the context, we use the abbreviation $\alpha_i = \alpha_i(S)$.

If $\alpha^\perp$ denotes the intersection numbers of the dual design, by the dimension formula and the properties of the dual we have

$$\alpha_i(S) = \begin{cases} 0, & \text{if } i > s \text{ or } k - i > v - s; \\ \alpha^\perp_{(v-s)-(k-i)}(S^\perp), & \text{otherwise.} \end{cases}$$

The following lemma shows that if the dimension or the codimension of $S$ in $V$ is at most $t$, the intersection numbers are uniquely determined and closely connected to the numbers $\lambda_{i,j}$.

**Lemma 6** ([31, Lemma 2.3])

$$\alpha_i(S) = \begin{cases} \begin{bmatrix} s \\ i \end{bmatrix}_q \cdot \lambda_{i,s-i}, & \text{if } \dim(S) \leq t; \\ q^{sk-iv} \begin{bmatrix} v-s \\ k-i \end{bmatrix}_q \cdot \lambda_{k-i,(v-s)-(k-i)}, & \text{if } \dim(S) \geq v-t. \end{cases}$$

*Proof* The first case is shown by double counting the set of all $(I, B) \in \begin{bmatrix} S \\ i \end{bmatrix}_q \times \mathscr{B}$ with $B \cap S = I$. The second case follows by making use of the dual design. $\qquad\square$

**Theorem 4** ([31, Theorem 2.4], $q$-analog of the Mendelsohn equations [29, Theorem 1]) *Let $\mathscr{D}$ be a $t$-$(v, k, \lambda)_q$ subspace design, $S$ a subspace of $V$ and $s = \dim(S)$. For $i \in \{0, \ldots, t\}$ we have the following equation on the intersection numbers of $S$ in $\mathscr{D}$:*

$$\sum_{j=i}^{s} \begin{bmatrix} j \\ i \end{bmatrix}_q \alpha_j = \begin{bmatrix} s \\ i \end{bmatrix}_q \lambda_i \tag{3}$$

*Proof* We count the set $X$ of all pairs $(I, B) \in \begin{bmatrix} V \\ i \end{bmatrix}_q \times \mathscr{B}$ with $I \leq B \cap S$ in two ways: On the one hand, there are $\begin{bmatrix} s \\ i \end{bmatrix}_q$ possibilities for the choice of $I \in \begin{bmatrix} S \\ i \end{bmatrix}_q$. By Lemma 3, there are $\lambda_i$ blocks $B$ such that $I \leq B$, which shows that $\#X$ equals the right hand side of Eq. (3). On the other hand, fixing a block $B$, the number of $i$-subspaces $I$ of $B \cap S$ is $\begin{bmatrix} \dim(B \cap S) \\ i \end{bmatrix}_q$. Summing over the possibilities for $j = \dim(B \cap S)$, we see that $\#X$ also equals the left hand side of Eq. (3). $\qquad\square$

The Mendelsohn equations can be read as a linear system of equations $Ax = b$ on the *intersection vector* $x = (\alpha_0, \alpha_1, \ldots, \alpha_k)$. The left $(t+1) \times (t+1)$ square part of the matrix $A = (\begin{bmatrix} j \\ i \end{bmatrix}_q)_{ij}$ ($i \in \{0, \ldots, t\}$ and $j \in \{0, \ldots, k\}$) is called the *upper triangular $q$-Pascal matrix*, which is known to be invertible with inverse matrix $\left( (-1)^{j-i} q^{\binom{j-i}{2}} \begin{bmatrix} j \\ i \end{bmatrix}_q \right)_{ij}$. Left-multiplication of the equation system with this inverse yields a parameterization of the intersection numbers $\alpha_0, \ldots, \alpha_t$ by $\alpha_{t+1}, \ldots, \alpha_k$. In this way, we get:

**Theorem 5** ([31, Theorem 2.6], $q$-analog of the Köhler equations [32, Satz 1]) *Let $\mathscr{D}$ be a $t$-$(v, k, \lambda)_q$ subspace design, $S$ a subspace of $V$ and $s = \dim(S)$. For $i \in \{0, \ldots, t\}$, a parametrization of the intersection number $\alpha_i$ by $\alpha_{t+1}, \ldots, \alpha_k$ is given by*

$$\alpha_i = \begin{bmatrix} s \\ i \end{bmatrix}_q \sum_{j=i}^{t} (-1)^{j-i} q^{\binom{j-i}{2}} \begin{bmatrix} s-i \\ j-i \end{bmatrix}_q \lambda_j$$

$$+ (-1)^{t+1-i} q^{\binom{t+1-i}{2}} \sum_{j=t+1}^{k} \begin{bmatrix} j \\ i \end{bmatrix}_q \begin{bmatrix} j-i-1 \\ t-i \end{bmatrix}_q \alpha_j.$$

## 3.4  Combinatorial Designs from Subspace Designs

By identifying the blocks of a subspace design with its set of points, we can construct combinatorial designs.

**Theorem 6** *Let* $\mathscr{D} = (V, \mathscr{B})$ *be a* $t$-$(v, k, \lambda)_q$ *subspace design. We define the following two combinatorial designs:*

- *Projective version: Let* $\mathscr{D}_p = \left( \begin{bmatrix} V \\ 1 \end{bmatrix}_q, \left\{ \begin{bmatrix} B \\ 1 \end{bmatrix}_q \mid B \in \mathscr{B} \right\} \right)$
- *Affine version: For any fixed hyperplane* $H \in \begin{bmatrix} V \\ v-1 \end{bmatrix}_q$ *let*

$$
\mathscr{D}_a = \left( \begin{bmatrix} V \\ 1 \end{bmatrix}_q \setminus \begin{bmatrix} H \\ 1 \end{bmatrix}_q, \left\{ \begin{bmatrix} B \\ 1 \end{bmatrix}_q \setminus \begin{bmatrix} H \\ 1 \end{bmatrix}_q \mid B \in \mathscr{B}, B \not\subseteq H \right\} \right)
$$

*If* $t = 2$, *then* $\mathscr{D}_p$ *is a combinatorial* $2$-$\left( \begin{bmatrix} v \\ 1 \end{bmatrix}_q, \begin{bmatrix} k \\ 1 \end{bmatrix}_q, \lambda \right)$ *design and* $\mathscr{D}_a$ *is a combinatorial* $2$-$(q^{v-1}, q^{k-1}, \lambda)$ *design. In the case* $q = 2$ *and* $t = 3$, $\mathscr{D}_a$ *is even a combinatorial* $3$-$(q^{v-1}, q^{k-1}, \lambda)$ *design.*

*Proof* Let $T$ be a 2-subset of $\begin{bmatrix} V \\ 1 \end{bmatrix}_q$. By the definition of $\mathscr{D}_p$, each block consists of the set of points of a certain $k$-subspace. Hence, any block containing the elements of $T$ has to contain all the points in the 2-subspace $\langle T \rangle$. By the design property of $\mathscr{D}$, there are exactly $\lambda$ blocks of $\mathscr{D}$ (and hence of $\mathscr{D}_p$) with this property.

   In the affine case, the proof is similar as by definition, any block of $\mathscr{D}_a$ containing a 2-subset $T$ of $\begin{bmatrix} V \\ 1 \end{bmatrix}_q \setminus \begin{bmatrix} H \\ 1 \end{bmatrix}_q$ has to contain the points in the affine line $\langle T \rangle \setminus \begin{bmatrix} H \\ 1 \end{bmatrix}_q$, too, and each block $B_a$ of $\mathscr{D}_a$ uniquely determines the corresponding block $B = \langle B_a \rangle$ of $\mathscr{D}$. For $q = 2$ and $t = 3$, a triple $T$ of points in $\begin{bmatrix} V \\ 1 \end{bmatrix}_q \setminus \begin{bmatrix} H \\ 1 \end{bmatrix}_q$ cannot be collinear, as affine lines consist of $q = 2$ points only. Hence $\dim(\langle T \rangle) = 3$ and the claim follows in an analogous way.                                                                             ☐

   The key of the above proof is that $\dim(\langle T \rangle)$ is uniquely determined independently of the choice of the $t$-subset $T$. This shows that the preconditions cannot be relaxed, as in the projective version, this property is not true any more for $t \geq 3$ (triples of points may be collinear or not), and in the affine situation the only exception is the special case $q = 2$, $t = 3$ of Theorem 6 (where triples cannot be collinear). The projective version of Theorem 6 is already mentioned in [33, Sect. 1]. For the case of Steiner systems, the affine version is found in [4, Theorems 7 and 8].

## 3.5  On Automorphism Groups of Designs

As the definition of a subspace design can entirely be expressed in terms of the subspace lattice $\mathscr{L}(V)$, the design property is invariant under the action of $\mathrm{Aut}(\mathscr{L}(V))$. Thus, $\mathrm{Aut}(\mathscr{L}(V))$ provides a notion of isomorphism of designs: Two designs

$\mathscr{D} = (V, \mathscr{B})$ and $\mathscr{D}' = (V, \mathscr{B}')$ on $V$ are called *isomorphic* if and only if they are contained in the same orbit of $\mathrm{Aut}(\mathscr{L}(V))$ or equivalently, if there exists an $\alpha \in \mathrm{Aut}(\mathscr{L}(V))$ such that $\alpha(\mathscr{B}) = \mathscr{B}'$. The stabilizer of $\mathscr{B}$ with respect to the action of $\mathrm{Aut}(\mathscr{L}(V))$ is called the *automorphism group* of $\mathscr{D}$, that is

$$\mathrm{Aut}(\mathscr{D}) = \mathrm{Aut}(\mathscr{L}(V))_{\mathscr{B}} = \{\alpha \in \mathrm{Aut}(\mathscr{L}(V)) \mid \alpha(\mathscr{B}) = \mathscr{B}\}.$$

For any subgroup $G$ of $\mathrm{Aut}(\mathscr{D})$, $\mathscr{D}$ is called *G-invariant*.

By Formula (1), the question for the existence of a $G$-invariant $t$-$(v, k, \lambda)_q$ design $\mathscr{D} = (V, \mathscr{B})$ only depends on the conjugacy class of $G$ in $\mathrm{Aut}(\mathscr{L}(V))$. Hence in order to choose possible groups of automorphisms for a $t$-$(v, k, \lambda)_q$ design, it is sufficient to take only one representative of each conjugacy class of subgroups of $\mathrm{Aut}(\mathscr{L}(V))$.

From the orbit-stabilizer-theorem of group actions, see [34, Sect. 1.2], we also get a one-to-one correspondence of the set of isomorphic designs of $\mathscr{B}$ with the set of cosets of the automorphism group of $\mathscr{B}$ within $\mathrm{Aut}(\mathscr{L}(V))$. More formally, the mapping

$$\mathrm{Aut}(\mathscr{L}(V))(\mathscr{B}) \to \mathrm{Aut}(\mathscr{L}(V))/\mathrm{Aut}(\mathscr{L}(V))_{\mathscr{B}}, \quad \alpha(\mathscr{B}) \mapsto \alpha\,\mathrm{Aut}(\mathscr{L}(V))_{\mathscr{B}}$$

is a bijection.

In chapter "Computational Methods in Subspace Designs" on computer construction of subspace designs it is explained in detail how to construct subspace designs with a prescribed group of automorphisms by computer. This is the so called *Kramer-Mesner method*.

**Theorem 7** (Kramer, Mesner [35]) *Let $G \leq \mathrm{P\Gamma L}(V)$. There exists a G-invariant $t$-$(v, k, \lambda)_q$ design if and only if there is a 0/1-vector $x$ satisfying*

$$A_{t,k}^G \cdot x = \begin{pmatrix} \lambda \\ \vdots \\ \lambda \end{pmatrix}. \tag{4}$$

Details on the matrix $A_{t,k}^G$ can be found in the chapter "Computational Methods in Subspace Designs". In general, our aim in that chapter is to prescribe certain groups of automorphisms which arise as subgroups $G$ of the general linear group $\mathrm{GL}(V)$. The corresponding images $\bar{G}$ in $\mathrm{PGL}(V) \leq \mathrm{Aut}(\mathscr{L}(V))$ are the groups to prescribed.

When it comes to the question which groups may occur as automorphism groups of designs, one group stands out, which is the *normalizer of a Singer cycle group*: At present, most of the successful computer searches for subspace designs have been performed by prescribing this group or a subgroup of small index, see Sect. 4.

Let $\phi \in \mathrm{GL}(V)$ and $\langle\phi\rangle = \{\mathrm{id}_V, \phi, \phi^2, \phi^3, \ldots\}$ be the cyclic subgroup of $\mathrm{GL}(V)$ generated by $\phi$. By Cayley–Hamilton, the minimal polynomial of $\phi$ has at most degree $v$, implying that the linear hull of $\langle\phi\rangle$ has at most dimension $v$. As $\mathbf{0} \notin \langle\phi\rangle$, we get that $\langle \phi \rangle$ contains at most $q^v - 1$ distinct elements. In other words, the element order of $\phi$ is at most $q^v - 1$.

An element of $\mathrm{GL}(V)$ of order $q^v - 1$ is called a *Singer cycle*, and a cyclic subgroup of order $q^v - 1$ is called a *Singer cycle group*. Indeed, Singer cycles and Singer cycle groups do always exist: To see this, we equip $V$ with a field multiplication, by setting $V = \mathbb{F}_{q^v}$. For any primitive element $\alpha$ of $\mathbb{F}_{q^v}$ (that is, $\alpha$ is a generator of the cyclic unit group $\mathbb{F}_{q^v}^\times$), $\sigma : V \to V$, $\mathbf{x} \mapsto \alpha \mathbf{x}$ is an $\mathbb{F}_q$-linear mapping of order $\#\mathbb{F}_q^\times = q^v - 1$ and thus a Singer cycle.

To get a matrix representation of $\sigma$ in $\mathrm{GL}(v, q)$, we can proceed as follows: Let $f = a_0 + a_1 x + a_2 x^2 + \ldots + a_{n-1}x^{n-1} + x^n \in \mathbb{F}_q[x]$ be the minimal polynomial of $\alpha$ over $\mathbb{F}_q$. Then $f$ is a primitive irreducible polynomial of degree $v$. With respect to the $\mathbb{F}_q$-basis $(1, \alpha, \alpha^2, \ldots, \alpha^{v-1})$ of $V = \mathbb{F}_{q^v}$, the transformation matrix of $\sigma$ is given by the *companion matrix* of $f$, which is

$$
\begin{pmatrix}
0 & 1 & 0 & \ldots & 0 \\
0 & 0 & 1 & \ldots & 0 \\
\vdots & \vdots & \vdots & \ddots & \vdots \\
0 & 0 & 0 & \ldots & 1 \\
-a_0 & -a_1 & -a_2 & \ldots & -a_{v-1}
\end{pmatrix}.
$$

By [36, Sect. 7], any two Singer cycle subgroups of $\mathrm{GL}(V)$ are conjugate. We point out that, in general, this is not true for Singer cycles: For example, the Singer cycles in $\mathrm{GL}(2, 3)$ and $\mathrm{GL}(3, 2)$ both fall into 2 conjugacy classes.[3]

The Galois group $\mathrm{Gal}(\mathbb{F}_{q^v}/\mathbb{F}_q)$ of the field extension $\mathbb{F}_{q^v}/\mathbb{F}_q$, i.e. the group of all field automorphisms of $\mathbb{F}_{q^v}$ element-wise fixing $\mathbb{F}_q$, is known to be cyclic of order $v$ and generated by the mapping $F : x \mapsto x^q$. $F$ is called *Frobenius automorphism*. It is easily seen that $F$ normalizes the Singer cycle group $S = \langle \sigma \rangle$, showing that $\langle \sigma, F \rangle$ is contained in the normalizer $N_{\mathrm{GL}(V)}(S)$ of $S$ in $\mathrm{GL}(V)$. In fact, both groups are equal [36, Satz 7.3], so

$$
N_{\mathrm{GL}(V)}(S) = \langle \sigma, F \rangle = \mathrm{Gal}(\mathbb{F}_{q^v}/\mathbb{F}_q) \rtimes S.
$$

In particular, the normalizer $N_{\mathrm{GL}(V)}(S)$ is of order $v(q^v - 1)$ and the quotient of $S$ in its normalizer is cyclic of order $v$.

In the following, $S(v, q)$ (and $N(v, q)$) will denote the conjugacy class of all (normalizers of) Singer cycle groups in $\mathrm{GL}(v, q)$.

---

[3]For a singer cyclic subgroup $S$, we have $[N_{\mathrm{GL}(V)}(S) : S] = v$ (see below), so under the action of $N_{\mathrm{GL}(V)}(S)$, the $\varphi(q^v - 1)$ Singer cycles in $S$ fall into $\frac{\varphi(q^v-1)}{v}$ orbits of length $v$ ($\varphi$ denoting the Euler's phi function). From this, we get that the number of conjugacy classes of Singer cycles in $\mathrm{GL}(v, q)$ equals $\frac{\varphi(q^v-1)}{v}$.

## 4 Examples

In order to represent a $G$-invariant design it is sufficient to list $G$ by its generators and the set of representatives of the selected orbits of $G$ on $\begin{bmatrix} V \\ k \end{bmatrix}$. For a compact representation we will write all $\alpha \times \beta$ matrices $X$ over $\mathbb{F}_q$ with entries $x_{i,j}$, whose indices are numbered from 0, as vectors of integers

$$\left[ \sum_j x_{0,j} q^j, \ldots, \sum_j x_{\alpha-1,j} q^j \right].$$

*Example 5* The first nontrivial simple subspace design for $t = 3$ was constructed by Braun, Kerber, and Laue [20]. It has parameters $3$-$(8, 4, 11)_2$ and is invariant under the normalizer of the Singer cycle group $N(8, 2) = \langle \sigma, F \rangle$. The order of the group is 2040, generators are

$$\sigma = [2, 4, 8, 16, 29, 32, 64, 128], \quad F = [1, 4, 16, 19, 29, 64, 116, 205].$$

The design can be constructed from the following orbit representatives.

| | | | | |
|---|---|---|---|---|
| [28, 32, 64, 128], | [2, 32, 64, 128], | [10, 32, 64, 128], | [6, 32, 64, 128], | [22, 32, 64, 128], |
| [5, 32, 64, 128], | [8, 16, 64, 128], | [6, 16, 64, 128], | [33, 16, 64, 128], | [3, 16, 64, 128], |
| [7, 16, 64, 128], | [39, 16, 64, 128], | [47, 16, 64, 128], | [2, 48, 64, 128], | [38, 48, 64, 128], |
| [14, 48, 64, 128], | [37, 48, 64, 128], | [13, 48, 64, 128], | [38, 8, 64, 128], | [22, 8, 64, 128], |
| [54, 8, 64, 128], | [54, 40, 64, 128], | [1, 24, 64, 128], | [51, 24, 64, 128], | [39, 24, 64, 128], |
| [6, 56, 64, 128], | [35, 56, 64, 128], | [59, 36, 64, 128], | [34, 20, 64, 128], | [59, 20, 64, 128], |
| [25, 12, 64, 128], | [34, 28, 64, 128], | [43, 28, 64, 128], | [61, 42, 64, 128], | [21, 62, 64, 128], |
| [105, 50, 84, 128] | | | | |

In Tables 1, 2, 3 and 4 we list sets of parameters for subspace designs which are known to be realizable. The subspace designs were either by the Kramer-Mesner method and by the methods described in Sects. 5 and 7 or by combining existing subspace designs to get new ones, as described in Theorem 3 and Corollary 1.

Due to the existence of dual and supplementary designs we restrict the lists to $t$-$(v, k, \lambda)_q$ designs satisfying $1 < t < k \le v/2$ and $1 \le \lambda \le \lambda_{\max}/2$.

Given a prime power $q$ and integers $0 < t < k < v$, the smallest positive integer value of $\lambda$ for which $t$-$(v, k, \lambda)_q$ is admissible is denoted by $\lambda_{\min}$. By Lemma 3, the set of all $\lambda$ such that $t$-$(v, k, \lambda)_q$ is admissible is given by $\{\lambda_{\min}, 2\lambda_{\min}, \ldots, \lambda_{\max}\}$.

As superscripts of the $\lambda$-values we give the earliest reference we are aware of. The references are encoded by the following letters:

**Table 1** Parameters of simple $t$-$(v, k, \lambda)_2$ designs known to be realizable

| $t$-$(v, k, \lambda)_q$ | $\lambda_{\min}$ | $\lambda_{\max}$ | $\lambda$ |
|---|---|---|---|
| $2$-$(6, 3, \lambda)_2$ | 3 | 15 | $3^a$, $6^b$. **Open:** – |
| $2$-$(7, 3, \lambda)_2$ | 1 | 31 | $3^c$, $4^a$, $5^c$, $6^a$, $7^d$, $8^a$, $9^b$, $10^c$, $11^b$, $12^c$, $14^c$, $15^b$. **Open:** 1, 2 |
| $2$-$(8, 3, \lambda)_2$ | 21 | 63 | $21^b$. **Open:** – |
| $2$-$(8, 4, \lambda)_2$ | 7 | 651 | $7^e$, $14^e$, $21^b$, $35^f$, $49^e$, $56^g$, $63^*$, $70^g$, $84^*$, $91^b$, $98^e$, $105^g$, $112^e$, $126^g$, $133^g$, $140^b$, $147^*$, $154^e$, $161^g$, $168^*$, $175^b$, $189^*$, $196^g$, $203^e$, $210^b$, $217^h$, $231^b$, $245^g$, $252^*$, $259^e$, $266^g$, $273^*$, $280^f$, $294^*$, $301^b$, $308^*$, $315^g$. **Open:** 28, 42, 77, 119, 182, 224, 238, 287, 322 |
| $3$-$(8, 4, \lambda)_2$ | 1 | 31 | $11^a$, $15^i$. **Open:** 1, ..., 10, 12, 13, 14 |
| $2$-$(9, 3, \lambda)_2$ | 1 | 127 | $7^e$, $12^e$, $19^e$, $21^g$, $22^g$, $24^e$, $31^e$, $36^e$, $42^g$, $43^g$, $48^e$, $49^m$, $55^e$, $60^e$, $63^g$ |
| $2$-$(9, 4, \lambda)_2$ | 7 | 2667 | $21^f$, $63^f$, $84^f$, $126^f$, $147^f$, $189^f$, $210^f$, $252^f$, $273^f$, $315^f$, $336^f$, $378^f$, $399^f$, $441^b$, $462^f$, $504^f$, $525^f$, $567^f$, $588^f$, $630^b$, $651^f$, $693^f$, $714^f$, $756^f$, $777^f$, $819^b$, $840^f$, $882^f$, $889^h$, $903^f$, $945^f$, $966^f$, $1008^f$, $1029^f$, $1071^f$, $1092^f$, $1134^f$, $1155^f$, $1197^f$, $1218^f$, $1260^b$, $1281^f$, $1323^f$ |
| $3$-$(9, 4, \lambda)_2$ | 21 | 63 | **Open:** 21 |
| $2$-$(10, 3, \lambda)_2$ | 3 | 255 | $15^b$, $30^b$, $45^k$, $60^j$, $75^j$, $90^j$, $105^j$, $120^b$ |
| $2$-$(10, 4, \lambda)_2$ | 5 | 10795 | $595^*$, $1020^*$, $1615^*$, $1785^*$, $1870^*$, $2040^*$, $2635^*$, $3060^*$, $3570^*$, $3655^*$, $4080^*$, $4165^*$, $4675^*$, $5100^*$, $5355^*$ |
| $3$-$(10, 4, \lambda)_2$ | 1 | 127 | – |
| $2$-$(10, 5, \lambda)_2$ | 15 | 97155 | $765^*$, $4590^*$, $5355^*$, $6885^*$, $7650^*$, $9180^*$, $9945^*$, $2295^*$, $3060^*$, $11475^*$, $12240^*$, $13770^*$, $14535^*$, $16065^*$, $16830^*$, $18360^*$, $19125^*$, $20655^*$, $21420^*$, $22950^*$, $23715^*$, $25245^*$, $26010^*$, $27540^*$, $28305^*$, $29835^*$, $30600^*$, $32130^*$, $32385^h$, $32895^*$, $34425^*$, $35190^*$, $36720^*$, $37485^*$, $39015^*$, $39780^*$, $41310^*$, $42075^*$, $43605^*$, $44370^*$, $45900^*$, $46665^*$, $48195^*$ |
| $3$-$(10, 5, \lambda)_2$ | 21 | 63 | **Open:** 21 |
| $2$-$(11, 3, \lambda)_2$ | 7 | 511 | $7^d$, $245^e$, $252^e$ |
| $2$-$(11, 4, \lambda)_2$ | 35 | 43435 | – |
| $2$-$(11, 5, \lambda)_2$ | 5 | 788035 | $43435^*$, $74460^*$, $117895^*$, $130305^*$, $136510^*$, $148920^*$, $192355^*$, $223380^*$, $260610^*$, $266815^*$, $297840^*$, $304045^*$, $341275^*$, $372300^*$, $390915^*$ |
| $2$-$(12, 3, \lambda)_2$ | 3 | 1023 | – |
| $2$-$(12, 4, \lambda)_2$ | 7 | 174251 | – |
| $2$-$(12, 5, \lambda)_2$ | 465 | 6347715 | – |
| $2$-$(12, 6, \lambda)_2$ | 31 | 53743987 | $2962267^*$, $5078172^*$, $8040439^*$, $8886801^*$, $9309982^*$, $10156344^*$, $13118611^*$, $15234516^*$, $17773602^*$, $18196783^*$, $20312688^*$, $20735869^*$, $23274955^*$, $25390860^*$, $26660403^*$ |
| $3$-$(12, k, \lambda)_2$ | | | – |
| $2$-$(13, 3, \lambda)_2$ | 1 | 2047 | $1^l$, $2^q$, ..., $6^q$, $7^d$, $8^q$ ..., $1023^q$. **Open:** – |

**Table 2** Parameters of simple $t$-$(v, k, \lambda)_3$ designs known to be realizable

| $t$-$(v, k, \lambda)_q$ | $\lambda_{min}$ | $\lambda_{max}$ | $\lambda$ |
|---|---|---|---|
| 2-$(6, 3, \lambda)_3$ | 4 | 40 | $8^e, 12^e, 16^f, 20^b$. **Open**: 4 |
| 2-$(7, 3, \lambda)_3$ | 1 | 121 | $5^c, 6^b, \ldots, 12^b, 13^n, 14^b, \ldots, 40^b, 41^e, 42^b, \ldots, 60^b$. **Open**: 1, …, 4 |
| 2-$(8, 3, \lambda)_3$ | 52 | 364 | $52^g, 104^f, 156^f$. **Open**: – |
| 2-$(8, 4, \lambda)_3$ | 13 | 11011 | $91 \cdot 5^*, 91 \cdot 6^*, \ldots, 91 \cdot 60^*$ |
| 2-$(10, 3, \lambda)_3$ | 4 | 3280 | $1640^o$ |
| 2-$(11, 3, \lambda)_3$ | 13 | 9841 | $13^n$ |
| 2-$(13, 3, \lambda)_3$ | 1 | 88573 | $13^n$ |

**Table 3** Parameters of simple $t$-$(v, k, \lambda)_4$ designs known to be realizable

| $t$-$(v, k, \lambda; q)$ | $\lambda_{min}$ | $\lambda_{max}$ | $\lambda$ |
|---|---|---|---|
| 2-$(6, 3, \lambda)_4$ | 5 | 85 | $10^e, 15^e, 25^e, 30^e, 35^e$. **Open**: 5, 40 |
| 2-$(7, 3, \lambda)_4$ | 1 | 341 | $21^p$ |
| 2-$(8, 4, \lambda)_4$ | 21 | 93093 | $5733^*$ |

**Table 4** Parameters of simple $t$-$(v, k, \lambda)_5$ designs known to be realizable

| $t$-$(v, k, \lambda; q)$ | $\lambda_{min}$ | $\lambda_{max}$ | $\lambda$ |
|---|---|---|---|
| 2-$(6, 3, \lambda)_5$ | 6 | 78 | $78^o$. **Open**: 6, 12, …, 72 |
| 2-$(7, 3, \lambda)_5$ | 1 | 781 | $31^n$ |
| 2-$(8, 4, \lambda)_5$ | 31 | 508431 | $20181^*$ |
| 2-$(10, 3, \lambda)_5$ | 6 | 97656 | $48828^o$ |

| | |
|---|---|
| a: | Braun, Kerber, Laue (2005) [20], |
| b: | Braun (2005) [37], |
| c: | Miyakawa, Munemasa, Yoshiara (1995) [38], |
| d: | Thomas (1987) [17], |
| e: | Braun (2015) presentation at ALCOMA 2015, |
| f: | S. Braun (2010) [39], |
| g: | S. Braun (2009) [40], |
| h: | Kiermaier, Laue, Wassermann (2016) [41], |
| i: | Braun (2013) [42], |
| j: | Braun (2011) presentation at Fq10, |
| k: | Braun (2016) [43], |
| l: | Braun, Etzion, Östergård, Vardy, Wassermann (2016) [44] |
| m: | Braun (2017) [45], |
| n: | Suzuki (1992) [19], |
| o: | Braun, Kiermaier, Kohnert, Laue (2017)[46], |
| p: | Suzuki (1990) [18], |
| q: | Braun, Wassermann (2017) [47], |
| *: | Kiermaier, Laue (2015) [25], see Theorem 3 and Corollary 1 |

In case there are only few remaining open cases, we list the $\lambda$-values in question.

## 4.1 Special Subspace Designs

Designs for $t = 2$

A considerable part of the literature on combinatorial designs studies *balanced incomplete block designs* (BIBD) which are 2-$(v, k, \lambda)$ designs. The number of blocks is denoted with $b = \lambda_0$ and each point appears in $r = \lambda_1$ blocks. Let $M$ be the $v \times b$ incidence matrix of the points and the blocks of a BIBD.

The results in the following theorem are classical and can be found e.g. in [2, Chap. II, Sects. 1 and 2].

**Theorem 8** *For a* 2-$(v, k, \lambda)$ *design it holds*

*(a) $bk = vr$,*
*(b) $b \geq \mathrm{rank}(M) = v$. (Fisher's inequality)*

From the definitions of $b$ and $r$ it follows immediately that the $q$-analog of equation (a) for 2-$(v, k, \lambda)_q$ subspace design is

$$[k]_q b = [v]_q r.$$

The incidence matrix $M$ between the points, i.e. the 1-subspaces, and the blocks of the designs has size $[v]_q \times b$. The $q$-analog of Fisher's inequality is

$$b \geq \mathrm{rank}(M) = [v]_q,$$

see [15, 48].

Symmetric Designs

If a (combinatorial) 2-$(v, k, \lambda)$ design attains the bound in Fisher's inequality, i.e. if $v = b$, the design is called *symmetric*. One consequence is that two different blocks of a symmetric 2-$(v, k, \lambda)$ design always intersect in exactly $\lambda$ points, compare [2, Chap. II, Sect. 3]. That is, for $B \in \mathscr{B}$, $\alpha_\lambda(B)$ is the only nonzero intersection number. Symmetric designs are especially interesting because of their relation to finite projective planes.

It was shown already in [49] that symmetric subspace designs are always trivial. To be precise, only the complete design $(V, \begin{bmatrix} V \\ v-1 \end{bmatrix}_q)$ is symmetric, the intersection of two different blocks always has dimension $\lambda$. We note however, that the resulting combinatorial design of a symmetric subspace design is in general nontrivial.

Quasi-symmetric Designs

As seen above, symmetric designs are not interesting in the context of subspace designs. Thus, one might relax the condition that two blocks always intersect in a $\lambda$-dimensional subspace and allow two intersection numbers. Combinatorial designs with this property are called *quasi-symmetric*. However, we are not aware of any work about quasi-symmetric subspace designs.

Signed Designs

Up to now, designs were either simple or non-simple, depending if the multiplicities of all blocks of the design are one or larger than one. If one allows the multiplicity of block to be also negative, the design is called a *signed design*. In [50] it is shown that all admissible designs are realizable as signed designs. This is the $q$-analog of a result by Wilson [51], which is a key ingredient in the proof of Keevash [52].

Generalized Subspace Designs

In [53, 54], Guruswami et al. give a more general definition of subspace designs. They define an $(s, \lambda)$-*subspace design* to be a collection of subspaces $B_1, \ldots, B_b$ of $V$ such that for every $t$-subspace $T \leq V$ at most $\lambda$ blocks $B_i$ intersect $T$ nontrivially.

## 4.2 Asymptotic Existence Results

For combinatorial designs, Teirlinck [55] showed that for all integers $t > 0$ there exist simple $t$-designs for $v, k, \lambda$ sufficiently large. In 2013, Fazeli, Lovett and Vardy [56] succeeded in proving a similar result for subspace designs, with the difference that the proof in [55] is constructive, but the proof in [56] is not.

**Theorem 9** (Fazeli, Lovett, Vardy) *Simple nontrivial* $t$-$(v, k, \lambda)_q$ *designs exist for all $q$ and $t$, and all $k > 12(t + 1)$ provided that $v \geq ckt$ for a large enough absolute constant $c$. Moreover, these $t$-$(v, k, \lambda)_q$ designs have at most $q^{12(t+1)v}$ blocks.*

In 2014, the long-standing question, whether Steiner systems over sets exist for all integers $t > 0$, could be answered affirmatively in a celebrated result by Keevash [52]. The non-constructive proof relies heavily on an asymptotic approximation by Frankl and Rödl [57].

For the time being, there is not such a result for $q$-Steiner systems. But in [57] the asymptotic approximation result is also shown in the subspace case.

**Theorem 10** (Frankl-Rödl) *Let $0 < t < k$ be integers. Then for $v \to \infty$ there exists a set $\mathscr{B} \subset \begin{bmatrix} V \\ k \end{bmatrix}$ such that*

$$\#\mathscr{B} = (1 - o(1)) \begin{bmatrix} v \\ t \end{bmatrix}_q \Big/ \begin{bmatrix} k \\ t \end{bmatrix}_q$$

*and* $\dim(K' \cap K'') < t$ *for all distinct* $K', K'' \in \mathscr{B}$.

# 5 Infinite Series of Subspace Designs

## 5.1 On Suzuki's Construction

In 1987 Thomas [17] constructed the first nontrivial simple *t*-subspace designs for $t = 2$. More precisely, for all $v \geq 7$ with $(v, 4!) = 1$ a Singer-invariant 2-$(v, 3, 7)_2$ design was constructed.

Suzuki [18, 19] extended Thomas' family using the following construction: For any natural number $r \geq 2$ we define the following map:

$$L_r : \begin{bmatrix} V \\ 2 \end{bmatrix} \to \mathscr{L}(V), \quad T \mapsto \langle b_0 \cdot b_1 \cdot \ldots \cdot b_{r-1} \mid \{b_0, \ldots, b_{r-1}\} \subset T \rangle.$$

The image of $L_r$ is the set of blocks

$$\mathscr{B}_r := \left\{ L_r(T) \mid T \in \begin{bmatrix} V \\ 2 \end{bmatrix} \right\}.$$

Suzuki proved that for $(v, (2r)!) = 1$ the mapping $T \mapsto L_r(T)$ defines an embedding from $\begin{bmatrix} V \\ 2 \end{bmatrix}$ into $\begin{bmatrix} V \\ r+1 \end{bmatrix}$ implying $\#\mathscr{B}_r = \begin{bmatrix} v \\ 2 \end{bmatrix}_q$. Furthermore, if $\mathscr{B}_r$ becomes a 2-$(v, r + 1, \lambda)_q$ design it has to satisfy $\lambda = \begin{bmatrix} r+1 \\ 2 \end{bmatrix}_q$. For $r = 2$, Suzuki could prove that $\mathscr{B}_2$ defines a design. In addition, Abe and Yoshiara [58] determined a nontrivial group of automorphisms of Suzuki's design.

**Theorem 11** (Suzuki, Abe, Yoshiara) *For any prime power q and any integer $v \geq 7$ satisfying $(v, 4!) = 1$ the set $\mathscr{B}_2$ of blocks defines an $N(v, q)$-invariant 2-$(v, 3, q^2 + q + 1)_q$ design.*

Abe and Yoshiara also tried to generalize Suzuki's construction to values $r > 2$. But so far, computer experiments for $q = 2$ and small parameters failed for $r \geq 3$. The only exception is the case $v = 7$ and $r = 3$ which yields a 2-$(7, 4, 35)_2$ design $\mathscr{B}_3$, the dual design to Thomas' 2-$(7, 3, 7)_2$ design.

## 5.2 On Itoh's Construction

In 1998 Itoh [59] gave a construction of a 2-$(\ell m, 3, \lambda)_q$ design invariant under the special linear group $G = \mathrm{SL}(m, q^\ell) \leq \mathrm{GL}(\ell m, q)$ for all $m \geq 3$, assuming the existence of certain $S(\ell, q)$-invariant 2-$(\ell, 3, \lambda)_q$ designs.

The major step Itoh did was the examination of the matrix $A_{2,3}^G$. The structure of this matrix turned out to be of the following shape consisting of four blocks of columns

$$A_{2,3}^{\text{SL}(m,q^\ell)} = \left[ \begin{array}{c|ccc} & \begin{matrix} a & & 0 \\ & \ddots & \\ 0 & & a \end{matrix} & 0 & 0 \\ A_{2,3}^{S(\ell,q)} & & & \\ \hline 0 & X & Y & Z \end{array} \right],$$

where $X, Y, Z$ are submatrices with exactly one row defined by

$$X = \begin{cases} ((q+1)^2, \ldots, (q+1)^2, q+1), & \text{if } 2 \mid \ell; \\ ((q+1)^2, \ldots, (q+1)^2), & \text{else,} \end{cases}$$

$$Y = \begin{cases} (q(q+1)(q^3-1), \ldots, q(q+1)(q^3-1), q(q^2-1)), & \text{if } 3 \mid \ell; \\ (q(q+1)(q^3-1), \ldots, q(q+1)(q^3-1)), & \text{else,} \end{cases}$$

$$Z = \begin{cases} (q^{2\ell-2}, \ldots, q^{2\ell-2}), & \text{if } m = 3; \\ (q^{2\ell-2} \frac{q^{(m-2)\ell}-1}{q-1}), & \text{if } m > 3. \end{cases}$$

Additionally, we have $a = q^{\ell-2}(q^{(m-1)\ell}-1)/(q-1)$.

Now, the following extension result is immediate from inspection of the matrix. If we have a selection of columns of $A_{2,3}^{S(\ell,q)}$ with constant row sum we can extend this to a selection of columns of $A_{2,3}^G$ with constant row sum by taking the right amount of columns of the third block of columns.

**Lemma 7** (Itoh) *If there exists a $S(\ell, q)$-invariant 2-$(\ell, 3, \lambda)_q$ design with*

$$\lambda = q(q+1)(q^3-1)s + q(q^2-1)t$$

*for some integer $s \geq 0$ and*

$$t \in \begin{cases} \{0, 1\}, & \text{if } 3 \mid \ell; \\ \{0\}, & \text{else,} \end{cases}$$

*then there exists an $\text{SL}(m, q^\ell)$-invariant 2-$(\ell m, 3, \lambda)_q$ design.*

Moreover, Itoh noticed that this lemma is also true for the $\text{GL}(m, q^\ell)$ action. Therefore $S(\ell, q)$-invariant designs also imply families of $\text{GL}(m, q^\ell)$-invariant designs.

Since the publication of Itoh's paper several $S(v, q)$-invariant subspace designs with index values $\lambda$ of the required form for Itoh's lemma have been constructed by computer. These are listed in the sequel.

In the binary case we need a 2-$(\ell, 3, \lambda)_q$ design with index

$$\lambda = 2(2+1)(2^3-1)s + 2(2^2-1)t = 42s + 6t$$

and over the ternary field we need

$$\lambda = 3(3+1)(3^3-1)s + 3(3^2-1)t = 312s + 24t,$$

**Table 5** Simple subspace designs from Itoh's lemma

| $t$-$(v, k, \lambda)_q$ | Parameters | Group of automorphisms |
|---|---|---|
| $2$-$(8m, 3, 42)_2$ | $m \geq 3$ | $\mathrm{SL}(m, 2^8)$ |
| $2$-$(9m, 3, 42s)_2$ | $m \geq 3, 1 \leq s \leq 2$ | $\mathrm{SL}(m, 2^9)$ |
| $2$-$(10m, 3, 210)_2$ | $m \geq 3$ | $\mathrm{SL}(m, 2^{10})$ |
| $2$-$(13m, 3, 42s)_2$ | $m \geq 3, 1 \leq s \leq 2$ | $\mathrm{SL}(m, 2^{13})$ |
| $2$-$(8m, 3, 312)_2$ | $m \geq 3$ | $\mathrm{SL}(m, 3^8)$ |

where in both cases $s \geq 0$ and $t$ is 0 or 1 if $\ell$ is divisible by 3 or $t$ is 0 otherwise.

All of the following given groups of automorphisms of the corresponding $2$-$(\ell, 3, \lambda)_q$ designs contain the Singer cycle group $S(\ell, q)$ as subgroup and therefore the designs also admit this group as group of automorphisms.

- In [60] three disjoint $N(8, 2)$-invariant $2$-$(8, 3, 21)_2$ designs are given. Combining two of them yields an $N(8, 2)$-invariant design of the supplementary parameters $2$-$(8, 3, 42)_2$.
- In Table 1 there are $N(3, 2^3)$-invariant $2$-$(9, 3, \lambda)_2$ designs. The supplementary design of the $2$-$(9, 3, 43)_2$ design has parameters $2$-$(9, 3, 42 \cdot 2)_2$.
- Also in Table 1 there is an $N(10, 2)$-invariant $2$-$(10, 3, 45)_2$ design. The supplementary design has parameters $2$-$(10, 3, 42 \cdot 5)_2$.
- In Table 2 there is a $N(8, 3)$-invariant $2$-$(8, 3, 52)_3$ design. The supplementary design is has the parameters $2$-$(8, 3, 312)_3$.
- There exist at least 586 pairwise disjoint $N(13, 2)$-invariant subspace designs with parameters $2$-$(13, 3, 1)_2$ [44]. The union of these designs yields $2$-$(13, 3, 42s)_2$ designs for $1 \leq s \leq 13$.

By Itoh's Lemma 7 these designs turn immediately into new families of infinite series [43] (Table 5).

## 6  *q*-Analogs of Steiner Systems

Among subspace designs the case $\lambda = 1$ is of particular interest. Such a design is called *q-Steiner system*. For $t = 1$, the existence of $q$-Steiner systems is already answered by Lemma 4, such designs are called *spreads*.

It was a long standing open question whether $q$-Steiner systems exist at all for $t \geq 2$. There was already the conjecture that such structures do not exist [61]. Especially, the smallest nontrivial set parameters $2$-$(7, 3, 1)_2$ (an $\mathrm{STS}(7)_q$) attracted quite a lot of interest.

## 6.1   q-Steiner Systems in Dimension 13

While the existence of a $q$-analog of the Fano plane is still undecided, meanwhile the first $q$-Steiner system for $t \geq 2$ has been found, namely an $S(2, 3, 13)_2$, see [44]. For the construction, the authors used the method of Kramer and Mesner. They prescribed the normalizer $N(13, 2)$ of a Singer subgroup of $\mathrm{Aut}(\mathscr{L}(V))$ as a group of automorphisms. The group order is $106\,483$ and the Kramer-Mesner matrix $A_{2,3}^{N(13,2)}$ has size $105 \times 30\,705$.

With the algorithm *dancing links* by Knuth [62] meanwhile 1316 *mutually disjoint* solutions have been determined [47]. That is, among $1 \leq \lambda \leq 1316$ mutually disjoint designs, no block appears twice. Therefore, the union of their block sets gives a 2-$(13, 3, \lambda)_2$ design.

Together with their supplementary designs and noting that $\lambda_{\max} = 2047$, this implies that all possible design parameters

$$2\text{-}(13, 3, \lambda)_2, \quad 1 \leq \lambda \leq 2047,$$

are realizable.

## 6.2   q-Analog of the Fano Plane

Arguably, one of the most tantalizing open problems in this field is the question for the existence of a $q$-analog of the Fano plane which was asked already in 1972 by Ray-Chaudhuri [14]. Its existence is still open over any finite base field $\mathbb{F}_q$. The most important single case is the binary case $q = 2$, as it is the smallest one. Nonetheless, so far the binary $q$-analog of the Fano plane has withstood all computational or theoretical attempts for its construction or refutation.

Following the approach for other notorious putative combinatorial objects as, e.g., a projective plane of order 10 or a self-dual binary [72, 36, 16] code, the possible automorphisms of a binary $q$-analog of the Fano plane have been investigated in [63, 64]. As a result, its automorphism group is at most of order 2.

For the groups of order 2, the above result was achieved as a special case of a more general result on restrictions of the automorphisms of order 2 of a binary $q$-analog of Steiner triple systems [63, Theorem 2].

Finally, the combination of the results of [63, 64] yields

**Theorem 12** *The automorphism group of a binary $q$-analog of the Fano plane is either trivial or of order 2. In the latter case, up to conjugacy in $\mathrm{GL}(7, 2)$ the automorphism group is represented by*

$$\left\langle \begin{pmatrix} 0&1&0&0&0&0&0 \\ 1&0&0&0&0&0&0 \\ 0&0&0&1&0&0&0 \\ 0&0&1&0&0&0&0 \\ 0&0&0&0&0&1&0 \\ 0&0&0&0&1&0&0 \\ 0&0&0&0&0&0&1 \end{pmatrix} \right\rangle.$$

# 7 "Large Sets" of Subspace Designs

In 1861 Sylvester [65] asked if it is possible to partition the set of 455 3-subsets on 15 points into 13 *resolvable* 2-(15, 3, 1) designs. His question was answered affirmatively by Denniston [66] as late as in 1974. Later on, a partition of the *k*-subsets into disjoint (combinatorial) *t*-(*v*, *k*, λ) designs was sometimes called *packing*. In the 1970s the term *large set of disjoint designs* evolved, nowadays such a partition into designs is simply called a *large set of designs*.

In the *q*-analog situation, a *large set of subspace designs*—denoted by $LS_q[N](t, k, n)$—is a partition of the whole set $\begin{bmatrix} V \\ k \end{bmatrix}$ of all *k*-dimensional subspaces of *V* into *N* disjoint *t*-(*v*, *k*, λ)$_q$ designs. Obviously, the complete *t*-(*v*, *k*, $\lambda_{max}$)$_q$ design forms an $LS_q[N](t, k, v)$ large set for $N = 1$ which is called *trivial*.

The parameter *N* determines the index λ, since it must be

$$\lambda \cdot N = \begin{bmatrix} v - t \\ k - t \end{bmatrix} = \lambda_{max}.$$

If $N = 2$ we speak of a *halving*.

As for subspace designs, we call a set of parameters $LS_q[N](t, k, v)$ *admissible* if the parameters are admissible, i.e. if *N* divides $\lambda_{max}$ and if the design parameters *t*-(*v*, *k*, $\lambda_{max}/N$)$_q$ are admissible. Using the second equation in Lemma 3 it is easy to see that this is equivalent to

$$\begin{bmatrix} v - s \\ k - s \end{bmatrix}_q \equiv 0 \pmod{N} \text{ for } 0 \leq s \leq t.$$

If a large set with this parameters exists, it is called *realizable* (Fig. 3).

Many results for subspace designs also hold true for their large sets. The following results are from [25, 46].

**Lemma 8** *If $LS_q[N](t, k, v)$ is realizable, then for each divisor $d \mid N$, the parameters $LS_q[d](t, k, v)$ are realizable, too.*

**Lemma 9** *If there exists an $LS_q[N](t, k, v)$ for $t \geq 1$ then there exists*

(a) *the dual large set $LS_q[N](t, v - k, v)$;*
(b) *the reduced large set $LS_q[N](t - 1, k, v)$;*
(c) *the derived large set $LS_q[N](t - 1, k - 1, v - 1)$;*
(d) *a residual large set $LS_q[N](t - 1, k, v - 1)$.*

**Lemma 10** ([25, Corollary 4]) *If there exists an $LS_q[N](t, k - 1, v - 1)$ and an $LS_q[N](t, k, v - 1)$, then there exists an $LS_q(t, k, v)$.*

A group $G \leq \text{Aut}(\mathcal{L}(V))$ is called the *automorphism group* of the large set $\{\mathcal{D}_1, \mathcal{D}_2, \ldots, \mathcal{D}_N\}$ if

$$G = \text{Aut}(\mathcal{L}(V))_{\{\mathcal{D}_1, \mathcal{D}_2, \ldots, \mathcal{D}_N\}}$$

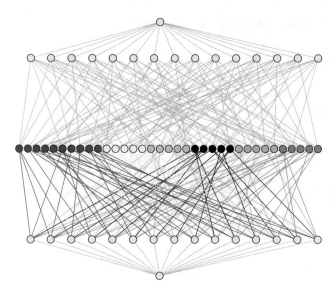

**Fig. 3** An $LS_2[7](1, 2, 4)$ in the subspace lattice of $\mathbb{F}_2^4$

**Table 6** Known parallelisms, given by their subspace design parameters

| $t$-$(v, k, \lambda)$ | Constraints | Source |
|---|---|---|
| $1$-$(2^i, 2, 1)_q$ | $i \geq 2$ | Denniston [69], Beutelspacher [69] |
| $1$-$(v, 2, 1)_q$ | $v \geq 3$ odd | Baker [70], Wettl [71] |
| $1$-$(6, 2, 1)_3$ | | Etzion, Vardy [72] |
| $1$-$(6, 3, 1)_2$ | | Hishida, Jimbo [73], Sarmiento [74] |

A large set $\{\mathscr{D}_1, \mathscr{D}_2, \ldots, \mathscr{D}_N\}$ is said to be uniformly-$G$ if the designs $\mathscr{D}_1, \mathscr{D}_2, \ldots, \mathscr{D}_N$ are $G$-invariant.

For ordinary combinatorial $t$-designs, large sets with $t = 1$ exist if and only if $k$ divides $v$ [67]. In the $q$-analog case, this question is wide open, as it includes the question for the existence of parallelisms in projective geometries: A large set of $1$-$(v, k, 1)_q$ designs (i.e. spreads) is known as a $(k-1)$-*parallelism* of the projective geometry $\mathrm{PG}(v-1, q)$. By admissibility, necessarily $k \mid v$. To our knowledge, the only known existence results are the following: If $v \geq 2$ is a power of 2, then all 1-parallelisms do exist [68, 69]. Furthermore, for $q = 2$ and $v$ even, all 1-parallelisms do exist [70, 71]. In [72], a 1-parallelism of $\mathrm{PG}(5, 3)$ is given. The only known parallelism for $k \geq 3$ is a 2-parallelism of $\mathrm{PG}(5, 2)$ [73, 74] (Table 6).

Chapter "Computational Methods in Subspace Designs" contains a summary of methods how to construct large sets of subspace designs by computer. For the time being, there are not many examples for $t \geq 2$ where this succeeded.

In [37] a 2-$(6, 3, 20)_3$ design and in [46] a 2-$(6, 3, 78)_5$ design has been presented which were detected by computer search. In both cases, the supplementary designs have exactly the same parameters. Thus, the design and it's supplementary designs form a large set for $N = 2$.

In [60] an $LS_2[3](2, 3, 8)$ was constructed. It is the first large set with $t \geq 2$ and $N \geq 3$. More large sets with the same parameters were constructed in [75]. Later on, in [41] an $LS_2[3](2, 4, 8)$ was presented (Table 7).

Table 8 contains the complete listing of these few large sets found by computer.

## 7.1 Infinite Series of Large Sets

Based on ideas of Teirlinck [76], Ajoodani-Namini and Khosrovshahi have developed a rich framework of recursive constructions of large sets of combinatorial designs [77–79]. For a survey see [80]. Recently, a $q$-analog version of parts of this theory was found [46]. The basic idea is to decompose the Graßmannian into *joins* of smaller large sets. In this way, given a suitable set of starting large sets, an infinite series of large sets with increasing values of $v$ and $k$ and constant values of $N$ and $t$. Principally, the method can be applied in a variety of situations. The problem is the current shortage of starting large sets on which the recursion can be founded.

As a detailed description of this theory is beyond the scope for this chapter, we confine ourselves to the presentation of the newly constructed large sets (and thus, subspace designs). The following theorems were tailored to fit the five known large sets from Table 8.

**Theorem 13** ([46, Theorem 6.4, Corollary 6.6]) *If there exists an* $LS_q[N](2, 3, 6)$, *then there exists an* $LS_q[N](2, k, v)$ *for all integers* $v$ *and* $k$ *with* $v \geq 6$, $v \equiv 2$ (mod 4), $3 \leq k \leq v - 3$ *and* $k \equiv 3$ (mod 4). *In particular, all these large sets exist for* $q \in \{3, 5\}$ *and* $N = 2$.

An overview of the resulting large sets is given in Table 9.

**Theorem 14** *Let* $q$ *and* $N$ *such that there exist large sets with parameters* $LS_2[N](2, 3, 8)$ *and* $LS_2[N](2, 4, 8)$. *Let* $v$ *and* $k$ *be integers with* $v \geq 8$ *and* $0 \leq k \leq v$ *such that*

*(a)* $v \equiv 2$ mod 6 *and* $k \equiv 3, 4, 5$ mod 6 *or*
*(b)* $v \equiv 3$ mod 6 *and* $k \equiv 4, 5$ mod 6 *or*
*(c)* $v \equiv 4$ mod 6 *and* $k \equiv 5$ mod 6.

**Table 7** Admissible large set parameters for small values. All possible values of $N > 1$ are given in brackets. Additionally, the design parameters for the largest given value of $N$ is listed. Large set parameters which are known to be are realizable are printed in boldface

| $LS_q[N](t, k, v)$ | $t\text{-}(v, k, \lambda)$ |
|---|---|
| $LS_2[5](2, 3, 6)$ | $2\text{-}(6, 3, 3)_2$ |
| $LS_2[31](2, 3, 7)$ | $2\text{-}(7, 3, 1)_2$ |
| $\mathbf{LS_2[3](2, 3, 8)}$ | $2\text{-}(8, 3, 21)_2$ |
| $\mathbf{LS_2[3, 31, 93](2, 4, 8)}$ | $2\text{-}(8, 4, 7)_2$ |
| $LS_2[31](3, 4, 8)$ | $3\text{-}(8, 4, 1)_2$ |
| $LS_2[127](2, 3, 9)$ | $2\text{-}(9, 3, 1)_2$ |
| $\mathbf{LS_2[3, 127, 381](2, 4, 9)}$ | $2\text{-}(9, 4, 7)_2$ |
| $LS_2[3](3, 4, 9)$ | $3\text{-}(9, 4, 21)_2$ |
| $LS_2[5, 17, 85](2, 3, 10)$ | $2\text{-}(10, 3, 3)_2$ |
| $LS_2[17, 127, 2159](2, 4, 10)$ | $2\text{-}(10, 4, 5)_2$ |
| $LS_2[127](3, 4, 10)$ | $3\text{-}(10, 4, 1)_2$ |
| $LS_2[73](2, 3, 11)$ | $2\text{-}(11, 3, 7)_2$ |
| $LS_2[17, 73, 1241](2, 4, 11)$ | $2\text{-}(11, 4, 35)_2$ |
| $LS_2[17](3, 4, 11)$ | $3\text{-}(11, 4, 15)_2$ |
| $LS_2[11, 31, 341](2, 3, 12)$ | $2\text{-}(12, 3, 3)_2$ |
| $LS_2[11, 31, 73, 341, 803, 2263, 24893](2, 4, 12)$ | $2\text{-}(12, 4, 7)_2$ |
| $LS_2[73](3, 4, 12)$ | $3\text{-}(12, 4, 7)_2$ |
| $LS_2[23, 89, 2047](2, 3, 13)$ | $2\text{-}(13, 3, 1)_2$ |
| $\mathbf{LS_3[2, 5, 10](2, 3, 6)}$ | $2\text{-}(6, 3, 4)_3$ |
| $LS_3[11, 121](2, 3, 7)$ | $2\text{-}(7, 3, 1)_3$ |
| $LS_3[7](2, 3, 8)$ | $2\text{-}(8, 3, 52)_3$ |
| $LS_3[7, 11, 77, 121, 847](2, 4, 8)$ | $2\text{-}(8, 4, 13)_3$ |
| $LS_3[11, 121](3, 4, 8)$ | $3\text{-}(8, 4, 1)_3$ |
| $LS_4[17](2, 3, 6)$ | $2\text{-}(6, 3, 5)_4$ |
| $LS_4[11, 31, 341](2, 3, 7)$ | $2\text{-}(7, 3, 1)_4$ |
| $LS_4[13](2, 3, 8)$ | $2\text{-}(8, 3, 105)_4$ |
| $LS_4[11, 13, 31, \ldots, 4433](2, 4, 8)$ | $2\text{-}(8, 4, 21)_4$ |
| $\mathbf{LS_5[2, 13, 26](2, 3, 6)}$ | $2\text{-}(6, 3, 6)_5$ |
| $LS_5[11, 71, 781](2, 3, 7)$ | $2\text{-}(7, 3, 1)_5$ |
| $LS_5[3, 7, 21](2, 3, 8)$ | $2\text{-}(8, 3, 186)_5$ |
| $LS_5[7, 11, \ldots, 71, \ldots, 16401](2, 4, 8)$ | $2\text{-}(8, 4, 31)_5$ |

**Table 8** Large sets of simple subspace designs by computer construction

| LS$[N]_q(t, k, v)$ | $G$ | $\#A^G_{t,k}$ |
|---|---|---|
| LS$[2]_3(2, 3, 6)$ | $\langle \sigma^2, F^2 \rangle$ | $25 \times 76$ |
| LS$[2]_3(2, 3, 6)$ | $S(5, 3) \times 1$ | $51 \times 150$ |
| LS$[2]_5(2, 3, 6)$ | $N(6, 5)$ | $53 \times 248$ |
| LS$[3]_2(2, 3, 8)$ | $S(8, 2)$ | $43 \times 381$ |
| LS$[3]_2(2, 4, 8)$ | $\langle \sigma^5, F^2 \rangle$ | $69 \times 1061$ |

*Then there exists an* LS$_q[N](2, k, v)$. *In particular, all these large sets exist for* $q = 2$ *and* $N = 3$.

An overview of the resulting large sets is given in Table 10.

# 8 Open Problems

The subject of subspace designs is still in its infancy and there are many questions which wait for an answer, see [81, 82]. The most tantalizing open problems seem to be:

- Is there a $q$-analog of the Fano plane? (subspace design with the parameters 2-$(7, 3, 1)_q$)
- Is there a geometric or algebraic description of the 2-$(13, 3, 1)_2$ designs?
- Find more $q$-analogs of Steiner systems!
- Find a concrete construction of a subspace design with $t \geq 4$!
- Do $q$-Steiner systems exist for all $t$, i.e. is there a $q$-analog of Keevash' theorem [52]?
- Are the parameters 2-$(7, 3, 2)_2$ realizable?
- Is LS$_2[5](2, 3, 6)$ realizable?
- Are the halvings LS$_q[2](2, 3, 6)$ realizable for odd $q > 5$?
- For combinatorial designs, the *halving conjecture* states that all admissible designs with $\lambda = \lambda_{\max}/2$ (i.e., halvings LS$[2](t, k, v)_q$) are realizable [83, Sect. 5]. For $t = 2$, the conjecture has been proven in [84]. Can anything be said about this conjecture in the $q$-analog case?

**Table 9** Admissibility and realizability of $\mathrm{LS}_q[2](2, k, v)$ for $q \in \{3, 5\}$. An integer value $k$ means $\mathrm{LS}_2[3](2, k, v)$ is realizable for these $k, v$. "-" means the parameter set is not admissible, and "?" means the parameter set is admissible but not known to be realizable

| k | | | | | | | | | | | | | | | | | v |
|---|---|---|---|---|---|---|---|---|---|---|---|---|---|---|---|---|---|
| | | | | | | | | | | | | | | | | 3 | 6 |
| | | | | | | | | | | | | | | | | - | 7 |
| | | | | | | | | | | | | | | | - | - | 8 |
| | | | | | | | | | | | | | | | - | - | 9 |
| | | | | | | | | | | | | | | 3 | ? | ? | 10 |
| | | | | | | | | | | | | | | - | ? | ? | 11 |
| | | | | | | | | | | | | | - | - | ? | ? | 12 |
| | | | | | | | | | | | | | - | - | - | ? | 13 |
| | | | | | | | | | | | | 3 | - | - | - | 7 | 14 |
| | | | | | | | | | | | | - | - | - | - | - | 15 |
| | | | | | | | | | | | - | - | - | - | - | - | 16 |
| | | | | | | | | | | | - | - | - | - | - | - | 17 |
| | | | | | | | | | | 3 | ? | ? | ? | 7 | ? | ? | 18 |
| | | | | | | | | | | - | ? | ? | ? | ? | ? | ? | 19 |
| | | | | | | | | | - | - | ? | ? | ? | ? | ? | ? | 20 |
| | | | | | | | | | - | - | - | ? | ? | ? | ? | ? | 21 |
| | | | | | | | | 3 | - | - | - | 7 | ? | ? | ? | 11 | 22 |
| | | | | | | | | - | - | - | - | - | ? | ? | ? | ? | 23 |
| | | | | | | | - | - | - | - | - | - | ? | ? | ? | ? | 24 |
| | | | | | | | - | - | - | ? | ? | - | - | - | - | ? | 25 |
| | | | | | | 3 | ? | ? | ? | 7 | - | - | - | 11 | ? | ? | 26 |
| | | | | | | - | ? | ? | ? | ? | - | - | - | - | ? | ? | 27 |
| | | | | | - | - | ? | ? | ? | - | - | - | - | - | ? | ? | 28 |
| | | | | | - | - | - | ? | ? | - | - | - | - | - | - | ? | 29 |
| | | | | 3 | - | - | - | 7 | - | - | - | 11 | - | - | - | 15 | 30 |
| | | | | - | - | - | - | - | - | - | - | - | - | - | - | - | 31 |
| | | | - | - | - | - | - | - | - | - | - | - | - | - | - | - | 32 |
| | | | - | - | - | - | - | - | - | - | - | - | - | - | - | - | 33 |
| | | 3 | ? | ? | ? | 7 | ? | ? | ? | 11 | ? | ? | ? | 15 | ? | ? | 34 |
| | | - | ? | ? | ? | ? | ? | ? | ? | ? | ? | ? | ? | ? | ? | ? | 35 |
| | - | - | ? | ? | ? | ? | ? | ? | ? | ? | ? | ? | ? | ? | ? | ? | 36 |
| | - | - | - | ? | ? | ? | ? | ? | ? | ? | ? | ? | ? | ? | ? | ? | 37 |
| 3 | - | - | - | 7 | ? | ? | ? | 11 | ? | ? | ? | 15 | ? | ? | ? | 19 | 38 |

**Table 10** Admissibility and realizability of $LS_2[3](2, k, v)$. An integer value $k$ means $LS_2[3](2, k, v)$ is realizable for these $k, v$. "-" means the parameter set is not admissible, and "?" means the parameter set is admissible but not known to be realizable

| 3 | 4 | 5 | 6 | 7 | 8 | 9 | 10 | 11 | 12 | 13 | 14 | 15 | 16 | 17 | 18 | 19 | 20 | $v$ |
|---|---|---|---|---|---|---|---|---|---|---|---|---|---|---|---|---|---|---|
| - |  |  |  |  |  |  |  |  |  |  |  |  |  |  |  |  |  | 6 |
| - |  |  |  |  |  |  |  |  |  |  |  |  |  |  |  |  |  | 7 |
| 3 | 4 |  |  |  |  |  |  |  |  |  |  |  |  |  |  |  |  | 8 |
| - | 4 |  |  |  |  |  |  |  |  |  |  |  |  |  |  |  |  | 9 |
| - | - | 5 |  |  |  |  |  |  |  |  |  |  |  |  |  |  |  | 10 |
| - | - | - |  |  |  |  |  |  |  |  |  |  |  |  |  |  |  | 11 |
| - | - | - | - |  |  |  |  |  |  |  |  |  |  |  |  |  |  | 12 |
| - | - | - | - |  |  |  |  |  |  |  |  |  |  |  |  |  |  | 13 |
| 3 | 4 | 5 | - | - |  |  |  |  |  |  |  |  |  |  |  |  |  | 14 |
| - | 4 | 5 | - | - |  |  |  |  |  |  |  |  |  |  |  |  |  | 15 |
| - | - | 5 | - | - | - |  |  |  |  |  |  |  |  |  |  |  |  | 16 |
| - | - | - | - | - | - |  |  |  |  |  |  |  |  |  |  |  |  | 17 |
| - | - | - | - | - | - | - |  |  |  |  |  |  |  |  |  |  |  | 18 |
| - | - | - | - | - | - | - |  |  |  |  |  |  |  |  |  |  |  | 19 |
| 3 | 4 | 5 | ? | ? | ? | 9 | 10 |  |  |  |  |  |  |  |  |  |  | 20 |
| - | 4 | 5 | ? | ? | ? | ? | 10 |  |  |  |  |  |  |  |  |  |  | 21 |
| - | - | 5 | ? | ? | ? | ? | ? | 11 |  |  |  |  |  |  |  |  |  | 22 |
| - | - | - | ? | ? | ? | ? | ? | ? |  |  |  |  |  |  |  |  |  | 23 |
| - | - | - | - | ? | ? | ? | ? | ? | ? |  |  |  |  |  |  |  |  | 24 |
| - | - | - | - | - | ? | ? | ? | ? | ? |  |  |  |  |  |  |  |  | 25 |
| 3 | 4 | 5 | - | - | - | 9 | 10 | 11 | ? | ? |  |  |  |  |  |  |  | 26 |
| - | 4 | 5 | - | - | - | - | 10 | 11 | ? | ? |  |  |  |  |  |  |  | 27 |
| - | - | 5 | - | - | - | - | - | 11 | ? | ? | ? |  |  |  |  |  |  | 28 |
| - | - | - | - | - | - | - | - | - | ? | ? | ? |  |  |  |  |  |  | 29 |
| - | - | - | - | - | - | - | - | - | - | ? | ? | ? |  |  |  |  |  | 30 |
| - | - | - | - | - | - | - | - | - | - | - | ? | ? |  |  |  |  |  | 31 |
| 3 | 4 | 5 | - | - | - | 9 | 10 | 11 | - | - | - | 15 | 16 |  |  |  |  | 32 |
| - | 4 | 5 | - | - | - | - | 10 | 11 | - | - | - | - | 16 |  |  |  |  | 33 |
| - | - | 5 | - | - | - | - | - | 11 | - | - | - | - | - | 17 |  |  |  | 34 |
| - | - | - | - | - | - | - | - | - | - | - | - | - | - | - |  |  |  | 35 |
| - | - | - | - | - | - | - | - | - | - | - | - | - | - | - | - |  |  | 36 |
| - | - | - | - | - | - | - | - | - | - | - | - | - | - | - | - |  |  | 37 |
| 3 | 4 | 5 | ? | ? | ? | 9 | 10 | 11 | ? | ? | ? | 15 | 16 | 17 | - | - |  | 38 |
| - | 4 | 5 | ? | ? | ? | ? | 10 | 11 | ? | ? | ? | ? | 16 | 17 | - | - |  | 39 |
| - | - | 5 | ? | ? | ? | ? | ? | 11 | ? | ? | ? | ? | ? | 17 | - | - | - | 40 |

# References

1. J. Tits, Sur les analogues algébriques des groupes semi-simples complexes, in *Colloque d'algèbre supérieure, tenu à Bruxelles du 19 au 22 décembre 1956*, Centre Belge de Recherches Mathématiques (Établissements Ceuterick, Louvain; Librairie Gauthier-Villars, Paris 1957), pp. 261–289
2. T. Beth, D. Jungnickel, H. Lenz, *Design Theory*, 2nd edn. (Cambridge University Press, Cambridge, 1999)
3. C.J. Colbourn, J.H. Dinitz, *Handbook of Combinatorial Designs*, 2nd edn., Discrete Mathematics and Its Applications (Chapman & Hall/CRC, Boca Raton, 2006)
4. M. Schwartz, T. Etzion, Codes and anticodes in the Grassman graph. J. Comb. Theory Ser. A **97**(1), 27–42 (2002). https://doi.org/10.1006/jcta.2001.3188
5. G, Birkhoff, Lattice theory, Colloquium Publications, vol 25, 3rd edn. American Mathematical Society (1967)
6. R.P. Stanley, *Enumerative Combinatorics*, 2nd edn., Cambridge Studies in Advanced Mathematics, vol. 1 (Cambridge University Press, Cambridge, 2012). https://doi.org/10.1017/CBO9781139058520
7. E. Artin, *Geometric Algebra* (Wiley classics library, Interscience Publishers Inc., New York, 1988). https://doi.org/10.1002/9781118164518.ch1
8. R.D. Fray, Congruence properties of ordinary and $q$-binomial coefficients. Duke Math. J. **34**(3), 467–480 (1967). https://doi.org/10.1215/S0012-7094-67-03452-7
9. F.H. Jackson, $q$-difference equations. Am. J. Math. **32**(3), 305–314 (1910). https://doi.org/10.2307/2370183
10. M. Ward, A calculus of sequences. Am. J. Math. **58**(2), 255–266 (1936). https://doi.org/10.2307/2371035
11. H. Cohn, Projective geometry over $\mathbb{F}_1$ and the Gaussian binomial coefficients. Am. Math. Mon. **111**(6), 487–495 (2004). https://doi.org/10.2307/4145067
12. J. Goldman, G.C. Rota, On the foundations of combinatorial theory IV finite vector spaces and Eulerian generating functions. Stud. Appl. Math. **49**(3), 239–258 (1970)
13. G. Pólya, G. Alexanderson, Gaussian binomial coefficients. Elem. Math. **26**, 102–109 (1971), http://eudml.org/doc/1410
14. C. Berge, D. Ray-Chaudhuri, Unsolved problems, in *Hypergraph Seminar: Ohio State University 1972*, Lecture Notes in Mathematics, ed. by C. Berge, D. Ray-Chaudhuri (Springer, Berlin, 1974), pp. 278–287. https://doi.org/10.1007/BFb0066199
15. P.J. Cameron, Generalisation of Fisher's inequality to fields with more than one element, in *Combinatorics - Proceedings of the British Combinatorial Conference 1973*, London Mathematical Society Lecture Note Series, ed. by T.P. McDonough, V.C. Mavron (Cambridge University Press, Cambridge, 1974), pp. 9–13. https://doi.org/10.1017/CBO9780511662072.003
16. P. Delsarte, Association schemes and $t$-designs in regular semilattices. J. Comb. Theory Ser. A **20**(2), 230–243 (1976). https://doi.org/10.1016/0097-3165(76)90017-0
17. S. Thomas, Designs over finite fields. Geom. Dedicata **24**(2), 237–242 (1987). https://doi.org/10.1007/BF00150939
18. H. Suzuki, 2-designs over GF($2^m$). Graphs Comb. **6**(3), 293–296 (1990). https://doi.org/10.1007/BF01787580
19. H. Suzuki, 2-designs over GF($q$). Graphs Comb. **8**(4), 381–389 (1992). https://doi.org/10.1007/BF02351594
20. M. Braun, A. Kerber, R. Laue, Systematic construction of $q$-analogs of $t$-$(v, k, \lambda)$-designs. Des. Codes Cryptogr. **34**(1), 55–70 (2005). https://doi.org/10.1007/s10623-003-4194-z
21. H. Suzuki, On the inequalities of $t$-designs over a finite field. Eur. J. Comb. **11**(6), 601–607 (1990). https://doi.org/10.1016/S0195-6698(13)80045-5
22. B. Segre, Teoria di galois, fibrazioni proiettive e geometrie non desarguesiane. Ann. Mat. Pura Appl. 4 **64**(1), 1–76 (1964). https://doi.org/10.1007/BF02410047
23. J. André, Über nicht-Desarguessche Ebenen mit transitiver Translationsgruppe. Math. Z. **60**(1), 156–186 (1954). https://doi.org/10.1007/BF01187370

24. D.K. Ray-Chaudhuri, R.M. Wilson, On *t*-designs. Osaka J. Math. **12**(3), 737–744 (1975)
25. M. Kiermaier, R. Laue, Derived and residual subspace designs. Adv. Math. Commun. **9**(1), 105–115 (2015). https://doi.org/10.3934/amc.2015.9.105
26. T.V. Trung, On the construction of *t*-designs and the existence of some new infinite families of simple 5-designs. Arch. Math. **47**(2), 187–192 (1986). https://doi.org/10.1007/BF01193690
27. D.C. van Leijenhorst, Orbits on the projective line. J. Comb. Theory Ser. A **31**(2), 146–154 (1981). https://doi.org/10.1016/0097-3165(81)90011-X
28. L.M.H.E. Driessen, *t*-designs, $t \geq 3$, Technical Report, Technische Universiteit Eindhoven, 1978
29. N.S. Mendelsohn, Intersection numbers of *t*-designs, in *Studies in Pure Mathematics*, ed. by L. Mirsky (Academic Press, London, 1971), pp. 145–150
30. W. Oberschelp, Lotto-Garantiesysteme und Blockpläne. Math.-Phys. Semesterber. **19**, 55–67 (1972)
31. M. Kiermaier, M.O. Pavčević, Intersection numbers for subspace designs. J. Comb. Des. **23**(11), 463–480 (2015). https://doi.org/10.1002/jcd.21403
32. E. Köhler, Allgemeine Schnittzahlen in *t*-Designs. Discret. Math. **73**(1–2), 133–142 (1988–1989). https://doi.org/10.1016/0012-365X(88)90141-0
33. S. Thomas, Designs and partial geometries over finite fields. Geom. Dedicata **63**, 247–253 (1996)
34. A. Kerber, *Applied Finite Group Actions*, Algorithms and Combinatorics, vol 19 (Springer, Berlin, 1999). https://doi.org/10.1007/978-3-662-11167-3
35. E.S. Kramer, D.M. Mesner, *t*-designs on hypergraphs. Discret. Math. **15**(3), 263–296 (1976). https://doi.org/10.1016/0012-365X(76)90030-3
36. B. Huppert, I. Endliche Gruppen, *Grundlehren der Mathematischen Wissenschaften* (Springer, Berlin, 1967), http://opac.inria.fr/record=b1108495
37. M. Braun, Some new designs over finite fields. Bayreuth. Math. Schr. **74**, 58–68 (2005)
38. M. Miyakawa, A. Munemasa, S. Yoshiara, On a class of small 2-designs over GF(*q*). J. Comb. Des. **3**(1), 61–77 (1995). https://doi.org/10.1002/jcd.3180030108
39. S, Braun, Construction of *q*-analogs of combinatorial designs, Presentation at the conference *Algebraic Combinatorics and Applications (ALCOMA10)*. (Thurnau, Germany, 2010)
40. S, Braun, Algorithmen zur computerunterstützten Berechnung von *q*-Analoga kombinatorischer Designs. Diplomathesis Universität Bayreuth (2009)
41. M. Kiermaier, R. Laue, A. Wassermann, A new series of large sets of subspace designs over the binary field. Des. Codes Cryptogr. (2016). https://doi.org/10.1007/s10623-017-0349-1
42. M. Braun, New 3-designs over the binary field. Int. Electron. J. Geom. **6**(2), 79–87 (2013)
43. M. Braun, New infinite series of 2-designs over the binary and ternary field. Des. Codes Cryptogr. **81**(1), 145–152 (2016). https://doi.org/10.1007/s10623-015-0133-z
44. M. Braun, T. Etzion, P.R.J. Östergård, A. Vardy, A. Wassermann, Existence of *q*-analogs of Steiner systems. Forum Math. **4**(7), (2016). https://doi.org/10.1017/fmp.2016.5
45. M. Braun, Designs over the binary field from the complete monomial group. Australas. J. Comb. **67**(3), 470–475 (2017)
46. M. Braun, M. Kiermaier, A. Kohnert, R. Laue, Large sets of subspace designs. J. Comb. Theory Ser. A **147**, 155–185 (2017). https://doi.org/10.1016/j.jcta.2016.11.004
47. M. Braun, A. Wassermann, Disjoint *q*-Steiner systems in dimension 13, Universität Bayreuth, Bayreuth, Technical Report, 2017
48. H. Suzuki, Five days introduction to the theory of designs (1989), http://subsite.icu.ac.jp/people/hsuzuki/lecturenote/designtheory.pdf
49. P.J. Cameron, Locally symmetric designs. Geom. Dedicata **3**, 65–76 (1974)
50. D. Ray-Chaudhuri, N. Singhi, *q*-analogues of *t*-designs and their existence. Linear Algebra Appl. **114–115**, 57–68 (1989). https://doi.org/10.1016/0024-3795(89)90451-5
51. R.M. Wilson, The necessary conditions for *t*-designs are sufficient for something. Util. Math **4**, 207–215 (1973)
52. P. Keevash, The existence of designs. arXiv identifier 1401.3665 (2014)

53. V. Guruswami, S. Kopparty, Explicit subspace designs, in *2013 IEEE 54th Annual Symposium on Foundations of Computer Science*, pp. 608–617 (2013). https://doi.org/10.1109/FOCS. 2013.71

54. V. Guruswami, C. Xing, List decoding Reed-Solomon, algebraic-geometric, and Gabidulin subcodes up to the singleton bound, in *Proceedings of the Forty-fifth Annual ACM Symposium on Theory of Computing, STOC '13* (ACM, New York, NY, USA, 2013), pp. 843–852. https://doi.org/10.1145/2488608.2488715

55. L. Teirlinck, Non-trivial $t$-designs without repeated blocks exist for all $t$. Discret. Math. **65**(3), 301–311 (1987). https://doi.org/10.1016/0012-365X(87)90061-6

56. A. Fazeli, S. Lovett, A. Vardy, Nontrivial $t$-designs over finite fields exist for all $t$. J. Comb. Theory Ser. A **127**, 149–160 (2014)

57. P. Frankl, V. Rödl, Near perfect coverings in graphs and hypergraphs. Eur. J. Comb. **6**(4), 317–326 (1985)

58. T. Abe, S. Yoshiara, On Suzuki's construction of 2-designs over GF($q$). Sci. Rep. Hirosaki Univ. **10**, 119–122 (1993)

59. T. Itoh, A new family of 2-designs over GF($q$) admitting $\mathrm{SL}_m(q^l)$. Geom. Dedicata **69**(3), 261–286 (1998). https://doi.org/10.1023/A:1005057610394

60. M. Braun, A. Kohnert, P.R.J. Östergård, A. Wassermann, Large sets of $t$-designs over finite fields. J. Comb. Theory Ser. A **124**, 195–202 (2014). https://doi.org/10.1016/j.jcta.2014.01. 008

61. K. Metsch, Bose–Burton Type Theorems for Finite Projective, Affine and Polar Spaces, in *Surveys in Combinatorics*, Lecture Notes Series, ed. by Lamb, Preece (London Mathematical Society, 1999)

62. D.E. Knuth, Dancing links, in *Millennial Perspectives in Computer Science, Cornerstones of Computing*, ed. by A.W. Roscoe, J. Davies, J. Woodcock (Palgrave, 2000), pp. 187–214

63. M. Braun, M. Kiermaier, A. Nakić, On the automorphism group of a binary $q$-analog of the Fano plane. Eur. J. Comb. **51**, 443–457 (2016). https://doi.org/10.1016/j.ejc.2015.07.014

64. M. Kiermaier, S. Kurz, A. Wassermann, The order of the automorphism group of a binary $q$-analog of the fano plane is at most two. Des. Codes Cryptogr. (2017). https://doi.org/10. 1007/s10623-017-0360-6

65. J.J. Sylvester, Note on the historical origin of the unsymmetrical six valued function of six letters. Philos. Mag. Ser. 4 **21**(141), 369–377 (1861). https://doi.org/10.1080/14786446108643072

66. R.H. Denniston, Sylvester's problem of the 15 schoolgirls. Discret. Math. **9**(3), 229–233 (1974). https://doi.org/10.1016/0012-365X(74)90004-1

67. Z. Baranyai, On the factorization of the complete uniform hypergraph, in *Infinite and finite Sets, Colloquia Mathematica Societatis János Bolyai*, vol. 1, ed. by A. Hajnal, R. Rado, V.T. Sós (Bolyai János Matematikai Társulat and North-Holland, Budapest and Amsterdam, 1975), pp. 91–107

68. A. Beutelspacher, On parallelisms in finite projective spaces. Geometriae Dedicata **3**(1), 35–40 (1974)

69. R.H.F. Denniston, Some packings of projective spaces. Atti Accad. Naz. Lincei Rend. Cl. Sci. Fis. Mat. Natur. **52**(8), 36–40 (1972)

70. R.D. Baker, Partitioning the planes of $AG_{2m}(2)$ into 2-designs. Discret. Math. **15**(3), 205–211 (1976)

71. F. Wettl, On parallelisms of odd dimensional finite projective spaces. Period. Polytech. Transp. Eng. **19**(1–2), 111 (1991)

72. T. Etzion, A. Vardy, Automorphisms of codes in the Grassmann scheme. arXiv identifier 1210.5724 (2012)

73. T. Hishida, M. Jimbo, Cyclic resolutions of the bib design in $PG(5, 2)$. Australas. J. Comb. **22**, 73–79 (2000)

74. J.F. Sarmiento, Resolutions of PG(5, 2) with point-cyclic automorphism group. J. Comb. Des. **8**(1), 2–14 (2000). https://doi.org/10.1002/(SICI)1520-6610(2000)8:1<2::AID-JCD2>3. 0.CO;2-H

75. M.R. Hurley, B.K. Khadka, S.S. Magliveras, Some new large sets of geometric designs of type $[3][2, 3, 2^8]$. J. Algebra Comb. Discret. Struct. Appl. **3**(3), 165–176 (2016). https://doi.org/10.13069/jacodesmath.40139

76. L. Teirlinck, Locally trivial $t$-designs and $t$-designs without repeated blocks. Discret. Math. **77**(1–3), 345–356 (1989). https://doi.org/10.1016/0012-365X(89)90372-5

77. G. B. Khosrovshahi, S. Ajoodani-Namini, Combining $t$-designs. JCTSA **58**(1), 26–34 (1991). https://doi.org/10.1016/0097-3165(91)90071-N

78. S. Ajoodani-Namini, Extending large sets of $t$-designs. J. Comb. Theory Ser. A **76**(1), 139–144 (1996). https://doi.org/10.1006/jcta.1996.0093

79. S. Ajoodani-Namini, G. Khosrovashahi, More on halving the complete designs. Discret. Math. **135**(1–3), 29–37 (1994). https://doi.org/10.1016/0012-365X(93)E0096-M

80. G.B. Khosrovshahi, B. Tayfeh-Rezaie, Trades and $t$-designs, *Surveys in Combinatorics 2009*, London Mathematical Society Lecture Note Series (Cambridge University Press, Cambridge, 2009), pp. 91–111. https://doi.org/10.1017/CBO9781107325975.005

81. T. Etzion, Problems on $q$-analogs in coding theory. arXiv identifier 1305.6126 (2013)

82. T. Etzion, L. Storme, Galois geometries and coding theory. Des. Codes Cryptogr. **78**(1), 311–350 (2016). https://doi.org/10.1007/s10623-015-0156-5

83. A. Hartman, Halving the complete design. Ann. Discret. Math. **34**, 207–224 (1987). https://doi.org/10.1016/S0304-0208(08)72888-3

84. S. Ajoodani-Namini, All block designs with $b = \binom{v}{k}/2$ exist. Discret. Math. **179**(1–3), 27–35 (1998). https://doi.org/10.1016/S0012-365X(97)00024-1

# Computational Methods in Subspace Designs

Michael Braun, Michael Kiermaier and Alfred Wassermann

**Abstract** Subspace designs are the $q$-analogs of combinatorial designs. Introduced in the 1970s, these structures gained a lot of interest recently because of their application to random network coding. Compared to combinatorial designs, the number of blocks of subspace designs are huge even for the smallest instances. Thus, for a computational approach, sophisticated algorithms are indispensible. This chapter highlights computational methods for the construction of subspace designs, in particular methods based on group theory. Starting from tactical decompositions we present the method of Kramer and Mesner which allows to restrict the search for subspace designs to those with a prescribed group of automorphisms. This approach reduces the construction problem to the problem of solving a Diophantine linear system of equations. With slight modifications it can also be used to construct large sets of subspace designs. After a successful search, it is natural to ask if subspace designs are isomorphic. We give several helpful tools which allow to give answers in surprisingly many situations, sometimes in a purely theoretical way. Finally, we will give an overview of algorithms which are suitable to solve the underlying Diophantine linear system of equations. As a companion to chapter "q-Analogs of Designs: Subspace Designs" this chapter provides an extensive list of groups which were used to construct subspace designs and large sets of subspace designs.

M. Braun (✉)
Faculty of Computer Science, Darmstadt University of Applied Sciences,
64295 Darmstadt, Germany
e-mail: michael.braun@h-da.de

M. Kiermaier · A. Wassermann
Department of Mathematics, University of Bayreuth, 95440 Bayreuth, Germany
e-mail: michael.kiermaier@uni-bayreuth.de

A. Wassermann
e-mail: alfred.wassermann@uni-bayreuth.de

© Springer International Publishing AG 2018                                213
M. Greferath et al. (eds.), *Network Coding and Subspace Designs*,
Signals and Communication Technology,
https://doi.org/10.1007/978-3-319-70293-3_9

# 1   Introduction

Combinatorial design theory has a long and venerable history starting in the middle of the 19th century. Combinatorial designs have many applications and there exists a huge amount of literature. For a comprehensive overview we recommend the two books [3, 29].

If one replaces in the definition of a combinatorial design *set of points* by *vector space* and *cardinality* by *dimension* one gets the definition of a *subspace design*.

Subspace designs, also called *q-analogs of designs*, *designs over finite fields*, or *designs in vector spaces* were introduced in the early 1970s independently by Ray-Chaudhuri [2], Cameron [25, 26] and Delsarte [31]. While receiving only occasionally attention by relatively few research groups until 10 years ago, the situation has changed a lot since then. Subspace designs have found much increased interest now due to the connection to *random network coding* pointed out by Kötter and Kschischang [64]. For an in-depth introduction to subspace designs the reader is referred to chapter "$q$-Analog of Designs: Subspace Design".

In terms of random network coding, one is mainly interested in very specific subspaces designs, namely $q$-Steiner systems. The reason is that $q$-Steiner systems are the best possible constant-dimension codes for random network coding. To be precise, they are *diameter-perfect* constant-dimension codes. This is analog to the situation in "classical" coding theory, where (combinatorial) Steiner systems are the best possible constant-weight codes for a given length and minimum distance. For more information on constant-dimension codes the reader is referred to chapters "Construction of Constant Dimension Codes" and "Construction of Cyclic Subspace Codes and Maximum Rank".

Subspace designs and subspace codes with $t = 1$ and $\lambda = 1$ are better known as (partial) spreads, which have been studied in finite geometry since many decades. However, the first subspace design for $t \geq 2$ was found only in 1987 [92], when the explicit question for their existence has been open for already about 15 years. For more information on geometric aspects the reader is referred to chapters "Geometrical Aspects of Subspace Codes" and "Partial Spreads and Vector Space Paritions".

Systematic computer search for subspace designs started in the 1990s [81]. Since then, quite a few subspace designs could be constructed, the most prominent being designs for $t = 3$ [17] and the first $q$-Steiner systems for $t = 2$ [16].

In this chapter we will study computational methods to construct subspace designs and so called *large sets of subspace designs*. For an in-depth study of subspace designs the reader is referred to chapter "$q$-Analog of Designs: Subspace Designs". In particular, chapter "$q$-Analog of Designs: Subspace Designs" contains comprehensive tables of all known simple subspace designs for small dimensions and $t \geq 2$.

Much more detailed information on construction and classification algorithms of combinatorial objects can be found in the book by Kaski and Östergård [53] and in the chapter on computer construction in the Handbook of combinatorial designs [42].

## 2  Preliminaries

This is an overview in compressed form about subspace designs. The definitions and results can be found in more details and together with references in chapter "*q*-Analogs of Designs:Subspace Design".

Throughout this chapter, $V$ will be a vector space of finite dimension $v$ over a finite field $\mathbb{F}_q$. Subspaces of dimension $k$ will be called $k$-*subspaces*. Sometimes, 1-subspaces will be denoted as *points*. If the dimension $k$ is clear from the context, $k$-subspaces will occasionally be called *blocks*.

For an integer $0 \le k \le v$ the set $\begin{bmatrix} V \\ k \end{bmatrix}_q$ consisting of all $k$-subspaces of $V$ is called *Graßmannian*. Its cardinality is determined by the $q$-*binomial coefficient*

$$\#\begin{bmatrix} V \\ k \end{bmatrix}_q = \begin{bmatrix} v \\ k \end{bmatrix}_q = \frac{(q^v - 1) \cdots (q^{v-k+1} - 1)}{(q^k - 1) \cdots (q - 1)}.$$

If clear from the context, we will omit the index $q$ from $\begin{bmatrix} V \\ k \end{bmatrix}_q$ and $\begin{bmatrix} v \\ k \end{bmatrix}_q$ for sake of readability. The $q$-analog of a combinatorial design is defined as follows.

**Definition 1** Let $q$ be a prime power, $V$ an $\mathbb{F}_q$-vector space of finite dimension $v$, $0 \le t \le k \le v$ integers and $\lambda$ a non-negative integer. A pair $\mathcal{D} = (V, \mathcal{B})$, where $\mathcal{B}$ is a collection of $k$-subspaces (*blocks*) of $V$, is called a $t$-$(v, k, \lambda)_q$ *subspace design* on $V$ if each $t$-subspace of $V$ is contained in exactly $\lambda$ blocks.

If $\mathcal{B}$ is a set, i.e. if every $k$-subset appears at most once in $\mathcal{B}$, the design is called *simple*.

A $t$-$(v, k, 1)_q$ design is also called $S(t, k, v)_q$ $q$-*Steiner system*.

If the strict equality conditions $\#\{B \in \mathcal{B} \mid T \subset B\} = \lambda$ for all $T \in \begin{bmatrix} V \\ t \end{bmatrix}_q$ in Definition 1 is relaxed to "$\le$", the structure is called *packing design*, whilst the relaxation to "$\ge$" is called *covering design*. Of course, the question is to maximize (minimize) the number of blocks in a packing (covering) design. From the viewpoint of random network coding, packing designs with $\lambda = 1$ are the same as *constant-dimension subspace codes* of minimum subspace distance $2(k - t - 1)$.

If not explicitly stated otherwise, for the rest of the chapter all designs will be simple designs. The empty set and the full Graßmannian are designs for all values of $0 \le t \le k$, called *trivial* designs. The parameters of the Graßmannian $\begin{bmatrix} V \\ k \end{bmatrix}_q$ as a $t$-design are $t$-$(v, k, \begin{bmatrix} v-t \\ k-t \end{bmatrix}_q)_q$. Thus, for fixed values of $t$, $v$, $k$, the largest possible value for $\lambda$ is $\lambda_{\max} := \begin{bmatrix} v-t \\ k-t \end{bmatrix}_q$.

**Lemma 1** ([91]) *Let $\mathcal{D} = (V, \mathcal{B})$ be a $t$-$(v, k, \lambda)_q$ subspace design. For each $s \in \{0, \ldots, t\}$, $\mathcal{D}$ is an $s$-$(v, k, \lambda_s)_q$ subspace design with*

$$\lambda_s = \frac{\begin{bmatrix} v-s \\ t-s \end{bmatrix}_q}{\begin{bmatrix} k-s \\ t-s \end{bmatrix}_q} \cdot \lambda = \frac{\begin{bmatrix} v-s \\ k-s \end{bmatrix}_q}{\begin{bmatrix} v-t \\ k-t \end{bmatrix}_q} \cdot \lambda.$$

*In particular, the number of blocks is given by*

$$\#\mathcal{B} = \lambda_0 = \lambda \cdot \begin{bmatrix} v \\ t \end{bmatrix}_q / \begin{bmatrix} k \\ t \end{bmatrix}_q.$$

A consequence of this lemma is that every 1-subspace is contained in

$$\lambda_1 = \lambda \cdot \frac{\begin{bmatrix} v-1 \\ t-1 \end{bmatrix}_q}{\begin{bmatrix} k-1 \\ t-1 \end{bmatrix}_q}$$

blocks of a $t$-$(v, k, \lambda)_q$ design.

A set $t$-$(v, k, \lambda)_q$ of design parameters is called *admissible* if for all $0 \le s \le t$ the *divisibility conditions*

$$\begin{bmatrix} v-s \\ t-s \end{bmatrix}_q \cdot \lambda \equiv 0 \pmod{\begin{bmatrix} k-s \\ t-s \end{bmatrix}_q}$$

are fulfilled. If a $t$-$(v, k, \lambda)_q$ design exists, the parameters are called *realizable*.

For fixed $t, k, v$ the divisibility conditions imply a minimum $\lambda_{\min}$ that fulfills the conditions. It is easy to see that all admissible values of $\lambda$ are multiples of $\lambda_{\min}$.

If a $t$-$(v, k, \lambda)_q$ design is realizable, then also the *dual design* with parameters $t$-$(v, v - k, \begin{bmatrix} v-t \\ k \end{bmatrix} / \begin{bmatrix} v-t \\ k-t \end{bmatrix} \lambda)_q$ and the *supplementary design* with parameters $t$-$(v, k, \lambda_{max} - \lambda)_q$ are realizable. Therefore, it is usually enough to restrict the search for designs to $2 \le k \le v/2$ and $1 \le \lambda \le \lambda_{max}/2$.

A celebrated, recent result by Fazeli, Lovett, and Vardy [36] is that there exist subspace designs for every value of $t$ and $q$ if the parameters $v, k, \lambda$ are large enough. However, their proof is not constructive.

An *automorphism* of a subspace design $\mathcal{D} = (V, \mathcal{B})$ is a bijective mapping $g \in P\Gamma L(V)$ such that $\mathcal{B}^g = \mathcal{B}$. Under the composition of mappings, the set of all automorphisms of a design forms a group, the *automorphism group* $\text{Aut}(\mathcal{D})$ of the design $\mathcal{D}$. For every subgroup $G$ of its automorphism group a design is said to be *G-invariant*.

The action of $P\Gamma L(V)$ on the subspace lattice of $V$ preserves containment, that is, for all $g \in P\Gamma L(V)$ and all subspaces $S, T \le V$,

$$S \le T \text{ if and only if } S^g \le T^g. \tag{1}$$

In particular, for $0 \le s \le v$ and $S \in \begin{bmatrix} V \\ s \end{bmatrix}$ the image $S^g$ is again in $\begin{bmatrix} V \\ s \end{bmatrix}$.

For computational purposes we will restrict ourselves to automorphisms from $GL(V)$ which is represented by the group $GL(v, q)$ of all invertible $v \times v$ matrices over $\mathbb{F}_q$, via assigning the map $\mathbf{x} \mapsto \mathbf{x}A$ to the matrix $A \in GL(v, q)$.

Let $G$ be a group acting on a set $X$ via $x \mapsto x^g$. The stabilizer of $x$ in $G$ is given by $G_x = \{g \in G \mid x^g = x\}$, and the $G$-orbit of $x$ is given by $x^G = \{x^g \mid g \in G\}$. For two elements $x, x' \in X$ in the same orbit, we write $x \sim x'$. Obviously, $x \sim x'$ if and only if there exists an element $g \in G$ with $x' = x^g$. The relation $\sim$ is an equivalence relation on $X$. Therefore, by the action of $G$, the set $X$ is partitioned into orbits. For all $x \in X$, there is the correspondence $x^g \mapsto G_x g$ between the orbit $x^G$ and the set $G_x \backslash G$ of the right cosets of the stabilizer $G_x$ in $G$. For finite orbit lengths, this implies the orbit-stabilizer theorem stating that $\#x^G = [G : G_x]$. In particular, the orbit lengths $\#x^G$ are divisors of the group order $\#G$. The stabilizers of elements in the same orbit are related by the formula

$$G_{x^g} = g^{-1} G_x g. \tag{2}$$

A partition of the Graßmannian into $N$ disjoint designs with the same parameters is called *large set of designs* and denoted by $LS_q[N](t, k, v)$. It is clear that

$$N \cdot \lambda = \lambda_{max}.$$

that is, $N$ has to divide $\lambda_{max}$ and $N$ determines $\lambda$. The large set parameters $LS_q[N](t, k, v)$ are called *admissible* if

$$\begin{bmatrix} v - s \\ k - s \end{bmatrix}_q \equiv 0 \pmod{N} \text{ for } 0 \leq s \leq t.$$

If an $LS_q[N](t, k, v)$ exists the parameters are called *realizable*.

If for a group $G \in P\Gamma L(v, q)$ and a large set with parameters $LS_q[N](t, k, v)$ all its designs—having parameters $t\text{-}(v, k, \lambda_{max}/N)_q$—are $G$-invariant, then the large set is called *uniformly-G*.

# 3 Computer Construction

The computational approaches to construct subspace designs and combinatorial designs are mostly the same. This is not a big surprise since both, combinatorial designs and subspace designs, are finite incidence structures.

The main difference to combinatorial designs is that the cardinalities of the objects which have to be considered for subspace designs are way larger than for combinatorial designs. As a consequence, so far only for comparatively small parameters subspace designs could be constructed by computer.

Now, the question is how can a computer be instructed to find $t\text{-}(v, k, \lambda)_q$ designs? The most naive approach would be the following algorithm.

**Algorithm 1** Naive algorithm to search for a simple $t$-$(v, k, 1)_q$ design.

$\mathcal{B} \leftarrow \emptyset, \mathcal{T} \leftarrow \begin{bmatrix} V \\ t \end{bmatrix}, \overline{\mathcal{T}} \leftarrow \emptyset$

success $\leftarrow$ true

**while** success **do**

  Search $B \in \begin{bmatrix} V \\ k \end{bmatrix}$ such that **for** $T \in \overline{\mathcal{T}}: T \not\subset B$

  **if** $B$ exists:

    $\mathcal{B} \leftarrow \mathcal{B} \cup \{B\}$

    **for** $T \subset B$:

      $\mathcal{T} \leftarrow \mathcal{T} \setminus \{T\}$

      $\overline{\mathcal{T}} \leftarrow \overline{\mathcal{T}} \cup \{T\}$

  **else**:

    success $\leftarrow$ false

**end while**

**return** $\mathcal{B}$ // return packing design

Thus, this simple, greedy algorithm packs blocks into the design until no more block can be found that consists solely of $t$-subspaces which are not covered yet. Surprisingly enough, an analysis by Frankl and Rödl [40] shows that asymptotically for $v \to \infty$, the set of blocks returned by Algorithm 1 is expected to come quite close to a Steiner system.

**Theorem 1** (Frankl and Rödl) *Suppose integers $t$, $k$ are given, $0 < t < k < v$. Then for $v \to \infty$ Algorithm 1 is expected to return a set $\mathcal{B} \subset \begin{bmatrix} V \\ k \end{bmatrix}$ such that*

$$\#\mathcal{B} = (1 - o(1)) \begin{bmatrix} v \\ t \end{bmatrix}_q \Big/ \begin{bmatrix} k \\ t \end{bmatrix}_q$$

*and $\dim(K' \cap K'') < t$ for all distinct $K', K'' \in \mathcal{B}$.*

An alternative description of Algorithm 1 is the following: We fix an order on all subspaces $T \in \begin{bmatrix} V \\ t \end{bmatrix}$ and $K \in \begin{bmatrix} V \\ k \end{bmatrix}$ and define the matrix $A_{t,k} = (a_{i,j})$ by

$$a_{i,j} = \begin{cases} 1, & \text{if } T_i \subset K_j; \\ 0, & \text{else.} \end{cases}$$

That is, the matrix $A_{t,k}$ is of size $\begin{bmatrix} v \\ t \end{bmatrix} \times \begin{bmatrix} v \\ k \end{bmatrix}$. Now, if the resulting selection of blocks in Algorithm 1 even forms a $q$-Steiner system then this selection can be expressed as a 0/1-vector $x$ of length $\begin{bmatrix} v \\ k \end{bmatrix}$ such that

$$A_{t,k} \cdot x = \begin{pmatrix} 1 \\ \vdots \\ 1 \end{pmatrix}. \tag{3}$$

*Example 1* For a really small example we look at combinatorial designs instead of subspace designs. That is, $V$ is a set of points and the blocks are $k$-element subsets of $V$.

In our example we want to cover every vertex (i.e. 1-subset) of the following graph by exactly one edge (i.e. 2-subset). The result is a 1-(4, 2, 1) (combinatorial)

design.

It is easy to see that there are the following three solutions:

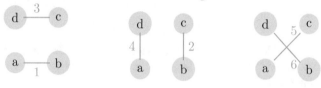

The matrix $A_{1,2}$ together with the three solution vectors $x^\top$ looks as follows.

| vertices\edges | 1 2 3 4 5 6 |
|:---:|:---|
| a | 1 0 0 1 1 0 |
| b | 1 1 0 0 0 1 |
| c | 0 1 1 0 1 0 |
| d | 0 0 1 1 0 1 |
| design1 | 1 0 1 0 0 0 |
| design2 | 0 1 0 1 0 0 |
| design3 | 0 0 0 0 1 1 |

The three designs consist of the block sets $\{\{a, b\}, \{c, d\}\}$, $\{\{a, d\}, \{b, c\}\}$, and $\{\{a, c\}, \{b, d\}\}$.

It turns out that problem (3) is well-known in computer science. It is called *exact cover* problem and is known to be *NP-complete* [52]. In the general case of searching for a $t$-$(v, k, \lambda)_q$ design we get a generalized exact cover problem of the form

$$
A_{t,k} \cdot x = \begin{pmatrix} \lambda \\ \vdots \\ \lambda \end{pmatrix}. \tag{4}
$$

For the construction of packing designs, the "=" in Eq. (4) has to be replaced by a "$\leq$", for covering designs it has to be replaced by a "$\geq$".

There is a plethora of algorithms available to solve the exact cover problem and its generalizations. We will discuss a few solving strategies in Sect. 6.

But even if there are powerful algorithms available to solve these problems, the biggest obstacle to this strategy for constructing subspace designs is that the sizes of the problem instances will be prohibitively large even for small parameters.

*Example 2* To find the binary analog of the Fano plane, i.e. a 2-$(7, 3, 1)_2$ design, the matrix $A_{2,3}$ has 2 667 rows and 11 811 columns. The design would consist of 381 3-subspaces.

Although being the smallest open parameter set for subspace designs, the problem is still out of reach for all known computer algorithms.

## 4   Using Groups

A possible strategy to tackle even very large instances is not to select individual blocks but to choose whole "chunks" of blocks which fit together well. For this, group theory comes in handy.

The idea to use groups to find combinatorial objects can be traced back at least to Dembowski [32], Parker [86] in the 1950s, and even to Moore [82] in 1896.

Dembowski [32] studied *tactical decompositions* via group actions and found additional necessary conditions for combinatorial objects having such groups as automorphism groups. Later on, Kramer and Mesner [65] had the idea to prescribe an automorphism group and search by computer for combinatorial designs with this automorphism group. The same approach has been used before the work of Kramer and Mesner at least by Magliveras [77] and Klin [57] in their PhD theses. In fact, the method is a systematic version of finding a tactical decomposition of a structure by means of the automorphism group of the structure.

So, with tactical decompositions as well as with the Kramer–Mesner method we choose a group $G$ and search for $G$-invariant designs. The resulting computer problem will be smaller and thus in many cases solvable by computer hardware. But we must be aware that those designs which are not $G$-invariant will not be found by this approach.

### 4.1   Tactical Decompositions

Tactical decompositions of combinatorial designs, mostly *symmetric designs*, were studied already by Dembowski [32, 33]. In [68, 83] tactical decompositions of subspace designs are introduced.

An *incidence structure* is a triple $(\mathscr{P}, \mathscr{B}, I)$ with $\mathscr{P} \cap \mathscr{B} = \emptyset$ and an incidence relation $I \subset \mathscr{P} \times \mathscr{B}$. The elements of $\mathscr{P}$ are called *points* and the elements of $\mathscr{B}$ are called *blocks*.

An incidence structure is called *tactical configuration* if there is a number $r'$ such that $\#\{B \in \mathscr{B} \mid (P, B) \in I\} = r'$ for all points $P \in \mathscr{P}$ and if there is a number $k'$ such that $\#\{P \in \mathscr{P} \mid (P, B) \in I\} = k'$ for all $B \in \mathscr{B}$. Counting the incidences $(P, B)$ in two ways, we get the equation $k'\#\mathscr{B} = r'\#\mathscr{P}$.

Therefore, $t$-$(v, k, \lambda)_q$ subspace designs are tactical configurations with the blocks of the design being the blocks of the tactical configuration. As points of the tactical configurations one can take one of the Graßmannians $\begin{bmatrix} V \\ s \end{bmatrix}$, $0 \le s \le t$. The incidence relation is given by "$\le$". Then, for the tactical configuration we have $k' = \begin{bmatrix} k \\ s \end{bmatrix}$ and $r' = \lambda_s$.

**Definition 2** A *decomposition* of a tactical configuration is a partition of the points $\mathscr{P} = \mathscr{P}_1 \sqcup \ldots \sqcup \mathscr{P}_m$ and the blocks $\mathscr{B} = \mathscr{B}_1 \sqcup \ldots \sqcup \mathscr{B}_n$.

A decomposition is *tactical* if there exist non-negative integers $\rho_{i,j}, \kappa_{i,j}, 1 \le i \le m, 1 \le j \le n$, such that

1. every $P \in \mathscr{P}_i$ is contained in exactly $\rho_{i,j}$ blocks in $\mathscr{B}_j$,
2. every block in $\mathscr{B}_j$ contains $\kappa_{i,j}$ points in $\mathscr{P}_i$.

The matrices $(\rho_{i,j})$ and $(\kappa_{i,j})$ are called *tactical decomposition matrices*.

In other words, a tactical decomposition is a partition $\mathscr{P} = \mathscr{P}_1 \sqcup \ldots \sqcup \mathscr{P}_m$ and $\mathscr{B} = \mathscr{B}_1 \sqcup \ldots \sqcup \mathscr{B}_n$ of the points and blocks of a tactical configuration such that each pair $(\mathscr{P}_i, \mathscr{B}_j)$ itself is a tactical configuration with regard to the incidence relation $I$.

A tactical configuration has always two trivial decompositions. One is the partition of $\mathscr{P}, \mathscr{B}$ into one set each, i.e. $m = n = 1$. The other one is the partition of $\mathscr{P}, \mathscr{B}$ into 1-element subsets.

Furthermore, the action of a subgroup $G \le \text{Aut}(\mathscr{D})$ yields a tactical decomposition which in general is nontrivial.

**Theorem 2** (Dembowski [32]) *Let $G$ be a group of automorphisms of an incidence structure $(\mathscr{P}, \mathscr{B}, I)$. Then the orbits of $G$ on the set of points $\mathscr{P}$ and the orbits of $G$ on the blocks $\mathscr{B}$ form a tactical decomposition.*

*Proof* The statement follows immediately from (1) and from the fact that automorphisms map points to points and blocks to blocks. □

*Remark 1* In [32] it was also observed that there are nontrivial tactical decomposition which do not stem from group actions.

The definition of the tactical decomposition matrices leads directly to the following equations for $t$-$(v, k, \lambda)_q$ designs and $0 \le s \le t$:

$$\sum_{i=1}^{m} \kappa_{i,j} = \begin{bmatrix} k \\ s \end{bmatrix}, \quad \sum_{j=1}^{n} \rho_{i,j} = \lambda_s. \tag{5}$$

Furthermore, for all $i \in \{1, \ldots, m\}$ and $j \in \{1, \ldots, n\}$ we have

$$\#\mathscr{P}_i \cdot \rho_{i,j} = \#\mathscr{B}_j \cdot \kappa_{i,j}, \tag{6}$$

since $(\mathscr{P}_i, \mathscr{B}_j)$ forms a tactical configuration.

In [83] the following further relations are stated.

**Theorem 3** (Nakić, Pavčević) *Let* $0 \le s \le t$. *Suppose that*

$$\begin{bmatrix} V \\ s \end{bmatrix} = \mathscr{P}_1 \sqcup \ldots \sqcup \mathscr{P}_m, \quad \mathscr{B} = \mathscr{B}_1 \sqcup \ldots \sqcup \mathscr{B}_n$$

*is a tactical decomposition of a* $2$-$(v, k, \lambda)_q$ *design. Let* $(\rho_{i,j})$ *and* $(\kappa_{i,j})$ *be the associated decomposition matrices. Then, for all pairs of rows* $1 \le l, r \le m$,

$$\sum_{j=1}^{n} \rho_{l,j}\kappa_{r,j} = \begin{cases} \lambda \cdot \#\mathscr{P}_r, & l \ne r; \\ \lambda_s + \lambda \cdot (\#\mathscr{P}_r - 1), & l = r. \end{cases} \tag{7}$$

In the same paper there are more, refined equations and inequalities for the special case $q = 2$. The preprint [30] generalizes the above equations for subspace designs with $t = 3$.

Algorithmic application

An algorithm to find subspace designs with the help of tactical decompositions can be the following approach. If we know the lengths of the block orbits then from Eq. (6) we see that it is enough to work with one of the matrices $(\rho_{i,j})$ or $(\kappa_{i,j})$.

**Algorithm 2** Choose a "promising" group $G \le P\Gamma L(v, q)$ and find $2$-$(v, k, \lambda)_q$ designs by using tactical decomposition matrices $(\rho_{i,j})$.

1. *Search for tentative tactical decomposition matrices*: Compute—up to permutations of rows and columns—all possible matrices $(\rho_{i,j})$ having nonnegative integer entries which satisfy the Eqs. (5) and (7) (and further equations or inequalities if available, see [83]).
2. *Indexing phase*:

   **for** all matrices $(\rho_{i,j})$ from 2.:
     **for** $1 \le i \le m, 1 \le j \le n$:
     replace all entries $(i, j)$ of $(\rho_{i,j})$ by 0/1-matrices of size $\#\mathscr{P}_i \times \#B_j$
       – having $\rho_{i,j}$ ones in each row and
       – being invariant under the simultaneous action of $G$ on its rows and
         columns.
     **if** successful:
     **output** incidence matrix of $2$-$(v, k, \lambda)_q$ design

*Remark 2* Not every orbit matrix from Step 1 will give rise to subspace designs. In contrast, a single matrix from Step 1 may produce more than one nonisomorphic designs.

*Example 3* We want to construct a combinatorial design from Example 1 by prescribing a cyclic symmetry, e.g. the mapping $\sigma : (a, b, c, d) \mapsto (b, c, d, a)$. The induced mapping on the edges maps $(1, 2, 3, 4, 5, 6) \mapsto (2, 3, 4, 1, 6, 5)$. Under the action of the group $\langle \sigma \rangle$ the set of vertices is a single orbit $\{a, b, c, d\}$ and the set of Edges

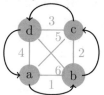

Is Partitioned Into the two orbits $\{1, 2, 3, 4\}$ and $\{5, 6\}$.

Recall that we are searching for a $1\text{-}(v, 2, 1)$ design. We know, that such a design consists of $\lambda_0 = 1\binom{4}{1}/\binom{2}{1} = 2$ blocks and every point appears in $\lambda_1 = 1\binom{3}{0}/\binom{1}{0} = 1$ blocks.

Therefore, we try to build our design by taking a block orbit of length two (there is only one choice in this small case). Since there is just one point orbit, the only choice in Step 1 is to take a trivial decomposition matrix consisting of the single entry, i.e. $n = m = 1$. Equation (5) for combinatorial designs give immediately $\kappa_{1,1} = 2$ and $\rho_{1,1} = 1$. Note, that we can not use the combinatorial design versions of Eq. (7) in this case, since we are searching for a 1-design. Thus, the only possible matrix in Step 1 is equal to (1).

In Step 2 we find that up to the action of $\langle \sigma \rangle$ just one $4 \times 2$-matrix having a single one in every row:

$$\begin{pmatrix} 1 & 0 \\ 0 & 1 \\ 1 & 0 \\ 0 & 1 \end{pmatrix}$$

It is the incidence matrix of the blocks points $a, b, c, d$ and the blocks 5, 6. Therefore, the blocks 5, 6 form a $\langle \sigma \rangle$-invariant $1\text{-}(4, 2, 1)$ design.

## 4.2 The Method of Kramer and Mesner

Similar as in the previous section we choose a group $G$ and search for $G$-invariant designs. Thus, instead of searching for individual $k$-subspaces as in (4) the design has to be composed of $G$-orbits on the $k$-subspaces. This approach applied to combinatorial designs is commonly attributed to Kramer and Mesner [65] and has been used there with great success.

The set of blocks $\mathscr{B}$ of a $G$-invariant $t\text{-}(v, k, \lambda)_q$ design ($G \leq \mathrm{Aut}(L(V))$) is the disjoint union of orbits of $G$ on the set $\begin{bmatrix} V \\ k \end{bmatrix}$ of $k$-dimensional subspaces of $V$.

To obtain an appropriate selection of orbits of $G$ on $\begin{bmatrix} V \\ k \end{bmatrix}$ we consider the incidence matrix $A^G_{t,k}$ whose rows are indexed by the $G$-orbits on the set of $t$-subspaces of $V$ and whose columns are indexed by the orbits on $k$-subspaces. The entry $a^G_{T,K}$ of $A^G_{t,k}$ corresponding to the orbits $T^G$ and $K^G$ is defined by

$$a^G_{T,K} := \#\{K' \in K^G \mid T \leq K'\}.$$

In other words, the matrix $A^G_{t,k}$ is the tactical decomposition matrix $(\rho_{i,j})$ for the tactical decomposition of the points $\begin{bmatrix} V \\ t \end{bmatrix}$ and the blocks $\begin{bmatrix} V \\ k \end{bmatrix}$ of the trivial $t$-$(v, k, \lambda_{\max})_q$ design into orbits on the $t$-subspaces and $k$-subspaces.

Since $A^G_{t,k}$ is a tactical decomposition of the complete design, i.e. a tactical configuration with $\#\{B \in \mathscr{B} \mid P \leq T\} = \lambda_{\max}$ for all $T \in \begin{bmatrix} V \\ t \end{bmatrix}$, we can search for a subset of $\mathscr{B}' \subset \mathscr{B}$ such that $\#\{B \in \mathscr{B}' \mid T \leq B'\} = \lambda < \lambda_{\max}$ for all $T \in \begin{bmatrix} V \\ t \end{bmatrix}$, which is a tactical decomposition of a $t$-$(v, k, \lambda)_q$ design. Therefore, the following theorem is obvious.

**Theorem 4** (Kramer, Mesner [65]) *Let $G \leq P\Gamma L(v, q)$. There exists a $G$-invariant $t$-$(v, k, \lambda)_q$ design if and only if there is a $0/1$-vector $x$ satisfying*

$$A^G_{t,k} \cdot x = \begin{pmatrix} \lambda \\ \vdots \\ \lambda \end{pmatrix}. \tag{8}$$

*Example 4* We look again at the combinatorial design from Example 1 and 3 and prescribe again the mapping $\sigma : (a, b, c, d) \mapsto (b, c, d, a)$. Counting how often a representative of the point orbit $\{a, b, c, d\}$ appears in each of the two orbits on lines, we get the following table.

|                | $\{1, 2, 3, 4\}$ | $\{5, 6\}$ |
|----------------|:----------------:|:----------:|
| $\{a, b, c, d\}$ | 2                | 1          |

That is, using the Theorem of Kramer and Mesner the matrix $A_{1,2}$ is condensed to the smaller matrix $A^{\langle \sigma \rangle}_{1,2} = (2 \ \ 1)$. Now, if we choose the orbit $\{5, 6\}$ as the set of blocks, we get a 1-$(4, 2, 1)$ design. If we choose the other orbit we get a 1-$(4, 2, 2)$ design.

Note that the other two designs in Example 1 are not found by this prescription of automorphisms.

*Remark 3* From Eq. (6) it is clear that $A^G_{t,k}$ can either be computed by determining $(\rho_{i,j})$ or by computing $(\kappa_{i,j})$.

In some cases it is faster to compute $(\kappa_{i,j})$ and then determine $(\rho_{i,j})$ by (6). The relation between these two matrices is also referred to as *Alltop's lemma* [1].

## 4.3 Promising Groups

In general, our approach in Sects. 4.1 and 4.2 is to prescribe certain groups of auto-morphisms which arise as subgroups $G$ of the general linear group $GL(V)$. When it comes to the question which groups may occur as automorphism groups of designs, one group stands out. At least up to now, most successful computer searches for sub-space designs with prescribed automorphism groups used this group or subgroups thereof, see Sect. 4.4. This group is the *normalizer of a Singer cycle group*. In the following we will describe this group in more detail and explain why it is a promising candidate for prescribing it as an automorphism group.

It will turn out to be helpful to apply the general linear group $GL(v, q)$ in matrix representation. In our notation, multiplication by group elements will be from the right. Further—when appropriate—we will switch between the vectors of the vector space $V = \mathbb{F}_q^v$ and the corresponding elements of the field $\mathbb{F}_{q^v}$. The multiplicative group $(\mathbb{F}_{q^v}^*, \cdot)$ consisting of the nonzero elements of $\mathbb{F}_{q^v}$ is known to be cyclic, generators are called primitive elements of $\mathbb{F}_{q^v}$.

Any cyclic subgroup of order $q^v - 1$ of $GL(V)$ is called a *Singer cycle group*. By setting $V = \mathbb{F}_{q^v}$ it can be seen that Singer cycle groups do always exist: For any primitive element $\alpha$ of $\mathbb{F}_{q^v}$, the mapping $\sigma : V \to V$, $\mathbf{x} \mapsto \alpha \mathbf{x}$ is $\mathbb{F}_q$-linear and of order $\#\mathbb{F}_q^* = q^v - 1$ and thus generates a Singer cycle group.

To get a matrix representation of $\sigma$ in $GL(v, q)$, we can proceed as follows: Let $f = a_0 + a_1 x + a_2 x^2 + \ldots + a_{n-1} x^{n-1} + x^n \in \mathbb{F}_q[x]$ be the minimal polynomial of $\alpha$ over $\mathbb{F}_q$. Then $f$ is a primitive irreducible polynomial of degree $v$. With respect to the $\mathbb{F}_q$-basis $(1, \alpha, \alpha^2, \ldots, \alpha^{v-1})$ of $V = \mathbb{F}_{q^v}$, the transformation matrix of $\sigma$ is given by the *companion matrix* of $f$, which is

$$\begin{pmatrix} 0 & 1 & 0 & \cdots & 0 \\ 0 & 0 & 1 & \cdots & 0 \\ \vdots & \vdots & \vdots & \ddots & \vdots \\ 0 & 0 & 0 & \cdots & 1 \\ -a_0 & -a_1 & -a_2 & \cdots & -a_{v-1} \end{pmatrix}.$$

By [49, Sect. 7], any two Singer cycle subgroups of $GL(V)$ are conjugate.

The Galois group $\mathrm{Gal}(\mathbb{F}_{q^v}/\mathbb{F}_q)$ of the field extension $\mathbb{F}_{q^v}/\mathbb{F}_q$, i.e. the group of all field automorphisms of $\mathbb{F}_{q^v}$ element-wise fixing $\mathbb{F}_q$, is known to be cyclic of order $v$ and generated by the mapping $F : x \mapsto x^q$. $F$ is called *Frobenius automorphism*.

It is shown in [49, Satz 7.3] that the normalizer $N_{GL(V)}(S)$ of the Singer cycle group $S = \langle \sigma \rangle$ is given by

$$N_{GL(V)}(S) = \langle \sigma, F \rangle = \mathrm{Gal}(\mathbb{F}_{q^v}/\mathbb{F}_q) \rtimes S$$

and is of order $v(q^v - 1)$.

In the following, $S(v, q)$ (and $N(v, q)$) will denote the conjugacy class of all (normalizers of) Singer cycle groups in $GL(v, q)$.

For a matrix group $G \leq \mathrm{GL}(v, q)$ we denote with $G \times n$ the group

$$\left\{ \begin{pmatrix} M & 0 \\ 0 & I_n \end{pmatrix} \mid M \in G \right\},$$

*embedded* into $\mathrm{GL}(v + n, q)$.

If one looks at the groups which have been used in the computer searches so far, then this *normalizer of a Singer cycle group*, subgroups thereof, and embeddings into higher dimensions are the predominant groups. Is there a reason for "the success" of the normalizer of a Singer cycle group?

A possible answer can be found when looking to the analog situation for combinatorial designs. There, many designs have been found which have a *transitive* automorphism group. In fact, a very successful approach is to search for designs having an automorphism group which is transitive on the $t$-subsets of the point set. If this is the case, then the Kramer–Mesner matrix of Sect. 4.2 shrinks down to one row and every collection of columns trivially fulfills the matrix equation of Theorem 4 and therefore gives a $t$-design.

But for subspace designs we face a slightly different situation as for combinatorial designs. We say that a group $G$ acts $t$-*transitively* on a vector space $V$ if the set of $t$-subspaces is a single orbit. Now, one might expect to construct a $t$-$(v, k, \lambda)_q$ design by prescribing a $t$-transitive group for $t \geq 2$. Unfortunately, by the following theorem from [27, Prop. 8.4] this is only possible for the trivial design.

**Theorem 5** (Cameron, Kantor) *Let $G \leq \mathrm{P\Gamma L}(v, q)$ be $t$-transitive with $2 \leq t \leq v - 2$. Then $G$ is $k$-transitive for all $1 \leq k \leq v - 1$.*

That is, every group which is at least 2-transitive would shrink down the Kramer–Mesner matrix not only to one row, but also to one column. But then, the only possible choices are the trivial designs consisting of all $k$-subspaces or the empty design.

Nevertheless, when prescribing an automorphism group in the search for subspace designs, choosing a 1-transitive group seems to be reasonable.

In [81] a list of all 1-transitive subgroups of $\mathrm{GL}(v, q)$ is given, based on the work of [46, 47, 76].

**Theorem 6** (Hering, Liebeck, see [81]) *If $G \leq \mathrm{GL}(v, q)$ acts 1-transitively on the 1-subspaces of $\mathbb{F}_q^v$ with $v \geq 6$, then one of the following holds:*

*(a)* $G \leq N(v, q)$,
*(b)* $\mathrm{SL}(a, q^{v/a}) \trianglelefteq G$, *where $a \mid v$, $a \leq 2$,*
*(c)* $\mathrm{Sp}(2a, q^{v/2a}) \trianglelefteq G$, *where $2a \mid v$,*
*(d)* $G_2(q^{v/6}) \trianglelefteq G < \mathrm{Sp}(6, q^{v/6})$, *where $q = 2^m$ and $6 \mid v$,*
*(e)* $G \cong U(3, 3)$ *if $q = 2$ and $v = 6$,*
*(f)* $G \cong \mathrm{SL}(2, 13) < \mathrm{Sp}(6, 3)$ *if $q = 3$ and $v = 6$.*

*Here, $\mathrm{Sp}(2n, q)$ denotes the symplectic group $\mathrm{Sp}(2n, \mathbb{F}_q)$, $U(n, q)$ are the unitary groups, and $G_2(2^n)$ are the Chevalley groups.*

In the special case that $v$ is prime, the possible 1-transitive automorphism groups are exactly the subgroups of the normalizer of a Singer cycle group.

**Corollary 1** *If $G$ is a transitive automorphism group of a non-trivial subspace design $(\mathbb{F}_q^v, \mathcal{B})$ with $v$ prime, then $G$ is a subgroup of $N(v, q)$.*

## 4.4 Results

In [16] the first $q$-Steiner system for $t \geq 2$, an $S(2, 3, 13)_2$, has been constructed. Since it was found with the method of Kramer and Mesner, this is a good chance to demonstrate the power of this approach.

Without using groups the system (3) would have $\begin{bmatrix} 13 \\ 2 \end{bmatrix}_2 = 11\,180\,715$ rows and $\begin{bmatrix} 13 \\ 3 \end{bmatrix}_2 = 3\,269\,560\,515$ columns. A $q$-Steiner system, i.e. a selection of columns such that every row is covered exactly once has to consist of precisely $\begin{bmatrix} 13 \\ 2 \end{bmatrix}_2 / \begin{bmatrix} 3 \\ 2 \end{bmatrix}_2 = 1\,597\,245$ columns. However, finding a selection of that many columns is by far out of reach of existing computer algorithms.

Instead, we prescribe the normalizer $N(13, 2)$ of a Singer cycle group as automorphism group. Specifically, in [16] the Singer cycle group $S(13, 2)$ for the polynomial $f(x) = x^{13} + x^{12} + x^{10} + x^9 + 1$ has been used.

In order to represent a $G$-invariant design it is sufficient to list $G$ by its generators and a set of representatives of the selected orbits of $G$ on $\begin{bmatrix} V \\ k \end{bmatrix}$. For a compact representation we will write all $n \times m$ matrices $X$ over $\mathbb{F}_q$ with entries $x_{i,j}$, whose indices are numbered from 0, as vectors of integers

$$\left[ \sum_{j=0}^{m-1} x_{0,j} q^j, \ldots, \sum_{j=0}^{m-1} x_{n-1,j} q^j \right].$$

The order of the normalizer of the Singer cycle groups is $\#N(13, 2) = 13(2^{13} - 1) = 106\,483$. Using compressed notation, one pair of generators is

$$[2, 4, 8, 16, 32, 64, 128, 256, 512, 1024, 2048, 4096, 27] \text{ and}$$
$$[1, 4, 16, 64, 256, 1024, 4096, 54, 216, 864, 3456, 5659, 6234].$$

The orbits of 2-subspaces and 3-subspaces under the action of $N(13, 2)$ are all full-length, resulting in a Kramer–Mesner matrix $A_{2,3}^{N(13,2)}$ with $\begin{bmatrix} 13 \\ 2 \end{bmatrix}_2 / 106\,483 = 105$ rows and $\begin{bmatrix} 13 \\ 3 \end{bmatrix}_2 / 106\,483 = 30\,705$ columns. Now, we need to find a solution of the Eq. (8) consisting of

$$\frac{\begin{bmatrix} 13 \\ 2 \end{bmatrix}_2 / \begin{bmatrix} 3 \\ 2 \end{bmatrix}_2}{\#N(13, 2)} = 15$$

columns of $A_{2,3}^{N(13,2)}$. Luckily, it turns out that this problem is in the reach of contemporary computer algorithms and hardware. It can be solved e.g. with the algorithm *dancing links* by Knuth [59]. One solution corresponds to the 15 orbit representatives listed below:

$$[416, 2048, 4096], \quad [32, 3072, 4096], \quad [1344, 512, 4096],$$
$$[3440, 1536, 4096], \quad [8, 3328, 4096], \quad [3284, 3840, 4096],$$
$$[3428, 128, 4096], \quad [617, 2176, 4096], \quad [1038, 3200, 4096],$$
$$[1113, 2688, 4096], \quad [1338, 576, 4096], \quad [3389, 2368, 4096],$$
$$[317, 2880, 4096], \quad [1448, 192, 4096], \quad [774, 3232, 4096].$$

Meanwhile, 1316 *mutually disjoint* solutions have been determined with the algorithm from Sect. 5.1 [23]. Therefore, the union of $1 \leq \lambda \leq 1316$ of their block sets gives a 2-$(13, 3, \lambda)_2$ design. Together with their supplementary designs and noting that $\lambda_{max} = 2047$, this implies that all possible design parameters

$$2\text{-}(13, 3, \lambda)_2, \quad 1 \leq \lambda \leq 2047,$$

are realizable.

Chapter "$q$-Analog of Designs: Subspace Designs" in this book contains tables of all known subspace designs with small parameters. We conclude this section with a list of those designs from these table which are constructed by computer and give the prescribed groups as well as the sizes of the resulting Kramer–Mesner matrices.

The list contains one group which is not a subgroup of $N(v, q)$, namely the monomial group $M(3, 2^3)$, see [15].

$q = 2$:

2-$(6, 3, \lambda)_2$: For all designs $\langle \sigma^7 \rangle$ was used. Matrix size: $77 \times 155$.

2-$(7, 3, \lambda)_2$: All known 2-$(7, 3, \lambda)_2$ designs can be constructed by prescribing $S(7, 2)$. Matrix size: $3 \times 15$.

$\quad$ $N(7, 2)$ for $\lambda = 3, 5, 7, 10, 12, 14$. Matrix size: $21 \times 93$.

2-$(8, 3, 21)_2$: $N(4, 2^2)$. Matrix size: $15 \times 105$.

2-$(8, 4, \lambda)_2$: $N(4, 2^2)$ for $\lambda = 21, 35, 56, 70, 91, 105, 126, 140, 161, 175, 196, 210, 231, 245, 266, 280, 301, 315$. Matrix size: $15 \times 217$.

$\quad$ $N(7, 2) \times 1$ for $\lambda = 7, 14, 49, 56, 63, 98$ $105, 112, 147, 154, 161, 196, 203, 210, 245, 252, 259, 294, 301, 308$. Matrix size: $13 \times 231$.

$\quad$ $\langle \sigma^5, \phi \rangle$ for $\lambda = 217, 224$. Matrix size: $35 \times 531$.

3-$(8, 4, \lambda)_2$: $N(4, 2^2)$ for $\lambda = 11, 15$. Matrix size: $105 \times 217$.

$\quad$ $N(8, 2)$: $\lambda = 11$. Matrix size: $53 \times 109$.

2-$(9, 3, \lambda)_2$: $N(3, 2^3)$ for $\lambda = 21, 22, 42, 43, 63$. Matrix size: $31 \times 529$.

$\quad$ $N(8, 2) \times 1$ for $\lambda = 7, 12, 19, 24, 31, 36, 43, 48, 55, 60$. Matrix size: $28 \times 408$.

$\quad$ $M(3, 2^3)$ for $\lambda = 49$. Matrix size: $40 \times 460$.

2-$(9, 4, \lambda)_2$: $N(9, 2)$. Matrix size: $11 \times 725$.

2-$(v, 3, \lambda)_2$, $v = 10, 11, 13$: $N(v, 2)$. Matrix sizes: $20 \times 633$ for $v = 10$, $31 \times 2263$ for $v = 11$, $105 \times 30705$ for $v = 13$.

$q = 3$:

2-$(6, 3, \lambda)_3$: $\langle \sigma^{13}, \phi \rangle$ for $\lambda = 16$. Matrix size: $93 \times 234$.
$\quad$ $S(5, 3) \times 1$ for $\lambda = 8, 16, 20$. Matrix size: $51 \times 150$.
$\quad$ $\langle \sigma^2 \rangle \times 1$ for $\lambda = 8, 12, 16, 20$. Matrix size: $91 \times 280$.
2-$(v, 3, \lambda)_3$, $v = 7, 8$: $N(v, 3)$. Matrix sizes: $13 \times 121$ for $v = 7, 41 \times 977$ for $v = 8$.

$q = 4$:

2-$(6, 3, \lambda)_4$: $\langle \sigma^3, \phi \rangle$ for $\lambda = 15, 35$. Matrix size: $51 \times 161$
$\quad$ $\langle \sigma^3, \phi \rangle \times 1$, designs for $\lambda = 10, 25, 30, 35$. Matrix size: $57 \times 229$.

$q = 5$:

2-$(6, 3, 78)_5$: $\langle \sigma^2, \phi \rangle$. Matrix size: $53 \times 248$.

## 4.5  Non-existence Results

In [16] a list of non-existence results for prominent design parameters is given. Their results show that $q$-Steiner systems with the following parameters and automorphisms do not exist:

$$
\begin{array}{ll}
S(2, 3, 7)_2, & \text{Galois group } F(7, 2) \text{ (order 7)} \\
S(3, 4, 8)_2, & \text{Singer subgroup (order 255)} \\
S(2, 4, 10)_2, & \text{normalizer of Singer subgroup (order 10 230)} \\
S(2, 4, 13)_2, & \text{normalizer of Singer subgroup (order 106 483)} \\
S(3, 4, 10)_2, & \text{normalizer of Singer subgroup (order 10 230)} \\
S(2, 3, 7)_3, & \text{Singer subgroup (order 2 186)} \\
S(2, 3, 7)_3, & \text{normalizer of Singer subgroup (order 546 868)}
\end{array}
$$

This extends upon the previous work on non-existence of $q$-Steiner systems [34, 63, 92]. For example, it was shown in [92] that there is no Singer-invariant $q$-Steiner system $S(2, 3, 7)_2$.

In [19, 55], a systematic exclusion of automorphisms of an $S(2, 3, 7)_2$ has been performed. For the question of the existence of such a $G$-invariant design, the subgroups of $P\Gamma L(7, 2) = GL(7, 2)$ only need to be considered up to conjugacy, see Eq. (2). The result is that the automorphism group of an $S(2, 3, 7)_2$ $q$-Steiner system is either trivial or of order two. In the latter case, group is conjugate to

$$
\left\langle \begin{pmatrix} 0100000 \\ 1000000 \\ 0001000 \\ 0010000 \\ 0000010 \\ 0000100 \\ 0000001 \end{pmatrix} \right\rangle.
$$

Interestingly, also a 2-$(7, 3, 2)_2$ design is not known yet. For the automorphism group of a design with these parameters, the authors were able to exclude all subgroups of $GL(7, 2)$ except

- three conjugacy classes of groups of order 2,
- two conjugacy classes of groups of order 3,
- two conjugacy classes of groups of order 7.

## 4.6 Generation of the Kramer–Mesner Matrix

Computing orbit representatives for matrix groups can be very time consuming even for small parameters. One way to do this is to keep the bases of the subspaces in *row reduced echelon form* and the group action is realized as multiplication of matrices followed by Gauss elimination.

A fast method relying on the subgroup lattice to compute orbit representatives was developed by [89] for combinatorial designs. This approach was generalized to subspace designs by [13, 17, 24]. Recently, Koch [62] improved this method considerably by removing the high memory consumption of the original algorithm.

In case the prescribed group is the Singer cycle group or its normalizer the computation of the Kramer–Mesner matrix can be sped up enormously by considering vectors in $\mathbb{F}_q^v$ as elements of the finite field $\mathbb{F}_{q^v}$. We use this to show how to classify the subspaces of $\mathbb{F}_q^v$ into orbits under the action of Singer cycle group or its normalizer, see [16, 35].

We fix a primitive element $\alpha$ of $\mathbb{F}_{q^v}$, and write a $k$-subspace $X$ of $\mathbb{F}_q^v$ as $X = \{0, \alpha^{x_1}, \alpha^{x_2}, \ldots, \alpha^{x_m}\}$, where $m = q^k - 1$ and $x_1, x_2, \ldots, x_m \in \mathbb{Z}_{q^v-1}$. In other words, we represent all nonzero vectors in $\mathbb{F}_q^v$ by their discrete logarithm with regard to basis $\alpha$.

For $x \in \mathbb{Z}_{q^v-1}$, let $\rho(x)$ be the minimal cyclotomic representative for $x$, that is $\rho(x) = \min\{xq^i \mod (q^v - 1) \mid 0 \leq i \leq v - 1\}$. We define

$$\text{inv}_F(X) := \{\rho(x_i) : 1 \leq i \leq m\},$$
$$\text{inv}_S(X) := \{x_i - x_j : 1 \leq i, j \leq m \text{ with } i \neq j\},$$
$$\text{inv}_N(X) := \{\rho(x_i - x_j) : 1 \leq i, j \leq m \text{ with } i \neq j\}.$$

**Lemma 2** *1. If two $k$-subspaces $X$, $Y$ of $\mathbb{F}_q^v$ are in the same orbit under the action of the Galois group $F(v, q)$ then $\text{inv}_F(X) = \text{inv}_F(Y)$.*
*2. If two $k$-subspaces $X$, $Y$ of $\mathbb{F}_q^v$ are in the same orbit under the action of the Singer cycle group $S(v, q)$ then $\text{inv}_S(X) = \text{inv}_S(Y)$.*
*3. If two $k$-subspaces $X$, $Y$ of $\mathbb{F}_q^v$ are in the same orbit under the action of the normalizer $N(v, q)$ of the Singer cycle group then $\text{inv}_N(X) = \text{inv}_N(Y)$.*

*Proof* Let $X = \{0, \alpha^{x_1}, \alpha^{x_2}, \dots, \alpha^{x_m}\}$ be a $k$-subspace of $\mathbb{F}_q^v$, with $x_1, x_2, \dots, x_m$ in $\mathbb{Z}_{q^v-1}$. The action of the generator of $S(v, q)$ on $X$ increases $x_1, x_2, \dots, x_m$ by one modulo $q^v - 1$, thereby preserving the differences between them.

The action of the Frobenius automorphism $\phi$ on $X$ multiplies each $x_i$ by $q$ modulo $q^v - 1$, thereby leaving it in the same cyclotomic coset.                                                                    $\square$

So, instead of using costly matrix multiplication we can compute the orbit representatives with simple arithmetics modulo $q^v - 1$.

In the special situation that $q^v - 1$ is a prime we can speed up the computation of $\text{inv}_N$ even further. One example is $q = 2$ and $v = 13$. Instead of working in the finite field $\mathbb{Z}_{q^v-1}$ we take once more the discrete logarithms and work modulo $q^v - 2$. By doing this, the action of the Frobenius automorphism $\phi$ reduces to an addition by $(q^v - 2)/v$. Thus, the numbers in $\text{inv}_N$ are considered modulo $(q^v - 2)/v$.

## 4.7  Isomorphism Problems for Subspace Designs

Whenever subspace designs are constructed, immediately the natural question arises which of them are isomorphic. The abstract formulation of this problem is the following: Given the action of a group $G$ on a set $X$, and $x, x' \in X$, how can we decide if $x \sim x'$? In our situation of subspace designs, the group $G$ typically is $\text{GL}(v, q)$ or $\text{PGL}(v, q)$ or $\text{P}\Gamma\text{L}(v, q)$ etc., and the set $X$ is the power set of $\begin{bmatrix} V \\ k \end{bmatrix}$.

A first approach is the computation of invariants. Invariants are functions $i : X \to \Omega$ with some set $\Omega$ which are constant on the orbits, that is, $x \sim x'$ implies $i(x) = i(x')$. In the case $i(x) \neq i(x')$, we know that $x \nsim x'$. However, in general an invariant cannot prove that $x \sim x'$, so we need further methods which we apply in the case $i(x) = i(x')$. A good invariant should be reasonably efficient to compute and not too often evaluate to the same expression if $x$ and $x'$ are from different orbits. A suitable source of invariants are the intersection numbers of blocks [56].

Sometimes, the following pragmatic approach can be applied: The isomorphism problem is transformed into a graph theoretic problem and then fed into the software nauty and its successors by McKay [80] to solve the graph isomorphism problem. Furthermore, we would like to mention the algorithm of [38] (based on [37], see also [39]), which is specialized for isomorphism problems on sets of subspaces of a vector space.

Now we focus on the quite common situation that our designs have been constructed by subscribing some subgroup $H$ of $G$, for example by the method of Kramer–Mesner. Compared to the above situation, now we have the additional information $H \subseteq G_x$ and $H \subseteq G_{x'}$. To take advantage of this extra knowledge, the theory of group actions comes in handy. By theoretical arguments, often it is possible to simplify or even solve the isomorphism problem.

For combinatorial designs, this approach was described by Laue et. al. [44, 71, 72], see also [53]. It involves the normalizers of certain groups. For a group $G$ with subgroup $H$, the normalizer of $H$ in $G$ will be denoted by $N_G(H)$.

**Lemma 3** *Let $G$ be a group acting on a set $X$. For any $x \in X$,*

$$N_G(G_x) = \{g \in G \mid G_x = G_{x^g}\}.$$

*Proof* This is a direct consequence of Eq. (2) $G_{x^g} = g^{-1}G_x g$. □

The above lemma has the following intuitive interpretation: The normalizer of the stabilizer of $x$ consists of all elements of $G$ which preserve the symmetries of $x$.

As a consequence, to investigate two elements $x$, $x'$ with the same stabilizer $H$ for $x \sim x'$, it is enough to consider the action of the subgroup $N_G(H)$ instead of the full group $G$. Depending on the group orders, this may be a significant reduction of the computational complexity. Unfortunately, in general this is not directly applicable in our situation: Typically, we only know that $x$ and $x'$ are $H$-invariant, but we don't know the full stabilizers $G_x$ and $G_{x'}$, which may well be larger and not identical to each other.

However, if we are in the comfortable situation that $H$ is a maximal (but proper) subgroup of $G$, then in all cases of interest, $H$ has to be the full automorphism group of $H$-invariant designs (as otherwise, the full automorphism group would be $G$, which acts transitively on $\binom{V}{k}$ and therefore only admits trivial solutions). If additionally $H$ is non-normal in $G$, then $N_G(H) = H$ and hence by Lemma 3, any two distinct $H$-invariant designs are non-isomorphic (with respect to the action of $G$). In the case that the maximal group $H$ is a normal subgroup of $G$, we have $N_G(H) = G$. Now by Lemma 3 we get that each $H$-invariant design will appear in exactly $[G : H]$ isomorphic copies among the set of all $H$-invariant designs.

*Example 5* The $q$-Steiner system $2\text{-}(13, 3, 1)_2$ has been found by prescribing the normalizer $N = N(13, 2)$ of a Singer cycle group in $GL(13, 2)$. As $N$ is a non-normal maximal subgroup of $GL(13, 2)$, any two different solutions of the Kramer–Mesner system describe non-isomorphic designs. In this way, we know that in fact there are thousands of non-isomorphic $q$-Steiner systems $2\text{-}(13, 3, 1)_2$.

If the prescribed group $H$ is a proper subgroup of its normalizer $N = N_G(H)$, this can be exploited algorithmically. For example, if we force one orbit $K^H$ to be in the design, and the solving algorithm shows that there is no solution which contains this orbit, all $K$-orbits in $(K^H)^N$ can be excluded from being part of a solution, i.e. the corresponding columns of the Kramer–Mesner matrix can be removed.

For algorithmic purposes also the Sylow subgroups of $G$ and $H$ are very valuable. The following theorem is a slight generalization of [49, Hilfssatz IV 2.5] and [10, Theorem 3.1].

**Theorem 7** *Let $G$ be a group acting on a set $X$, $x, x' \in X$ with $x \sim x'$ and $P$ a common Sylow subgroup of $G_x$ and $G_{x'}$. Then there exists an $n \in N_G(P)$ with $x' = x^n$.*

*Proof* Let $g \in G$ with $x' = x^g$. From $g^{-1}Pg \subseteq g^{-1}G_x g = G_{x^g} = G_{x'}$, both groups $P$ and $g^{-1}Pg$ are Sylow subgroups of $G_{x'}$. As any two Sylow subgroups of the same order are conjugate, there is a $h \in G_{x'}$ with

$$P = h^{-1}(g^{-1}Pg)h = (gh)^{-1}P(gh).$$

Now $n = gh \in N_G(P)$ satisfies $x^n = (x^g)^h = (x')^h = x'$.                                              □

By Lemma 3, we saw that in the case $G_x = G_{x'}$, the investigation of $x \sim x'$ may be done using the action of the subgroup $N(G_x)$ instead of $G$, which potentially provides a huge improvement in terms of computational complexity. Now by Theorem 7, the knowledge of a common Sylow subgroup $P$ of $G_x$ and $G_{x'}$ is enough to provide a similar simplification. In this case, the acting group $G$ may be replaced by the subgroup $N_G(P)$.

*Example 6* We illustrate the above theorem with 2-$(9, 4, 21)_2$ designs.

Prescribing $N(9, 2)$ in the Kramer–Mesner method gives many solutions, which are $N(9, 2)$-invariant 2-$(9, 4, 21)_2$ designs. We can prove that all these designs are mutually non-isomorphic without having knowledge of their full automorphism groups. Let us point out that $N(9, 2)$ is not maximal in $\mathrm{GL}(9, 2)$, so the argument of Example 5 doesn't work in this case.

We have $\#N(9, 2) = (2^9 - 1) \cdot 9 = 3^2 \cdot 7 \cdot 73$. Let $P$ be a Sylow-73 subgroup of $N(9, 2)$. Then $\#P = 73$, and by $\#\mathrm{GL}(9, 2) = \prod_{i=0}^{8}(2^9 - 2^i) = 2^{36} \cdot 3^5 \cdot 5^2 \cdot 7^3 \cdot 17 \cdot 31 \cdot 73 \cdot 127$, $P$ is also a Sylow-73 subgroup of $\mathrm{GL}(9, 2)$. Hence, $P$ is a Sylow-73 subgroup of the full automorphism group of any $N(9, 2)$-invariant design. With the above theorem we can conclude that if two $N(9, 2)$-invariant designs $\mathcal{D}_1 = (V, \mathcal{B}_1)$, $\mathcal{D}_2 = (V, \mathcal{B}_2)$ are isomorphic there would be an element $n \in N_{\mathrm{GL}(9,2)}(P)$ with $\mathcal{B}_1^n = \mathcal{B}_2$. But $N_{\mathrm{GL}(9,2)}(P) = N(9, 2)$ and we know that both designs are $N(9, 2)$-invariant. Therefore, $\mathcal{D}_1$ and $\mathcal{D}_2$ are non-isomorphic.

For the cases where Theorem 7 is not applicable, further theoretical results exist, see [72]. One example is the following theorem.

**Theorem 8** *Let $G \leq \mathrm{GL}(v, q)$ and let $\Delta$ be the set of $G$-invariant $t$-$(v, k, \lambda)_q$ designs. Let $P$ be a Sylow $p$-subgroup of $G$. Then after removing every design from $\Delta$ that is invariant under any subgroup $H$ such that either*

- $G < H < N_{\mathrm{GL}(v,q)}(G)$ *or*
- $H = \langle G, G^n \rangle$ *for some $n \in N_{\mathrm{GL}(v,q)}(P)$ with $G^n \neq G$*

*two designs in $\Delta$ are isomorphic if and only if some element from $N_{\mathrm{GL}(v,q)}(G)$ maps one onto the other.*

## 5 Constructing "Large Sets of Designs" by Computer

The method of Kramer–Mesner can also be adapted to construct large sets of designs. This is explained for combinatorial designs in Chee [28]. In fact, the authors describe three approaches based on the Kramer–Mesner theorem.

1. Use $t$-homogeneous groups, i.e. the Kramer–Mesner matrix shrinks down to a single row. This is not possible for subspace designs with $t \geq 2$ because of Theorem 5.
2. *Recursive approach.* Construct a design from the Kramer–Mesner matrix, remove the columns of the matrix which correspond to the design orbits and recurse.
3. *Isomorphic designs.* Search for uniformly-$G$ large set consisting of $N$ isomorphic designs.

For the time being, large sets with $t \geq 2$ could only be constructed by computer with the second approach. Sarmiento [88] used an hybrid approach mixing methods 2 and 3 to classify all $LS_3[155](1, 3, 6)$ consisting of 155 point-transitive, uniformly-$G$ $1$-$(6, 3, 1)_3$ designs.

## 5.1 Recursive Approach

The following algorithm describes a basic approach to find large sets. A version of this algorithm for large sets of combinatorial designs can be found in [28, 73, 74].

The algorithm computes an $LS_q[N](t, k, v)$ large set $\mathscr{L}$ consisting of $N$ $G$-invariant $t$-$(v, k, \lambda)_q$ designs. Either the algorithm terminates with a large set or it ends without any statement about the existence.

**Algorithm 3** 1. [*Initialize.*] Set $\Omega$ as the complete set of $G$-orbits on $\begin{bmatrix} V \\ k \end{bmatrix}$ and set $\mathscr{L} := \emptyset$.
2. [*Solve.*] Find a random $t$-$(v, k, \lambda)_q$ design $\mathscr{D} = (V, \mathscr{B}))$ consisting of orbits of $\Omega$. If such a $t$-design exists insert $\mathscr{D}$ into $\mathscr{L}$ and continue with 3. Otherwise terminate without a large set.
3. [*Remove.*] Remove the selected orbits in $\mathscr{B}$ from $\Omega$. If $\Omega = \emptyset$ then terminate with a large set $\mathscr{L}$. Otherwise go to 2.

The described algorithm can be implemented by a slight modification of the Kramer–Mesner approach. We just have to add a further row to the Diophantine system of Eq. (8) in the following way:

$$\begin{pmatrix} A_{t,k}^G \\ \hline \cdots y \cdots \end{pmatrix} \cdot x = \begin{pmatrix} \lambda \\ \vdots \\ \lambda \\ \hline 0 \end{pmatrix}$$

The vector $y$ is indexed by the $G$-orbits on $\begin{bmatrix} V \\ k \end{bmatrix}$ corresponding to the columns of $A_{t,k}^G$. The entry indexed by the $G$-orbit containing $K$ is defined to be one if the orbit of $K$ has already been covered by a selected $t$-$(v, k, \lambda)_q$ design. Otherwise it is zero. In every iteration step the vector $y$ has to be updated.

*Remark 4* Another approach is to compute all $0/1$-vectors which are solutions of the Kramer–Mesner system (8) for a given group $G$. Every solution vector corresponds to a design.

In a second step we try to find a subset of $N$ disjoint solution vectors. This is again an exact cover problem.

## 5.2 Large Sets from Isomorphic Designs

This approach is described in detail in [28] for combinatorial designs. The task will be to search for a uniformly-$G$ large set $LS_q[N](t, k, v)$, where $G \leq H \leq P\Gamma L(v, q)$. Moreover, the designs of the large set will not only be $G$-invariant, but also mutually isomorphic.

Let $n_k^G$ and $n_k^H$ be the number of orbits of $G$ and $H$ acting on $\begin{bmatrix} V \\ k \end{bmatrix}$ and fix some order $K_j^G$, $1 \leq j \leq n_k^G$, and $K_i^H$, $1 \leq i \leq n_k^H$, on the $G$-orbits and $H$-orbits acting on $\begin{bmatrix} V \\ k \end{bmatrix}$.

The *fusion matrix* $F_k^{G,H} = (f_{i,j})$ is the $n_k^H \times n_k^G$ matrix defined by

$$f_{i,j} = \begin{cases} 1, & \text{if } K_j^G \subseteq K_i^H; \\ 0, & \text{else.} \end{cases}$$

We want to find a large set consisting of designs $\mathcal{D}_1, \mathcal{D}_2, \ldots, \mathcal{D}_N$ such that each design $\mathcal{D}_i$ is $G$-invariant. Suppose further that we want an isomorphism $g \in P\Gamma L(V)$ of order $N$ such that

$$\mathcal{D}_1^{g^i} = \mathcal{D}_i, \quad 0 \leq i < N.$$

Let $H = \langle G, \sigma \rangle$. Then, an orbit of $H$ on $k$-subspaces is the union of (disjoint) $G$-orbits on $k$-subspaces.

As a consequence, if we are able to find a design $(V, \mathcal{B}_1)$ that contains exactly one $G$-orbit from every $H$-orbit, then $\{\mathcal{B}_1^{g^i} \mid 0 \leq i < N\}$ gives rise to disjoint designs, i.e. a large set of designs. This can be summarized in the following theorem.

**Theorem 9** (Chee [28]) *Suppose that* $\sigma \in P\Gamma L(V)$ *with* $\text{ord}(\sigma) = N$ *and* $G \leq P\Gamma L(V)$. *Let* $H = \langle G, \sigma \rangle$.

*There exists a uniformly-$G$ large set* $LS_q[N](t, k, v)$ *if there is a $0/1$-vector $x$ satisfying*

$$\begin{pmatrix} A_{t,k}^G \\ F_k^{G,H} \end{pmatrix} \cdot x = \begin{pmatrix} \lambda \\ \vdots \\ \lambda \\ 1 \\ \vdots \\ 1 \end{pmatrix}.$$

## 6   Solving Algorithms

Solving Eq. (8) is a special instance of the *multi-dimensional subset sum problem* which is known to be NP-complete [41]. Since problem (8) can be reduced to many other NP-hard problems it is no surprise that there are many solving algorithms available. In this section we will give an overview of the so far most promising strategies to find subspace designs. For a survey, see also [42, 53, 78].

### 6.1   Backtracking

The problem (8) for $\lambda = 1$ is known as the *exact cover problem*. A common approach to find all solution of an exact cover problem is to systematically test all combinations of block orbits. Walker [93] was the first to call such an approach *back-track*.

A backtracking algorithm for solving the system (8) for $\lambda = 1$ and Kramer–Mesner matrix $A = (a_{i,j})$ is quite simple and straight forward to describe:

**Algorithm 4**  Choose columns of $A$ to find a solution of (8) for $\lambda = 1$.

1.  If $A$ has no columns, the problem is solved; terminate successfully.
2.  Otherwise choose a row $r$ with the least number of nonzero entries.
3.  **for each** column $c$ such that $a_{r,c} = 1$:
    include $c$ in the partial solution
    **for each** $i$ such that $a_{i,c} = 1$
    delete row $i$ from the matrix $A$;
    **for each** $j$ such that $a_{i,j} = 1$,
    delete column $j$ from the matrix $A$.
4.  Repeat this algorithm recursively on the reduced matrix $A$.

In [58] the strategy of Step 2 to choose those rows first which have the least number of nonzero entries is justified. If the goal is to find all solutions then the order in which the columns in Step 3 taken is irrelevant. This may be different if one wants to find one solution at all.

The speed of this algorithm is largely determined by the choice of the data structures. In [59] Knuth uses doubly linked lists, all the navigation through the matrix is done via pointers. By a trick due to Hitotumatu and Noshita [48], the use of pointers enables a very fast recovering of the original data after stepping back from recursion. The algorithm is called *dancing links*.

In [95] a parallel version of dancing links is described. [55] uses a brute force parallelization which is better suited for batch system on computing clusters.

For problem instances with $\lambda > 1$ the situation changes. The dancing links algorithm can be adapted to this case. The library `libexact` [54] contains an implementation, see also [79].

## 6.2 Maximum Clique Algorithms

A *weighted graph* $G = (V, E)$ is a set of vertices $V$ together with a set of edges $E$ which are 2-element subsets of $V$. Additionally, every vertex carries a (nonzero) weight, i.e. there is a mapping wgt $: V \to \mathbb{Z}_{\geq 0}$.

A *clique* of a graph $G$ is a set of vertices $C \subset V$ such that $\{v, w\} \in E$ for all $v, w \in C, v \neq w$.

The following decision problem is NP-complete: Given a weighted graph $G$ and an integer $k$, is there a clique in $G$ with weight at least $k$? Although no polynomial-time algorithm is known for the maximum weight clique problem, various algorithms have been developed as these have many important applications. `cliquer` [85] is one freely available software package that solves instances of the maximum weight clique problem.

For $\lambda = 1$ problem (8) can be formulated as maximum weight clique problem. The vertices of the graph are the block orbits. Their weights are given by the orbit lengths. Two orbits share an edge if and only if they cover disjoint sets of $t$-subspaces.

This approach is promising if the cardinality of a maximum clique can be expected to be reasonably small. The advantage of the method is that it can tackle instances with very many block orbits.

If one is interested in improving the known bounds for packing designs, i.e. constant dimension subspace codes, then it is sometimes good enough to find approximate solutions of the maximum clique problem. In such a situation one may resort to stochastic methods.

Stochastic algorithms for finding cliques have been thoroughly studied, but most of the studies consider unweighted graphs. For some recent results, see [87]. Typically, a stochastic algorithm proceeds by adding and removing single vertices in a specific manner, when building up a large clique.

Lower bounds for constant dimension subspace codes could be improved by such a stochastic algorithm in [21].

For the time being, no $q$-Steiner systems could be found by this approach. In [35] it is reported that with a weighted clique approach 14 of the necessary 15 orbits could be packed together in the search for 2-$(13, 3, 1)_2$ design.

## 6.3 Lattice Point Enumeration

When one attempts to solve systems of type (8) for large values of $\lambda$ a detour via lattices proves to be worthwhile.

Suppose that $b_1, b_2, \ldots, b_n \in \mathbb{Q}^m$. The integer span of these vectors, i.e.

$$L = \{\sum_{i=1}^{n} u_i b_i \mid u_i \in \mathbb{Z}\}$$

is called *lattice*. Given a lattice, central problems are to find shortest nonzero lattice vectors with regard to various norms $\|.\|$ and to find a basis of $L$ consisting of short vectors with regard to $\|.\|_2$. We will not dive further into this subject and not give a precise definition what a "basis of short vectors" exactly is. The reader is referred to [84] for an extensive overview.

In 1982, Lenstra, Lenstra and Lovász [75] gave a celebrated method—*LLL algorithm*—which produces approximate solutions to both problems in polynomial time. The LLL algorithm does *lattice basis reduction* – it takes a lattice basis as input and outputs a reduced basis of the lattice, hopefully consisting of short vectors. The amazing power of the LLL algorithm is that it performs much better than the theoretical analysis predicts and in many practical cases the output bases are already solutions of the above problems.

The first ones to use lattice basis reduction for the search of combinatorial designs were Kreher and Radziszowski [66, 67]. They used the original LLL algorithm as proposed in [75] and the formulation of problem (8) as lattice problem from Lagarias and Odlyzko [69].

Since then, this approach could be improved in many aspects. With improved variants of the LLL algorithm much smaller basis vectors can be achieved, see [84]. In [94] an improved lattice formulation of problem (8) is given. For this lattice, a solution of (8) corresponds to a shortest nonzero lattice vector in $\|.\|_\infty$ norm. This was generalized in [96] to non-simple designs.

Further, the lattice basis reduction algorithms behave somewhat random. But [90] proposes a method to find a shortest vector for the Euclidean norm from an LLL reduced lattice basis by exhaustive enumeration. This was generalized by [51] to arbitrary norms.

Combining these improvements many combinatorial designs for $t = 6, 7, 8, 9$ were found [4, 5, 7–12, 70, 71, 94]. The algorithm is used in a software system called DISCRETA [6] which allows the user to easily prescribe automorphism groups and try to solve the corresponding Kramer–Mesner system of equations.

The algorithm was also the method of choice to construct subspace designs in many publications, see [13, 14, 17–20, 24] for an incomplete list.

Let $A$ be an $l \times s$ Kramer–Mesner matrix from (8). The corresponding lattice proposed in [94] is generated by the linearly independent columns of the matrix

$$
L = \begin{pmatrix}
 & & c \cdot \lambda \\
 & c \cdot A & \vdots \\
 & & c \cdot \lambda \\
\hline
2 & 0 & 1 \\
 & \ddots & \vdots \\
 & 0 & 2 & 1 \\
0 & \cdots & 0 & 1
\end{pmatrix},
\tag{9}
$$

consisting of $s + 1$ column vectors with $l + s + 1$ rows. The constant $c$ is chosen large enough such that the output of the lattice basis reduction looks like

$$\begin{pmatrix} 0 & * \\ B & * \end{pmatrix}$$

and $B$ has size $(s + 1) \times (s + 1 - \mathrm{rk}(A))$.

In the second phase of the algorithm we search with the enumeration algorithm from [51] for all nonzero lattice vectors $b$ of the lattice spanned by the columns of $B$ such that $\|b\|_\infty = 1$. The integer vectors $b = (b_1, \ldots, b_s, b_{s+1})^\top$ of the lattice spanned by the columns of $B$ fulfill the equation

$$A \cdot \begin{pmatrix} (b_1 - b_{s+1})/2 \\ \vdots \\ (b_s - b_{s+1})/2 \end{pmatrix} = -b_{s+1} \begin{pmatrix} \lambda \\ \vdots \\ \lambda \end{pmatrix}.$$

The condition $\|b\|_\infty = 1$ ensures that $b_{s+1} = \pm 1$ and that $b_{s+1} \cdot (b_i - b_{s+1})/2 \in \{0, 1\}$ for $1 \leq i \leq s$.

The big advantage of this method is that the size of $\lambda$ has not much influence on the runtime. This is in strong contrast with the backtracking approach above which is best for $\lambda = 1$.

The height of the search tree in the second phase is determined by the number of columns of $B$, i.e. it is roughly the difference between the number of unknowns and the number of equations. As a consequence, having many equations is good for the runtime of the algorithm and the number of orbits which have to be selected has not much influence on the runtime.

The disadvantage of this lattice based approach is that the lattice basis reduction in the first phase—albeit having polynomial runtime—takes prohibitively long time for systems when the number of unknowns exceeds 3 000 (as a rule of thumb).

## 6.4 Counting Algorithms

If the design parameters are small enough it may be easy to find one solution and one is tempted to try to count all designs having automorphisms. Schmalz [89] developed a graph theoretical approach to enumerate all solutions of (8) implicitly.

A similar approach is known in computer science as *binary decision diagrams* and related data structures, see [60]. In [61, Exercise 50] Knuth describes a binary method to count all solutions.

## 6.5 Integer Linear Programming

The system of Eq. (8) can be regarded as *integer linear programming* problem. One possible formulation is

$$\max \sum_i x_i \quad \text{such that}$$

$$A_{t,k}^G \cdot x = \begin{pmatrix} \lambda \\ \vdots \\ \lambda \end{pmatrix},$$

$$x_i \in \{0, 1\}.$$

There are a many software systems available to solve integer linear programming problems. Among the most powerful and popular are CPLEX [50] and Gurobi [43].

It seems that for problems of type (8) integer linear programming algorithms are still inferior to lattice basis reduction algorithms. But often the linear programming part—i.e. using the relaxation $x_i \in [0, 1] \subset \mathbb{R}$—of these solvers is sufficient to show nonexistence of solutions.

Moreover, these solving algorithms show their power when it comes to find packing designs, i.e. replacing "=" by "≤". The online tables for the best bounds on subspace codes[1] contain many example where lower bounds were constructed by using integer linear programming, compare [45, 63].

## 6.6 Other Algorithms

As already mentioned, since problem (8) is an NP-complete problem it can be reduced to other NP-hard problems. Recently, randomized algorithms were successful for related packing problems, e.g. [22, 97]. Other randomized algorithms are *simulated annealing, tabu search, hill climbing*, see [42]. But also *constraint logic algorithms* and *SAT algorithms* are candidates to be tried out.

## References

1. W.O. Alltop, On the construction of block designs. J. Comb. Theory **1**, 501–502 (1966)
2. C. Berge, D. Ray-Chaudhuri, in *Unsolved problems*, ed. by C. Berge, D. Ray-Chaudhuri. Hypergraph Seminar: Ohio State University 1972, in Lecture Notes in Mathematics, vol. 411 (Springer, Berlin, 1974), pp. 278–287. https://doi.org/10.1007/BFb0066199

---

[1] http://subspacecodes.uni-bayreuth.de.

3. T. Beth, D. Jungnickel, H. Lenz, *Design Theory*, vol. 1, 2, 2nd edn. (Cambridge University Press, London, 1999)
4. A. Betten, A. Kerber, A. Kohnert, R. Laue, A. Wassermann, The discovery of simple 7-designs with automorphism group PΓL(2,32), in *AAECC 11*, Lecture Notes in Computer Science, vol. 948 (Springer, Heidelberg, 1995), pp. 131–145
5. A. Betten, A. Kerber, R. Laue, A. Wassermann, Simple 8-designs with small parameters. Des. Codes Cryptogr. **15**(1), 5–27 (1998). https://doi.org/10.1023/A:1008263724078
6. A. Betten, R. Laue, A. Wassermann, DISCRETA – A tool for constructing $t$-designs. Lehrstuhl II für Mathematik, Universität Bayreuth, http://www.mathe2.uni-bayreuth.de/discreta/
7. A. Betten, R. Laue, A. Wassermann, Simple 6 and 7-designs on 19 to 33 points. Congr. Numer. **123**, 149–160 (1997)
8. A. Betten, R. Laue, A. Wassermann, Some simple 7-designs, eds. by. J.W.P. Hirschfeld, S.S. Magliveras, M.J. de Resmini Geometry, Combinatorial Designs and Related Structures, Proceedings of the First Pythagorean Conference, London Mathematical Society Lecture Notes, vol. 245, pp. 15–25 (1997)
9. A. Betten, R. Laue, A. Wassermann, New $t$-designs and large sets of $t$-designs. Discret. Math. **197/198**, 83–109 (1999). Also appeared in the special volume Discrete Mathematics, Editor's Choice, Edition 1999
10. A. Betten, R. Laue, A. Wassermann, Simple 7-designs with small parameters. J. Comb. Des. **7**, 79–94 (1999)
11. A. Betten, R. Laue, A. Wassermann, Simple 8-(40, 11, 1440) designs. Discret. Appl. Math. **95**, 109–114 (1999)
12. A. Betten, R. Laue, A. Wassermann, A Steiner 5-design on 36 points. Des. Codes Cryptogr. **17**, 181–186 (1999)
13. M. Braun, Konstruktion diskreter Strukturen unter Verwendung von Operationen linearer Gruppen auf dem linearen Verband, Ph.D. thesis, University of Bayreuth, Germany (2004)
14. M. Braun, Some new designs over finite fields. Bayreuth. Math. Schr. **74**, 58–68 (2005)
15. M. Braun, Designs over the binary field from the complete monomial group. Australas. J. Comb. **67**(3), 470–475 (2017)
16. M. Braun, T. Etzion, P.R.J. Östergård, A. Vardy, A. Wassermann, Existence of $q$-analogs of steiner systems. Forum Math. Pi **4**(e7), 14 (2016). https://doi.org/10.1017/fmp.2016.5
17. M. Braun, A. Kerber, R. Laue, Systematic construction of $q$-analogs of $t$-$(v, k, \lambda)$-designs. Des. Codes Cryptogr. **34**(1), 55–70 (2005). https://doi.org/10.1007/s10623-003-4194-z
18. M. Braun, M. Kiermaier, A. Kohnert, R. Laue, Large sets of subspace designs. J. Comb. Theory Ser. A **147**, 155–185 (2017). https://doi.org/10.1016/j.jcta.2016.11.004
19. M. Braun, M. Kiermaier, A. Nakić, On the automorphism group of a binary $q$-analog of the fano plane. Eur. J. Comb. **51**, 443–457 (2016). https://doi.org/10.1016/j.ejc.2015.07.014
20. M. Braun, A. Kohnert, P.R.J. Östergård, A. Wassermann, Large sets of $t$-designs over finite fields. J. Comb. Theory Ser. A **124**, 195–202 (2014). https://doi.org/10.1016/j.jcta.2014.01.008
21. M. Braun, P.R.J. Östergård, A. Wassermann, New lower bounds for binary constant-dimension subspace codes. Exp. Math. 1–5 (2016). https://doi.org/10.1080/10586458.2016.1239145
22. M. Braun, J. Reichelt, $q$-analogs of packing designs. J. Comb. Des. **22**(7), 306–321 (2014). https://doi.org/10.1002/jcd.21376
23. M. Braun, A. Wassermann, *Disjoint q-Steiner systems in dimension 13* Universität Bayreuth, Bayreuth, Technical Report (2017)
24. S. Braun, Algorithmen zur computerunterstützten Berechnung von $q$-Analoga kombinatorischer Designs. Diplomathesis Universität Bayreuth (2009)
25. P.J. Cameron, Generalisation of Fisher's inequality to fields with more than one element. eds. by T.P. McDonough, V.C. Mavron. Combinatorics - Proceedings of the British Combinatorial Conference 1973, London Mathematical Society Lecture Note Series, vol. 13 (Cambridge University Press, Cambridge, 1974), pp. 9–13. https://doi.org/10.1017/CBO9780511662072.003
26. P.J. Cameron, Locally symmetric designs. Geom. Dedicata **3**, 65–76 (1974)

27. P.J. Cameron, W.M. Kantor, 2-transitive and antiflag transitive collineation groups of finite projective spaces. J. Algebra **60**(2), 384–422 (1979). https://doi.org/10.1016/0021-8693(79)90090-5

28. Y.M. Chee, C.J. Colbourn, S.C. Furino, D.L. Kreher, Large sets of disjoint $t$-designs. Australas. J. Comb. **2**, 111–119 (1990)

29. C.J. Colbourn, J.H. Dinitz, in *Handbook of Combinatorial Designs*, 2nd edn, Discrete Mathematics and Its Applications. (Chapman and Hall/CRC , 2006)

30. M. De Boeck, A. Nakić, Necessary conditions for the existence of 3-designs over finite fields with nontrivial automorphism groups. ArXiv e-prints arXiv:1509.09158 (2015)

31. P. Delsarte, Association schemes and $t$-designs in regular semilattices. J. Comb. Theory Ser. A **20**(2), 230–243 (1976). https://doi.org/10.1016/0097-3165(76)90017-0

32. P. Dembowski, Verallgemeinerungen von Transitivitätsklassen endlicher projektiver Ebenen. Math. Z. **69**, 59–89 (1958)

33. P. Dembowski, Finite Geometries: Reprint of the 1968 Edition. (Springer, 2012)

34. T. Etzion, A. Vardy, On $q$-analogs of Steiner systems and covering designs. Adv. Math. Commun. **5**(2), 161–176 (2011). https://doi.org/10.3934/amc.2011.5.161

35. T. Etzion, A. Vardy, Automorphisms of codes in the Grassmann scheme. ArXiv e-prints arXiv:1210.5724 (2012)

36. A. Fazeli, S. Lovett, A. Vardy, Nontrivial $t$-designs over finite fields exist for all $t$. J. Comb. Theory Ser. A **127**, 149–160 (2014)

37. T. Feulner, The automorphism groups of linear codes and canonical representatives of their semilinear isometry classes. Adv. Math. Commun. **3**(4), 363–383 (2009). https://doi.org/10.3934/amc.2009.3.363

38. T. Feulner, Canonical forms and automorphisms in the projective space (2013)

39. T. Feulner, Eine kanonische Form zur Darstellung äquivalenter Codes – Computergestützte Berechnung und ihre Anwendung in der Codierungstheorie, Kryptographie und Geometrie. Ph.D. thesis, Universität Bayreuth (2013)

40. P. Frankl, V. Rödl, Near perfect coverings in graphs and hypergraphs. Eur. J. Comb. **6**(4), 317–326 (1985)

41. M.R. Garey, D.S. Johnson, *Computers and Intractability: A Guide to the Theory of NP-Completeness* (W.H Freeman and Company, New York, 1979)

42. P.B Gibbons, P.R.J Östergård, in *Computational methods in design theory*,eds. by C.J. Colbourn, J.H. Dinitz. Handbook of Combinatorial Designs, 2 edn., chap. VII.6, (Chapman and Hall/CRC, 2007), pp. 755–783

43. I. Gurobi Optimization, Gurobi optimizer reference manual (2016), http://www.gurobi.com

44. E. Haberberger, A. Betten, R. Laue, Isomorphism classification of $t$-designs with group theoretical localisation techniques applied to some Steiner quadruple systems on 20 points. Congr. Numer. 75–96 (2000)

45. D. Heinlein, M. Kiermaier, S. Kurz, A. Wassermann, Tables of subspace codes. ArXiv e-prints arXiv:1601.02864 (2016)

46. C. Hering, Transitive linear groups and linear groups which contain irreducible subgroups of prime order. Geom. Dedic. **2**(4), 425–460 (1974). https://doi.org/10.1007/BF00147570

47. C. Hering, Transitive linear groups and linear groups which contain irreducible subgroups of prime order. II. J. Algebra **93**(1), 151–164 (1985). https://doi.org/10.1016/0021-8693(85)90179-6

48. H. Hitotumatu, K. Noshita, A technique for implementing backtrack algorithms and its application. Inf. Process. Lett. **8**(4), 174–175 (1979). https://doi.org/10.1016/0020-0190(79)90016-4

49. B. Huppert, Endliche Gruppen I, in *Grundlehren der mathematischen Wissenschaften*, vol. 134 (Springer, Heidelberg, 1967)

50. IBM: ILOG CPLEX Optimizer (2010), http://www-01.ibm.com/software/integration/optimization/cplex-optimizer/

51. M. Kaib, H. Ritter, *Block reduction for arbitrary norms* Universität Frankfurt, Preprint (1995)

52. R.M. Karp, Reducibility among combinatorial problems, eds. by R.E. Miller, J.W. Thatcher, J.D. Bohlinger. Complexity of Computer Computations: Proceedings of a symposium on the Complexity of Computer Computations, March 20–22, 1972, (Springer, Boston, 1972), pp. 85–103. https://doi.org/10.1007/978-1-4684-2001-2_9

53. P. Kaski, P.R. Östergård, *Classification Algorithms for Codes and Designs* (Springer, Berlin, 2006). https://doi.org/10.1007/3-540-28991-7

54. P. Kaski, O. Pottonen, libexact user's guide version 1.0. Technical Report 2008-1, Helsinki University of Technology (2008)

55. M. Kiermaier, S. Kurz, A. Wassermann, The order of the automorphism group of a binary q-analog of the fano plane is at most two. Designs, Codes and Cryptography (2017). To appear https://doi.org/10.1007/s10623-017-0360-6

56. M. Kiermaier, M.O. Pavčević, Intersection numbers for subspace designs. J. Comb. Des. **23**(11), 463–480 (2015). https://doi.org/10.1002/jcd.21403

57. Klin, M.H.: Investigations of algebras of invariant relations of certain classes of permutation groups. Ph.D. thesis, Nikolaev (1974). In russian

58. D.E. Knuth, Estimating the efficiency of backtrack programs. Math. Comp. **29**(129), 121–136 (1975)

59. D.E Knuth, Dancing links, eds. by A.W. Roscoe, J. Davies, J. Woodcock. Millennial perspectives in computer science, Cornerstones of computing, (Palgrave, 2000), pp. 187–214

60. D.E. Knuth, *The art of computer programming*, vol. 4A (Addison-Wesley, New Jersey, 2011)

61. D.E Knuth, Dancing links. Technical Report Fasc 5c, Stanford University (2017)

62. M. Koch, Neue Strategien zur Lösung von Isomorphieproblemen. Ph.D. thesis, University of Bayreuth, Germany (2016)

63. A. Kohnert, S. Kurz, Construction of large constant dimension codes with a prescribed minimum distance, eds. by J. Calmet, W. Geiselmann, J. Müller-Quade. Mathematical Methods in Computer Science: Essays in Memory of Thomas Beth, (Springer, Heidelberg, 2008), pp. 31–42. https://doi.org/10.1007/978-3-540-89994-5_4

64. R. Kötter, F.R. Kschischang, Coding for errors and erasures in random network coding. IEEE Trans. Inf. Theory **54**(8), 3579–3591 (2008). https://doi.org/10.1109/TIT.2008.926449

65. E.S. Kramer, D.M. Mesner, t-designs on hypergraphs. Discret. Math. **15**(3), 263–296 (1976). https://doi.org/10.1016/0012-365X(76)90030-3

66. D.L. Kreher, S.P. Radziszowski, The existence of simple 6-(14, 7, 4) designs. J. Comb. Theory Ser. A **43**, 237–243 (1986)

67. D.L. Kreher, S.P. Radziszowski, Constructing 6-(14,7,4) designs. Contemp. Math. **111**, 137–151 (1990)

68. V. Krčadinac, A. Nakić, M.O. Pavčević, The Kramer–Mesner method with tactical decompositions: some new unitals on 65 points. J. Comb. Des. **19**(4), 290–303 (2011). https://doi.org/10.1002/jcd.20277

69. J.C. Lagarias, A.M. Odlyzko, Solving low-density subset sum problems. J. Assoc. Comp. Mach. **32**, 229–246 (1985). Appeared already in Proc. 24th IEEE Symp. Found. Comp. Sci. (1983), 1–10

70. R. Laue, Halvings on small point sets. J. Comb. Des. **7**, 233–241 (1999)

71. R. Laue, Constructing objects up to isomorphism, simple 9-designs with small parameters, in Algebraic Combinatorics and Applications, (Springer, New York, 2001), pp. 232–260

72. R. Laue, Solving isomorphism problems for t-designs, ed by W.D. Wallis. Designs 2002: Further Computational and Constructive Design Theory (Springer US, Boston, MA , 2003), pp. 277–300. https://doi.org/10.1007/978-1-4613-0245-2_11

73. R. Laue, S. Magliveras, A. Wassermann, New large sets of t-designs. J. Comb. Des. **9**, 40–59 (2001)

74. R. Laue, G.R. Omidi, B. Tayfeh-Rezaie, A. Wassermann, New large sets of t-designs with prescribed groups of automorphisms. J. Comb. Des. **15**(3), 210–220 (2007). https://doi.org/10.1002/jcd.20128

75. A.K. Lenstra, H.W. Lenstra Jr., L. Lovász, Factoring polynomials with rational coefficients. Math. Ann. **261**, 515–534 (1982)

76. M.W. Liebeck, The affine permutation groups of rank three. Proc. Lond. Math. Soc. (3) **54**(3), 477–516 (1987). https://doi.org/10.1112/plms/s3-54.3.477
77. S.S Magliveras, The subgroup structure of the Higman-Sims simple group. Ph.D. thesis, University of Birmingham (1970)
78. R. Mathon, Computational methods in design theory, ed. by A.D. Keedwell. Surveys in combinatorics, Proceeding 13th Br. Combinatorial Conference, London Mathematical Society Lecture Notes, vol. 166 (Guildford/UK, 1991), pp. 101–117
79. R. Mathon, Searching for spreads and packings, eds by J.W.P. Hirschfeld, S.S. Magliveras, M.J. de Resmini. Geometry, Combinatorial Designs and Related Structures, Proceedings of the first Pythagorean conference, Mathematical Society Lecture Notes, vol. 245, (London, 1997), pp. 161–176
80. B.D. McKay, A. Piperno, Practical graph isomorphism, II. J. Symb. Comput. **60**, 94–112 (2014). https://doi.org/10.1016/j.jsc.2013.09.003
81. M. Miyakawa, A. Munemasa, S. Yoshiara, On a class of small 2-designs over GF($q$). J. Comb. Des. **3**(1), 61–77 (1995). https://doi.org/10.1002/jcd.3180030108
82. E.H. Moore, Tactical memoranda i-iii. Am. J. Math. **18**(4), 264–303 (1896)
83. A. Nakić, M.O. Pavčević, Tactical decompositions of designs over finite fields. Des. Codes Cryptogr. **77**(1), 49–60 (2015). https://doi.org/10.1007/s10623-014-9988-7
84. P.Q Nguyen, B. Vallée, in *The LLL Algorithm: Survey and Applications*, 1st edn. Information Security and Cryptography. (Springer, Heidelberg, 2009). https://doi.org/10.1007/978-3-642-02295-1
85. S. Niskanen, P.R.J Östergård, Cliquer user's guide, version 1.0. Technical Report T48, Helsinki University of Technology (2003)
86. E.T. Parker, On collineations of symmetric designs. Proc. Am. Math. Soc. **8**(2), 350–351 (1957). http://www.jstor.org/stable/2033742
87. W. Pullan, Optimisation of unweighted/weighted maximum independent sets and minimum vertex covers. Discret. Optim. **6**(2), 214–219 (2009). https://doi.org/10.1016/j.disopt.2008.12.001
88. J.F. Sarmiento, Resolutions of PG(5, 2) with point-cyclic automorphism group. J. Comb. Designs **8**(1), 2–14 (2000). https://doi.org/10.1002/(SICI)1520-6610(2000)8:1<2::AID-JCD2>3.0.CO;2-H
89. B. Schmalz, $t$-Designs zu vorgegebener automorphismengruppe. Bayreuth. Math. Schr **41**, 1–164 (1992). Ph.D thesis, Universität Bayreuth
90. C.P Schnorr, M. Euchner, Lattice basis reduction: Improved practical algorithms and solving subset sum problems, in Proceedings of Fundamentals of Computation Theory '91, Lecture Notes in Computer Science, vol. 529, (Springer, Heidelberg, 1991), pp. 68–85
91. H. Suzuki, On the inequalities of $t$-designs over a finite field. Euro. J. Comb. **11**(6), 601–607 (1990). https://doi.org/10.1016/S0195-6698(13)80045-5
92. S. Thomas, Designs over finite fields. Geom. Dedic. **24**(2), 237–242 (1987). https://doi.org/10.1007/BF00150939
93. R.J. Walker, An enumerative technique for a class of combinatorial problems. Proc. Sympos. Appl. Math. **10**, 91–94 (1960). American Mathematical Society, Providence, R.I. (1960)
94. A. Wassermann, Finding simple $t$-designs with enumeration techniques. J. Comb. Des. **6**(2), 79–90 (1998). https://doi.org/10.1002/(SICI)1520-6610(1998)6:2<79::AID-JCD1>3.0.CO;2-S
95. A. Wassermann, Covering the Aztec diamond with one-sided tetrasticks. Bull.Inst. Comb. Appl. (ICA) **32**, 70–76 (2001)
96. A. Wassermann, Attacking the market split problem with lattice point enumeration. J. Comb. Optim. **6**(1), 5–16 (2002)
97. J. Zwanzger, A heuristic algorithm for the construction of good linear codes. IEEE Trans. Inf. Theory **54**(5), 2388–2392 (2008). https://doi.org/10.1109/TIT.2008.920323

# Part III
# Application of Network Coding

# Index Coding, Network Coding and Broadcast with Side-Information

**Eimear Byrne and Marco Calderini**

**Abstract** Index coding, the problem of efficient broadcast to many receivers with side-infomation, is a rich and active research area. It has applications to a range of multi-user broadcast scenarios such as video-on-demand and satellite communications. It has attracted significant theoretical interest both as a hard problem in its own right and due to its connections to other network capacity problems. The central problem of index coding, that of determining the optimal rate of an index code, is still open. We describe recent advances on the index coding problem and its generalizations in the context of broadcast with side-information. The two main approaches to bounding the optimal rate of an index code, namely rank-minimization methods and linear programming models, are discussed in detail. The latter of these, based on graph-theoretical ideas in the classical case, can be extended even to generalizations of the index coding problem for which there is no associated side-information hypergraph. We discuss error-correction in the index coding problem, the corresponding bounds on the optimal transmission rate and decoding algorithms. We also illustrate the connections to network coding, interference alignment and coded caching.

## 1 Introduction

Broadcast with side-information describes a number of problems in network information theory, including index coding, network coding, coded caching and interference alignment. The equivalences and connections between these different topics has been observed in the literature [18, 19, 27, 28]. The problem has several applications, such as for satellite communications, distribution of media files (eg. video-on-demand), topological interference alignment and distributed caching.

E. Byrne (✉)
School of Mathematics and Statistics, University College Dublin, Dublin, Ireland
e-mail: ebyrne@ucd.ie

M. Calderini
Deptartment of Mathematics, University of Trento, Trento, Italy
e-mail: marco.calderini@unitn.it

© Springer International Publishing AG 2018
M. Greferath et al. (eds.), *Network Coding and Subspace Designs*,
Signals and Communication Technology,
https://doi.org/10.1007/978-3-319-70293-3_10

The canonical example of the broadcast with side-information problem is provided by *index coding* and is the main focus of this chapter. Birk and Kol in [6] introduced the index coding (with side-information) problem as an aspect of *informed source coding on demand*. One of first appearances of the term 'index coding' was in [2]. It relates to a problem of source coding with *side-information*, in which receivers have partial information of a broadcast prior to its transmission. The problem for the sender is to exploit knowledge of the users' side-information in order to optimize the transmission rate. The problem has since become a subject of several studies and generalizations, including error-correction, privacy and secrecy [1, 3–5, 12–14].

The index coding problem may be described informally as follows. A single sender has a set of data packets. There is a set of some $m$ clients each of whom already possesses some packets of the sender's data. Each receiver sends a request to the sender for a single packet and these requests may differ from client to client. The task for the sender is to satisfy all users' demands using a minimum number of transmissions. If the sender only transmits uncoded data packets, then the number of transmissions required is simply the number of different requests of all users, and all users will receive the requests of the other users. However, if data coding is permitted, that is if the broadcaster transmits functions of its packets, then the number of required transmissions can be greatly reduced. The main problem of index coding is the determination of this minimum transmission rate, or the minimum number of packet transmissions required by the sender in order to satisfy all users' requests. Given an ICSI instance, we also seek to obtain an explicit optimal encoding function, which is computationally very hard. Therefore we seek bounds on the optimal transmission rate and algorithms that generate good but possibly sub-optimal encoding functions.

It was shown in [2] that the best rate of a *scalar linear* binary index code is characterized by the *minrank* of a graph, which is NP-hard to compute [29]. Clearly, any encoding function for an instance necessarily gives an upper bound on the optimal length of an index code for that instance. There have been a number of papers addressing this aspect of the problem, in fact finding sub-optimal but feasible solutions, using linear programming methods to obtain partitions of the users into solvable subsets. Such solutions involve obtaining *clique covers, partial-clique covers, multicast partitions* and some variants of these [5, 30, 33, 36]. Other than these LP approaches, low-rank matrix completion methods may also be applied. This was considered for index coding over the real numbers in [22], but the problem is essentially still open for index coding over finite fields.

In Sect. 2 we describe the general problem of broadcast with side information. This is a problem with two fundamental aspects, namely that of delivery and placement. The delivery problem is essentially a generalization of index coding, while the placement problem is a generalized coded-caching problem. In Sect. 3 we introduce the classical index coding problem and describe the known efforts to estimate the optimal transmission rate for an arbitrary instance, including approaches from graph theory and algebraic approaches. In Sect. 4 we describe a generalized version of the index coding problem, namely that for coded side-information. We outline how many of the graph-theoretic bounds can be extended to this more general case. In addition, we consider the problem of error correction, when the sender's transmission is subject to noise. We give bounds on the transmission rate in this context and describe a decoding method for Hamming like errors. Finally, in Sect. 5 we discuss connections of the classical index coding problem to other problems of network communications, such as its equivalence to network coding, relation to interference alignment and to the coded-caching problem.

## 1.1  Notation

For any positive integer $n$, we let $[n] := \{1, \ldots, n\}$. We write $\mathbb{F}_q$ to denote the finite field of order $q$ and use $\mathbb{F}_q^{n \times t}$ to denote the vector space of all $n \times t$ matrices over $\mathbb{F}_q$. Given a matrix $X \in \mathbb{F}_q^{n \times t}$ we write $X_i$ and $X^j$ to denote the $i$th row and $j$th column of $X$, respectively. More generally, for subsets $\mathscr{S} \subset [n]$ and $\mathscr{T} \subset [t]$ we write $X_{\mathscr{S}}$ and $X^{\mathscr{T}}$ to denote the $|\mathscr{S}| \times t$ and $n \times |\mathscr{T}|$ submatrices of $X$ comprised of the rows of $X$ indexed by $\mathscr{S}$ and the columns of $X$ indexed by $\mathscr{T}$ respectively. We write $\langle X \rangle$ to denote the row space of $X$.

## 2  Broadcast with Side-Information

We present a general scenario, by which we define an *instance* of the broadcast with side-information problem over $\mathbb{F}_q$. The main ingredients are as follows.

- There is a single sender and $m$ receivers (or users).
- $X \in \mathbb{F}_q^{n \times t}$ is the uncoded data held by the sender.
- User $i$ has side information $(V^{(i)}, V^{(i)}X)$, for some matrix $V^{(i)} \in \mathbb{F}_q^{v_i \times n}$ of rank $v_i$.
- User $i$ has request matrix $R_i$, for some $R_i \in \mathbb{F}_q^{r_i \times n}$ of rank $r_i$.
- User $i$ demands the request packet $R_i X \in \mathbb{F}_q^{r_i \times t}$.

The task of the sender is to ensure that each user $i$ can combine its side-information with the sender's broadcast to decode to its demanded packet $R_i X$. We assume that each $i$th user can generate all linear combinations of the rows of $V^{(i)}$, so it effectively possesses this $v_i$-dimensional space. The sender, after receiving each request $R_i$, may send $Y \in \mathbb{F}_q^{N \times t}$, a function of $X$, say $Y = E(X)$ for some map $E : \mathbb{F}_q^{n \times t} \longrightarrow \mathbb{F}_q^{N \times t}$. We say that the encoding $E$ *realizes* a length $N$ code for this problem if indeed each user can retrieve its demand $R_i X$ for any source data matrix $X$, given knowledge of $E, Y, V^{(i)}, V^{(i)}X$. Therefore, the source data matrix $X$ should be thought of as a variable of the above instance. Unless stated otherwise, we will assume that $E$ is $\mathbb{F}_q$-linear, so that for all $X$, $E(X) = LX$ for some $N \times n$ matrix $L$ over $\mathbb{F}_q$. Then we say that $L$ realizes the given instance if each user can retrieve its demand $R_i X$ for any source data matrix $X$, given knowledge of $L, Y = LX, V^{(i)}, V^{(i)}X$. If such an $L$ exists, we say that the length $N$ is *achievable* for the given instance. Finding such an $L$ (and indeed the length $N$) is computationally hard and the central problem of broadcast with side-information.

User $i$ can retrieve its demand $R_i X$, for all possible choices of $X$ if and only if there exist matrices $A_i$, $B_i$ satisfying

$$R_i = A_i V^{(i)} + B_i L, \tag{1}$$

from which it computes $R_i X$, knowing $V^{(i)} X$ (as its side information) and $LX$ (which was transmitted) [4]. It is generally assumed that a user does not demand $R_i X$ if it already has it in its cache. Therefore, we assume that no row of $R_i$ is contained in the row space of $V^{(i)}$.

*Remark 1* The uncoded downlink cost is $nt$, which is the cost of sending the full data matrix $X$. The total coded downlink cost is $N(n + t)$, which is the cost of transmitting $L$ and $Y = LX$. Therefore there is a coding gain only if $t > \frac{Nn}{n-N}$.

We now describe this in terms of a matrix code. First let $r = \sum_{j=1}^{m} r_j$ and let $R$ be the $r \times n$ matrix

$$R = [R_1^T, R_2^T, ..., R_m^T]^T.$$

We call $R$ the request matrix. For each $i$, let

$$\mathscr{C}_i = \{AV^{(i)} : A \in \mathbb{F}_q^{r_i \times v_i}\} \subset \mathbb{F}_q^{r_i \times n}.$$

$\mathscr{C}_i$ is a vector space of $r_i \times n$ matrices for the scalars $\mathbb{F}_q$. It can be thought of as $r_i$ copies of the length $n$ linear code generated by the rows of $V^{(i)}$. Now define

$$\mathscr{C} = \{[U_1^T, U_2^T, ..., U_m^T]^T : U_i \in \mathscr{C}_i\} \subset \mathbb{F}_q^{r \times n},$$

which is a vector space of $r \times n$ matrices for the scalars $\mathbb{F}_q$.

We say that the pair $(\mathscr{C}, R)$ is an instance of the broadcast with side-information problem and we call the matrix code $\mathscr{C}$ the side-information code of the instance. The problem of determining the optimal code length of the instance $(\mathscr{C}, R)$ and a corresponding encoding matrix $L$ is a *delivery* problem.

It can be shown, using (1), that the minimum length of a code for $(\mathscr{C}, R)$ is

$$\kappa(\mathscr{C}, R) := \min\{\text{rank}(R + C) : C \in \mathscr{C}\},$$

which is called the *minrank* of the instance $(R, \mathscr{C})$ [4]. The block length $t$ does not affect the minrank parameter, however, due to overhead transmission costs, the gains of coding are greater as $t$ increases. The set $R + \mathscr{C} := \{R + C : C \in \mathscr{C}\}$ is a coset or translate of $\mathscr{C}$, so $\kappa(R + \mathscr{C})$ is the minimum rank of any member of this coset. It is also the rank distance of the matrix $R$ to the side information code $\mathscr{C}$. This generalizes the *minrank* of a *side-information graph* or hypergraph, as it arises in the *index coding problem* (cf. [3, 4, 13, 25]). Implicit in this is the fact that any full-rank matrix $L$ that realizes the instance $(\mathscr{C}, R)$ can be obtained by rank-factorization of a member of $R + \mathscr{C}$. Note that

$$\dim \mathscr{C} = \sum_{i \in [m]} r_i v_i$$

over $\mathbb{F}_q$, so $|R + \mathscr{C}| = q^s$ where $s = \sum_{i \in [m]} r_i v_i \le rn$.

For a given side-information code $\mathscr{C}$, the sender can satisfy any set of requests in at most

$$\rho(\mathscr{C}) := \max\{\kappa(\mathscr{C}, R) : R \in \mathbb{F}_q^{m \times n}\}$$

transmissions, which is the *rank-metric covering radius* of the code $\mathscr{C}$. So if the side-information code $\mathscr{C}$ has low covering radius, then all instances $(\mathscr{C}, R)$ require a small number of transmissions.

This relates to a *placement* problem: that of determining side-information codes $\mathscr{C}$ with least maximal minrank $\kappa(\mathscr{C}, R)$ for some fixed set of request matrices $R$ in $\mathbb{F}_q^{r \times n}$.

The delivery problem of broadcast with side-information is essentially the index coding problem and its generalizations. The placement problem is central to coded-caching. The main focus of this chapter will be an account of the former. We remark that with respect to delivery, there is no loss of generality in assuming that each request matrix $R_i$ is a single row vector; for example, if the user $m$ has $2 \times n$ request matrix $R_m$ then the instance can be equivalently represented by one for which user $m$ is now represented by two users $m$ and $m + 1$, each with the same side-information $\mathscr{C}_m$. User $m$'s request vector is the first row of $R_m$ and the request vector of user $m + 1$ is the 2nd row of $R_m$. Observe that both instances are represented by the same pair $(R, \mathscr{C})$, and have the same minrank.


## 3  Index Coding with Side-Information

We start with the classical index coding (ICSI) with side information problem. In this case it is assumed that the side-information held by the users is *uncoded*, so each user has a subset of the data packets held by the broadcaster, and each user requests a single other packet from the sender. As we will see below, a prominent characteristic of the ICSI problem is that it can be identified with a unique directed hypergraph.

The unique sender has a data matrix $X \in \mathbb{F}_q^{n \times t}$. There are $m$ receivers, each with a request for a data packet $X_i$, and it is assumed that each receiver has a subset of messages $X_{\mathscr{X}_i}$, for a subset $\mathscr{X}_i \subseteq [n]$. The requested packets of the users are described by a surjection $f : [m] \to [n]$, such that the packet requested by $i$ is denoted by $X_{f(i)}$, and it is assumed that $f(i) \notin \mathscr{X}_i$ for all $i \in [m]$. With respect to this viewpoint, the side-information and requested data are represented as subsets of $[n]$, the index set of the data matrix $X$. Hence the term *index coding*. This description (see [13]), is indeed a special case of the broadcast with side-information problem as given in Sect. 2. To translate back to this setting, set the side information codes $\mathscr{C}_i$ to be generated by matrices $V^{(i)}$ whose rows are standard basis vectors and the request matrices to be the standard basis vectors $R_i = \mathbf{e}_{f(i)}$.

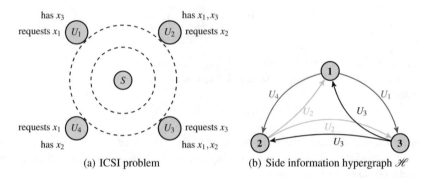

(a) ICSI problem                          (b) Side information hypergraph $\mathscr{H}$

**Fig. 1** The ICSI problem of Example 1

For the remainder, let us fix $t, m, n$ to denote those parameters as described above. Then for any $\mathscr{X} = (\mathscr{X}_1, \ldots, \mathscr{X}_n)$, $\mathscr{X}_i \subset [n]$ and map $f : [m] \to [n]$, the corresponding instance of the ICSI problem (or the ICSI instance) is denoted by $\mathscr{I} = (\mathscr{X}, f)$.

The index coding problem given by an instance $\mathscr{I}$ can be described by a side-information (directed) hypergraph [1]. We define a directed hypergraph $\mathscr{H} = (\mathscr{V}, \mathscr{E})$, where the set of vertices is $\mathscr{V} = [n]$. Each vertex $i$ of $\mathscr{H}$ corresponds to the data packet $X_i$. The set $\mathscr{E}$ is composed by the hyper-edges of type $(f(i), \mathscr{X}_i)$.

*Example 1* Consider the ICSI instance with $n = 3$ (three messages), $m = 4$ (four users), $f(1) = 1$, $f(2) = 2$, $f(3) = 3$, $f(4) = 1$, $\mathscr{X}_1 = \{3\}$, $\mathscr{X}_2 = \{1, 3\}$, $\mathscr{X}_3 = \{1, 2\}$, and $\mathscr{X}_4 = \{2\}$. The hypergraph $\mathscr{H}$ that describes this instance has three vertices $1, 2, 3$, and has four hyperarcs. These are $e_1 = (1, \{3\})$, $e_2 = (2, \{1, 3\})$, $e_3 = (3, \{1, 2\})$, and $e_4 = (2, \{2\})$. This hypergraph is depicted in Fig. 1.

When $m = n$ we assume that $f(i) = i$ for all $i \in [n]$, and the corresponding side information hypergraph has precisely $n$ hyperarcs, each with a different origin vertex. It is simpler to describe such an ICSI instance as a digraph $\mathscr{G} = ([n], \mathscr{E})$, the so-called *side information (di)graph* [2]. For each hyperarc $(i, \mathscr{X}_i)$ of $\mathscr{H}$, are $|\mathscr{X}_i|$ arcs $(i, j)$ of $\mathscr{G}$, for $j \in \mathscr{X}_i$. Equivalently, $\mathscr{E} = \{(i, j) : i, j \in [n], j \in \mathscr{X}_i\}$.

We now formally define what is meant by a code for an instance of the ICSI problem.

**Definition 1** Given an instance of the ICSI problem described by an hypergraph $\mathscr{H}$. Let $N$ be a positive integer. We say that the map

$$E : \mathbb{F}_q^{n \times t} \to \mathbb{F}_q^N,$$

is an $\mathbb{F}_q$-code of length $N$ for the instance described by $\mathscr{H}$ if for each $i \in [m]$ there exists a decoding map

$$D_i : \mathbb{F}_q^N \times \mathbb{F}_q^{|\mathscr{X}_i|} \to \mathbb{F}_q^t,$$

satisfying

$$\forall X \in \mathbb{F}_q^{n \times t} : D_i(E(X), X_{\mathscr{X}_i}) = X_{f(i)},$$

in which case we say that $E$ is an $\mathscr{H}$-IC. $E$ is called an $\mathbb{F}_q$-linear $\mathscr{I}$-IC if $E(X) = LX$ for some $L \in \mathbb{F}_q^{N \times n}$, in which case we say that $L$ represents the code $E$. If $t = 1$ it is called *scalar* linear.

## 3.1 Bounding the Optimal Rate of the Index Coding Problem

Bar-Yossef et al. showed [2, 3] that (for a given field size), the optimal length of a scalar linear index code is equal to the minrank of the associate side-information graph. The notion of minrank for an undirected graph $\mathscr{G}$ was first considered by Haemers [21] in 1978 to obtain a bound for the Shannon graph capacity.

Such a result was extended to the more general case with $m \geq n$ in [13]. Therefore, minrank characterizes the best possible scalar linear index code for a given finite field.

Let $p$ be a fixed prime. The broadcast rate of an IC-instance $\mathscr{I}$ is defined as follows [1].

**Definition 2** Let $\mathscr{H}$ be a side-information hypergraph. We denote by $\beta_t(\mathscr{H})$ the minimal number of symbols required to realize an index coding instance $\mathscr{I}$ associated to $\mathscr{H}$ for block length $t$, over all possible extensions of $\mathbb{F}_p$. That is,

$$\beta_t(\mathscr{H}) = \inf_q \{N \mid \exists \text{ a } q\text{-ary index code of length } N \text{ for } \mathscr{I}\}.$$

Moreover we denote by $\beta(\mathscr{H})$ the limit

$$\beta(\mathscr{H}) = \lim_{t \to \infty} \frac{\beta_t(\mathscr{H})}{t} = \inf_t \frac{\beta_t(\mathscr{H})}{t}.$$

In the following we will consider the scalar case, i.e. $t = 1$, moreover we report the results on the minrank in the more general case of the hypergraphs.

### 3.1.1 Algebraic Methods

**Definition 3** Let $\mathscr{H}$ be the side-information hypergraph of an instance of the IC problem. A matrix $M = (m_{i,j}) \in \mathbb{F}_q^{m \times n}$ *fits* the hypergraph if

$$m_{i,j} = \begin{cases} 1 & \text{if } j = f(i) \\ 0 & \text{if } j \text{ does not lie in } \mathscr{X}_i \end{cases}$$

The min-rank of $\mathcal{H}$ over $\mathbb{F}_q$ is defined to be

$$\mathrm{minrk}_q(\mathcal{H}) = \min\{\mathrm{rank}_q(M) : M \text{ fits } \mathcal{H}\}$$

*Example 2* Consider the ICSI instance given in Example 1. Then we have that a matrix $M$ that fits the hypergraph $\mathcal{H}$ is of type

$$M = \begin{array}{c} \\ U_1 \\ U_2 \\ U_3 \\ U_4 \end{array} \begin{array}{c} X_1\ X_2\ X_3 \\ \begin{bmatrix} 1 & 0 & * \\ * & 1 & * \\ * & * & 1 \\ 1 & * & 0 \end{bmatrix} \end{array}$$

where the symbol "$*$" may be replace by an arbitrary element of the field $\mathbb{F}_q$.

The following lemma specifies a sufficient condition on a matrix $L$ to correspond to a $\mathcal{H}$-IC. This result was implicitly formulated by Bar-Yossef et al. [2, 3] for the case where $m = n$, $f(i) = i$ for all $i \in [n]$, and $q = 2$, then generalized to the case $m \geq n$ for any $q$ by [13].

Let $\mathrm{Supp}(\mathbf{v})$ denotes the support of a vector $\mathbf{v} \in \mathbb{F}_q^n$.

**Lemma 1** *A $\mathscr{I}(\mathscr{X}, f)$-IC of length $N$ over $\mathbb{F}_q$ has a linear encoding map if and only if there exists a matrix $L \in \mathbb{F}_q^{N \times n}$ such that for each $i \in [m]$, there exists a vector $\mathbf{u}^{(i)} \in \mathbb{F}_q^n$ satisfying*

$$\mathrm{Supp}(\mathbf{u}^{(i)}) \subseteq \mathscr{X}_i \tag{2}$$

$$\mathbf{u}^{(i)} + \mathbf{e}_{f(i)} \in \langle L \rangle. \tag{3}$$

The lemma above implies the existence of a vector $\mathbf{b}^{(i)} \in \mathbb{F}_q^N$ such that $\mathbf{b}^{(i)} L = \mathbf{u}^{(i)} + \mathbf{e}_{f(i)}$, in which case the receiver at $i$ retrieves

$$x_{f(i)} = \mathbf{e}_{f(i)} X = \mathbf{b}^{(i)} L X - \mathbf{u}^{(i)} X = \mathbf{b}^{(i)} Y - \mathbf{u}^{(i)} X_{\mathscr{X}_i}, \tag{4}$$

where $Y$ is the message sent by the source over the broadcast channel.

As consequence we obtain the following

**Theorem 1** *Let $\mathscr{I} = (\mathscr{X}, f)$ be an instance of the ICSI problem, and $\mathcal{H}$ its side information hypergraph. Then the optimal length of a $q$-ary linear $\mathcal{H}$-IC is $\mathrm{minrk}_q(\mathcal{H})$.*

Therefore, min-rank characterizes the best possible scalar linear index code for a given finite field.

**Theorem 2** *Let $\mathcal{H}$ be a side-information hypergraph, for any $q$ we have*

$$\beta(\mathcal{H}) \leq \mathrm{minrk}_q(\mathcal{H}).$$

*Example 3* Consider the instance given by Example 1. It is easy to check that a matrix that fits the hypergraph has at least rank 2 (see Example 2). Thus

$$M = \begin{bmatrix} 1 & 0 & 1 \\ 1 & 1 & 0 \\ 1 & 0 & 1 \\ 1 & 1 & 0 \end{bmatrix}$$

achieves the minimum rank possible 2. Now if we consider 2 linear independent rows of $M$, suppose

$$L = \begin{bmatrix} 1 & 0 & 1 \\ 1 & 1 & 0 \end{bmatrix}.$$

Take into account receiver 1, that requested $X_1$ and knows $X_3$. Encoding $X$ with $L$ we obtain

$$LX = [X_1 + X_3, \ X_1 + X_2]^T.$$

So, deleting $X_3$ from $X_1 + X_3$ receiver 1 obtain $X_1$. Similarly for the other receivers.

Thus, do find an encoded matrix we need to select the linear independents rows of a fitting matrix of the hypergraph.

The authors of [2] proved that in various cases, linear codes are optimal, so their main conjecture was that linear index coding is always optimal, that is $\beta(\mathscr{G}) = \text{minrk}_2(\mathscr{G})$ for any graph $\mathscr{G}$. This conjecture was later disproved by Lubetzky and Stav in [26]. The authors show that for any positive $\varepsilon > 0$ there exist graphs, of order $n$, where every linear index code requires at least $n^{1-\varepsilon}$ bits, whereas a given non-linear index code utilizes only $n^\varepsilon$ bits.

However, as shown by Peeters [29], computing the minrank of a general graph is a hard task. More specifically, Peeters showed that deciding whether a graph has min-rank three is an NP-complete problem.

Due to the result given in Theorem 1, we have that the problem of find a linear index code is equivalent to a rank minimization problem over a finite field. An approach to the minimum rank matrix completion problem over finite fields representing the linear index coding problem was studied in [17].

In [17] first the complete sub-matrix of highest rank is identified, using a heuristic scheme of polynomial complexity and then by using row (column) projection to expand the complete sub-matrix iteratively. The goal of the projection step is to find possible completions of an incomplete row or column such that they are in the span of the complete sub-matrix. The steps of row and column projections are administered over a decision tree.

To identify the maximal complete sub-matrix of M of highest rank, a heuristic algorithm, starting with an initial complete sub-matrix, tries to improve it iteratively.

The row-improving step is performing in the following way, supposing that we have the complete sub-matrix $M_I^J$:

- Let $\bar{I} = [m] \setminus I$, select a row of $M_{\bar{I}}^J$ with fewer erasures (that is, stars "$*$").
- Suppose $i$ is selected, consider $M_{I \cup \{i\}}^J$ and remove the incomplete columns in it.
- If the rank of the resulting matrix is greater than $M_I^J$, update the select $I$ and $J$.

A similar approach is applied to improve the set of columns.

---

**Data**: Incomplete matrix $M$
**Result**: $I$, $J$ such that the complete sub-matrix $M_I^J$ is of maximal rank
$I \leftarrow \{1, ..., m\}$;
$J \leftarrow \emptyset$;
$k \leftarrow 0$;
$N \leftarrow 100$;
**while** *No change detected over N iterations in sequence* **do**
    $th \leftarrow |I|/(|I| + |J|)$;
    **if** $\text{rand}([0, 1]) > th$ **then**
       | perform rows improvement
    **else**
       | perform columns improvement
    **end**
    $M' \leftarrow$ the matrix obtain from the improvement;
    **if** $\text{rk}(M') > k$ **then**
       | Update $I$ and $J$
    **end**
**end**

**Algorithm 1:** Complete sub-matrix

---

It is preferred to identify the maximal complete sub-matrix of highest rank within the matrix $M$, however we can proceed to the subsequent steps for matrix completion even with a sub-optimal choice. Thus, we may limit the number of iterations in Algorithm 1 to $N$ iterations.

The completion of the matrix $M$ is done by alternating projection steps on rows and columns. The goal of this step is to find possible completions of an incomplete row (column) such that it in the span of the complete sub-matrix. The row and column projections are administered over a decision tree.

Here we describe the horizontal projection (that is, on the rows) in a branch. It is similar for the columns. There are three possible cases (Fig. 2).

1. There are one or more incomplete rows that may be completed in a unique way in the subspace spanned by the rows of the current complete sub-matrix. In this case we complete these rows and update the complete sub-matrix. Then we switch the projection direction over the next branch.
2. There is no row that may be completed uniquely, but there are some rows that are in the subspace spanned by the rows of the complete sub-matrix. In this case, we choose the incomplete row with minimum possible solutions. Then, for each solution we analyze the consequent matrix completion, over multiple subsequent branches, continuing the procedure in the alternate direction of projection.

**Fig. 2** A sample structure
for the decision tree

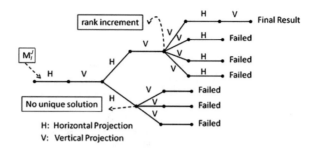

3. The last case is when no row can be completed with a vector of the span of
the complete sub-matrix. Thus, the rank of the solution is to be increased. If the
increased rank is larger than the rank of the previously completed branches, the
current branch is eliminated; otherwise, all the possible solutions are examined
over multiple subsequent branches.

Step 2, described above, can be performed using erasure decoding techniques. Consider a generator matrix $G$ of the code spanned by the complete submatrix and its
parity check matrix $H$. Then suppose to have selected the row $M_i$ restricted to the
column of the complete sub-matrix (the symbols stars now represent our variables).
We verify the solution of the system $M_i \cdot H = 0$.

At each iteration, the branch with 'maximum opportunity' is selected, which
is quantified by a metric defined by the ratio of the completion percentage of the
matrix with respect to the rank of its complete sub-matrix. The branch with minimum
achieved rank identifies the solution.

Note that the growth of the tree due to a rank increment, that is the number of
new branches, is bounded above by $q^e$, where $e$ is the maximum number of erasures
that we have in the rows or in the columns. Moreover, in a path of the tree we could
increment the rank at most $\text{minrk}(\mathcal{H}) - k$ times, where $k$ is the rank of the starting
complete submatrix.

Furthermore, the growth due to the multiple allowed combinations for completing
a row or (column) is bounded above by $q^{\max(n,m)-k}$ for each projection.

For large incomplete matrices the authors in [17] propose a sub-optimum algorithm. Specifically, whenever the number of branches goes beyond a pre-set threshold, we prune the branches having a small value with respect to the metric described
above.

In [31] the authors propose a linear algebraic algorithm to complete $M$ (over $\mathbb{F}_2$)
given a random subset of its entries. They establish some conditions on the row and
column spaces of $M$ which guarantee that the algorithm runs in polynomial time.
Moreover a linear programming-based extension is proposed.

In particular, given a matrix $M \in \mathbb{F}_2^{n \times n}$ of rank $r$ and a random subset of its
entries $\Omega$, we can complete successfully $M$ with high probability in time $O(n^{2r+3})$,
whenever the cardinality of the set $\Omega$ is at least $\widetilde{\Omega}\left(n^{2-\frac{1}{r+1}}\right)$.

**Data**: Incomplete matrix $M$
**Result**: Complete matrix with minimum possible rank
Find maximal complete sub-matrix $M_I^J$;
**while** *incomplete branches* $> 0$ **do**
    Choose the branch with maximum opportunity;
    Perform projection in the proper direction;
    Based on projection results add more branches or
    eliminate current one if necessary ;
    **if** *matrix is complete in this branch* **then**
        **if** *achieved rank < minimum achieved rank* **then**
          | Update minimum achieved rank
        **end**
    **end**
**end**

**Algorithm 2:** Complete matrix

There have been several hardness results for matrix completion over finite fields. Tan et al. [34] studied the more general problem where, instead of entries, random linear combinations of the entries are observed. They give various information-theoretic bounds on the number of measurements necessary for low rank matrix recovery.

The matrix completion problem over the reals seems to behave rather differently and techniques do not appear to transfer to the finite field case. A method for matrix completion over the real numbers is given in [22]. Note that linear index codes over the reals have applications to topological interference management in wireless networks [23, 27].

The method developed in [22] is based on an alternating projection (AP) approach. As it is shown in the paper, completing the index coding matrix $M$ by choosing values for the symbols "$*$" such that $M$ has a low rank $r$ can be thought of as finding the intersection of two regions $\mathscr{C}$ and $\mathscr{D}$ in $\mathbb{R}^{n \times n}$ defined by,

$$\mathscr{C} = \{M \in \mathbb{R}^{n \times n} \mid \mathrm{rk}(M) \le r\},$$

and

$$\mathscr{D} = \{M \in \mathbb{R}^{n \times n} \mid m_{ij} = 0 \text{ if } (i, j) \in \mathscr{E}(\mathscr{G}) \text{ and } m_{ii} = 1, i \in [n]\}.$$

$\mathscr{C}$ is not convex and therefore convergence of the AP method is not guaranteed. However, the AP method can give a certificate that a certain rank $r$ is achievable. Therefore we can use the AP method as a heuristic. This AP method, in some cases, leads to up to 13% average savings in broadcast messages compared to graph coloring (see next section).

The authors in [22] compare the AP methods also to the well studied Alternating Minimization (AltMin) algorithm [20, 41, 42]. Jain et al. [41] gave one of the first performance guarantees of AltMin, in particular the authors proved that by observing $|\Omega| = O\left(\frac{\sigma_1}{\sigma_r} n \log n \log(r\|M\|_F/\varepsilon)\right)$ random entries of an incoherent $M$, AltMin can recover $M$ in $O(\log(1/\varepsilon))$ steps. Here $\sigma_1$ and $\sigma_r$ denote the largest and smallest singular values of $M$, respectively, and $|| \cdot ||_F$ is the Frobenius norm. However in this context the authors in [22] show that AltMin does not perform as well as AP.

Candès and Recht in [10] showed that replacing the rank function by the nuclear norm leads to finding the minimum rank with high probability (under certain conditions). However, these results do not carry over directly to the index coding problem because the model in [10] assumes the location of the fixed entries is chosen uniformly at random. In the index coding problem the matrix $M$ has a specific structure, that is, all the diagonal entries have to be equal to one. Indeed, as noted in [22] the approach in [10] always output the maximum rank $n$.

In the analysis of the performance of these algorithms it is always assumed that the set $\Omega$ is given at random. However, in the index coding problem, also in the case of random graph, the diagonals entries are fixed to one, so it is not guaranteed that this algorithms performs well.

### 3.1.2 Graph Theoretic Methods

Graph theoretic methods start from the well-known fact that all the users forming a clique in the side information digraph can be simultaneously satisfied by transmitting the XOR of their packets. Indeed, for such a graph we have that $\mathscr{X}_i = [n] \setminus \{i\}$. Thus sending $Y = \sum_i X_i$, any receiver retrieves its requested packet.

Moreover, from the fact that for a graph $\mathscr{G}$ that is a union of disjoint graphs, i.e. $\mathscr{G} = \mathscr{G}_1 \cup \cdots \cup \mathscr{G}_s$, it holds that $\mathrm{minrk}(\mathscr{G}) = \sum_i \mathrm{minrk}(\mathscr{G}_i)$, we have that an achievable scheme for index coding on graphs is the number of disjoint cliques required to cover $\mathscr{G}$ [2]. This number is called the clique-covering number $\mathbf{cc}(\mathscr{G})$, which is equal to the chromatic number of the complement graph $\chi(\overline{\mathscr{G}})$. This is because all the vertices assigned to the same color cannot share an edge and hence must form a clique on the complement graph.

In [2] the authors prove that for an acyclic graph $\mathscr{G}$ of order $n$ the optimal broadcast rate is $n$. Therefore if $\alpha(\mathscr{G})$ is the order of the maximum acyclic induced subgraph of $\mathscr{G}$ then $\alpha(\mathscr{G}) \leq \beta(\mathscr{G})$. In particular, when $\mathscr{G}$ is symmetric, $\alpha(\mathscr{G})$ is the independence number of $\mathscr{G}$. These two results give the so-called sandwich property on the optimal broadcast rate.

**Theorem 3** ([2])
$$\alpha(\mathscr{G}) \leq \beta(\mathscr{G}) \leq \mathbf{cc}(\mathscr{G}).$$

*Example 4* Consider the graph $\mathscr{G}$ in Fig. 3. We can see that any clique partition of $\mathscr{G}$ is composed of at least 3 cliques, e.g. $\{\{1, 2\}, \{3, 4\}, \{5\}\}$. Thus $\mathbf{cc}(\mathscr{G}) = 3$, and an encoded scheme for the instance associated to $\mathscr{G}$ is $Y = [X_1 + X_2, X_3 + X_4, X_5]$.

An other scheme given in [2] is based on the concept of partial-clique.

**Definition 4** A graph $\mathscr{G}$ is called a $k$-partial clique if for all node $i$ we have $\deg_{Out}(i) \geq n - k - 1$ and there exists at least one node for which the equality

**Fig. 3** The graph $\mathscr{G}$ of
Example 4

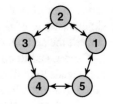

holds. Here $\deg_{Out}(i)$ denotes the out-degree of the node $i$, that is, the edges outgoing from $i$.

Note that a 0-partial clique is a usual clique.

The nodes contained in a partial clique correspond to a set $n$ of clients, each missing at most $k$ packets of the $n-1$ packets requested by the other receivers of the set, and at least one client missing exactly $k$ of those blocks. To satisfy the requests of a $k$-partial clique we can use an $(n, k+1)$-MDS code, e.g. Reed-Solomon codes. Indeed suppose $G$ is generator matrix, over a sufficient large field, of an MDS code of length $n$ and dimension $k+1$. Now we can broadcast the message $Y = GX$. Now any receiver is able to delete at least $n-k-1$ columns of $G$ from $Y$. So since every set of $k+1$ columns of $G$ are linearly independent, we can retrieve the requested packet. So, instead of partitioning a graph in clique we can use partial-clique.

**Theorem 4** ([2]) *Let $\mathscr{G}$ a side-information graph and let $\mathscr{G}_1, \ldots, \mathscr{G}_r$ be a partition in partial clique whose parameters are $k_1, \ldots, k_r$. Then*

$$\beta(\mathscr{G}) \leq \sum_{i=1}^{r} k_i + 1.$$

*Example 5* Consider the instance given by the side information graph $\mathscr{G}$ in Fig. 4. We can see that using the scheme based on the clique cover we need 3 transmissions. Using the partial clique scheme we can use 2 transmissions. In fact each receiver knows one packet, so $\mathscr{G}$ is a 1-partial clique. Consider the matrix

$$G = \begin{bmatrix} 1 & 0 & 1 \\ 0 & 1 & 1 \end{bmatrix}$$

**Fig. 4** The graph $\mathscr{G}$ of
Example 5

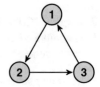

which represents a $(3, 2)$-MDS code over $\mathbb{F}_2$. We can encode $X = [X_1, X_2, X_3]^T$ using $G$, that is $Y = GX = [X_1 + X_3, X_2 + X_3]$. It is easy to check that each receiver can retrieve the requested packet.

It turns out that the idea based on partitioning with cliques leads to a family of stronger bounds, starting with an LP relaxation called the fractional clique covering number. It can be readily checked that this rate is the solution to the integer program

$$\min \sum_{C \in \mathscr{C}} y_C$$

$$\text{s.t.} \sum_{C:j \in C} y_C = 1 \text{ for all } j \in [n] \tag{5}$$

$$y_C \in \{0, 1\} \text{ for all } C \in \mathscr{C}.$$

where $\mathscr{C}$ is the collection of all cliques in $\mathscr{G}$.

Blasiak, Kleinberg, and Lubetzky [7] extended the clique covering bound to the fractional clique covering bound, namely, the solution to the linear program obtained by relaxing the integer constraint $y_C \in \{0, 1\}$ to $y_C \in [0, 1]$ in (5). This scheme corresponds to an achievable (vector-linear) index code.

*Example 6* Consider the graph $\mathscr{G}$ given in Example 4. A possible fractional partition of the nodes is given by $\{\{1, 2\}, \{2, 3\}, \{3, 4\}, \{4, 5\}, \{5, 1\}\}$. Any node is contained in two sets of the fractional partition. So if we consider a data matrix $X$ composed by two sub-packets for each packet, i.e.

$$X \begin{bmatrix} X_{11} & X_{12} \\ X_{21} & X_{22} \\ X_{31} & X_{32} \\ X_{41} & X_{42} \\ X_{51} & X_{52} \end{bmatrix}.$$

Sending $Y = [X_{11} + X_{21}, X_{31} + X_{41}, X_{12} + X_{51}, X_{22} + X_{32}, X_{42} + X_{52}]$ we satisfy all the requests, achieving a rate of $5/2$.

**Theorem 5** ([7]) *The optimal broadcast rate is upper bounded by the optimal solution* $\text{cc}_f(\mathscr{G})$ *of the LP relaxation of (5),*

$$\beta(\mathscr{G}) \leq \text{cc}_f(\mathscr{G}) \leq \text{cc}(\mathscr{G}).$$

In [40], Shanmugam, Dimakis, and Langberg showed, extending the clique covering scheme, that if the users are partitioned in cliques and the packets encoded for each clique with the XOR as before, then we can reduce the number of transmissions by applying an MDS code and using side information to recover them. Just as Birk and Kol reduced the number of message transmissions with partial-clique [2].

The coding scheme given in [40] achieves the local clique covering number of $\mathscr{G}$, denoted by $\mathbf{cc}_l(\mathscr{G})$ [16]. Further extending this scheme with fractional coloring, we can establish the following.

**Theorem 6** ([40]) *The optimal broadcast rate is upper bounded by the optimal solution* $\mathbf{cc}_{lf}(\mathscr{G})$ *of the LP relaxation of*

$$\min \quad k$$
$$s.t. \sum_{C:C\cap\mathscr{X}_j\neq\emptyset} y_C \leq k \text{ for all } j \in [n]$$
$$\sum_{C:j\in C} y_C = 1 \text{ for all } j \in [n] \tag{6}$$
$$y_C \in \{0, 1\} \text{ for all } C \in \mathscr{C}.$$

*Example 7* Consider the instance with $n = m = 6$, $f(i) = i$ for all $i$ and

$$\mathscr{X}_1 = \{2, 3, 4\}, \quad \mathscr{X}_2 = \{1, 3, 4\},$$

$$\mathscr{X}_3 = \{4, 5, 6\}, \quad \mathscr{X}_4 = \{3, 5, 6\},$$

$$\mathscr{X}_5 = \{1, 2, 6\}, \quad \mathscr{X}_6 = \{1, 2, 5\}.$$

We associate the graph $\mathscr{G}$ given in Fig. 5 with this instance, where an edge from one clique to another means that all the nodes in a clique are connected to all the nodes in the other clique. Using the scheme based on clique covering requires us to use at least three transmissions. With local clique covering, we first encode using clique covering to obtain $Y' = [X_1 + X_2, X_3 + X_4, X_5 + X_6]$. Then, exploiting the side information, we can encode $Y'$ using the MDS code also used in Example 5 to obtain $Y = [X_1 + X_2 + X_5 + X_6, X_3 + X_4 + X_5 + X_6]$. Then all receivers are able to decode ther requested packets.

This linear programming approach was then used to extend these bounds to the more general case of the hypergraphs. Blasiak, Kleinberg, and Lubetzky in [7] introduced the concept of a hyperclique.

**Fig. 5** The graph $\mathscr{G}$ for the ICSI instance in Example 7

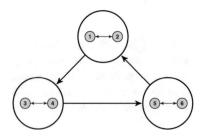

**Definition 5** Let $\mathcal{H}$ a side-infromation hypergraph. A subset $S \subseteq [m]$ is a hyper-clique if for every pair of distinct elements $i, j \in S$ $f(i) \in \mathcal{X}_j \cup \{f(j)\}$.

As in the graph case we can satisfy the requests of all the receivers in a hyperclique by just sending the XOR of all packets. We define the hyperclique covering number $\psi(\mathcal{H})$ as the optimal solution of

$$\min \sum_{C \in \mathcal{C}} y_C$$

$$\text{s.t.} \sum_{C:j \in C} y_C = 1 \text{ for all } j \in [m] \tag{7}$$

$$y_C \in \{0, 1\} \text{ for all } C \in \mathcal{C}.$$

The solution of the LP relaxation of (7) is denoted by $\psi_f(\mathcal{H})$.

**Theorem 7** ([7]) *The optimal broadcast rate is upper bounded by $\psi_f(\mathcal{H})$*

$$\beta(\mathcal{G}) \le \psi_f(\mathcal{H}) \le \psi(\mathcal{H}).$$

In the work of Tehrani, Dimakis and Neely [36] the authors extend the idea of the partial clique to the case $m \ge n$, introducing the multicast partition.

As noted in [36], if we have a side-information hypergraph $\mathcal{H}$ and each receiver knows at least $k$ packets (i.e. $|\mathcal{X}_i| \ge k$), then we can use the same approach as in the partial-clique case, namely, to use the generator matrix of an MDS code to encode the packets.

**Theorem 8** ([36]) *Let $\psi^p(\mathcal{H})$ the optimal solution of*

$$\min \sum_M a_M d_m$$

$$\text{s.t.} \sum_{M:j \in M} a_M = 1 \text{ for all } j \in [m] \tag{8}$$

$$a_M \in \{0, 1\} \text{ for all } M \subseteq [m], \quad d_m = |R(M)| - \max_{j \in M} |R(M) \cap \mathcal{X}_j|,$$

*where $R(M) = \{f(i) \mid i \in M\}$.*
*Then $\beta(\mathcal{H}) \le \psi^p(\mathcal{H})$.*

Later, Shanmugam, Dimakis, and Langberg [30] proved that the fractional version of this parameter provides an upper bound for the optimal rate.

In [30] the authors extended their previous result, defining the local hyperclique covering number $\psi_l(\mathcal{H})$ of an hypergraph as the optimal solution of the integer program

$$\min \quad k$$

$$\text{s.t.} \sum_{C:C\cap U_j\neq\emptyset} y_C \leq k \ \text{ for all } j \in [m]$$

$$\sum_{C:j\in C} y_C = 1 \text{ for all } j \in [m] \tag{9}$$

$$y_C \in \{0, 1\} \text{ for all } C \in \mathscr{C},$$

where $U_j = \{i \in [m] \mid f(i) \notin \mathscr{X}_j\}$. The LP relaxation of (9) is defined to be the fractional local hyperclique cover, denoted $\psi_{lf}(\mathscr{H})$.

**Theorem 9** ([30]) *The optimal broadcast rate is upper bounded by* $\psi_{lf}(\mathscr{H})$

$$\beta(\mathscr{G}) \leq \psi_{lf}(\mathscr{H}) \leq \psi_l(\mathscr{H}).$$

In [30], another parameter, called the partitioned local hyperclique cover and its fractional version for the groupcast setting is defined. This pararameter is stronger than those based on local hyperclique covering and partition multicast.

**Definition 6** The partitioned local hyperclique cover number of $\mathscr{H}$, denoted $\psi_l^P(\mathscr{H})$, is given by the following integer program:

$$\min \quad \sum_M a_M k_M$$

$$\text{s.t.} \sum_{C:C\cap U_j\cap M\neq\emptyset} y_C \leq k_M \text{ for all } j \in M$$

$$\sum_{M:j\in M} a_M = 1 \text{ for all } j \in [m] \tag{10}$$

$$\sum_{C:j\in C} y_C = 1 \text{ for all } j \in [m]$$

$$y_C \in \{0, 1\} \text{ for all } C \in \mathscr{C}, \ a_M \in \{0, 1\} \text{ for all } M \subseteq [m].$$

The fractional version $\psi_{lf}^P(\mathscr{H})$ is given by the LP relaxation of (10).

**Theorem 10** ([30])
$$\beta(\mathscr{G}) \leq \psi_{lf}^P(\mathscr{H}) \leq \psi_l^P(\mathscr{H}).$$

However this new scheme is within a factor $e$ from the fractional hyperclique cover (implying the same for all previous bounds as well).

**Theorem 11** ([30])
$$\psi_f(\mathscr{H}) \leq e\psi_{lf}^P(\mathscr{H}).$$

In Fig. 6 we report the comparison of all these parameters.

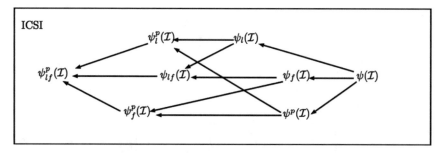

**Fig. 6** Comparison of graph theoretic bounds for the ICSI problem. $u \leftarrow v$ means $u \leq v$

## 4 Generalizations of the Index Coding Problem - Coded-Side Information

In [12, 33] the authors give a generalization of the index coding problem in which both demanded packets and locally cached packets may be linear combinations of some set of data packets. We refer to this as the *index coding with coded side information* problem (ICCSI). This represents a significant departure from the ICSI problem in that an ICCSI instance no longer has an obvious association to a graph, digraph or hypergraph, as in the ICSI case. However, as we show here, it turns out that many of the results for index coding have natural extensions in the ICCSI problem.

One motivation for the ICCSI generalization is related to the coded-caching problem. The method in [28] uses uncoded cache placement, but the authors give an example to show that coded cache placement performs better in general. In [8], it is shown that in a small cache size regime, when the number of users is not less than the number of files, a scheme based on coded cache placement is optimal. Moreover in [39] the authors show that the only way to improve the scheme given in [28] is by coded cache placement.

Another motivation is toward applications for wireless networks with relay helper nodes and cloud storage systems [12] (Fig. 7).

**Fig. 7** State of the network after the 6th time slot

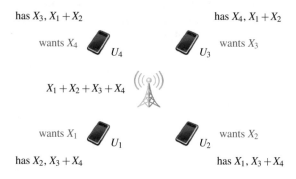

**Table 1** Illustration of utilizing coded packets as side information

| Time slot | Packet sent | Received by $U_1$? | Received by $U_2$? | Received by $U_3$? | Received by $U_4$? |
|---|---|---|---|---|---|
| 1 | $X_1$ | No | Yes | No | No |
| 2 | $X_2$ | Yes | No | No | No |
| 3 | $X_3$ | No | No | No | Yes |
| 4 | $X_4$ | No | No | Yes | No |
| 5 | $X_1 + X_2$ | No | No | Yes | Yes |
| 6 | $X_4 + X_3$ | Yes | Yes | No | No |
| 7 | $X_1 + X_2 + X_3 + X_4$ | Yes | Yes | Yes | Yes |

Consider the example in Table 1. We have a scenario with one sender and four receivers $U_1$, $U_2$, $U_3$ and $U_4$. The source node has four packets $X_1$, $X_2$, $X_3$ and $X_4$ and for $i = 1, ..., 4$ user $U_i$ wants packet $X_i$. The transmitted packet is subject to independent erasures. It is assumed that there are feedback channels from the users, informing the transmitting node which packets are successfully received. At the beginning, in time slot 1, 2, 3 and 4 the source node transmits packets $X_1$, $X_2$, $X_3$ and $X_4$, respectively. After time slot 4 we have the following setting: $U_1$ has packet $X_2$, $U_2$ has packet $X_1$, $U_3$ has packet $X_4$ and $U_4$ has packet $X_3$. Now from the classical ICSI problem we have that receivers $U_1$ and $U_2$ form a clique, in the associated graph, and then we can satisfy their request sending $X_1 + X_2$. Similarly for $U_3$ and $U_4$ we can use $X_3 + X_4$. So, the source node in time slot 5 and 6 transmits the coded packet $X_1 + X_2$ and $X_3 + X_4$, intending that users receive the respective packet. However, $U_1$ and $U_2$ receive the coded packet $X_3 + X_4$ and $U_3$ and $U_4$ receive $X_1 + X_2$. At this point if only the uncoded packets in their caches are used, we still need to send two packets. If all packets in their caches are used, the source only needs to transmit one coded packet $X_1 + X_2 + X_3 + X_4$ in time slot 7. If all four users can receive this last transmission successfully, then all users can decode the required packets by linearly combining with the packets received earlier.

We now describe an instance of index coding with coded-side information. As for the uncoded case we have a data matrix $X \in \mathbb{F}_q^{n \times t}$ and a set of $m$ receivers. For each $i \in [m]$, the $i$th user seeks some linear combination of the rows of $X$, say $R_i X$ for some $R_i \in \mathbb{F}_q^n$. We will refer to $R_i$ as the request vector and to $R_i X$ as the request packet of User $i$. In this scenario a user's cache is represented by a pair of matrices

$$V^{(i)} \in \mathbb{F}_q^{d_i \times n} \text{ and } \Lambda^{(i)} \in \mathbb{F}_q^{d_i \times t}$$

related by the equation

$$\Lambda^{(i)} = V^{(i)} X.$$

It is assumed that any vector in the row spaces of $V^{(i)}$ and $\Lambda^{(i)}$ can be generated at the $i$th receiver. We denote these respective row spaces by $\mathscr{X}^{(i)} := \langle V^{(i)} \rangle$ and $\mathscr{L}^{(i)} :=$

$\langle \Lambda^{(i)} \rangle$ for each $i$. The side information of the $i$th user is $(\mathscr{X}^{(i)}, \mathscr{L}^{(i)})$. Similarly, the sender $S$ has the pair of row spaces $(\mathscr{X}^{(S)}, \mathscr{L}^{(S)})$ for matrices

$$V^{(S)} \in \mathbb{F}_q^{d_S \times n} \text{ and } \Lambda^{(S)} = V^{(S)} X \in \mathbb{F}_q^{d_S \times t}$$

and does not necessarily possess the matrix $X$ itself.

The $i$th user requests a coded packet $R_i X \in \mathscr{L}^{(S)}$ with $R_i \in \mathscr{X}^{(S)} \setminus \mathscr{X}^{(i)}$. We denote by $R$ the $m \times n$ matrix over $\mathbb{F}_q$ with each $i$th row equal to $R_i$. The matrix $R$ thus represents the requests of all $m$ users.

*Example 8* Consider the instance given by the example represented in Fig. 4. Then we have $q = 2$, $m = n = 4$, $t = 1$ and $R_i = \mathbf{e}_i$ for all $i \in [4]$ and $\mathscr{X}^{(S)} = \mathbb{F}_2^4$. The side information are given by the following matrices

$$V^{(1)} = \begin{bmatrix} 0 & 1 & 0 & 0 \\ 0 & 0 & 1 & 1 \end{bmatrix}, \ V^{(2)} = \begin{bmatrix} 1 & 0 & 0 & 0 \\ 0 & 0 & 1 & 1 \end{bmatrix},$$

$$V^{(3)} = \begin{bmatrix} 1 & 1 & 0 & 0 \\ 0 & 0 & 0 & 1 \end{bmatrix}, \ V^{(4)} = \begin{bmatrix} 1 & 1 & 0 & 0 \\ 0 & 0 & 1 & 0 \end{bmatrix}.$$

*Remark 2* The reader will observe that the classical ICSI problem is indeed a special case of the index coding problem with coded side information. Setting $V^{(S)}$ to be the $n \times n$ identity matrix, $R_i = \mathbf{e}_{f(i)} \in \mathbb{F}_q^n$ and $V^{(i)}$ to be the $d_i \times n$ matrix with rows $V_j^{(i)} = \mathbf{e}_{i_j}$ for each $i_j \in \mathscr{X}_i$, yields $\mathscr{X}^{(i)} = \langle \mathbf{e}_j : j \in \mathscr{X}_i \rangle$. Then User $i$ has the rows of $X$ indexed by $\mathscr{X}_i$ and requests $X_{f(i)}$.

*Remark 3* The case where the sender does not necessarily possess the matrix $X$ itself can be applied to the *broadcast relay channel*, as described in [33]. The authors consider a channel as in Fig. 8, and assume that the relay is close to the users and far away from the source, and in particular that all relay-user links are erasure-free. Each node is assumed to have some storage capacity and stores previously received data in its cache. The packets in the cache of the relay node are obtained as previous broadcasts, hence it may contain both coded and uncoded packets. The relay node, playing the role of the sender, transmits packets obtained by linearly combining

**Fig. 8** A schematic for the broadcast relay channel

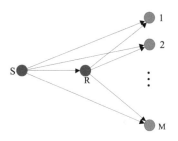

the packets in its cache, depending on the requests and coded side information of all users. It seeks to minimize the total number of broadcasts such that all users' demands are met.

We denote by

$$\mathcal{X} := \{A \in \mathbb{F}_q^{m \times n} : A_i \in \mathcal{X}^{(i)}, i \in [m]\},$$

so that $\mathcal{X} = \oplus_{i \in [m]} \mathcal{X}^{(i)}$ is the direct sum of the $\mathcal{X}^{(i)}$ as a vector space over $\mathbb{F}_q$. For the remainder, we let $\mathcal{X}$, $\mathcal{X}^{(S)}$, $R$ be as defined above and write $\mathscr{I} = (\mathcal{X}, \mathcal{X}^{(S)}, R)$ to denote an instance of the ICCSI problem for these parameters.

The definition of an index code is similar to that of the ICSI case.

**Definition 7** Let $N$ be a positive integer. We say that the map

$$E : \mathbb{F}_q^{n \times t} \to \mathbb{F}_q^{N \times t},$$

is an $\mathbb{F}_q$-code for $\mathscr{I}$ of length $N$ if for each $i$th receiver, $i \in [m]$ there exists a decoding map

$$D_i : \mathbb{F}_q^{N \times t} \times \mathcal{X}^{(i)} \to \mathbb{F}_q^t,$$

satisfying

$$\forall X \in \mathbb{F}_q^{n \times t} : D_i(E(X), A) = R_i X,$$

for some vector $A \in \mathcal{X}^{(i)}$, in which case we say that $E$ is an $\mathscr{I}$-IC. $E$ is called an $\mathbb{F}_q$-linear $\mathscr{I}$-IC if $E(X) = L V^{(S)} X$ for some $L \in \mathbb{F}_q^{N \times ds}$, in which case we say that $L$ represents the code $E$, or that the matrix $L$ realizes $E$. If $t = 1$, we say that $L$ represents a scalar linear index code. If $t > 1$ we say that the code is vector linear. We write $\mathscr{L}$ to denote the space $\langle L V^{(S)} \rangle$.

As before, for the ICCSI instance $\beta_t(\mathscr{I})$ denotes the minimum broadcast rate for block-length $t$ where the encoding is over all possible extensions of $\mathbb{F}_p$. That is, for $\mathscr{I} = (\mathcal{X}, \mathcal{X}^{(S)}, R)$

$$\beta_t(\mathscr{I}) = \inf_q \{N \mid \exists \text{ a } q\text{-ary index code of length } N \text{ for } \mathscr{I}\}.$$

and the optimal broadcast rate is given by the limit

$$\beta(\mathscr{I}) = \lim_{t \to \infty} \frac{\beta_t(\mathscr{I})}{t} = \inf_t \frac{\beta_t(\mathscr{I})}{t}.$$

## 4.1  Bounding the Optimal Rate with Coded-Side Information

As shown in [4], it turns out that also for the case of coded side information the optimal length of a scalar linear index code is linked to a problem of rank minimization.

The result of Lemma 1 is generalized by the following

**Lemma 2** ([4]) *Let* $L \in \mathbb{F}_q^{N \times ds}$. *Then* $L$ *represents an* $\mathbb{F}_q$-*linear* $\mathscr{I}$-*IC index code of length* $N$ *if and only if for each* $i \in [m]$, $R_i \in \mathscr{L} + \mathscr{X}^{(i)}$.

So Lemma 2 gives necessary and sufficient conditions for a matrix $L$ to represent a linear code of the instance $\mathscr{I}$. The sufficiency of the statement of Lemma 2 has already been noted in [33]).

*Remark 4*  If the equivalent conditions of the above lemma hold we have that for each $i \in [m]$, $R_i = \mathbf{b}^{(i)} L V^{(S)} + \mathbf{a}^{(i)} V^{(i)}$ for some vectors $\mathbf{a}^{(i)}, \mathbf{b}^{(i)}$. So User $i$ decodes its request by computing

$$R_i X = \mathbf{b}^{(i)} L V^{(S)} X + \mathbf{a}^{(i)} V^{(i)} X = \mathbf{b}^{(i)} Y + \mathbf{a}^{(i)} \Lambda^{(i)},$$

where $Y$ is the received message.

The analogue of the min-rank is as follows:

**Definition 8** ([4]) The min-rank of the instance $\mathscr{I} = (\mathscr{X}, \mathscr{X}^{(S)}, R)$ of the ICCSI problem over $\mathbb{F}_q$ is

$$\kappa(\mathscr{I}) = \min \left\{ \mathrm{rank}(A + R) : \begin{matrix} A \in \mathbb{F}_q^{m \times n}, \\ A_i \in \mathscr{X}^{(i)} \cap \mathscr{X}^{(S)}, \forall i \in [m] \end{matrix} \right\}.$$

Let $\tilde{\mathscr{X}} = \{Z \in \mathbb{F}_q^{m \times n} : Z_i \in \mathscr{X}^{(S)}\}$. We observe that the quantity $\kappa(\mathscr{I})$ is $d_{\mathrm{rk}}$ $(R, \mathscr{X} \cap \tilde{\mathscr{X}})$, which is the rank-distance of $R \in \mathbb{F}_q^{m \times n}$ to the $\mathbb{F}_q$-linear code $\mathscr{X} \cap \tilde{\mathscr{X}}$, or equivalently the minimum rank-weight of the coset $R + (\mathscr{X} \cap \tilde{\mathscr{X}}) \subset \mathbb{F}_q^{m \times n}$.

As consequence of Lemma 2 we have

**Theorem 12** ([4]) *The length of an optimal* $\mathbb{F}_q$-*linear* $\mathscr{I}$-*IC is* $\kappa(\mathscr{I})$. *In particular*

$$\beta(\mathscr{I}) \leq \kappa(\mathscr{I}).$$

Similarly to the classical ICSI, we have that optimal linear index coding matrix can be obtained by solving a matrix completion problem over a finite field. However, this matrix completion problem is different from the conventional matrix completion problem. This comes from the fact that, in this more general problem, the rows of the matrix have to lie in the spaces $\mathscr{X}^{(i)}$'s. In [25] it is proposed a random greedy algorithm (over $\mathbb{F}_2$) that minimizes the rank of the derived matrix.

As said before an ICCSI instance no longer has an obvious association to a combinatorial structure such as a graph or hypergraph. However, all the bounds given in Sect. 3.1 can be generalized to the ICCSI case adopting a Linear Programming approach.

We start with the following definition, introduced in [33] as a *coding group*, wherein a procedure to detect such as subset is given. It is easy to see that this definition generalizes the definition of a hyperclique given in Sect. 3.1.

**Definition 9** Let $\mathscr{I} = (\mathscr{X}, \mathscr{X}^{(S)}, R)$ be an instance of the ICCSI problem. A subset of receivers $C \subseteq [m]$ is called *generalized clique* if there exists $\mathbf{v} \in \mathscr{X}^{(S)}$ such that $R_i \in \langle \mathbf{v} \rangle + \mathscr{X}^{(i)}$ for all $i \in C$.

For simplicity in the following we refer to a generalized clique just as a clique.

Note that demand $R_i X$ of each user $i$ of a clique can be met by sending the message $\mathbf{v}X$, that is one packet as for the classical case. Hence a set of $\ell$ cliques that partitions the set $[m]$ ensures that all requests can be delivered in at most $\ell$ transmissions.

We denote by $\mathscr{C}$ the set of all cliques of $\mathscr{I} = (\mathscr{X}, \mathscr{X}^{(S)}, R)$.

**Definition 10** We define the *generalized clique cover number* of $\mathscr{I}$, denoted by $\varphi(\mathscr{I})$, to be the optimal solution of the following integer program:

$$\min \sum_{C \in \mathscr{C}} y_C$$

$$\text{s.t. } \sum_{C: j \in C} y_C = 1 \text{ for all } j \in [m] \tag{11}$$

$$y_C \in \{0, 1\} \text{ for all } C \in \mathscr{C}.$$

The LP relaxation of (11) (so with the relaxed constraint $0 \le y_C \le 1$ for all $C$) is the *fractional generalized clique cover number* $\varphi_f(\mathscr{I})$.

**Theorem 13** ([5]) *Let $\mathscr{I} = (\mathscr{X}, \mathscr{X}^{(S)}, R)$. There exist achievable $\mathbb{F}_q$-linear index codes corresponding to $\varphi(\mathscr{I})$ and $\varphi_f(\mathscr{H})$. In particular, we have*

$$\beta(\mathscr{I}) \le \varphi_f(\mathscr{I}) \le \varphi(\mathscr{I}).$$

For each clique $C \in \mathscr{C}$ define the set

$$\mathscr{R}(C) := \{\mathbf{v} \in \mathbb{F}_q^n \mid R_i \in \langle \mathbf{v} \rangle + \mathscr{X}^{(i)} \ \forall i \in C\}.$$

That is, $\mathscr{R}(C)$ contains all the vectors that we can use to encode $X$ in order that all receivers in $C$ can decode correctly.

**Definition 11** For each $C \in \mathscr{C}$ fix a vector $\mathbf{v}_C \in \mathscr{R}(C)$. We define the following integer program with respect to the vectors $\mathbf{v}_C$.

min $k$

s.t. $\displaystyle\sum_{C:\mathbf{v}_C\notin\mathcal{X}^{(j)}} y_C \leq k$ for all $j \in [m]$ (12)

$\displaystyle\sum_{C:j\in C} y_C = 1$ for all $j \in [m]$     $y_C \in \{0, 1\}$ for all $C \in \mathcal{C}$ and $k \in \mathbb{N}$.

We denote by $\phi_l(\mathcal{I}, (\mathbf{v}_C \in \mathcal{R}(C) : C \in \mathcal{C}))$ the optimal solution of (12), depending on the fixed $\mathbf{v}_C$'s. The minimum over all possible $\mathbf{v}_C$'s is called the *local generalized clique cover number*

$$\varphi_l(\mathcal{I}) = \min_{(\mathbf{v}_C\in\mathcal{R}(C):C\in\mathcal{C})} \phi_l(\mathcal{I}, (\mathbf{v}_C : C \in \mathcal{C})).$$

This is an extension of the local hyperclique cover: for a set of fixed $\mathbf{v}_C$, given user $j \in [m]$ and some feasible solution to (11), count number of cliques $C$ in that generalized clique cover such that $\mathbf{v}_C$ is not contained in the side-information $\mathcal{X}^{(j)}$ and let $k$ be the maximum number of such cliques for each $j$. The optimal solution of (12) is the minimum value of $k$ over all possible solutions of (11) and all choices of $\mathbf{v}_C$. The minimum of the LP relaxation of (12) over all possible $\mathbf{v}_C$'s is called the fractional local generalized clique cover number $\varphi_{lf}(\mathcal{I})$. Clearly the optimal solution of (12) depends on the choice of vectors $\mathbf{v}_C$.

**Theorem 14** ([5]) *Let $\mathcal{I} = (\mathcal{X}, \mathcal{X}^{(S)}, R)$. There are achievable linear index codes corresponding to $\varphi_l(\mathcal{I})$ and $\varphi_{lf}(\mathcal{I})$ implying $\beta(\mathcal{I}) \leq \varphi_{lf}(\mathcal{I}) \leq \varphi_l(\mathcal{I})$.*

In the multicast partition scheme for the ICSI case we want to find a partition where the knowledge, i.e. the cardinality of $\mathcal{X}_i$, of the users among the sets of the partition is maximized. Now, the knowledge of a user is given by the dimension of the space $\mathcal{X}^{(i)}$, so we obtain the following

**Definition 12** We define the *partition generalized multicast number*, $\varphi^P(\mathcal{I})$ to be the optimal solution of the following integer program

$$\min \sum_{M\subset[m]} a_M d_M$$

s.t. $\displaystyle\sum_{M:j\in M} a_M = 1$ for all $j \in [m]$ (13)

$$a_M \in \{0, 1\} \text{ for all } M \subset [m], M \neq \emptyset.$$

and $d_M = \dim(\langle R_M\rangle) - \displaystyle\min_{j\in M}\dim(\langle R_M\rangle \cap \mathcal{X}^{(j)}).$

The LP relaxation of (13) is called the *fractional partition generalized multicast number*, $\varphi_f^P(\mathcal{I})$.

We briefly justify the above: each user is assigned to exactly one multicast group $M$, so the selected groups $M$ form a partition of $[m]$. Each member $j$ of a multicast group $M \subset [m]$ already has access to at least $\dim(\langle R_M \rangle \cap \mathscr{X}^{(j)})$ independent vectors in $\langle R_M \rangle$, so a coding scheme can be applied to ensure delivery of all remaining requests within a group using at most $d_M$ transmissions. In fact as it is shown in [4] we have

**Proposition 1** *Let* $\mathscr{I} = (\mathscr{X}, \mathscr{X}^{(S)}, R)$. *If* $q > m$ *then* $\kappa(\mathscr{I}) \leq \max\{n - d_i : i \in [m]\}$. *For any* $q$, $\kappa(\mathscr{I}) \leq \operatorname{rank}(R)$.

The essential content of the proof of Proposition 1 is that there exists an $N \times n$ matrix $L$ realizing $\mathscr{I}$ for $N \leq \max\{n - d_i : i \in [m]\}$, which corresponds to a multicast solution, so every user can retrieve any linear combination of the $X_i$. In this case the matrix $L$ is such that $\mathscr{L} + \mathscr{X}^{(i)} = \mathbb{F}_q^n$ for each $i$.

**Theorem 15** ([5]) *Let* $\mathscr{I} = (\mathscr{X}, \mathscr{X}^{(S)}, R)$. *There are achievable linear index codes of lengths* $\varphi^p(\mathscr{I})$ *and* $\varphi_f^p(\mathscr{I})$, *which implies that* $\beta(\mathscr{I}) \leq \varphi_f^p(\mathscr{I}) \leq \varphi^p(\mathscr{I})$.

The final approach considered combines partition multicast and local clique covering. The users are partitioned into multicast groups and independently covered by generalized cliques. Each multicast group offers a reduced ICCSI problem to which a restricted local clique cover is applied.

**Definition 13** Define the following integer program

$$\min \sum_{M \subset [m]} a_M k_M$$

$$\text{s.t.} \sum_{\substack{C : v_C \notin \mathscr{X}^{(j)} \\ C \cap M \neq \emptyset}} y_C \leq k_M \text{ for all } j \in M \tag{14}$$

$$\sum_{M : j \in M} a_M = 1, \quad \sum_{C : j \in C} y_C = 1 \text{ for all } j \in [m]$$

$$a_M, y_C \in \{0, 1\} \text{ for all } C \in \mathscr{C}, M \subset [m] \text{ and } t_M \in \mathbb{N}.$$

We denote by $\phi_l^p(\mathscr{I}, (v_C \in \mathscr{R}(C) : C \in \mathscr{C}))$ the optimal solution of (14) with respect to $(v_C \in \mathscr{R}(C) : C \in \mathscr{C})$ fixed. The minimum over all possible choices of $v_C$ is called the *partitioned local generalized clique cover number*

$$\varphi_l^p(\mathscr{I}) = \min_{(v_C \in \mathscr{R}(C) : C \in \mathscr{C})} \phi_l^p(\mathscr{I}, (v_C \in \mathscr{R}(C) : C \in \mathscr{C})).$$

The minimum of the LP relaxation of (14) over all possible choices of $v_C$ is called the fractional partitioned local generalized clique cover number $\varphi_{lf}^p(\mathscr{I})$.

**Theorem 16** ([5]) *There are achievable linear index codes corresponding to* $\varphi_l^p(\mathscr{I})$ *and* $\varphi_{lf}^p(\mathscr{I})$ *implying* $\beta(\mathscr{I}) \leq \varphi_{lf}^p(\mathscr{I}) \leq \varphi_l^p(\mathscr{I})$.

The comparison between the parameter introduced in Sect. 3.1 (see Fig. 6) are no more valid in this contest, as shown in the following remarks.

*Remark 5* The parameters $\varphi^P$ and $\varphi_l^P$ are not comparable. From the parameters given in [30] we have that there exist instances of the ICSI problem for which $\varphi^P(\mathscr{I}) \geq \varphi_l^P(\mathscr{I})$. Now, consider the ICCSI instance with $m = n = 3$, $q = 2$, $\mathscr{X}^{(S)} = \mathbb{F}_2^3$.

$$V^{(1)} = [0\ 1\ 1]\quad V^{(2)} = [1\ 1\ 1]\quad V^{(3)} = [1\ 1\ 1],$$

and $R_1 = 100$, $R_2 = 010$, $R_3 = 001$.

In order to satisfy the request of a receiver using only one vector then the coding vectors should be

- $\mathbf{v}_1 = 100$ or $\mathbf{v}_1' = 111$ for User 1;
- $\mathbf{v}_2 = 010$ or $\mathbf{v}_2' = 101$ for User 2;
- $\mathbf{v}_3 = 001$ or $\mathbf{v}_3' = 110$ for User 3.

Then the set of all cliques is $\mathscr{C} = \{\{1\}, \{2\}, \{3\}\}$. Moreover we can see that $\mathbf{v}_i, \mathbf{v}_i' \notin \mathscr{X}^{(1)}$ for all $i$. Now, if we consider the multicast group $M = \{1, 2, 3\}$ we can note that $d_M = 2$ and that $k_M = 3$ because none of the six vectors above is in the space $\mathscr{X}^{(1)}$. Then we have $2 = \varphi^P(\mathscr{I}) \leq \varphi_l^P(\mathscr{I}) = 3$.

*Remark 6* The parameters $\varphi^P$ and $\varphi$ are not comparable. From the parameters given in [30], there exist instances of the ICSI problem for which $\varphi(\mathscr{I}) \geq \varphi^P(\mathscr{I})$. Now consider the ICCSI instance with $m = n = 2$, $q = 2$, $\mathscr{X}^{(S)} = \mathbb{F}_2^2$.

$$V^{(1)} = [1\ 1]\quad V^{(2)} = [0\ 0],$$

and $R_1 = 10$, $R_2 = 01$. It is easy to check that using the multicast group partition we need two transmissions, but it can be seen that $\{1, 2\}$ is a clique and that $\mathbf{v}_{\{1,2\}} = 01 \in \mathscr{R}(\{1, 2\})$, yielding $1 = \varphi(\mathscr{I}) \leq \varphi^P(\mathscr{I}) = 2$.

*Remark 7* We have $\varphi_l^P(\mathscr{I}) \leq \varphi_l(\mathscr{I}) \leq \varphi(\mathscr{I})$. It is easy to check that $\varphi_l(\mathscr{I}) \leq \varphi(\mathscr{I})$ as $k$ is at most equal to the number of cliques that form a partition of $[m]$. Then we have also $\varphi_l^P(\mathscr{I}) \leq \varphi_l(\mathscr{I})$. In fact, among the possible optimal solutions we have those where $M = [m]$ and in that case we obtain exactly $\varphi_l(\mathscr{I})$.

It is possible to introduce a weak definition of clique. $C \subseteq [m]$ is called weak clique if for all $i, j \in C$ we have $R_j \in \mathscr{X}^{(i)}$ or $\langle R_j \rangle = \langle R_i \rangle$. Using this definition, it is possible to introduce the notion of a *weak clique cover, local weak clique cover* and *partitioned local weak clique cover* with respective corresponding parameters $_w\varphi(\mathscr{I})$, $_w\varphi_l(\mathscr{I})$ and $_w\varphi_l^P(\mathscr{I})$ along with their fractional counterparts. Note that if $C$ is a weak clique then it is also a generalized clique. In fact we can encode the message using the sum of distinct requests as vector $\mathbf{v}_C$.

Moreover from the definition of weak clique, if we consider a clique as a multicast group $M$ then it results $d_M = 1$. Therefore $\varphi^P(\mathscr{I}) \leq {}_w\varphi(\mathscr{I})$ and the same holds for the fractional parameters. However, also in this case, the partitioned local weak clique cover and the partitioned multicast cover are not comparable (see example in Remark 5) (Fig. 9).

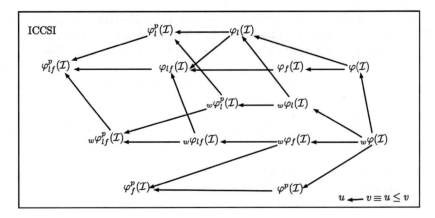

**Fig. 9** ICCSI bounds introduced in this section. Smaller quantities are placed to the left and the weakest bound is placed to the rightmost of the figure. Arrows indicate the relationship they satisfy

## 4.2  Error Correction

Error correction in the index coding problem has been considered in [4, 14]. In this model it is assumed that some erroneous packets have been transmitted among $N$ coded packets $E(X)$, that is the users receive a transmission $Y = E(X) + W$, for some error matrix $W$. The question is now to determine the shortest possible length $N$ of a linear index code such that each receiver can recover its demanded packet, assuming that $E(X)$ has been corrupted by noise.

This problem was introduced in [14] for the case of uncoded side information and extended in [4] to include the coded-side information case. For the case of Hamming errors, it is shown that ideas from classical coding theory can be adapted to the linear index coding problem to obtain bounds on the transmission rate where at most $\delta$ rows of $E(X)$ have been affected by noise. For linear index codes, generic decoding algorithms based on syndrome decoding can be applied. Rank-metric error correction was also considered in [4]. In this exposition, for brevity we will describe the Hamming case only.

The Hamming distance $d_H(Z, Z')$ between a pair of matrices $Z, Z'$ is the number of non-zero rows of their difference $Z - Z'$, which is also the Hamming weight $w_H(Z - Z')$.

The formal notion of a $\delta$-error correcting index code is as follows.

**Definition 14** Let $\mathscr{I}$ be an instance of an ICCSI problem and let $N$ be a positive integer. We say that the map

$$E : \mathbb{F}_q^{n \times t} \to \mathbb{F}_q^{N \times t},$$

is a $\delta$-error correcting code for $\mathscr{I}$ of length $N$, and say that $E$ is an $(\mathscr{I}, \delta)$-ECIC, if for each $i$th receiver there exists a decoding map

$$D_i : \mathbb{F}_q^{N \times t} \times \mathscr{X}^{(i)} \to \mathbb{F}_q^t,$$

satisfying

$$D_i(E(X) + W, A) = R_i X$$

for all $X, W \in \mathbb{F}_q^{N \times t}$, $w_H(W) \leq \delta$ for some vector $A \in \mathscr{X}^{(i)}$. $E$ is called a linear code for $\mathscr{I}$, or an $\mathbb{F}_q$-linear $\mathscr{I}$-ECIC if $E(X) = LV^{(S)}X$ for some $L \in \mathbb{F}_q^{N \times ds}$, in which case we say that $L$ represents the linear $(\mathscr{I}, \delta)$-ECIC $E$.

**Theorem 17** *Let $\mathscr{I}$ be an instance of an ICCSI problem and let $N$ be a positive integer. A matrix $L \in \mathbb{F}_q^{N \times ds}$ represents a linear $(\mathscr{I}, \delta)$-ECIC if and only if for all $i \in [m]$ it holds that*

$$w_H\left(LV^{(S)}(X - X')\right) \geq 2\delta + 1,$$

*for all $X, X' \in \mathscr{M}$ such that $V^{(i)}X = V^{(i)}X'$, $R_i X \neq R_i X'$.*

### 4.2.1 Bounds on the Transmission Rate of an $(\mathscr{I}, \delta)$-ECIC

We denote by $\mathscr{N}(\mathscr{I}, \delta)$ the optimal length of an $\mathbb{F}_q$-linear Hamming metric $(\mathscr{I}, \delta)$-ECIC and by $N(k, d)$ the optimal length $\ell$ of an $\mathbb{F}_q$-$[\ell, k, d]$ code, i.e. a $k$-dimensional $\mathbb{F}_q$-linear code in $\mathbb{F}_q^\ell$ of minimum Hamming distance $d$.

**Definition 15** We denote by $\alpha(\mathscr{I})$ the maximum dimension of any subspace $U$ of $\mathbb{F}_q^n$ such that every member $\mathbf{u}$ of $U$ satisfies $V^{(i)}\mathbf{u} = 0$ and $R_i\mathbf{u} \neq 0$ for some $i \in [m]$. That is,

$$\alpha(\mathscr{I}) := \max\{\dim U \mid U < \mathbb{F}_q^n : \forall \mathbf{u} \in U \setminus \{0\} \exists i \in [m] \text{ s.t. } V^{(i)}\mathbf{u} = 0, R_i\mathbf{u} \neq 0\}$$

*Example 9* Let $q = 2, m = 6, n = 4$. Let

$$V^{(1)} = \begin{bmatrix} 1 & 0 & 1 & 0 \\ 0 & 0 & 0 & 1 \end{bmatrix}, V^{(2)} = \begin{bmatrix} 1 & 0 & 0 & 0 \\ 0 & 0 & 1 & 1 \end{bmatrix}, V^{(3)} = \begin{bmatrix} 1 & 0 & 0 & 0 \\ 0 & 1 & 0 & 0 \end{bmatrix},$$

$$V^{(4)} = \begin{bmatrix} 0 & 1 & 0 & 1 \\ 0 & 0 & 1 & 0 \end{bmatrix}, V^{(5)} = \begin{bmatrix} 1 & 0 & 1 & 0 \\ 0 & 0 & 0 & 1 \end{bmatrix}, V^{(6)} = \begin{bmatrix} 0 & 1 & 0 & 0 \\ 0 & 0 & 0 & 1 \end{bmatrix},$$

and let the request vectors be given by

$$R_1 = 1100, R_2 = 0111, R_3 = 1010, R_4 = 1001, R_5 = 0100, R_6 = 1111.$$

Then the set of vectors of $\mathbb{F}_q^4$ such that $V^{(i)}\mathbf{u} = 0$ and $R_i\mathbf{u} \neq 0$ for some $i \in [4]$ is given by

$$\{1000, 0100, 0010, 1010, 0101, 0011, 1101, 1110, 0111\},$$

which, after including the zero vector, contains the subspace $\{0000, 1010, 0100, 1110\}$, and no larger subspace. It follows that $\alpha(\mathscr{I}) = 2$.

**Theorem 18** ($\alpha$-bound) *Let $\mathscr{I}$ be an instance of the ICCSI problem. Then*

$$N(\alpha(\mathscr{I}), 2\delta + 1) \leq \mathscr{N}(\mathscr{I}, \delta).$$

The argument used in the proof of the $\alpha$-bound may be sketched as follows. We let $C_U := \{LV^{(S)}\mathbf{u} : \mathbf{u} \in U\}$ for $U$ of maximum dimension as described above. If $L$ realizes an $(\mathscr{I}, \delta)$-ECIC then $C_U$ is an $\mathbb{F}_q$-linear $[N, \alpha, 2\delta + 1]$ code, so that the ECIC necessarily has length at least $\alpha(\mathscr{I})$. We give sufficient conditions for tightness of the $\alpha$-bound.

Let $U^\perp$ denotes the orthogonal space of $U$ with respect to the usual scalar product in $\mathbb{F}_q^n$.

**Corollary 1** *Let $\mathscr{I}$ be an instance of the ICCSI problem. If there exists a matrix $B \in \mathbb{F}_q^{\alpha(\mathscr{I}) \times d_S}$ satisfying $BV^{(S)\perp} \cap V^{(i)\perp} \subset R_i^\perp$ for all $i \in [m]$ then*

$$N(\alpha(\mathscr{I}), 2\delta + 1) = \mathscr{N}(\mathscr{I}, \delta).$$

Setting $\delta = 0$ gives the following lower bound on the min-rank of an instance as an immediate consequence of the $\alpha$-bound and Corollary 1.

**Corollary 2** *Let $\mathscr{I}$ be an instance of the ICCSI problem. Then*

$$\alpha(\mathscr{I}) \leq \kappa(\mathscr{I}),$$

*with equality occurring if there exists $L \in \mathbb{F}_q^{\alpha(\mathscr{I}) \times d_S}$ satisfying $LV^{(S)\perp} \cap V^{(i)\perp} \subset R_i^\perp$ for all $i \in [m]$.*

**Theorem 19** ($\kappa$-bound) *Let $\mathscr{I}$ be an instance of the ICCSI problem. Then*

$$\mathscr{N}(\mathscr{I}, \delta) \leq N(\kappa(\mathscr{I}), 2\delta + 1).$$

In the proof of the $\kappa$ bound, a matrix realizing an $(\mathscr{I}, \delta)$-ECIC is found as the product $L = L_2L_1$ where $L_1$ is a $\kappa(\mathscr{I}) \times d_S$ matrix realizing an optimal $\mathscr{I}$-IC and the column space of $L_2$ is an optimal $\mathbb{F}_q$-$[N, \kappa(\mathscr{I}), 2\delta + 1]$ code.

A Singleton-like bound is deduced by observing that deleting any $2\delta$ rows of a matrix $L$ that realizes an $(\mathscr{I}, \delta)$-ECIC results in an $\mathscr{I}$-IC.

**Theorem 20** (Singleton bound) *Let $\mathscr{I}$ be an instance of the ICCSI problem. Then*

$$\kappa(\mathscr{I}) + 2\delta \leq \mathscr{N}(\mathscr{I}, \delta).$$

*Example 10* Let $m = 6, n = 5, q = 2$ and let $\mathscr{X}^{(S)} = \mathbb{F}_q^5$. Suppose that $\mathscr{X}^{(i)}$ has dimension $d_i = 2$ for each $i \in \{1, ..., 6\}$. Let $\mathscr{I}$ be the instance defined by user side-information

$$V^{(1)} = \begin{bmatrix} 0 & 1 & 1 & 1 & 0 \\ 0 & 0 & 1 & 1 & 1 \end{bmatrix}, \ V^{(2)} = \begin{bmatrix} 1 & 0 & 0 & 0 & 1 \\ 0 & 0 & 1 & 1 & 0 \end{bmatrix}, \ V^{(3)} = \begin{bmatrix} 1 & 1 & 1 & 1 & 0 \\ 0 & 0 & 0 & 1 & 1 \end{bmatrix},$$

$$V^{(4)} = \begin{bmatrix} 1 & 0 & 0 & 1 & 0 \\ 0 & 1 & 1 & 1 & 1 \end{bmatrix}, \ V^{(5)} = \begin{bmatrix} 0 & 0 & 1 & 1 & 0 \\ 0 & 0 & 0 & 1 & 1 \end{bmatrix}, \ V^{(6)} = \begin{bmatrix} 1 & 0 & 0 & 1 & 0 \\ 0 & 0 & 1 & 1 & 0 \end{bmatrix},$$

and request vectors

$$R_1 = 10000, \ R_2 = 10000, \ R_3 = 00101, \ R_4 = 10001, \ R_5 = 11000, \ R_6 = 00111.$$

It can be checked that $\kappa(\mathscr{I}) = 3$ and that $\alpha(\mathscr{I}) = 3$. It follows from the $\alpha$-bound that $6 = N(3, 3) = N(\alpha(\mathscr{I}), 3) \leq \mathscr{N}(\mathscr{I}, 1)$. From the $\kappa$-bound we have $6 = N(3, 3) = N(\kappa(\mathscr{I}), 3) \geq \mathscr{N}(\mathscr{I}, 1)$.

We construct a matrix $L$ that realizes a 1-error correcting ECIC for this instance as

$$L = \begin{bmatrix} 10000 \\ 01000 \\ 00011 \\ 01011 \\ 11000 \\ 11011 \end{bmatrix} = \begin{bmatrix} 100 \\ 010 \\ 001 \\ 011 \\ 110 \\ 111 \end{bmatrix} \begin{bmatrix} 10000 \\ 01000 \\ 00011 \end{bmatrix} = L_2 L_1,$$

where $L_2$ has column space a $[6, 3, 3]$ binary linear code and $L_1$ realizes an optimal code $\mathscr{I}$-IC.

If $q \geq \kappa(\mathscr{I}) + 2\delta - 1$ then there exists an $\mathbb{F}_q$-linear $[q + 1, \kappa(\mathscr{I}), 2\delta + 1]$ optimal code (in fact a maximum distance separable code), so the $\kappa$ and Singelton bounds combine to give the following result.

**Corollary 3** *Let $\mathscr{I}$ be an instance of the ICCSI problem. If $q \geq \kappa(\mathscr{I}) + 2\delta - 1$ then*

$$\mathscr{N}(\mathscr{I}, \delta) = \kappa(\mathscr{I}) + 2\delta.$$

We let $V_q(n, r)$ denote the size of a Hamming sphere of radius $r$ in $\mathbb{F}_q^n$.

**Theorem 21** *Let $\mathscr{I}$ be an instance of the ICCSI problem. Let $L \in \mathbb{F}_q^{N \times d_S}$ be selected uniformly at random over $\mathbb{F}_q$. The probability that $L$ corresponds to a Hamming metric $\mathbb{F}_q$-linear $(\mathscr{I}, \delta)$-ECIC is at least*

$$1 - \sum_{i=1}^{m} q^{n-d_i-1}(q - 1) \frac{V_q(N, 2\delta)}{q^N}.$$

*In particular there exists an $\mathbb{F}_q$-linear $(\mathscr{I}, \delta)$-ECIC of length $N$ if*

$$N > n - d - 1 + \log_q(m(q-1)V_q(N, 2\delta)),$$

*where $d = \min\{d_i : i \in [m]\}$.*

Let $H_q$ denote the $q$-ary entropy function:

$$H_q : (0, 1) \to \mathbb{R} : x \mapsto x \log_q(q-1) - x \log_q(x) - (1-x)\log_q(1-x).$$

It is well known that the function $H_q(x)$ is continuous and increasing on $(0, 1 - (1/q))$ and that $V_q(n, \lambda n) \leq q^{H_q(\lambda)n}$ [24].

**Corollary 4** *Let $\mathscr{I}$ be an instance of the ICCSI problem with. Let $\lambda \in \mathbb{Q}$ such that $0 < \lambda < 1 - 1/q$ and let $N \in \mathbb{Z}$ satisfy $\lambda N \in \mathbb{Z}$ Then, choosing the entries of $L \in \mathbb{F}_q^{N \times d_S}$ uniformly at random over the field $\mathbb{F}_q$, the probability that $L$ corresponds to a Hamming metric $\mathbb{F}_q$-linear $(\mathscr{I}, \delta)$-ECIC, with $\delta = \lfloor \frac{\lambda N}{2} \rfloor$, is at least*

$$1 - (q-1) \sum_{i \in \hat{m}} \frac{q^{(n-d_i-1)}}{q^{N(1-H_q(\lambda))}}.$$

*In particular there exists an $\mathbb{F}_q$-linear Hamming metric $(\mathscr{I}, \delta)$-ECIC if*

$$m < \frac{q^{N(1-H_q(\lambda))-(n-d-1)}}{q-1},$$

*where $d = \min\{d_i : i \in [m]\}$.*

*Remark 8* The $\alpha$ and Singleton bounds have rank-metric analogues, as does Theorem 21. The interested reader is referred to [4] for further details.

### 4.2.2  Decoding an $(\mathscr{I}, \delta)$-ECIC

For the remainder of this section, we let $L \in \mathbb{F}_q^{N \times d_S}$ be a matrix corresponding to an $(\mathscr{I}, \delta)$-ECIC. Suppose that for some $i \in [m]$ the $i$th user, receives the message

$$Y_{(i)} = LV^{(S)}X + W_{(i)} \in \mathbb{F}_q^{N \times t},$$

where $LV^{(S)}X$ is the codeword transmitted by $S$ and $W_{(i)}$ is the error vector in $\mathbb{F}_q^N$. Since $R_i \notin \mathscr{X}^{(i)}$, there exists an invertible matrix $M_{(i)} \in \mathbb{F}_q^{n \times n}$ such that

$$V^{(i)}M_{(i)} = [I|0] \text{ where } I \text{ is the identity matrix in } \mathbb{F}_q^{d_i \times d_i}, \text{ and } R_i M_{(i)} = \mathbf{e}_{d_i+1}.$$

Now define $X' := M_{(i)}^{-1} X \in \mathbb{F}_q^{n \times t}$. Then we have

$$V_j^{(i)} X = \mathbf{e}_j M_{(i)}^{-1} X = X'_j \text{ for } j \in [d_i]$$

and

$$R_i X = \mathbf{e}_{d_i+1} M_{(i)}^{-1} X = X'_{d_i+1}.$$

Let $L' = L V^{(S)} M_{(i)}$ and let $\overline{[s]} := [n] \setminus [s]$. Consider the following two codes. We define $\mathscr{C}^{(i)} \subset \mathbb{F}_q^N$ to be the column space of the matrix $[L'^{d_i+1} | L'^{\overline{[d_i+1]}}] \in \mathbb{F}_q^{N \times n}$ and we define $\mathscr{C}_{(i)} \subset \mathbb{F}_q^N$ to be the subspace of $\mathscr{C}^{(i)}$ spanned by the columns of $L'^{\overline{[d_i+1]}}$. For each $i \in [m]$, we have $\mathscr{C}_{(i)} \subseteq \mathscr{C}^{(i)}$ with $\dim(\mathscr{C}^{(i)}) = \dim(\mathscr{C}_{(i)}) + 1$. As usual, for an $\mathbb{F}_q$-linear code $C \in \mathbb{F}_q^N$ we write $C^\perp := \{y \in \mathbb{F}_q^N : x \cdot y = 0\}$ to denote its dual code. Then we have $\mathscr{C}^{(i)\perp} \subseteq \mathscr{C}_{(i)}^\perp$ with $r_i = \dim(\mathscr{C}_{(i)}^\perp) = \dim(\mathscr{C}^{(i)\perp}) + 1$ for some $r_i$. Let $H_{(i)}$ be a parity check matrix of $\mathscr{C}_{(i)}$ of the form

$$H_{(i)} = \begin{bmatrix} h_{(i)} \\ \hline H^{(i)} \end{bmatrix} \in \mathbb{F}_q^{r_i \times N}, \tag{15}$$

where $H^{(i)}$ is a parity check matrix of $\mathscr{C}^{(i)}$ and $h_{(i)} \in \mathscr{C}_{(i)}^\perp \setminus \mathscr{C}^{(i)\perp}$.

Then

$$H_{(i)} L'^{d_i+1} = [s_{d_i+1}, 0, \ldots, 0]^T$$

for some $s_{d_i+1} \in \mathbb{F}_q \setminus \{0\}$.

We now outline a procedure for decoding the demand $R_i X$ at the $i$th receiver, which is based on syndrome decoding. In the first step we compute syndrome, of $H_{(i)}$, in which is embedded a syndrome of $H^{(i)}$. In the second step a table of syndromes is computed for the code $\mathscr{C}^{(i)}$. Finally, in the third step the output $R_i X$ is computed.

**Theorem 22** *If $w_H(W_{(i)}) \leq \delta$ then Algorithm 3 has output $\hat{X}_{d_i+1} = X'_{d_i+1} = R_i X$.*

*Example 11* Let $q = 2$, $m = n = 4$, $t = 1$, $\delta = 1$ and $R_i = \mathbf{e}_i$ for all $i \in [4]$ and $\mathscr{X}^{(S)} = \mathbb{F}_2^4$.

Let the side-information for this instance be encoded according to the matrices:

$$V^{(1)} = \begin{bmatrix} 0 & 1 & 1 & 0 \\ 0 & 0 & 1 & 0 \end{bmatrix}, V^{(2)} = \begin{bmatrix} 1 & 0 & 0 & 0 \\ 0 & 0 & 1 & 1 \end{bmatrix}, V^{(3)} = \begin{bmatrix} 1 & 0 & 0 & 0 \\ 0 & 0 & 0 & 1 \end{bmatrix}, V^{(4)} = \begin{bmatrix} 1 & 1 & 0 & 1 \\ 0 & 1 & 1 & 0 \end{bmatrix}.$$

The vectors in $\mathbb{F}_2^4$ in $V^{(i)\perp} \setminus R_i^\perp$ for some $i \in [4]$ comprise the set

$$\mathscr{S} = \{1001, 1000, 0111, 0100, 0010, 0110\}.$$

**Data**: Erroneous Transmission $Y_{(i)} = LV^{(S)}X + W_{(i)} \in \mathbb{F}_q^{N \times t}$
**Output**: Requested data $R_i X$
**Step** *I*:      Compute

$$H_{(i)}(Y_{(i)} - L'^{[d_i]}X'_{[d_i]}) = \begin{bmatrix} \alpha_i \\ \beta_i \end{bmatrix} \in \mathbb{F}_q^{r_i \times t} \tag{16}$$

**Step** *II*:    Find $\varepsilon \in \mathbb{F}_q^{N \times t}$ with $w_H(\varepsilon) \leq \delta$ such that

$$H^{(i)}\varepsilon = \beta_i \in \mathbb{F}_q^{(r_i - 1) \times t}. \tag{17}$$

**Step** *III*: Compute

$$\hat{X}_{d_i + 1} = (\alpha_i - h_{(i)}\varepsilon)/s_{d_i + 1}. \tag{18}$$

**Result**: $R_i X = \hat{X}_{d_i + 1}$

**Algorithm 3:** Correcting $\delta$ erroneous packets

It is easy to check that $\alpha(\mathscr{I}) = 2$ and $\kappa(\mathscr{I}) = 2$, so from the $\alpha$ and $\kappa$ bounds we have

$$\mathscr{N}(\mathscr{I}, 1) \leq N(\kappa(\mathscr{I}), 3) = 5 = N(2, 3) = N(\alpha(\mathscr{I}), 3) \leq \mathscr{N}(\mathscr{I}, 1).$$

We claim that the matrix

$$L = \begin{bmatrix} 1 & 0 & 1 & 0 \\ 0 & 1 & 1 & 1 \\ 1 & 1 & 0 & 0 \\ 1 & 1 & 1 & 0 \\ 0 & 0 & 1 & 0 \end{bmatrix}$$

represents an optimal linear $(\mathscr{I}, 1)$-ECIC. Since no element of $\mathscr{S}$ is in the null space of $L$, any $4 \times t$ matrix $Z$ satisfying $V^{(i)}Z = 0$ and $R_i Z \neq 0$ has at least one column from $\mathscr{S}$. $L\mathbf{v}$ has weight at Hamming weight least 3 for any $\mathbf{v} \in \mathscr{S}$, and hence $LZ$ has minimum Hamming weight at least 3.

Now suppose $t = 1$ and let $X = [1111]^T$. The sender broadcasts $LX$. Suppose one Hamming error occurs and $U_4$ receives the vector

$$Y_4 = LX + W_{(4)} = [01010]^T + [00010]^T = [01001]^T.$$

Let

$$M_{(4)} = \begin{bmatrix} 1 & 0 & 1 & 1 \\ 0 & 0 & 0 & 1 \\ 0 & 1 & 0 & 1 \\ 0 & 0 & 1 & 0 \end{bmatrix} \quad \text{and} \quad L' = LM_{(4)} = \begin{bmatrix} 1 & 1 & 1 & 0 \\ 0 & 1 & 1 & 0 \\ 1 & 0 & 1 & 0 \\ 1 & 1 & 1 & 1 \\ 0 & 1 & 0 & 1 \end{bmatrix}$$

We obtain a parity check matrix of $\mathscr{C}_{(4)}$, as in (15)

$$H_{(4)} = \begin{bmatrix} 0\ 0\ 0\ 1\ 1 \\ 1\ 0\ 0\ 1\ 1 \\ 0\ 1\ 0\ 1\ 1 \\ 0\ 0\ 1\ 1\ 1 \end{bmatrix}.$$

Applying Step I of Algorithm 3 we obtain

$$H_{(4)}(Y - L'^{[2]}X'_{[2]}) = \begin{bmatrix} 0\ 0\ 0\ 1\ 1 \\ 1\ 0\ 0\ 1\ 1 \\ 0\ 1\ 0\ 1\ 1 \\ 0\ 0\ 1\ 1\ 1 \end{bmatrix}\begin{bmatrix} 1 \\ 1 \\ 1 \\ 1 \\ 1 \end{bmatrix} = \begin{bmatrix} 0 \\ 1 \\ 1 \\ 1 \end{bmatrix}.$$

Therefore $\alpha_4 = 0$ and $\beta_4 = [111]^T$. Now from Step II, we obtain that the vector $\varepsilon = [00001]^T$ is a solution of (3) and in Step III we obtain

$$\hat{X}_3 = (0 - [00011] \cdot [00001])/1 = 1 = X_4.$$

## 5 Connections

The simplicity of the ICSI problem as a problem of multi-user communication, the fact that it is a network with a single coding point or common link, has encouraged many researchers to study capacity problems from the index coding point of view [17, 18, 23, 27, 35, 38]. In this section we consider some connections of index coding to other network communication problems.

### 5.1 Equivalence to Network Coding

The equivalence of the index and network coding problems is important, particularly in relation to the fundamental question of determining the capacity of an arbitrary network coding instance.

An instance of the index coding problem may be viewed as an instance of network coding in which there is a single internal node of in-degree greater than one, i.e. a single coding point, whose outgoing edge has some capacity $c_B$, which represents the broadcast rate. We illustrate this connection by an example shown in the Fig. 5.1 below, where the index coding instance on the left is the same as that of Fig. 1 and the figure on the right is the corresponding network coding instance. The edges outgoing from the source nodes all have unit capacity, whilst all other edges have capacity $c_B$. An edge joining an $i$th source node with a $j$th sink node (shown below in red) exists

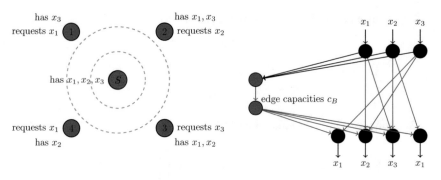

**Fig. 10** Index coding as an instance of network coding

in the network coding representation if Receiver $j$ possesses source $x_i$ as part of its side-information (Fig. 10).

This equivalence was described in [18, 19]. In [19] the authors establish this equivalence for both scalar linear and vector linear coding, while in [18] is shown that this equivalence holds in general. Explicitly, given a network coding instance, the authors show how to *reduce* this to an index coding problem, showing that a solution to the index coding problem also yields one for the network coding problem.

Recall that in describing an instance of index coding we have fixed $n$ as the number of sources and $m$ as the number of users. We give a slight generalization of the index coding problem to describe this equivalence: we will assume now that each receiver may demand more than one row of $X$. As noted before, this can be easily translated to the problem of single requests by adding new users with identical side information.

An instance of the index coding problem is given by $\mathscr{I} = (\mathscr{X}, \mathscr{R})$ where $\mathscr{X} = (\mathscr{X}_i : i \in [n])$ for $\mathscr{X}_i \subset [n]$ corresponding to the side infomation held by each $i$th receiver and $\mathscr{R} = (\mathscr{R}_i : i \in [m])$ for $\mathscr{R}_i \subset [n]$ corresponding to the requests of each receiver.

Then $\mathscr{I}$-IC is realized by $(E, D)$ with $D = \{D_i : i \in [m]\}$ if

$$E : \mathbb{F}_q^{n \times t} \to \mathbb{F}_q^N \text{ and } D_i : \mathbb{F}_q^N \times \mathbb{F}_q^{|\mathscr{X}_i|} \to \mathbb{F}_q^t$$

are maps such that

$$D_i(E(X), X_{\mathscr{X}_i}) = (X_j : j \in \mathscr{R}_i)$$

for each $X \in \mathbb{F}_q^{n \times t}$ and $i \in [m]$.

We now describe an instance of the network coding problem as a tuple $\mathscr{I}' = (G, S, U, B)$, where $G = (\mathscr{V}, \mathscr{E})$ is a directed acyclic graph, $S$ is a set of source nodes (those with in-degree 0), $U$ is a set of receiver nodes (those with out-degree 0) and $B$ is an $|S| \times |U|$ matrix with entry $B_s^u = 1$ if receiver $u$ requests data from source $s$ and 0 otherwise.

We let $\text{In}(v)$ and $\text{In}(e)$ deonte the edges incoming to a node $v$ or an edge $e$. Define some *local* encoding functions $F = \{F_e : e \in \mathscr{E}\}$ and decoding functions

$D' = \{D'_u : u \in U\}$ such that the arguments of $F_e$ and $D'_u$ respectively are the outputs of edges incoming to $e$ and $u$. The acyclic structure of $G$ means that $F$ yields a well-defined set of *global* encoding functions $P = \{P_e : e \in \mathscr{E}\}$, so that $P_e(X)$ is the output of edge $e$ for any transmitted $X = (X_s : s \in S)$. More precisely, for any $e \in \mathscr{E}$, we have

$$P_e(X) = F_e(P_{e'}(X) : e' \in \text{In}(e)).$$

A network code for the instance $\mathscr{I}'$ is realized by $(F, D')$ if

$$D'_u(P_e(X) : e \in \text{In}(u)) = (X_s : B^u_s = 1),$$

that is, if the output of $D'_u$ is $(X_s : B^u_s = 1)$ for any transmitted $X$.

We now outline the reduction of a network coding instance $\mathscr{I}'$ to an index coding instance $\mathscr{I}$.

- $n := |S| + |\mathscr{E}|$.
  For any $X \in \mathbb{F}^{n \times t}_q$, the first $|S|$ rows of $X$ are enumerated by $S$ and the remaining rows are enumerated by $|\mathscr{E}|$.
- $m := |U| + |\mathscr{E}| + 1$
  The $m$ receivers are given by the set $\{t_u : u \in U\} \cup \{t_e : e \in E\} \cup \{t_0\}$.
- For each $u \in U$, assign the side-information $\mathscr{X}_u := \{e : e \in \text{In}(u)\}$ and the requests $R_u := \{s : B^u_s = 1\}$ to receiver $t_u$.
- For each $e \in \mathscr{E}$, assign the side-information $\mathscr{X}_e := \{e' : e' \in \text{In}(e)\}$ and the request $R_e = \{e\}$ to receiver $t_e$.
- Assign the side-information $\mathscr{X}_0 := S$ and the requests $R_0 = \mathscr{E}$ to receiver $t_0$.
- $N := |\mathscr{E}|$.

The main result of [18] shows that a network code $(F, D')$ exists for $\mathscr{I}'$ if and only if an index code $(E, D)$ of length $N$ exists for $\mathscr{I}$ constructed from $\mathscr{I}'$ as outlined above.

Let $(F, D')$ be network code for the instance $\mathscr{I}'$ with corresponding matrix $B$, indicating the requests of each user $u \in U$. For each edge $e \in \mathscr{E}$, the global encoding functions $P_e$ have argument $X_S$ and

$$X_e = P_e(X_S) = F_e(P_{e'}(X_S) : e' \in \text{In}(e)).$$

The decoding functions $D'_u$ satisfy

$$D'_u(P_e(X_S) : e \in \text{In}(u)) = (X_s : B^u_s = 1),$$

for each $u \in U$. Define a map by

$$E : \mathbb{F}^{n \times t}_q \longrightarrow \mathbb{F}^{N \times t}_q : X \mapsto E(X) := (E_e(X) : e \in \mathscr{E}) = (X_e + P_e(X_S) : e \in \mathscr{E}).$$

We claim that for each receiver of the corresponding instance $\mathscr{I}$ there have a decoding map that outputs its request vector.

- For each $u \in U$, define

$$
\begin{aligned}
D_u(E(X), X_{\mathscr{X}_u}) &:= D'_u((E_e(X) : e \in \mathscr{X}_u) - X_{\mathscr{X}_u}) = D'_u((E_e(X) : e \in \mathrm{In}(u)) - X_{\mathrm{In}(u)}) \\
&= D'_u(X_e + P_e(X_S) - X_e : e \in \mathrm{In}(u)) = D'_u(P_e(X_S) : e \in \mathrm{In}(u)) \\
&= (X_s : B^u_s = 1) = X_{R_u},
\end{aligned}
$$

so receiver $t_u$ recovers its requested data.

- For each $e \in \mathscr{E}$, define

$$
\begin{aligned}
D_e(E(X), X_{\mathscr{X}_e}) &:= F_e((E_{e'}(X) : e' \in \mathscr{X}_e) - X_{\mathscr{X}_e}) \\
&= F_e(P_{e'}(X_S) : e' \in \mathrm{In}(e)) \\
&= P_e(X_S) = X_e,
\end{aligned}
$$

which is the data requested by $t_e$.

- For terminal $t_0$, define

$$
\begin{aligned}
D_0(E(X), X_{\mathscr{X}_0}) &:= (E_e(X) : e \in \mathscr{E}) - (P_e(X_{\mathscr{X}_0}) : e \in \mathscr{E}) \\
&= (X_e + P_e(X_S) : e \in \mathscr{E}) - (P_e(X_S) : e \in \mathscr{E}) \\
&= X_{\mathscr{E}},
\end{aligned}
$$

as requested by user $t_0$.

Therefore the network code $(F, D')$ for $\mathscr{I}'$ yields an index code for the instance $\mathscr{I}$.

Conversely, suppose there is a code $(E, D)$ of length $N = |\mathscr{E}|$ for the index coding instance $\mathscr{I}$. We will construct a code $(F, D')$ for the network coding instance $\mathscr{I}'$. Given any $Y \in \mathrm{Im}\, E \subset \mathbb{F}_q^{N \times t}$, the existence of $D_0$ means that for each $W \in \mathbb{F}_q^{|S| \times t}$ there exists unique $Z \in \mathbb{F}_q^{N \times t}$ satisfying $D_0(Y, W) = Z$; equivalently, such that $E(W, Z) = Y$.

Then choose $Y \in \mathrm{Im}\, E$. Let $X_S$ be the source data to be transmitted. Then $X_{\mathscr{E}} = (X_e : e \in \mathscr{E}) := D_0(Y, X_S)$. For each $e \in \mathscr{E}$, define recursively a local and global encoding functions as follows.

- If $e$ is incident with a source node $s \in S$ then

$$
X_e = P_e(X_S) = F_e(X_s) := D_e(Y, X_s) = X_s.
$$

- For every $e \in \mathscr{E}$, define

$$
X_e = P_e(X_S) = F_e(X_{\mathrm{In}(e)}) := D_e(Y, X_{\mathrm{In}(e)}).
$$

So $F_e(X_{\mathrm{In}(e)}) = F_e(P_{e'}(X_S) : e' \in \mathrm{In}(e))$ for each $e \in \mathscr{E}$ not incident with a source node.

**Fig. 11** The butterfly
network

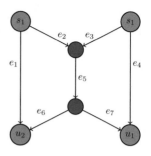

For each $u \in U$, define a decoding function:

$$D'_u(X_{\text{In}(u)}) := D_u(Y, X_{\text{In}(u)}) = (X_s : B^u_s = 1).$$

The index code $(E, D)$ for the instance $\mathscr{I}$ thus yields the network code $(F, D')$ for the network coding instance $\mathscr{I}'$.

We summarize this in the following theorem, which is a simplified version of [18, Theorem 1], in which we assume constant rate of all sources and a zero-error solution. The reader is referred to [18] for further details.

**Theorem 23** *Let $\mathscr{I}'$ be an instance of the network coding problem. Let $\mathscr{I}$ be the corresponding instance of the index coding problem of length $N$. Then there exists a network code for $\mathscr{I}'$ if and only if there exists an index code for $\mathscr{I}$ for the same packet length $t$.*

*Example 12* We construct an index coding instance $\mathscr{I}$ from the instance of the Butterfly Network, as shown below (Fig. 11).

User 1 wants $Y_1$ and User 2 wants $Y_2$ for the sources transmitting $Y_1$ and $Y_2$ respectively, so $B = I_2$.

- $n := 2 + 7 = 9$.
  For any $X \in \mathbb{F}_q^{7 \times t}$, the first 2 rows of $X$ are enumerated by $\{1, 2\}$ and the remaining rows are enumerated by $\{e_1, ..., e_7\}$.
- $m := 2 + 7 + 1 = 10$
  We label the 10 receivers by the set $\{t_{u_1}, t_{u_2}, t_{e_1}, ..., t_{e_7}, t_0\}$.
- The side infomation sets are given by:

$$\mathscr{X}_{u_1} = \{e_4, e_7\}, \ \mathscr{X}_{u_2} = \{e_1, e_6\}, \ \mathscr{X}_{e_1} = \mathscr{X}_{e_2} = \{1\}, \ \mathscr{X}_{e_3} = \mathscr{X}_{e_4} = \{2\},$$
$$\mathscr{X}_{e_5} = \{e_2, e_3\}, \ \mathscr{X}_{e_6} = \mathscr{X}_{e_7} = \{e_5\}, \ \mathscr{X}_0 = \{1, 2\}.$$

- The request sets are given by:

$$\mathscr{R}_0 = \{e_1, ..., e_7\}, \ \mathscr{R}_{u_1} = \{1\}, \ \mathscr{R}_{u_2} = \{2\}, \ \mathscr{R}_{e_i} = \{e_i\}, i \in [7].$$

- $N := 7$.

We remark that the set of all $\mathbb{F}_2$-linear feasible solutions for this index coding problem are matrices with the same row space as one of the form

$$
\begin{bmatrix}
1 & 0 & 0 & 0 & 0 & * & 0 & 0 & * \\
0 & 1 & * & 0 & 0 & 0 & * & 0 & 0 \\
* & 0 & 1 & 0 & 0 & 0 & 0 & 0 & 0 \\
* & 0 & 0 & 1 & 0 & 0 & 0 & 0 & 0 \\
0 & * & 0 & 0 & 1 & 0 & 0 & 0 & 0 \\
0 & * & 0 & 0 & 0 & 1 & 0 & 0 & 0 \\
0 & 0 & 0 & * & * & 0 & 1 & 0 & 0 \\
0 & 0 & 0 & 0 & 0 & 0 & * & 1 & 0 \\
0 & 0 & 0 & 0 & 0 & 0 & * & 0 & 1 \\
* & * & 1 & 1 & 1 & 1 & 1 & 1 & 1
\end{bmatrix}
\tag{19}
$$

where $*$ may be replace by an arbitrary element of $\mathbb{F}_2$. It can be checked that the least $\mathbb{F}_2$-rank any such matrix may have is 7, so the optimal length of an $\mathbb{F}_2$-linear index code of $\mathscr{I}$ is 7.

Of course there exists a linear network code over $\mathbb{F}_2$ for the instance $\mathscr{I}'$ illustrated by the butterfly network. For $Y \in \mathbb{F}_2^2$, the global encoding functions given by:

$$
P_{e_1}(Y) = P_{e_2}(Y) = Y_1, \; P_{e_3}(Y) = P_{e_4}(Y) = Y_2, \; P_{e_5}(Y) = P_{e_6}(Y) = P_{e_7}(Y) = Y_1 + Y_2
$$

realize a a network code, with the decoding functions:

$$
D'_{u_1}(Y) = P_{e_7}(Y) - P_{e_4}(Y) = Y_1 \text{ and } D'_{u_2}(Y) = P_{e_6}(Y) - P_{e_1}(Y) = Y_2.
$$

The claim is now that there exists a length 7 index code for the instance described. Define a map $E : \mathbb{F}_2^9 \longrightarrow \mathbb{F}_2^7$ by

$$
E(X) = (E_{e_1}(X), ..., E_{e_7}(X)), \text{ with } E_{e_i}(X) := X_{e_i} + P_{e_i}(X_1, X_2).
$$

A matrix representing the linear map $E$ is given by

$$
L = \begin{bmatrix}
1 & 0 & 1 & 0 & 0 & 0 & 0 & 0 & 0 \\
1 & 0 & 0 & 1 & 0 & 0 & 0 & 0 & 0 \\
0 & 1 & 0 & 0 & 1 & 0 & 0 & 0 & 0 \\
0 & 1 & 0 & 0 & 0 & 1 & 0 & 0 & 0 \\
1 & 1 & 0 & 0 & 0 & 0 & 1 & 0 & 0 \\
1 & 1 & 0 & 0 & 0 & 0 & 0 & 1 & 0 \\
1 & 1 & 0 & 0 & 0 & 0 & 0 & 0 & 1
\end{bmatrix},
$$

which can be checked to be an $\mathbb{F}_2$-linear $\mathscr{I}$-IC. For example,

$$
\begin{aligned}
D_{u_1}(LX, X_{\{e_4, e_7\}}) &= D'_{u_1}(L_{e_4} \cdot X - X_{e_4}, L_{e_7} \cdot X - X_{e_7}) \\
&= D'_{u_1}(X_2 + X_{e_4} - X_{e_4}, X_1 + X_2 + X_{e_7} - X_{e_7}) \\
&= D'_{u_1}(X_2, X_1 + X_2) = X_1 + X_2 - X_2 = X_1,
\end{aligned}
$$

which is the demand of $t_{u_1}$.

$$
\begin{aligned}
D_0(LX, X_{\{1,2\}}) &= LX - (P_e(X_{\{1,2\}}) : e \in \{e_1, ..., e_7\}) \\
&= LX - (X_1, X_1, X_2, X_2, X_1 + X_2, X_1 + X_2, X_1 + X_2) \\
&= (X_{e_1}, X_{e_2}, X_{e_3}, X_{e_4}, X_{e_5}, X_{e_6}, X_{e_7}),
\end{aligned}
$$

which is the demand of $t_0$.

Conversely, we know that the instance of the index coding problem $\mathscr{I}$ is realized by a code of length 7; a linear one in fact. Let $E(X) = LX$ for the matrix $L$ given above. Then the column space of $[L^1, L^2]$ is contained in the column space of $[L^{e_1}, ..., L^{e_7}]$, so for any choice of $X_1, X_2$ there exist $X_{e_1}, ..., X_{e_7}$ such that $[L^1, L^2]X_{\{1,2\}} = [L^{e_1}, ..., L^{e_7}]X_{\{e_1,...,e_7\}}$, i.e. such that $LX = 0$. Moreover, the existence of the decoding function $D_0$ for $t_0$ ensures that the $X_{e_1}, ..., X_{e_7}$ are uniquely determined by $X_1, X_2$. The decoding functions for the index coding instance are given by:

$$
D_{u_1}(LX, X_{e_4, e_7}) = X_1, D_{u_2}(LX, X_{e_1, e_6}) = X_2, D_0(LX, X_{1,2}) = X_{\{e_1,...,e_7\}}
$$
$$
D_{e_1}(LX, X_1) = X_{e_1}, D_{e_2}(LX, X_1) = X_{e_2}, D_{e_3}(LX, X_2) = X_{e_3}, D_{e_4}(LX, X_2) = X_{e_4}
$$
$$
D_{e_5}(LX, X_{e_2, e_3}) = X_{e_5}, D_{e_6}(LX, X_{e_5}) = X_{e_6}, D_{e_7}(LX, X_{e_5}) = X_{e_7}.
$$

If we set $LX = 0$, then any pair $X_1, X_2 \in \mathbb{F}_2$ determines a unique set of values $X_{e_1}, ... X_{e_7}$, so we may define the local encoding functions for the edges $e_1, ..., e_7$ by

$$
F_{e_1}(X_1) = D_{e_1}(0, X_1) = X_{e_1}, F_{e_2}(X_1) = D_{e_2}(0, X_1) = X_{e_2}, F_{e_3}(X_2) = D_{e_3}(0, X_2) = X_{e_3},
$$
$$
F_{e_4}(X_2) = D_{e_4}(0, X_2) = X_{e_4}, F_{e_5}(X_{\{e_2, e_3\}}) = D_{e_5}(0, X_{\{e_2, e_3\}}) = X_{e_5},
$$
$$
F_{e_6}(X_{e_5}) = D_{e_6}(0, X_{e_5}) = X_{e_6}, F_{e_7}(X_{e_5}) = D_{e_7}(0, X_{e_5}) = X_{e_7}.
$$

The decoding functions for $u_1$ and $u_2$ are given by

$$
D'_{u_1}(X_{\{e_4, e_7\}}) := D_{u_1}(0, X_{\{e_4, e_7\}}) = X_1 \text{ and } D'_{u_2}(X_{\{e_1, e_6\}}) := D_{u_2}(0, X_{\{e_1, e_6\}}) = X_2.
$$

The stronger version of the equivalence between network and index coding is given by the following.

**Theorem 24** ([18, Theorem 1]) Let $\mathscr{I}'$ be an instance of the network coding problem. Let $\mathscr{I}$ be the corresponding instance of the index coding problem with broadcast rate $c_B$. For any rate vector $R$, any integer $t$, and any $\varepsilon \geq 0$ it holds that $\mathscr{I}'$ is $(\varepsilon, R', t)$ feasible if and only if $\mathscr{I}$ is $(\varepsilon, R, c_B, t)$ feasible.

Here $R' = (R_s : s \in S) \in \mathfrak{R}^{|S|}$ is a rate vector for $\mathscr{I}'$, indicating a transmission rate of $R_s$ for each source $s \in S$, and $R = (R', c_e : e \in \mathscr{E}) \in \mathfrak{R}^n$ represents a rate vector for $\mathscr{I}$, where $c_e$ is the capacity of edge $e$ in the network. The quantity $c_B$ is the sum of the edge capacities: $c_B = \sum_{e \in \mathscr{E}} c_e$. As before $t$ is the block length. The instance $\mathscr{I}'$ (respectively $\mathscr{I}$) is called $(\varepsilon, R', t)$-feasible (respectively $(\varepsilon, R, c_B, t)$-feasible) if each terminal node can retrieve its required data with probability at least $1 - \varepsilon$. The *capacity region* of the instance $\mathscr{I}'$ is the set of $R'$ such that for any $\varepsilon, \delta \geq 0$ the instance $\mathscr{I}'$ is $(\varepsilon, (1 - \delta)R', t)$-feasible for some blocklength $t$.

In spite of the equivalence demonstrated in Theorem 24, it is not known whether or not the network coding capacity region of $\mathscr{I}'$ can be obtained by solving the capacity region of the index coding instance $\mathscr{I}$. This is referred to as *capacity equivalence*. While this equivalence holds for linear codes, and for certain network topologies the question of capapcity equivalence for general networks is unknown. The question has an interesting connection to the *edge removal problem*.

## 5.2   Interference Alignment

Interference alignment is a technique used to manage interfering signals between multiple sender-receiver pairs sharing the same channel. It has applications to distributed storage problems in network coding [9]. Such interference is common in wireless networks, where topological interference management (TIM) may be viewed as an index coding problem, with multiple data streams competing for a common communication link [23, 27]. The term *alignment* comes from the fact that the method exploits communication differences between clients, so that alignment of signals for one user may yield orthogonality at another. The index coding analogue is that diversity of side information among different users means that even though alignment occurs at the broadcast, different users can still decode their own different demands. In several cases, as shown in [27], optimal solutions of the index coding problem provide optimal solutions of the TIM problem. We briefly outline how the scalar-linear ($t = 1$) ICSI problem may be viewed as a TIM problem, as described in [27].

A scalar-linear index coding scheme of length $N$ for an instance $\mathscr{I} = (\mathscr{X}, f)$ can be described as follows. There are

- *precoding vectors $L^1, \ldots, L^n \in \mathbb{F}_q^{N \times 1}$* and
- for each pair $(k, r = f(k))$, *receiver combining vectors $U_{r,k} \in \mathbb{F}_q^{1 \times N}$*, satisfying $U_{r,k}L^i = 0$ if $i \neq r$ and $i \notin \mathscr{X}_k$ and is non-zero if $i = r$.

The sender then transmits

$$LX = [L^1, \ldots, L^n][X_1^T, \ldots, X_n^T]^T.$$

If $r = f(k)$ then User $k$ retrieves its demand $X_r$ via

$$(U_{r,k}L^r)^{-1}U_{r,k}(LX - \sum_{i \in \mathcal{X}_k} L^i X_i) = (U_{r,k}L^r)^{-1}U_{r,k}(\sum_{i \notin \mathcal{X}_k} L^i X_i)$$

$$= \sum_{i \notin \mathcal{X}_k}(U_{r,k}L^r)^{-1}(U_{r,k}L^i)X_i$$

$$= X_r + \sum_{i \notin \mathcal{X}_k, i \neq r}(U_{r,k}L^r)^{-1}(U_{r,k}L^i)X_i$$

$$= X_r.$$

Choosing $\mathbf{b}^{(k)} = (U_{f(k),k}L^{f(k)})^{-1}U_{f(k),k}$ and $\mathbf{u}^{(k)} = (U_{f(k),k}L^{f(k)})^{-1}U_{f(k),k}L^{\mathcal{X}_k}$ yields a solution to the index coding problem as given in (4).

The *alignment* property corresponds to linear dependence among the columns of $L$, which results in reduction of broadcast rate. If all columns of $L$ are linearly independent, $N = n$ and the transmission rate is the same as for routing, that is, where every transmitted symbol occupies a different time-slot. In TIM, the side-information, $\mathcal{X}_i$, of the $i$th user is called an *antidote* for the messages $\{X_j, j \in \mathcal{X}_i\}$. User $i$ uses its antidotes to effectively cancel $|\mathcal{X}_i|$ columns of $L$ and to hence observe a linear combination of $n - |\mathcal{X}_i| - 1$ *interfering* vectors in the column space of the remaining columns of $L$, i.e. in the column space of $L_{[n]\setminus(\mathcal{X}_i \cup f(i))}$. Therefore, a necessary condition for decoding at the $i$th receiver is that the column space of $L_{f(i)}$ has trivial intersection with the column space of $L_{[n]\setminus(\mathcal{X}_i \cup f(i))}$. In particular, resolvability for User $i$ requires the dimension of the space spanned by the interfering columns of $L$, those indexed by $[n]\setminus(\mathcal{X}_i \cup f(i))$ should be no more than $n - 1$, in which case the interfering vectors must align in an $n - 1$ dimensional space.

In Fig. 5.2, the diagram on the left is the index coding instance shown already in Fig. 5.1. The image on the right is a related interference alignment problem. The upper nodes represent transmitters and the lower nodes are receivers. In the TIM graph shown a receiver node is connected to a sender node if the receiver lies within the transmission range of the sender. In the corresponding index coding problem, there is a link from a source node to a receiver if and only if it is absent in the TIM graph (Fig. 12).

An optimal solution to the toy index coding instance shown above (for which receivers 1, ..., 4 demand $X_1, X_2, X_3, X_1$ respectively) is given by

$$L = [L^1, L^2, L^3] = \begin{bmatrix} 100 \\ 011 \end{bmatrix}.$$

In the language of TIM, we say that there is alignment with respect to $L^2$ and $L^3$. The vectors

$$U_{1,1} = [1\ 0], U_{2,2} = [0\ 1], U_{3,3} = [0\ 1], U_{4,1} = [1\ 0],$$

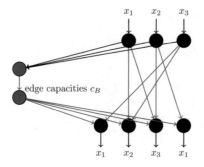 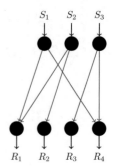

**Fig. 12** Topological interference management and index coding

satisfy the conditions as described above, so each $k$th user can decode its demand $X_{f(k)}$. For example,

$$X_{f(1)} = \left( [1\ 0] \begin{bmatrix} 1 \\ 0 \end{bmatrix} \right) [1\ 0] \left( \begin{bmatrix} 100 \\ 011 \end{bmatrix} \begin{bmatrix} X_1 \\ X_2 \\ X_3 \end{bmatrix} - X_3 \begin{bmatrix} 0 \\ 1 \end{bmatrix} \right) = X_1$$

## 5.3 The Coded Caching Problem

The *canonical coded-caching* problem [28] corresponds to the placement phase of a special class of the instances $(\mathscr{C}, R)$, as introduced in Sect. 2. First it is assumed that the $n$ packets of $X$ comprise $k$ blocks $X^{(i)}, ..., X^{(k)}$ of size $\ell$ (so that $n = k\ell$) and that each $i$th user wants some complete block, say $X^{(j)}$, after delivery. So $R_i$ is an $r_i \times n$ matrix of the form

$$R_i = [O \cdots H \cdots O]$$

for some $r_i \times \ell$ matrix $H$ with standard basis vectors of length $\ell$ as rows. Each user has a subset of some number of packets from each block and the same number of packets $v$ in total. In terms of $(\mathscr{C}, R)$, this imposes the constraints that for each $i$,

- $V^{(i)}$ is an $v \times n$ matrix,
- $R_i$ and $V^{(i)}$ have standard basis vectors as rows,
- the $j$th block of $\ell$ columns of $V^{(i)}$ has some $\ell - r_i$ standard basis vectors of length $\ell$ as columns that complete $H$ to a basis of $\mathbb{F}_q^\ell$.

If a subset of users wish to receive the same block, the delivery to that set of users becomes a local multicast problem.

The main difference between the coded-caching problem and the index coding problem is the role of the sender in the placement phase. The index coding problem is essentially one of delivery for given $(\mathscr{C}, R)$; the placement is not necessarily controlled by the sender. However, in the coded caching problem, we seek an optimal placement of the side-information code $\mathscr{C}$ in advance of knowing the users' request matrix $R$. The sender tries to choose $\mathscr{C}$ in such a way that for any $R$ all users' demands can be met with a small number of transmissions. More precisely, the coded caching problem seeks to find, or obtain bounds on:

$$\min\{\max\{\kappa(R, \mathscr{C}) : R \in \mathbb{F}_q^{r \times n}\} : \mathscr{C} = \mathscr{C}_1 \oplus \cdots \mathscr{C}_m, \dim\mathscr{C}_i = r_i v\}.$$

In [28] the authors use the cut-set bound to derive a lower bound on the optimal storage memory rate trade-off. Furthermore, they devise a scheme that achieves this rate within a constant factor. So asymptotically, the canonical coded caching problem is solved. Moreover, it was shown in [39] that improvements to the scheme presented in [28] can only be achieved by considering caching schemes with coded side-information.

For example, the matrices $V^{(i)}$ may have rows that are not standard basis vectors, which corresponds to the cache data (the side-information) being encoded. Then $\mathscr{C}_i$ is an arbitrary $nv$-dimensional matrix code for each $i$. In [35] the authors propose a scheme for coded-caching with coded side-information using MDS and rank metric codes. Their scheme delivers an improvement in the memeory-rate trade-off of several known schemes and are in some cases optimal. However, their scheme requires large field sizes.

# References

1. N. Alon, A. Hassidim, E. Lubetzky, U. Stav, A. Weinstein, Broadcasting with side information, in *Proceedings 49th Annual IEEE Symposium on Foundation of Computer Science (FOCS)* (2008), pp. 823–832
2. Z. Bar-Yossef, Z. Birk, T.S. Jayram, T. Kol, Index coding with side information, in *Proceedings of 47th Annual IEEE Symposium Foundations of Computer Science* (2006), pp. 197–206
3. Z. Bar-Yossef, Z. Birk, T.S. Jayram, T. Kol, Index coding with side information. IEEE Trans. Inf. Theory **57**(3), 1479–1494 (2011)
4. E. Byrne, M. Calderini, Error correction for index coding with coded side information, in *IEEE Transactions on Information Theory*, vol. 63(6), published online, 27 March 2017, pp. 3712–3728. https://doi.org/10.1109/TIT.2017.2687933
5. E. Byrne, M. Calderini, Bounding the Optimal Rate of the ICSI and ICCSI Problems, SIAM J. Discret. Math. **31**(2), 1403–1427 (2017). arXiv:1604.05991
6. Y. Birk, T. Kol, Informed source coding on demand (ISCOD) over broadcast channels, in *Proceedings of IEEE Conference on Computer Communication* (San Francisco, CA, 1998), pp. 1257–1264
7. A. Blasiak, R. Kleinberg, E. Lubetsky, Broadcasting with side information: bounding and approximating the broadcast rate. IEEE Trans. Inf. Theory **59**(9), 5811–5823 (2013)
8. Z. Chen, P. Fan, K.B. Letaief, Fundamental limits of caching: improved bounds for users with small buffers. IET Commun. **10**(17), 2315–2318 (2016)
9. V.R. Cadambe, S.A. Jafar, H. Maleki, K. Ramchandran, C. Suh, Asymptotic interference alignment for optimal repair of MDS codes in distributed storage. IEEE Trans. Inf. Theory **59**(5), 2974–2987 (2013)

10. E.J. Candès, B. Recht, Exact matrix completion via convex optimization. Found. Comput. Math. **9**(6), 717–772 (2009)
11. T.H. Cormen, C. Stein, R.L. Rivest, C.E. Leiserson, *Introduction to Algorithms*, 2nd edn. (McGraw-Hill Higher Education, New York, 2001)
12. M. Dai, K.W. Shum, C.W. Sung, Data dissemination with side information and feedback. IEEE Trans. Wireless Commun. **13**(9), 4708–4720 (2014)
13. S.H. Dau, V. Skachek, Y.M. Chee, On the security of index coding with side information. IEEE Trans. Inf. Theory **58**(6), 3975–3988 (2012)
14. S.H. Dau, V. Skachek, Y.M. Chee, Error correction for index coding with side information. IEEE Trans. Inf. Theory **59**(3), 1517–1531 (2013)
15. S.H. Dau, V. Skachek, Y.M. Chee, Optimal index codes with near-extreme rates. IEEE Trans. Inf. Theory **60**, 1515–1527 (2014)
16. P. Erdös, Z. Füredi, A. Hajnal, P. Komja'th, V. Rödl, A. Seress, Coloring graphs with locally few colors. Discret. Math. **59**(1), 21–34 (1986)
17. H. Esfahanizadeh, F. Lahouti, B. Hassibi, A matrix completion approach to linear index coding problem, in *Information Theory Workshop (ITW)* (IEEE, 2014), pp. 531–535
18. M. Effros, S. El Rouayheb, M. Langberg, An equivalence between network coding and index coding. IEEE Trans. Inf. Theory **61**(5), 2478–2487 (2015)
19. A. El Rouayheb, A. Sprintson, C. Georghiades, On the index coding problem and its relation to network coding and matroid theory. IEEE Trans. Inf. Theory **56**(7), 3187–3195 (2010)
20. M. Fazel, H. Hindi, S. Boyd, Rank minimization and applications in system theory in *American Control Conference*, vol. 4 (IEEE, 2004), pp. 3273–3278
21. W. Haemers, An upper bound for the Shannon capacity of a graph. Colloq. Math. Soc. Jànos Bolyai **25** (1978)
22. X. Huang, S. El Rouayheb, Index coding and network coding via rank minimization, in *IEEE Information Theory Workshop (ITW)* (2015), pp. 14–18
23. S.A. Jafar, Topological interference management through index coding. IEEE Trans. Inf. Theory **60**(1), 529–568 (2014)
24. H. Loeliger, An upper bound on the volume of discrete spheres. IEEE Trans. Inf. Theory **40**(6), 2071–2073 (1994)
25. N. Lee, A.G. Dimakis, R.W. Heath Jr, Index coding with coded side-information. IEEE Commun. Lett. **19**(3)(2015)
26. Eyal Lubetzky, Uri Stav, Nonlinear index coding outperforming the linear optimum. IEEE Trans. Inf. Theory **55**(8), 3544–3551 (2009)
27. H. Maleki, V.R. Cadambe, S.A. Jafar, Index coding - an interference alignment perspective. IEEE Trans. Inf. Theory **60**(9), 5402–5432 (2014)
28. M. Maddah-Ali, U. Niesen, Fundamental limits of caching. IEEE Trans. Inf. Theory **60**(5), 2856–2867 (2014)
29. R. Peeters, Orthogonal representations over finite fields and the chromatic number of graphs. Combinatorica **16**(3), 417–431 (1996)
30. K. Shanmugan, A. Dimakis, M. Langberg, Graph theory versus minimum-rank for index coding, in *Proceedings of the 2014 IEEE International Symposium on Information Theory (ISIT)* (2014), pp. 291–295. arXiv:1402.3898.v1
31. J. Saunderson, M. Fazel, B. Hassibi, Simple algorithms and guarantees for low rank matrix completion over $\mathbb{F}_2$. IEEE Int. Symp. Inf. Theory (ISIT). Barcelona **2016**, 86–90 (2016)
32. D. Silva, F.R. Kschischang, R. Koetter, Communication over finite-field matrix channels. IEEE Trans. Inf. Theory **56**(3), 1296–1305 (2010)
33. K.W. Shum, D. Mingjun, C. Sung, Broadcasting with coded side information. 2012 IEEE 23rd Int. Symp. Pers. Indoor Mobile Radio Commun. (PIMRC) **89**(94), 9–12 (2012)
34. V.Y.F. Tan, L. Balzano, S.C. Draper, Rank minimization over finite fields: fundamental limits and coding-theoretic interpretations. IEEE Trans. Inf. Theory **58**(4), 2018–2039 (2012)
35. C. Tian, J. Chen, Caching and delivery via interference elimination, in *2016 IEEE International Symposium on Information Theory (ISIT)* (2016), pp. 830–834

36. A.S. Tehrani, A.G. Dimakis, M.J. Neely, Bipartite index coding, in *Proceedings of the IEEE 2012 International Symposium on Information Theory (ISIT)* (Boston, 1-6 Jul 2012), pp. 2246–2250

37. J.I. Tamir, E.R. Elenberg, A. Banerjee, S. Vishwanath, Wireless index coding through rank minimization, in *IEEE ICC 2014 - Wireless Communications Symposium* (2014), pp. 5209–5214

38. M.F. Wong, M. Langberg, M. Effros, On a capacity equivalence between network and index coding and the edge removal problem, in *2013 IEEE International Symposium on Information Theory* (2013), pp. 972–976

39. K. Wan, D. Tuninetti, P. Piantanida, On the optimality of uncoded cache placement, in *2016 IEEE Information Theory Workshop (ITW)* (2016), pp. 161–165

40. K. Shanmugam, A.G. Dimakis, M. Langberg, Local graph coloring and index coding, in *2013 IEEE International Symposium on Information Theory Proceedings (ISIT)* (2013), pp. 1152–1156

41. P. Jain, P. Netrapalli, S. Sanghavi, Low-rank matrix completion using alternating minimization, in *Proceedings of the forty-fifth annual ACM symposium on Theory of computing* (ACM, 2013), pp. 665–674

42. M. Hardt, Understanding alternating minimization for matrix completion, in *IEEE 55th Annual Symposium on Foundations of Computer Science (FOCS)* (2014), pp. 651–660

# Implementation of Network Coding in Wireless Systems

Semiha Tedik Basaran, Ali Reza Heidarpour, Selahattin Gokceli,
Gunes Karabulut Kurt, Murat Uysal and Ibrahim Altunbas

**Abstract** In this chapter, we target to give extensive performance analyses about application of network coding (NC) in wireless systems, referred to as network coded cooperation (NCC), brings both diversity and multiplexing gains. We use the diversity-multiplexing trade-off (DMT) to determine performance bounds of NCC systems. Within the scope of this study, NCC is integrated with orthogonal frequency division multiple access (OFDMA) and the corresponding system model is characterized by specifically focusing on frequency diversity gain. DMT expressions of the NCC-OFDMA system is given. A real-time implementation of the NCC-OFDMA system is presented by creating a testbed NI USRP-2921, NI PXIe-5644R, NI PXI-6683H software defined radio (SDR) modules and LabVIEW software. Obtained real-time performance measurements are essential to demonstrate the practical advantages or disadvantages of the usage of the NCC-OFDMA system. Overall, we aim to present a detailed overview of the fundamental performance bounds of NCC and its extension to the practical applicability of NCC in wireless networks.

S. Tedik Basaran (✉) · S. Gokceli · G. K. Kurt · I. Altunbas
Department of Communications and Electronics Engineering, Istanbul Technical University,
3469 Istanbul, Turkey
e-mail: tedik@itu.edu.tr

S. Gokceli
e-mail: gokcelis@itu.edu.tr

G. K. Kurt
e-mail: gkurt@itu.edu.tr

I. Altunbas
e-mail: ibraltunbas@itu.edu.tr

A. R. Heidarpour
Department of Electrical and Computer Engineering, University of Alberta, Edmonton,
AB T6G 2R3, Canada
e-mail: alirezaheidarpour@ualberta.ca

M. Uysal
Department of Electrical and Electronics Engineering, Ozyegin University,
34794 Istanbul, Turkey
e-mail: murat.uysal@ozyegin.edu.tr

© Springer International Publishing AG 2018 295
M. Greferath et al. (eds.), *Network Coding and Subspace Designs*,
Signals and Communication Technology,
https://doi.org/10.1007/978-3-319-70293-3_11

# 1 Introduction

In classical routing applications, intermediate nodes simply store and forward the received symbol to the destinations or other intermediate nodes without modifying the content of received packets. As 5G and beyond 5G communication technologies target high data rate and low delay demands based on real-time applications in dense network scenarios, the efficient usage of network resources (bandwidth, power, and time) becomes critical at intermediate nodes to jointly increase the throughput and reduce the delay. Network coding (NC) is proposed as a smart routing solution which meets high throughput and low latency demands [1].

The transmission process of NC systems is completed in two orthogonal phases; the broadcast phase and the relaying phase. In the broadcast phase, information symbols are radiated from source nodes and received by intermediate nodes (referred to as relay nodes hereafter). Relay nodes typically decode the received symbols and combine symbols from multiple sources to obtain network coded symbols. In the relaying phase, each relay node determines the NC coefficients based on the selected coding type for all symbols and encodes the detected source symbols. Then, relay nodes forward the encoded symbols to destination nodes. Each destination node needs to receive several coded symbols, at least as many as the number of source nodes, to successfully decode the transmitted symbols.

In classical routing schemes, relay nodes sequentially send the received symbols, which may cause a high delay and a low data rate. Hence, by using NC, the data rate can be increased while obtaining low transmission latency. NC has a flexible nature for an extension to multiple source and multiple relay cases by determining different code sets for each relay. Hence, we can use NC as a smart routing tool, by efficiently serving dense network users, furthermore it can be used as a solution for the scalability problem in dense networks. We focus on linear NC in this chapter, but there are also some works about non-linear mapping schemes on received bits [2].

The application of NC in wireless networks inherently holds one advantage and one disadvantage. Firstly, there may be direct links that may be used to provide an improved error performance, between source and destination pairs due to the broadcast nature of wireless channel. On the other hand, error propagation may emerge at destination node due to wireless channel impairments. Whether a direct link, also termed as cooperation link, is present or not depends on the corresponding link qualities. These links provide additional cooperative diversity to improve the error performance [3]. Hence, when the implementation of NC in wireless networks is designed by considering cooperation links, the system is named as network coded cooperation (NCC). NCC has higher diversity gain and increased spectral efficiency when compared to NC with the help of cooperation links.

Wireless networks are more prone to transmission errors when compared to wired counterparts due to channel impairments. In the early works on NC, only error-free cases are considered [1]. Although this is a valid assumption for wired networks, it is not realistic for wireless networks. As mentioned above, different from NC, in NCC, destination nodes use the source symbols that are received through direct links in the broadcast phase to improve the error performance. Hence, destination nodes have multiple received symbols in both broadcast and relaying phases. The symbols received in the relaying phase are encoded by using the selected network code. A combining procedure at destination nodes needs to be used to exploit full diversity. As the combining procedure, there are two frequently used detector types; the rank-based detector [4] and the maximum likelihood (ML) [5] detector.

Diversity-multiplexing trade-off (DMT) is a tool that determines the set of diversity and multiplexing gain pairs that can be obtained simultaneously for communication systems. DMT can be used to assess the performance of NCC systems. The asymptotic DMT of NCC systems is studied in [6, 7]. These works focus on time-division multiple access (TDMA) to preserve orthogonality among different source and relay transmissions. On the other hand, [8] considers an efficient multiple access technique, orthogonal-frequency division multiple access (OFDMA), to provide orthogonality while providing frequency diversity gain. This system model is called as NCC-OFDMA.

The theoretical DMT results of NCC-OFDMA system which is superior according to benchmark systems is presented in [8]. In [9], an NCC-OFDMA framework is established to evaluate its performance by using software defined radio (SDR) nodes. The results show that NCC-OFDMA system enables easy implementation and provides higher reliability against wireless channel errors, leading to higher throughput when compared to its NC counterpart. Through implementation, real-time issues are analyzed and insights about a more comprehensive deployment are shared.

The aim of this chapter is to show with theoretical analysis and practical application scenarios, that NCC is a strong candidate for next generation wireless systems. We address the theoretical performance bounds of NCC through DMT analyses and highlight practical application challenges with real-time implementation by using SDR nodes. In order to evaluate the system performance, bit error rate (BER), error vector magnitude (EVM) and signal-to-noise ratio (SNR) metrics are used.

The rest of this chapter is organized as follows. In Sect. 2, the signalling model of NCC-OFDMA system and coding details are given. After that, DMT results of NCC system are given in comparison with the state-of-the-art studies in Sect. 3. In Sect. 4, implementation and test results of NCC-OFDMA system are presented. The benefits of NCC-OFDMA are emphasized in Sect. 5.

## 2 Basic Principles

System models of NC and NCC are given in Fig. 1 a, b, respectively. Both two systems include $P$ source nodes, $M$ relay nodes, and a single destination node. The main advantage of NCC different from NC is the exploitation of direct transmission links between source and destination pairs, as shown in the figure. In the linear NC system, the destination node has $M$ network-coded symbols, composed of linear combinations of $P$ source symbols. In order to recover all $P$ source symbols, the destination node needs at least $P$ symbols, hence, we need to satisfy $P \leq M$ the

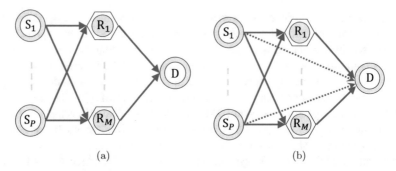

Fig. 1  **a** Block diagram of NC system. **b** Block diagram of NCC system with the presence of direct source to destination links

constraint in an NC system. On the other hand, this necessity is eliminated due to the presence of direct transmission links in the NCC system. The direct transmission links assist the relay-aided communication by providing additional cooperative diversity gain.

NCC systems using TDMA strategy are generally planned for multiple users. To exploit frequency diversity gain while ensuring orthogonality among multiple users, OFDMA technique is considered for multiple access for NCC systems, referred to as NCC-OFDMA. The transmission process of the NCC-OFDMA system is also completed in two orthogonal phases: broadcast and relaying phases. In the broadcast phase, all source nodes emit their own signals. The signals of the sources can be captured jointly by the relay nodes and the destination node because of the broadcast nature of wireless channel. In the wireless systems, the transmission qualities of source-relay links are generally considered to be higher than the source-destination links. Therefore, the subcarrier assignment process is performed according to the attributes of the source-destination links. Following the broadcast phase, the relay nodes demodulate the signals when they received during the broadcast phase and combine them by using the coefficients of the selected network code. The combined symbols formed at the relay nodes are called network coded symbols. After that, in the relaying phase, relay nodes transmit the network coded symbols to the destination nodes.

Let the frequency domain representation of the $n$th subcarrier channel gain of link $t$ be given by $H_t[n]$, where $t \in \{s_i d,\ s_i r_j,\ r_j d\}, i = 1, \ldots, P,\ j = 1, \ldots, M,\ n \in F_t$ with $F_t$ is the set of the assigned subcarriers of the transmitter of link $t$. The number of subcarriers is equal to $N$. All transmissions are carried out with unit power in the system. The received signal from $j$th relay node of $i$th source node in the broadcast phase is given as:

$$Y_{s_i r_j}[n] = H_{s_i r_j}[n]X_i[n] + W_{s_i r_j}[n], \tag{1}$$

where $W_{s_i r_j}[n]$ represents the additive white Gaussian noise (AWGN) at the $j$th relay node. $j$th relay node detects $X_i[n]$ by using $Y_{s_i r_j}[n]$, and the detected symbol is denoted by $\tilde{X}_{ij}[n]$. The network code coefficients indicated by $\alpha_{ij}$ can be selected from a maximum distance separable (MDS) code satisfying the Singleton bound [10], maximum rank distance (also called Gabidulin) codes [11] or can be generated randomly in random network coding (RNC). By using $\tilde{X}_{ij}[n]$ and $\alpha_{ij}$ values the network coded symbol at $j$th relay node is calculated as:

$$C_j[n] = \alpha_{1j}\tilde{X}_{1j}[n] \boldsymbol{+} \alpha_{2j}\tilde{X}_{2j}[n] \ldots, \boldsymbol{+} \alpha_{Pj}\tilde{X}_{Pj}[n] \tag{2}$$

where $\boldsymbol{+}$ denotes the summation operator in the field, $\mathcal{GF}(q)$, where $q$ is the size of the field. Source and network coded signals received by the destination node in two phases from the source and relay nodes are defined as follows:

$$Y_{s_id}[n] = H_{s_id}[n]X_i[n] + W_{s_id}[n]$$
$$Y_{r_jd}[n] = H_{r_jd}[n]C_j[n] + W_{r_jd}[n], \tag{3}$$

where AWGN components at the destination node in the broadcast and relaying phases are denoted by $W_{s_id}[n]$ and $W_{r_jd}[n]$, respectively.

If we assume error-free transmission among source-relay pairs, $(\tilde{X}_{ij} = X_i)$, to simplify the expressions, the vector notation of received signals can be given as:

$$
\begin{bmatrix} Y_{s_1d} \\ Y_{s_2d} \\ \vdots \\ Y_{s_pd} \\ Y_{r_1d} \\ \vdots \\ Y_{r_Md} \end{bmatrix}
=
\begin{bmatrix} a_1 & 0 & \cdots & 0 & \alpha_{11} & \cdots & \alpha_{1M} \\ 0 & a_2 & \cdots & 0 & \alpha_{21} & \cdots & \alpha_{2M} \\ \vdots & \vdots & \ddots & \vdots & \vdots & \ddots & \vdots \\ 0 & 0 & \cdots & a_P & \alpha_{P1} & \cdots & \alpha_{PM} \end{bmatrix}^T
\begin{bmatrix} X_1 \\ X_2 \\ \vdots \\ X_P \end{bmatrix},
\tag{4}
$$

which can be also expressed as $\mathbf{y} = \mathbf{Zx}$. The first $P$ entries of $\mathbf{y}$ represent the received symbols of the direct transmission links among source and destination nodes. If the $i$th direct transmission link is available, $a_i$ is equal to 1, otherwise it is equal to 0. $\alpha_{ij}$ is generated from $\mathcal{GF}(q)$ according to the current coding scheme in the relay nodes. The last $M$ entries of $\mathbf{y}$ denote the received encoded symbols from $M$ relay nodes. $\mathbf{Z}$ indicates the global coding matrix of NCC, consisting of direct link and relay link coefficients. The aim of destination node is to recover $\mathbf{x}$ from $\mathbf{y}$ thanks to $\mathbf{Z}$. Detailed explanations and comparisons about signal modeling of NC and NCC can be found in [12]. We explain the detection methods of NCC in detail to show effect of the detector on the system in the next subsection.

## 2.1 Detector Design

There are two receiver strategies of NC depending on the availability of the global coding matrix ($\mathbf{Z}$) at destination node: coherent and non-coherent NC. Most of NC and NCC works assume that destination node knows the coding coefficients [4, 13], and this strategy is named as coherent NC. On the other hand, if coding coefficients are not available at the destination node, the destination node can obtain the coefficients through transmitting pilot symbols. This second approach is called non-coherent NC [14]. In order to recover source symbols, a new coding metric based on subspaces is proposed against erasures and errors of non-coherent RNC. To obtain $\mathbf{x}$, Gauss–Jordan elimination method can be used to solve the set of linear equations of NCC over $\mathcal{GF}(q)$ [4, 13]. To apply Gauss–Jordan elimination for obtaining $P$ source symbols, we need at least $P$ independent equations. This condition corresponds to $\mathbf{Z}$ to be of full rank. In addition to performance corruptions due to wireless channels as mentioned above, incorrect estimations of $X_i$ at relay nodes may cause error propagation at destination node.

## 2.2 Code Design

Coding coefficients, $\alpha_{ij}$, can be selected from different linear code sets like as MDS [10], maximum rank distance (also called Gabidulin) codes [11] or be produced randomly [4, 13]. MDS codes achieve equality in Singleton bound [11], and are preferred in different applications as error correction and storage. MDS codes are accepted as an important coding class due to their capabilities about error detection and error correction based on the Hamming distance criterion. The redundancy-reliability ratio of MDS codes is the optimal.

The other class of coding includes rank distance approach. In the rank distance approach, the code construction is based on detecting and correcting of rank errors by supplementing redundancy [11]. Instead of selecting from predetermined code sets, in RNC, network code coefficients can be generated randomly. In RNC, coding coefficients have a uniform distribution and independently chosen from $\mathcal{GF}(q)$. RNC yields flexible scheduling opportunity to NCC systems based on dynamically changing network components with improving the efficiency of the NCC systems. To obtain more information about the dynamic coding according to wireless channel can be found in chapter "Opportunistic Network Coding".

## 2.3 On the Resource Usage of NCC

In both NC and NCC, broadcast and relaying phases are completed in $P$ and $M$ orthogonal transmission blocks, respectively. TDMA, which causes undesired delay for dense networks, can be used to provide orthogonality between transmission links. On the other hand, multiplexing gain can be obtained by allocating transmission links in frequency domain thanks to OFDMA. At the same time, OFDMA provides additional frequency diversity gain through assigning different subcarriers to users according to channel quality of each subcarrier to the corresponding user. There are also joint resource allocation schemes to unify assigning of various limited resources as power, subcarrier, and bits. Employing subcarrier allocation processes can improve the efficiency of the system. The comparative theoretical performance bounds of various NCC schemes will be assessed in the following section by providing DMT results.

# 3 Performance Limitations and DMT Analysis of NCC

Due to occurence at the same time, there is a fundamental trade-off referred to as DMT between diversity and multiplexing gains of a communication systems. DMT is an appropriate tool to evaluate performance of NCC-OFDMA system.

## 3.1 DMT of NCC Systems and Comparison with Cooperative Communication Protocols

Figure 2 depicts diversity gain $d$ versus multiplexing gain $r$ for various cooperative communication protocols. It is known that conventional cooperative (CC) systems including space-time coding [3] and opportunistic relaying [15] achieve the same DMT of $d(r) = (M + 1)(1 - 2r)$, $r \in (0, 0.5)$.

In [6], Peng et al. consider a NCC system which consists of $P$ source-destination pairs and $M$ relay nodes with dynamic coding (DC-NCC). Particularly, the "best" relay (which has the best end-to-end path between source and destination) among the set of $M$ available relay nodes are selected and then the relay dynamically employs XOR operation on the source packets based on instantaneous source to relay channel quality. It is shown that the DC-NCC system can achieve a full diversity gain of $M + 1$. However, diversity gain of $M + 1$ can only be obtained under the assumption where the destination can successfully overhear the data from other source nodes. If this optimistic assumption is removed, the achievable diversity gain of the DC-NCC system reduces to only two and does not improve by increasing the number of relay candidates.

In an effort to improve the diversity gain, random NCC (RNCC) system and deterministic NCC (DNCC) system are presented in [7], where relay nodes encode the sources' packets using an encoding matrix of size $(P + M) \times P$. As described in the previous section, in this setup, the first $P$ rows form an identity sub-matrix, corresponding to the direct transmissions in the broadcasting phase. In addition, the remaining columns and rows correspond to the packets transmitted by relays in the relaying phase. In DNCC system the coefficients in the encoding matrix are preset while in RNCC system are drawn randomly from finite field. The associated DMT

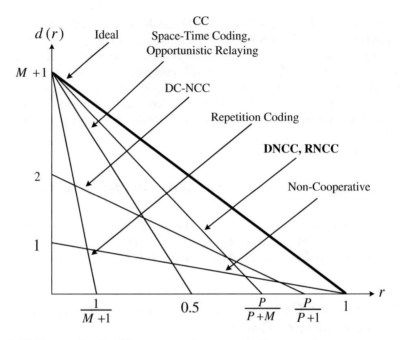

**Fig. 2** DMT comparison for different cooperative communication protocols

analysis reveals that a full diversity gain of $M + 1$ can be maintained through the use of MDS codes typically used for point to point channels.

The DMT expressions of DC-NCC and DNCC/RNCC systems are given in [7]. As can be seen from Fig. 2, CC and RNCC/DNCC systems can achieve full diversity gain of $M + 1$ when $r = 0$. However, DNCC/RNCC systems outperform CC systems in terms of multiplexing gain and offer higher diversity gain than that of CC systems for the same spectral efficiency. On the other hand, in comparison to other schemes, DC-NCC systems have the highest multiplexing gain and offer more diversity gain as $r$ increases. This is due to the fact that only one relay XOR'es source packets and therefore the overall transmission takes place in $P + 1$ orthogonal time slots. It is worth mentioning that as $P$ increases the spectral efficiency of the NCC systems moves towards the ideal case. Therefore, NCC systems have better DMT performance and outperform CC systems. On the other hand, repititon coding which is based on simply repeating the data bits, has the worst DMT performance.

## 3.2 Extension to CC-OFDMA

In CC systems, relay nodes either amplify and transmit (amplify-and-forward relaying) or decode and retransmit (decode-and-forward relaying) the source signals to the destination without any combining operation. It is shown that both protocols can achieve full diversity gains [16] and therefore CC with OFDMA, referred to as CC-

OFDMA systems are expected to extract both frequency and spatial diversity. In the literature, DMT performance of CC-OFDMA system has been studied in [17]. The underlying assumption in [17] is that the subcarriers are independent. Furthermore, just one subcarrier is allocated to each source node. The derived DMT expression is given by [17]

$$d(r) = N(M + 1)(1 - 2r), r \in (0, 0.5). \qquad (5)$$

The maximum diversity gain is obtained when multiplexing gain $r = 0$, i.e., $N(M + 1)$. This means that both frequency and spatial diversity can be achieved in CC-OFDMA. On the other hand, the maximum multiplexing gain can be obtained when $d(r) = 0$, i.e., $r = 0.5$.

## 3.3  Extension to NCC-OFDMA

The aforementioned NCC systems [6, 7] build upon the assumption of TDMA. To exploit the multipath diversity gains and have further gains over NCC systems, the combination of OFDM with NCC has been investigated in the literature. More specifically, in [8] the DMT performance of NCC-OFDMA system is studied. In the system under consideration, the transmission rate for all nodes is equal to $R_0$. There is $L$ number of coherence bandwidths due to the frequency selective channel. Furthermore, $K_1 = N/P$ and $K_2 = N/M$ are the numbers of subcarriers assigned to each source and each relay in the broadcasting phase and the relaying phase, respectively. This leads to different transmission rates for subcarriers in these two phases i.e., $R_{s_1} = R_0/K_1$ for the broadcasting phase and $R_{s_2} = R_0/K_2$ for the relaying phase. By using Marcum $Q$-function of order one represented by $Q_1(\cdot, \cdot)$, the outage probabilities of subcarriers in the broadcasting phase for $S \rightarrow D$ links over Rician fading channels are given by [8]

$$P_{out_{sub}^{SD}}(K_1) = 1 - Q_1(\alpha, \beta_{SD}), \qquad (6)$$

where $\alpha = \sqrt{2k}$ and $\beta_{SD} = \sqrt{2(1 + k)\left(2^{R_{s_1}\hat{N}} - 1\right) \Big/ \rho\, G_{SD}}$. $\rho$ is SNR and $G_{XY}$ is normalized path loss expressions with respect to $S \rightarrow D$ links. Similarly, for $S \rightarrow R$ links, the outage probability of the subcarriers is given by

$$P_{out_{sub}^{SR}}(K_1) = 1 - Q_1(\alpha, \beta_{SR}), \qquad (7)$$

where $\beta_{SR} = \sqrt{2(1 + k)\left(2^{R_{s_1}\hat{N}} - 1\right) \Big/ \rho\, G_{SR}}$.

In Eqs. (6) and (7), $k$ is Rician factor, $\hat{N}$ is the number of subcarriers within the coherence bandwidth while calculating as $\hat{N} = N/L$. On the other hand, the outage probability of subcarriers in the relaying phase (i.e., $R \rightarrow D$ links) can be obtained as

$$P_{out_{sub}^{RD}}(K_2) = 1 - Q_1(\alpha, \beta_{RD}),\tag{8}$$

where $\beta_{RD} = \sqrt{2(1+k)\left(2^{R_{s_2}\hat{N}} - 1\right) \Big/ \rho\, G_{RD}}$.

## 3.4 Relaying Strategies

Relay nodes which successfully decode all of their received packets from $P$ source nodes are allowed to participate in the relaying phase and the rows of the encoding matrix corresponding to the relays with erroneous decisions are discarded. The DMT expression of NCC-OFDMA system is given by [8]

$$d(r) = L(M+1)\left(1 - \frac{P + M\max\{P,M\}}{P}\frac{L}{L}r\right),\ r \in \left(0, \frac{P}{P+M\max\{P,M\}}\frac{L}{L}\right).\tag{9}$$

As expected, NCC-OFDMA systems are capable to fully extract both frequency and spatial diversity and significantly outperform NCC systems which build upon TDMA. The DMT of CC-OFDMA system with the same system configuration is given by [8]

$$d(r) = L(M+1)\left(1 - (M+1)\frac{\max\{P,M\}}{L}r\right),\ r \in \left(0, \frac{1}{M+1}\frac{L}{\max\{P,M\}}\right)$$

where all relay nodes transmit in the relaying phase or

$$d(r) = L(M+1)\left(1 - 2\frac{\max\{P,M\}}{L}r\right),\ r \in \left(0, 0.5\frac{L}{\max\{P,M\}}\right)$$

where selection relaying is considered. As can be seen from Fig. 3, both CC-OFDMA and NCC-OFDMA systems achieve full diversity gain of $L(M+1)$ when $r = 0$. However, maximum multiplexing gain of CC-OFDMA systems is always lower than that of NCC-OFDMA systems indicating multiplexing gain loss in CC systems.

As another transmission strategy, if relay can decode even one packet correctly from one of $P$ source packets, it is allowed to participate in the relaying phase. In this case, only the coefficients in the encoding matrix corresponding to the erroneous packets would be zero instead of discarding the whole row. Let $P_{out}^{opt}$ denote the outage probability of the system under this assumption. A lower bound on the outage probability $P_{out}^{low}$ is the case when all the relays decode all $P$ packets correctly and can participate in the relaying phase i,e., $m = 0$ (see Eq. (14) of [8]) and the probability that any relay successfully decodes all $P$ packets received during the broadcasting phase $P_R$ is equal to 1. Therefore, $P_{out}$ in (16) of [8] is reduced to

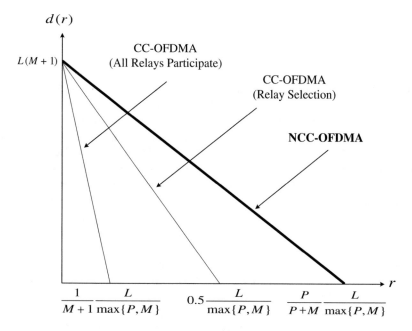

**Fig. 3** DMT comparison for CC-OFDMA and NCC-OFDMA systems

$$P_{out}^{low} = \sum_{j=0}^{M} \binom{M}{j} P_{out_{RD}}^{M-j} \left(1 - P_{out_{RD}}\right)^{j} \sum_{i=0}^{P-1-j} \binom{P}{i} P_{out_{SD}}^{P-i} \left(1 - P_{out_{SD}}\right)^{i}. \quad (10)$$

The lower bound of the outage probability in high SNR regime can be obtained as

$$\lim_{\beta \to 0} P_{out}^{low} = C' \left( \exp\left(-\frac{\alpha^2}{2}\right) \frac{\beta^2}{2} \right)^{L(M+1)}, \quad (11)$$

where $C' = \sum_{j=0}^{M} \binom{M}{j}\binom{P}{P-1-j} G_{RD}^{-L(M-j)} G_{SD}^{-L(j+1)}$.

Note that, $P_{out}^{low} < P_{out}^{opt} < P_{out}$. Using squeezing theorem [18], we have

$$L(M+1) < -\lim \frac{\log(P_{out}^{opt})}{\log(\rho)} < L(M+1). \quad (12)$$

Therefore, one can conclude that the slopes of outage probability in log-log scale for $P_{out}^{low}$, $P_{out}^{opt}$, and $P_{out}$ are identical and the achievable diversity gain of the system remains the same regardless of which relaying strategy is used. Therefore, the DMTs under two relaying strategies are identical.

# 4 Practical Implementation of NCC

NCC-OFDMA system gives full diversity gain when multiplexing gain is equal to zero and maximum multiplexing gain performance among the other fundamental systems while no diversity gain. In order to test its high theoretical capability, the real time implementation of NCC-OFDMA system is necessary. To perform the implementation of NCC-OFDMA system, SDR implementation is used. SDR implementations of various wireless communication techniques have been realized recently. This trend continues to grow because of the benefits of SDR implementations over simulations. Modeling a transmission environment in simulation is not realistic in most cases because hardware related issues show random characteristics at each transmission and appropriate modeling of these becomes impossible.

CC structure that is inherent to NCC systems, may complicate the implementation of NCC models in real-time. Such an implementation requires the realization of multiple nodes in addition to proper communication between these nodes. As a natural consequence of these requirements, accurate synchronization of these nodes is a must in order to operate them cooperatively. Beyond these necessities, a flexible design is also necessary to deploy algorithms to hardware without any problem.

In order to create a functional NCC testbed, an SDR based testbed is implemented in [9] by considering the mentioned aspects. A testbed was created for the real-time implementation of proposed NCC model in order to observe performance from a different perspective and validate simulation results. As SDR components, LabVIEW, NI USRP-2921 nodes, NI PXI-6683H module, and NI PXIe-5644R Vector Signal Transceiver (VST) are used. NCC testbed is implemented with 5 NI USRP-2921 SDR nodes where three of them are used as source nodes and other two nodes are used as destination. We used two physically separated receivers to seperately examine their performances. For relay node, NI PXIe-5644R VST module is utilized which can operate in 65 MHz to 6 GHz frequency range and can support 80 MHz instantaneous bandwidth. Moreover, NI PXI-6683H timing and synchronization module which includes a TCXO oscillator is used as synchronization source. All PXI modules are managed via NI PXIe-1082 chassis. LabVIEW is used as software tool due to its easy programming and SDR compatibility features.

As an expected necessity, NCC testbed needs a proper synchronization solution for the robust performance due to its cooperative setup. Main component of this solution is NI PXI-6683H module which provides 10 MHz clock signal. This signal is distributed to one destination and two source nodes via RF cables. Other nodes receive this signal with MIMO cables from synchronized nodes. Due to initial connection of VST module to the chassis, relay node is also synchronized with the same synchronization resource. One drawback of this setup is the limitation of distances between nodes due to limited lengths of cables. These cables are used because of their sufficient RF performances, longer cables have higher losses. This limitation makes observing the effects of different distances on the system performance difficult, however we overcome this drawback by configuring source gains properly. By configuring the hardware in LabVIEW code, synchronization solution is generated.

In hardware side, RF cables are used to distribute synchronization signals to two sources and one destination. Remaining pairs receive these signals via pair-wise cables. In the software side, these cable connections are indicated by configuring corresponding settings and a code is used in order to start and manage the distribution of synchronization signals. For source, relay and destination functionalities, the codes are prepared for each operation in LabVIEW. NCC related functions are also implemented and necessary codes are prepared. In the end, a flexible combination of codes is obtained, where possible modifications about algorithm can easily be realized.

A physical view of the testbed is shown in Fig. 4. The measurement parameters of the given implementation setup are defined in Table 1. It can be seen from Fig. 4 that physical distances between nodes are limited in the testbed. However, configured distances as well as transmission gains provide realistic transmission environment. Therefore, a proper balance between distance and realistic transmission is empirically determined and nodes are located accordingly. It is noteworthy to state that during experiments, distances between nodes were longer, distances between source and destination nodes were around 250 cm. Particularly low transmission gains provide more realistic transmission circumstances. In this case, a limited distance up to 1 m is sufficient to separate nodes and obtain realistic transmission qualities which are also verified with real-time experiments. It should be highlighted that two experimentation setups are considered. These two setups are differentiated with the link quality between source and destination nodes, and relay's transmission quality. In the first setup, source nodes have a lower path loss, and transmission quality between source and relay as well as destination nodes is sufficient, as will be shown via measurement results. However, in the second setup source nodes cannot properly communicating

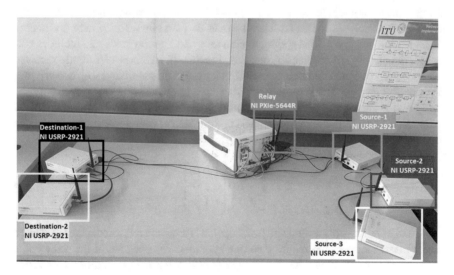

**Fig. 4** Created testbed of NCC system

**Table 1** The measurement parameters of the given setup

| | |
|---|---|
| Carrier frequency | 2.45 GHz |
| I/Q data rate | Megasamples/sec |
| Number of bits used in one frame | 2080 bits |
| Number of 4-QAM symbols | 1040 samples |
| Total number of subcarriers of the one user data portion | 320 samples |
| Number of reference subcarriers | 40 samples |
| Number of source nodes | 3 |
| Number of relay node | 1 |
| Number of destination nodes | 2 |
| Zero padding length | 120 samples |
| DFT length ($N$) | 1200 samples |
| CP length samples | 300 |

with destination nodes. Moreover, relay's link quality is subject to a higher path loss, when compared to the first setup. When experiment results are evaluated, transmission gain levels also need to be considered with distance configurations.

## 4.1 Software Description

As mentioned, code of NCC-OFDMA system is prepared by using LabVIEW. In Lab-VIEW, smaller code fragments known as subVIs are used to increase functionality. Hence, the code includes several SubVI components for source, relay and destination nodes which are managed from a timed flad structure. Relay network coding SubVI implements network coding to data received by relay node and includes corresponding $\mathcal{GF}(4)$ multiplication and addition SubVI. In this SubVI, firstly received source symbols from source nodes by the relay node are multiplied with corresponding coefficients $i_{SR}$ that represent whether data is received properly or not by the relay node. Then these multiplication outputs are multiplied with $\alpha_{ij}$ in $\mathcal{GF}(4)$ by using $\mathcal{GF}(4)$ multiplication SubVI. The coding matrix consists of $\alpha_{ij}$ which can be defined as:

$$\mathbf{Z} = \begin{bmatrix} 1\ 0\ 0 & 1\ 0\ 0 \\ 0\ 1\ 0 & 0\ 1\ 0 \\ 0\ 0\ 1 & 1\ 1\ 1 \end{bmatrix}^T. \tag{13}$$

Note that although only a single physical device is used as a relay node, there are three linear components of the received source symbols are generated at the relay node. Then, results of these are summed in $\mathcal{GF}(4)$ by using $\mathcal{GF}(4)$ addition SubVI for each data loop. In result, three separate network coded symbols are obtained which are transmitted by relay node's transmitter in corresponding frequency resources.

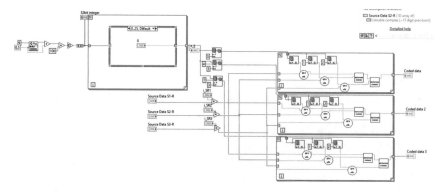

**Fig. 5** Block diagram of the relay network coding SubVI

In multiplication and addition SubVI, $\mathcal{GF}(4)$ multiplication and addition tables are implemented by using multiple array functions and case structures. An exemplary view of relay network coding SubVI can be seen in Fig. 5. With corresponding input-output configurations, NCC-OFDMA code is obtained.

In order to measure the performance of our system, real-time tests are experimented. As explained earlier, two different setups are created, where difference between the setups are the different distance configurations between nodes. The quality of each link between nodes is measured. Both NC and NCC results are measured to compare their performances. The results of all types of links are shown in this section, except source-relay links because of their superior performances. For measuring the link performance, EVM, SNR and BER metrics are measured for each link. EVM and SNR metrics are measured in terms of percentage and dB respectively, and given with these units in the tables. The EVM value of the $i$th source for single OFDMA frame can be calculated as:

$$EVM_i = \frac{\sqrt{\frac{1}{|F_{s_id}|}\sum_{n\in F_{s_id}}\left[\left(I_i[n]-\tilde{I}_i[n]\right)^2 + \left(Q[n]-\tilde{Q}[n]\right)^2\right]}}{|v_{max}|}, \qquad (14)$$

where $I_i[n]$ ($\tilde{I}_i[n]$) and $Q_i[n]$ ($\tilde{Q}_i[n]$) denote the real and imaginary components of the $X_i[n]$ ($\tilde{X}_i[n]$). $|F_{s_id}|$ represents the cardinality of the subcarrier set, where $|v_{max}|$ denotes the maximum absolute value of $X_i[n]$.

For the first setup, three different relay transmit gain levels are experimented for each destination nodes. Because of directional differences between source-destination pairs due to location differences, error performance of the sources can be quite different from each other. In order to approximate error performances, source gains are selected differently for each source node. However, diversified source performances can provide more comprehensive results. Therefore, locations and gains of sources are organized in a way that clear performance differences can be observed.

The source transmission gains of $S_1$, $S_2$ and $S_3$ are set as 14, 8 and 6 dB respectively. These gains have not been changed during the experiments. As stated, the effect of source gain changes on the overall performance is observed with two setups. Each one has a different location-gain adjustment. As the first step, first setup is experimented. Firstly, source-destination link performances are evaluated to analyze performance differences between source nodes. For $D_1$, $S_2$ has the worst transmission performance where the $S_1$ is the best one, as demonstrated in Table 3. EVM and BER results can be used for this analysis. EVM and BER results are obtained as 26.71% and $1.5 \times 10^{-3}$ for the data transmitted by the $S_2$ where results are lower for the $S_1$ node's transmitted data, which are 15.09% and $1 \times 10^{-6}$ respectively. SNR results show similar differences, which can be observed from the Table 3. EVM, BER and SNR metrics are obtained as 18.73%, $4.5 \times 10^{-4}$ and 14.82 dB for the data transmitted by $S_3$. As explained, transmission gains of source nodes are not changed during each setup, however, usage of two different setups allows us to understand the effect of different source node gains on the performance. This effect will be discussed when source-destination link performances of the second setup will be given. If similar error performance evaluation is done for the $D_2$, it can be observed from the Table 3 that $S_2$ has the worst transmission performance and $S_1$ has the best transmission performance similar to results obtained for the $D_1$. One difference is the fact that error performance is slightly better for the $S_2$, however, this is not valid for other source nodes and their transmission performances slightly worse for the $D_2$. EVM and BER results for the transmission of $S_2$ are obtained as 25.11% and $1.3 \times 10^{-3}$ respectively. These values are obtained as 16.15% and $6 \times 10^{-5}$ for the transmission of $S_1$ and 19.02% and $7 \times 10^{-4}$ for the transmission of $S_3$. Secondly, error performance results of relay-destination links are obtained which are shown in Table 2, and performance results for the overall NC and NCC operations are obtained as in Table 4.

**Table 2** Relay-destination performances for the first setup

| Destination | Source | BER | EVM | SNR |
|---|---|---|---|---|
| $D_1$ | $S_1$ | $1 \times 10^{-6}$ | 15.09 | 15.72 |
| | $S_2$ | $1.5 \times 10^{-3}$ | 26.71 | 13.39 |
| | $S_3$ | $4.5 \times 10^{-4}$ | 18.73 | 14.82 |
| $D_2$ | $S_1$ | $6 \times 10^{-5}$ | 16.15 | 15.31 |
| | $S_2$ | $1.3 \times 10^{-3}$ | 25.11 | 13.40 |
| | $S_3$ | $7 \times 10^{-4}$ | 19.02 | 14.38 |

**Table 3** Source-destination link performances for the first setup

| Destination | Relay gain (dBm) | BER | EVM | SNR |
|---|---|---|---|---|
| $D_1$ | $-12$ | $1 \times 10^{-7}$ | 13.60 | 18.80 |
| | $-16$ | $1.6 \times 10^{-4}$ | 15.59 | 15.57 |
| | $-20$ | $4.2 \times 10^{-3}$ | 33.01 | 10.40 |
| $D_2$ | $-12$ | 0 | 8.91 | 18.13 |
| | $-16$ | $9 \times 10^{-5}$ | 14.71 | 16.50 |
| | $-20$ | $1 \times 10^{-3}$ | 25.60 | 12.12 |

**Table 4** NC and NCC performances for the first setup

| Destination | Source | Relay gain (dBm) | NC-BER | NCC-BER |
|---|---|---|---|---|
| $D_1$ | $S_1$ | −12 | $1.4 \times 10^{-4}$ | 0 |
| | | −16 | $5.5 \times 10^{-4}$ | $2 \times 10^{-5}$ |
| | | −20 | $3.6 \times 10^{-3}$ | $2.7 \times 10^{-4}$ |
| | $S_2$ | −12 | $8.6 \times 10^{-4}$ | $4.7 \times 10^{-4}$ |
| | | −16 | $1.3 \times 10^{-3}$ | $8 \times 10^{-4}$ |
| | | −20 | $6.5 \times 10^{-3}$ | $3.3 \times 10^{-3}$ |
| | $S_3$ | −12 | $3.3 \times 10^{-4}$ | $1.9 \times 10^{-4}$ |
| | | −16 | $5.5 \times 10^{-4}$ | $3.4 \times 10^{-4}$ |
| | | −20 | $6 \times 10^{-3}$ | $8 \times 10^{-4}$ |
| $D_2$ | $S_1$ | −12 | 0 | 0 |
| | | −16 | $2.9 \times 10^{-5}$ | $1 \times 10^{-6}$ |
| | | −20 | $7 \times 10^{-4}$ | $4 \times 10^{-5}$ |
| | $S_2$ | −12 | $6 \times 10^{-4}$ | $6 \times 10^{-4}$ |
| | | −16 | $8 \times 10^{-4}$ | $1 \times 10^{-4}$ |
| | | −20 | $1.7 \times 10^{-3}$ | $8 \times 10^{-4}$ |
| | $S_3$ | −12 | 0 | 0 |
| | | −16 | $7.9 \times 10^{-5}$ | $1.4 \times 10^{-5}$ |
| | | −20 | $1.1 \times 10^{-3}$ | $4 \times 10^{-4}$ |

For the $D_1$, at −12 dBm relay transmit gain, EVM results for the relay-destination links are obtained as 13.29, 13.74 and 13.78% respectively. BER results for the same components are obtained as $1 \times 10^{-7}$ on average. At the same relay transmit gain, NC operation's BER results for the transmissions of $S_1$, $S_2$ and $S_3$ are obtained as $1.4 \times 10^{-4}$, $8.6 \times 10^{-4}$ and $3.3 \times 10^{-4}$ respectively. When compared to the source-destination link results, slight performance improvement can be observed. However, NCC operation improves error performances further, where these error performances for the source nodes are obtained as 0, $4.7 \times 10^{-4}$ and $1.9 \times 10^{-4}$ respectively. As shown, NCC operation is quite effective and error performance for the transmission of first source node decreases to 0.

For the −16 dBm relay transmit gain, relay-destination link EVM results are obtained as 15.14, 16.07 and 15.57% respectively. BER performance of the same links are obtained as $1.6 \times 10^{-4}$ on average. At the same configuration, NC operation's BER results increase and obtained as $5.5 \times 10^{-4}$, $1.3 \times 10^{-3}$ and $5.5 \times 10^{-4}$ respectively. BER values decrease with the NCC operation, obtained as $2 \times 10^{-5}$, $8 \times 10^{-4}$ and $3.4 \times 10^{-4}$ respectively.

For the −20 dBm relay transmit gain, error performances decrease further. For this gain, BER results for the relay-destination links are obtained as $4.2 \times 10^{-3}$ on average and EVM results are obtained as 32.38, 32.66 and 34% respectively. At this gain, BER results for the NC operation are obtained as $3.6 \times 10^{-3}$, $6.5 \times 10^{-3}$ and $6 \times 10^{-3}$ respectively. Better performances are observed with NCC operation where

BER results are obtained as $2.7 \times 10^{-4}$, $3.3 \times 10^{-3}$ and $8 \times 10^{-4}$ respectively. It is clear that performances of relay transmissions are critical for robust NC or NCC operation and if low relay transmit gain is used, reliable performance cannot be obtained.

For the $D_2$, at $-12$ dBm relay transmit gain, BER results of NC operation for the transmissions of $S_1$ and $S_3$ are obtained as 0, where related value is $6 \times 10^{-4}$ for the $S_2$. NCC operation does not create any improvement, as the same BER values are obtained. Error performance results are better for the $D_2$, compared to the $D_1$. This is due to better relay-destination link performances which create effective improvements. EVM results of relay-destination links are obtained as on average 8.91% and transmissions are completed without any bit error in these links. At the $-16$ dBm relay transmit gain, BER results of NC operation are obtained as $2.9 \times 10^{-5}$, $8 \times 10^{-4}$ and $7.9 \times 10^{-5}$ for source nodes, respectively. For these relay experiments, NCC operation creates clear performance improvements as corresponding BER values are obtained as $10^{-6}$, $1 \times 10^{-4}$ and $1.4 \times 10^{-5}$ respectively. EVM results are obtained as 14.71% and BER results are obtained as $9 \times 10^{-5}$ on average for relay-destination link transmissions. At the $-20$ dBm relay transmit gain, for relay-destination links, EVM results and BER results are obtained as 25.6% and $1 \times 10^{-3}$ on average, respectively. At the same gain, BER results of NC operation are obtained as $7 \times 10^{-4}$, $1.7 \times 10^{-3}$ and $1.1 \times 10^{-3}$ for source nodes, respectively. Moreover, BER results of NCC operation are obtained as $4 \times 10^{-5}$, $8 \times 10^{-4}$ and $4 \times 10^{-4}$ respectively and clear performance improvements are observed. In result, when compared to the previous results, relay-destination links are important especially for NCC operation's performance and certain improvements can be observed. As a common observation, as relay transmission performance becomes better, NC and NCC operations can decrease source-destination errors effectively.

As mentioned, one more setup is configured in order to create worse source-destination and relay-destination link performances compared to the first setup. For this setup, two relay transmit gains are experimented because, performance results of good relay transmit gain levels are evaluated in the first setup and relay transmit gain can be decreased to the minimum of $-15$ dBm, configurations below this value results in poor performance results. It is noteworthy that gain level is a parameter configured in the LabVIEW, performance of relay or source nodes should also be evaluated by considering distances to destination nodes.

As first step, source-destination link performances are measured which are demonstrated in Table 5. For the $D_1$, again, worst performance is performed by the $S_2$ and the $S_1$ has the best performance. BER results for source nodes are obtained as $7 \times 10^{-4}$, $2.2 \times 10^{-3}$ and $1 \times 10^{-3}$ respectively. Similarly, EVM results for source nodes are obtained as 29.20, 35.64 and 34.97% respectively. SNR results verify error performances and obtained as 11.41, 10.61 and 10.66 dB respectively. Same measurements are also performed for the $D_2$. BER results for source nodes are obtained as $2.3 \times 10^{-3}$, $3.6 \times 10^{-3}$ and $4.6 \times 10^{-3}$ respectively. EVM measurement is also performed and EVM results are obtained as 31.05, 35.45 and 39.11% respectively. Moreover, SNR values are measured as 11.16, 10.09 and 9.67 dB respectively. Therefore, it is clear that $S_3$ has the worst performance where the $S_1$ has the best performance.

**Table 5** Source-destination link performances for the second setup

| Destination | Source | BER | EVM | SNR |
|---|---|---|---|---|
| $D_1$ | $S_1$ | $7 \times 10^{-4}$ | 29.20 | 11.41 |
| | $S_2$ | $2.2 \times 10^{-3}$ | 35.64 | 10.61 |
| | $S_3$ | $1 \times 10^{-3}$ | 34.97 | 10.66 |
| $D_2$ | $S_1$ | $2.3 \times 10^{-3}$ | 31.05 | 11.16 |
| | $S_2$ | $3.6 \times 10^{-3}$ | 35.45 | 10.09 |
| | $S_3$ | $4.6 \times 10^{-3}$ | 39.11 | 9.67 |

**Table 6** Relay-destination performances for the second setup

| Destination | Relay gain (dBm) | BER | EVM | SNR |
|---|---|---|---|---|
| $D_1$ | $-13$ | $3 \times 10^{-5}$ | 14.64 | 17.08 |
| | $-15$ | $7 \times 10^{-4}$ | 18.59 | 14.55 |
| $D_2$ | $-13$ | $8 \times 10^{-5}$ | 19.29 | 14.29 |
| | $-15$ | $2.5 \times 10^{-3}$ | 22.05 | 13.13 |

**Table 7** NC and NCC performances for the second setup

| Destination | Source | Relay gain (dBm) | NC-BER | NCC-BER |
|---|---|---|---|---|
| $D_1$ | $S_1$ | $-13$ | $3 \times 10^{-5}$ | $1 \times 10^{-5}$ |
| | | $-15$ | $7 \times 10^{-4}$ | $3 \times 10^{-4}$ |
| | $S_2$ | $-13$ | $5.3 \times 10^{-4}$ | $5 \times 10^{-4}$ |
| | | $-15$ | $1.7 \times 10^{-3}$ | $5 \times 10^{-4}$ |
| | $S_3$ | $-13$ | $5 \times 10^{-4}$ | $4.6 \times 10^{-4}$ |
| | | $-15$ | $1.6 \times 10^{-3}$ | $1 \times 10^{-3}$ |
| $D_2$ | $S_1$ | $-13$ | $3 \times 10^{-4}$ | $1.7 \times 10^{-4}$ |
| | | $-15$ | $2.5 \times 10^{-3}$ | $1 \times 10^{-3}$ |
| | $S_2$ | $-13$ | $5 \times 10^{-4}$ | $2.5 \times 10^{-4}$ |
| | | $-15$ | $3.5 \times 10^{-3}$ | $1.2 \times 10^{-3}$ |
| | $S_3$ | $-13$ | $6.5 \times 10^{-4}$ | $3 \times 10^{-4}$ |
| | | $-15$ | $4.2 \times 10^{-3}$ | $2.4 \times 10^{-3}$ |

As second step, performances of relay-destination links with error performance results for the overall NC and NCC operations are measured as shown in Tables 6 and 7. For the $D_1$, at $-13$ dBm relay transmit gain, BER results of the NC operation are obtained as $3 \times 10^{-5}$, $5.3 \times 10^{-4}$ and $5 \times 10^{-4}$ respectively. With NCC operation, these performances are improved and results are obtained as $1 \times 10^{-5}$, $5 \times 10^{-4}$ and $4.6 \times 10^{-4}$ respectively. BER results and EVM results for relay-destination links are measured as $3 \times 10^{-5}$ and 14.64% on average. For the same setup, at $-15$ dBm relay transmit gain, BER results of the NC operation are obtained as $7 \times 10^{-4}$, $1.7 \times 10^{-3}$ and $1.6 \times 10^{-3}$ respectively. Similarly, NCC operation's BER results are measured as $3 \times 10^{-4}$, $5 \times 10^{-4}$ and $1 \times 10^{-3}$ respectively. When relay-destination link performances are measured, BER and EVM results are obtained as $7 \times 10^{-4}$ and 18.59% on average. Same observations are also done for the $D_2$. At $-13$ dBm relay transmit gain, BER results for NC operation are measured as $3 \times 10^{-4}$, $5 \times 10^{-4}$ and $6.5 \times 10^{-4}$ respectively. Accordingly, BER results for NCC operation are obtained as $1.7 \times 10^{-4}, 2.5 \times 10^{-4}$ and $3 \times 10^{-4}$ respectively. Furthermore, BER and EVM results for relay-destination links are obtained as $8 \times 10^{-5}$ and 19.29% on average. Similarly, at $-15$ dBm relay transmit gain, BER results for NC operation are measured as $2.5 \times 10^{-3}, 3.5 \times 10^{-3}$ and $4.2 \times 10^{-3}$ respectively. Accordingly, BER results for NCC operation are obtained as $1 \times 10^{-3}, 1.2 \times 10^{-3}$ and $2.4 \times 10^{-3}$ respectively. Thus, improvement that is obtained with NCC operation, is observed similar to previous results. Furthermore, BER and EVM results for relay-destination links are obtained as $2.5 \times 10^{-3}$ and 22.05% on average.

In order to verify results demonstrated in the tables, individual BER and EVM results obtained at each of 20 tests are given by bar charts. Firstly, BER results obtained for the link between $S_2$ and $D_1$ that are obtained with first setup at $-16$ dBm relay transmit gain, are given in Fig. 6. Secondly, BER results obtained for the link between $S_3$ and $D_2$ that are obtained with the second setup at $-15$ dBm relay transmit gain, are given in Fig. 7. It can be observed that individual results also represent overall results given in the tables.

Moreover, EVM results obtained for the links between each source node and relay nodes, are shown in Fig. 8. As mentioned, these links have very good performances. As the last observation, individual EVM results obtained at each test are shown in Fig. 9, where links between source nodes and $D_1$ are considered with the second setup. Similar to BER charts, EVM charts also show consistency of results shown in the tables.

The usage of OFDMA is vital to analyze realistic usage scenarios of NCC operation. By evaluating multiple user scenarios with OFDMA, this NCC implementation

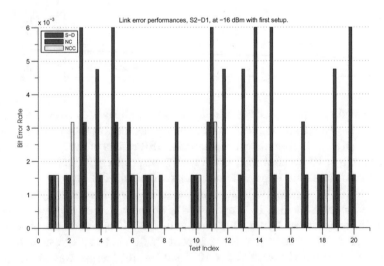

**Fig. 6** BER results obtained during consecutive tests for the first setup

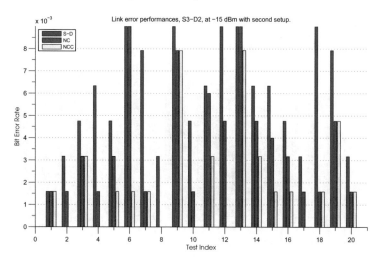

**Fig. 7** BER results obtained at consecutive tests for the second setup

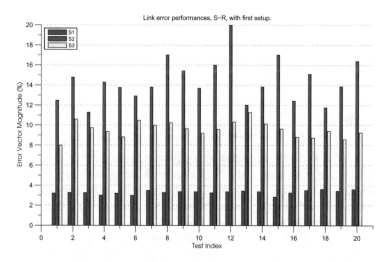

**Fig. 8** EVM results obtained for the source-relay links

shows how resources can be divided and managed. Validating theoretical components in real-time with realistic transmission environment and hardware issues is the most crucial outcome of the implementation. It is shown that with a flexible design of such a testbed, NCC operation in real-time can be easily realized and error performance can be improved significantly. Beyond NCC, NC performance is also measured in real-time. NC also improves error performance, but NCC has clear benefits especially in terms of error performance. As future mobile networks will include massive number of devices, it will be more challenging to obtain desired error performances. Thus, relay-based communication would be very beneficial in terms of the error per-

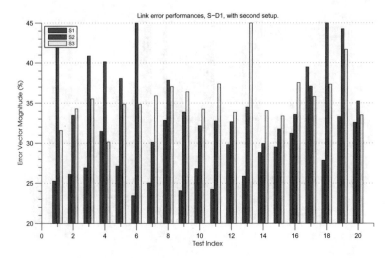

**Fig. 9** EVM results obtained at consecutive tests

formances. NCC is a suitable candidate for such dense network scenarios and tight system requirements resulting from increasing usage rates can be overcome with easy deployment and cost efficiency of NCC.

## 5 Conclusion

NCC is a candidate for smart routing application, that can be used in 5G or beyond technologies. Due to its natural characteristics as allowing scalability and providing robustness to channel impairments, NCC can be easily adapted to the target network, alleviating the expected implementation issues in dense network deployments. In this chapter, we present the theoretical DMT bounds in comparison with the state-of-the-art design solutions and details of implementation setup of an NCC system. We show that NCC is a convenient smart routing approach to satisfy requirements of next generation communication systems by performing both of theoretical analyses and real-time implementation.

## References

1. R. Ahlswede, N. Cai, S.Y. Li, R.W. Yeung, Network information flow. IEEE Trans. Inf. Theory **46**(4), 1204–1216 (2000)
2. R. Dougherty, C. Freiling, K. Zeger, Insufficiency of linear coding in network information flow. IEEE Trans. Inf. Theory **51**(8), 2745–2759 (2005)
3. J.N. Laneman, G.W. Wornell, Distributed space-time-coded protocols for exploiting coopera-tive diversity in wireless networks. IEEE Trans. Inf. Theory, **49**(10), 2415–2425 (2003)

4. T. Ho, M. Mdard, R. Koetter, D.R. Karger, M. Effros, J. Shi, B. Leong, A random linear network coding approach to multicast. IEEE Trans. Inf. Theory, **52**(10), 4413–4430 (2006)
5. M. Di Renzo, M. Iezzi, F. Graziosi, On diversity order and coding gain of multisource multirelay cooperative wireless networks with binary network coding. IEEE Trans. Veh. Tech. **62**(3), 1138–1157 (2013)
6. C. Peng, Q. Zhang, M. Zhao, Y. Yao, W. Jia, On the performance analysis of network-coded cooperation in wireless networks. IEEE Trans. Wireless Commun. **7**, 3090–3097 (2008)
7. H. Topakkaya, Z. Wang, Wireless network code design and performance analysis using diversity-multiplexing tradeoff. IEEE Trans. Commun. **59**(2), 488–496 (2011)
8. A.R. Heidarpour, G.K. Kurt, M. Uysal, Finite-SNR diversity-multiplexing tradeoff for network coded cooperative OFDMA Systems. IEEE Trans. Wirel. Commun. **16**(3), 1385–1396 (2017)
9. S. Gokceli, H. Alakoca, S.T. Basaran, et al., EURASIP J. Adv. Sig. Process. **2016**(8) (2016)
10. F.J. MacWilliams, N.J.A. Sloane, *The Theory of Error-Correcting Codes.* North-Holland Mathematical Library (North-Holland Publishing Company, Amsterdam, 1977)
11. E.M. Gabidulin, Theory of codes with maximum rank distance. Problemy Peredachi Informatsii, **21**(1), 3–16 (1985)
12. S.T. Basaran, G.K. Kurt, M. Uysal, I. Altunbas, A tutorial on network coded cooperation. IEEE Commun. Surv. Tutor. **18**(4), 2970–2990 (2016)
13. P.A. Chou, Y. Wu, K. Jain, Practical network coding. in *Proceedings of the Annual Allerton Conference on Communication Control and Computing*, vol. 41, No. 1, (2003) pp. 40–49
14. R. Koetter, F.R. Kschischang, Coding for errors and erasures in random network coding. IEEE Trans. Inf. Theory, IEEE, **54**, 3579–3591 (2008)
15. A. Bletsas, A. Khisti, D.P. Reed, A. Lippman, A simple cooperative diversity method based on network path selection. IEEE J. Sel. Areas Commun. **24**(3), 659–672 (2006)
16. J.N. Laneman, Network coding gain of cooperative diversity. in *Proceedings of IEEE MILCOM*, vol. 1, (2004) pp. 106–112
17. B. Bai, W. Chen, K.B. Letaief, Z. Cao, Joint relay selection and sub-channel allocation for amplify-and-forward OFDMA cooperative networks. in *Proceedings of IEEE International Communications Conference (ICC'2012)* (2012) pp. 4192–4196
18. M. Gromov, Pseudo holomorphic curves in symplectic manifolds. Invent. Math. **82**(2), 307–347 (1985)

# Opportunistic Network Coding

**Kemal Alic and Ales Svigelj**

**Abstract** In this chapter we describe a practical view of the usage of the network coding theory over realistic communication networks. In particular, we introduce opportunistic network coding, which can be applied to wireless networks with mesh topologies and multiple unicast streams. With the term opportunistic we describe the opportunistic nature of this type of coding, as packets are encoded only if the opportunity arises and there are no mechanisms within it to increase the number of coding opportunities. Different coding approaches have been proposed that cover different network configurations and traffic patterns. In particular, we focused on the main design aspects of BON and COPE, the two opportunistic network coding procedures that perform network coding on the packet level and can significantly improve the network throughput. In addition, we are also describing performance metrics, which are found suitable for performance evaluation of network coding algorithms. For illustration purposes we also show typical results for the above mentioned algorithms. The results show that opportunistic network coding can significantly improve the wireless mesh network performance in terms of throughput and delay.

## Introduction

The network coding benefits are not limited to either wireline networks or multicast applications. The opportunistic network coding which is investigated in this chapter shows, how benefits can be achieved also in the wireless networks and with multiple unicast streams.

The term opportunistic network coding describes the opportunistic nature of this type of coding, as packets are encoded only if the opportunity arises and there are no mechanisms to increase the number of coding opportunities. Packets are coded for one hop only, where they are decoded. In general coding of different and same streams is possible, though we can expect only coding of different streams.

K. Alic (✉) · A. Svigelj
Jozef Stefan Institute, Ljubljana, Slovenia
e-mail: kemal.alic@ijs.si

A. Svigelj
e-mail: ales.svigelj@ijs.si

© Springer International Publishing AG 2018
M. Greferath et al. (eds.), *Network Coding and Subspace Designs*,
Signals and Communication Technology,
https://doi.org/10.1007/978-3-319-70293-3_12

**Fig. 1** Example of how
opportunistic network coding
increases the throughput

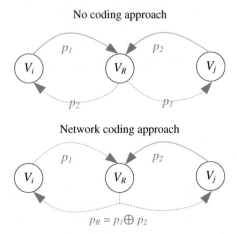

Opportunistic network coding benefits can be best explained with the help of the simple example depicted in Fig. 1. Consider a three-node chain topology where node $V_1$ has a packet $p_1$ destined for node $V_2$, and node $V_2$ has a packet $p_2$ for the destination node $V_1$. The two nodes can exchange their packets through an intermediate relay node $V_R$. Hence, nodes $V_1$ and $V_2$ send packets $p_1$ and $p_2$ to $V_R$. In a conventional system where network coding is not used, $V_R$ first transmits $p_1$, which is then followed by the $p_2$. In the case where a network coding procedure is applied, the $V_R$ performs an algebraic operation over the two packets (e.g., *XOR*) and sends out an encoded packet $p_R = p_1 \oplus p_2$. Upon reception, node $V_1$ *XORs* the received encoded packet $p_R$ with the sent packet $p_1$ and obtains the packet $p_2$. The same procedure is used on node $V_2$. $V_2$ is familiar with the content of $p_2$ and *XORs* it with the just-received packet $p_R$, thus obtaining $p_1$ officially destined to it. By using network coding the number of transmissions that all three nodes need to perform in order to deliver both packets to their destinations has been reduced from 4 to 3.

The example in Fig. 1 presents a topology that is suitable for the chain-structure coding, i.e., a topology where the two packets are travelling in the opposite directions and get encoded in the intermediate node(s). In such a case, all the recipient nodes have sufficient information for successful packet decoding without packet overhearing. Furthermore, packet retransmission may appear only due to the unsuccessful delivery of the encoded packet to one of the recipient nodes.

A more interesting case arises with a more general network deployment. In such a case, the question is which packets the coding node should encode together in such a way that the recipient nodes will be able to decode the encoded packet. When a node has a packet to forward, it needs to know whether encoding of this packet with other queued packet(s) might save bandwidth, i.e., it needs to determine whether the receivers can decode the encoded packets. Consider the situation in Fig. 2. Nodes $V_1$ and $V_2$ are sending packets $p_{13}$ and $p_{24}$ via the relay node $V_R$ to the nodes $V_3$ and $V_4$, respectively. If $V_R$ creates an encoded packet from $p_{13}$ and $p_{24}$, e.g., $p_R = p_{13} \oplus p_{24}$, the receiving node $V_3$ will be able to decode the encoded packet $p_R$ if $V_3$ has overheard

**Fig. 2** General relaying
scenario in which network
coding can be used

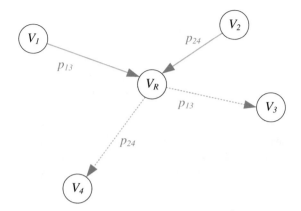

the $p_{24}$ transmission from $V_2$ to $V_R$. The situation is similar for $V_4$, which needs to overhear the $p_{13}$ transmission from $V_1$ to $R$.

Let us highlight why the coding packets that travel in the opposite directions are preferred. Assume the situation presented in Fig. 1 where packets $p_n$ and $p_m$ travel in the opposite directions, i.e. $p_n$ from $V_j$ via $V_R$ to node $V_i$ and $p_m$ via versa. If we describe the links between nodes with packet delivery probabilities, i.e. probability that packets are successfully transmitted between two links, then we can estimate the chances that the encoded packet $p_n \oplus p_m$ will be successfully received and decoded on the two nodes is

$$P = P_{V_R V_j} P_{V_R V_i}, \tag{1}$$

where $P_{V_R V_j}$ and $P_{V_R V_i}$ are the probabilities that packets are successfully transmitted from node $V_R$ to node $V_j$ and to node $V_i$, respectively.

In a more general situation where packet $p_n$ travels from $V_i$ to $V_j$ and $p_m$ from $V_k$ to $V_l$, then the probability that the encoded packet $p_n \oplus p_m$ will be successfully received and decoded on the two recipient nodes is:

$$P = P_{V_R V_j} P_{V_R V_i} P_{V_i V_i} P_{V_k V_j}. \tag{2}$$

The difference to the first case is that also two overhearing events need to take place for each of the receiving nodes to have sufficient information to perform decoding.

### Information Required in the Opportunistic Coding Process

The more packets we encode together the looser the conditions in coding process are, and with that the higher the possibility that receiving nodes will not be able to decode those packets. However, if coding conditions are hard to meet, then there are only few coding opportunities and low bandwidth saving benefits. Good opportunistic coding algorithm does not only attempt to encode as many packets as possible, but rather makes sure that the encoded packets are very likely to be decoded. Making incorrect coding decisions leads to unnecessary retransmissions, which results in wasting of the wireless resources. Therefore, an important development issue is finding a balance between the number of coded packets and the successful decoding rate, by using small network coding overhead.

Coding decisions can be based on knowledge, e.g. each node records the state of the traffic for its neighbouring nodes, although as shown in [1–3] this is only sufficient for a portion of coding decisions. Thus, guessing mechanism needs to be used in the coding decision process. The guessing can be based on the information that is already available on the node, and was collected by some protocol. For example, the COPE build on the information already available on the node and acquire the information gathered by the routing protocol. This is an excellent option as no new additional overhead is introduced to the network [4, 5]. The drawback with this solution is that the routing information has to be available in the layer where coding procedure is implemented. With communication components available on the market this is not always possible. An alternative is to acquire an estimation of coding possibilities through own network coding evaluation. This could be performed through (i) transmitting the probe (measurement) packets; or (ii) by estimating the coding success of the regularly encoded packets. The first option introduces additional overhead to the network. By collecting the required information through measurements scheduled periodically we introduce additional overhead, thus lowering goodput and benefits of network coding. With the second option the data is collected by just observing the traffic going through the network coding module e.g. collecting the statistics on which packets have been encoded and how many retransmissions were needed. The main drawback of such an approach is that there is a rather small portion of packets that get encoded. Collecting them over longer periods of time is in general not acceptable as this would make the system slow and unable to adapt to the fast changes on wireless links. Practical value of such approach is low for general network performance since the number of coding options (combinations of packet's previous hop and packet's next hop) grows with the number of neighbours a node has i.e. faster than exponentially (i.e. $N!/2$). In practice, there are too many possibilities to provide a reliable estimation on which node pairs provide good coding matches. In addition, the ongoing measurements should be able to detect the changes in wireless links conditions and react fast in case of their degradation. The subset of possible coding matches should be made in order to provide reliable decision process and react to changes in wireless environment.

### Living Environment for Opportunistic Network Coding

Opportunistic network coding works well in wireless mesh networks. With this term we describe networks where nodes connect to each other via multi-hop wireless links. Usually we consider them to be self-sustainable, i.e. adaptive and self-organised in order to be able to maintain connectivity.

Wireless mesh networks are very limited in capacity. E.g. the 802.11 family of products are advertised to support data rate of up to 54 Mbps. Still, "protection" mechanisms such as binary exponential backoff, rate adaptation, and protocol overheads, cut the throughput by 50%. Moreover, owing to backward compatibility with

802.11b and 802.11g is encumbered with legacy issues that reduce throughput by an additional ~20%. Besides, the actual bandwidth available to individual clients can be even much lower due to the shared nature of the wireless medium. Also, mesh networks usually operate in the unlicensed 2.45 GHz Industrial, Scientific and Medical (ISM) frequency band. Hence, bandwidth is shared also with other networks or network devices, e.g. Bluetooth peripheral devices, spread-spectrum cordless phones, or microwave ovens. Interference affects the quality of a wireless link and, consequently, its error rate and achievable capacity.

Regardless of their capacity limitations wireless mesh networks are appealing in areas with scarce wired infrastructure. Such places can be found in less developed areas or in places such as tunnels, battlefields, on board of public transportation etc. Furthermore, due to low operational and deployment costs commercial deployment can also be found in urban areas with the main application of offering cheap Internet connectivity.

In a mesh network, each router can be seen as part of a backbone infrastructure and also as an access point for end users. In this chapter we see mesh routers as access points that offer end-users access, hence traffic generated on nodes can be considered aggregated from multiple users.

The network coding procedures discussed in this work are positioned between the network and the link layer.

Routing-path selection also influences the performance of network coding. Routing that takes into account also network coding can further boost the network performance, as can poorly selected routing in combination with network coding result even in poorer network performance. The proposed BON procedure works as an independent layer, but its performance is indirectly still related to the performance of routing [1, 2].

In this chapter we consider the link-state routing protocol that is widely adopted in the wireless mesh networks. We adopted symmetric routing (with symmetric we refer to path selection of traffic flows that travel in the opposite directions).

Wireless mesh networks were initially based on the IEEE 802.11 standard. Therefore, we provide a brief overview of general properties of the standard important from the opportunistic network coding perspective.

The standard offers mechanisms for packet distribution to the nodes within their communication range. Unicast inherently assure reliable delivery of packets to the next hop while broadcast offers distribution of data to multiple users. With network coding we want all the nodes to listen to all the transmissions. To this end, the broadcast mechanism in the link layer seems a suitable solution at a first glance. However, due to the absence of collision detection and backoff on the source node the reliability of broadcast is questionable. For this purpose we adopt pseudo broadcast first presented in [1] and discussed in this chapter.

**The Coding Algorithms**

The general COPE and BON architecture is shown in Fig. 3. Packets arriving from the network layer are placed into the network coding queues. Both coding procedures

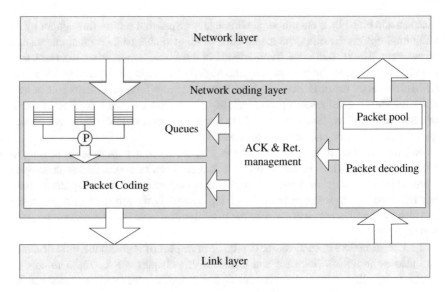

**Fig. 3** Opportunistic network coding architecture

foresee use of FIFO queues, but also other queue scheduling can be used. When the node is granted access to the wireless channel the coding process takes the first packer from the queues and searches for coding opportunities amongst remaining packets in the output queue. If there are no coding opportunities identified, the packets are sent out as they are. None of the two algorithms deliberately delays packets for the purpose of finding additional coding opportunities. If packet coding finds coding opportunities, the outgoing packets are coded into one encoded packet and retransmissions are scheduled for each individual native packet.

All the received packets go first through the decoding process. If packets are successfully decoded each native packet coded in the encoded packet is stored into the packet pool. If any of the native packets are destined to the recipient node acknowledgement message is sent out. If received packet is acknowledgement type of packet the scheduled retransmission is cancelled. Packets with next hop address the same as the recipient node are also sent to network layer for further processing.

### *COPE*

All the received packets go first through the decoding process. If packets are successfully decoded each native packet coded in the encoded packet is stored into the packet pool. If any of the native packets are destined to the recipient node acknowledgement message is sent out. If received packet is acknowledgement type of packet the scheduled retransmission is cancelled. Packets with next hop address the same as the recipient node are also sent to network layer for further processing.

In COPE the coding process depends on the nodes' knowledge on what information (which packets) its neighbouring nodes have. Based on this knowledge the coding process is straightforward and the decoding process will have a high success rate. Due to the nature of wireless communications and imposed delays that comes with this knowledge can be used only to find a small portion of all possible coding opportunities. Hence, the COPE coding process also uses guessing which is done through the delivery probability that is calculated in all of the ETX (Expected Transmissions Count) based routing protocols. COPE incorporates two techniques to support such coding process, namely opportunistic listening and learning neighbour state:

- opportunistic listening and
- learning neighbour state
- guessing

**Opportunistic Listening**

In COPE the coding process depends on the nodes' knowledge on what information (which packets) its neighbouring nodes have. Based on this knowledge the coding process is straightforward.

COPE foresees that all the nodes listen to all transmissions. As already mentioned all the packets carry potentially useful information that can be used in the decoding process. COPE also uses information obtained for the coding process. When a packet is received on the node the sender and recipient node information is extracted and noted. The node that overheard the transmission now knows that the two nodes have the packet.

**Learning Neighbour State**

COPE procedure also foresees the use of the reception reports. All the nodes in the network send out reports in which information on the received packets is stored. Reports are broadcasted periodically or when opportunity arises they are attached to the regular outgoing packets.

**Guessing**

In a network where we do not want to delay packets and Quality of Service (QoS) should be at least considered, relying solely on the knowledge on the state of the network would miss a great deal of coding opportunities. Hence, COPE foresees also guessing, when information on the state of the network, i.e. whether the packet under the question was received by the node.

The well-known COPE algorithm guessing relies on the information gathered in the network layer, i.e. the expected transmission count matrix (ETX):

$$ETX = \frac{1}{P_{V_i V_l} P_{V_l V_i}} \tag{3}$$

With this metric each link between two nodes is described with the inverse of the packet delivery probability and its successful acknowledgement. Information can be used in a way that it reflects the probability that a node received certain packet.

## The COPE Coding Algorithm

The node estimates probability that the node $N$ has packet $p$ by looking at the delivery probability for the link between packet's previous hop and node $N$.

With all the needed information, the node can code together as many packets $(p_1, \ldots, p_n)$ as possible provided that none of the packets have been created on the coding node, all the packets have different next hops and that there is a strong possibility that all next hops will be able to decode the packet. The next hop can decode the packet if it has already received all but one of the packets coded together. Let the probability that a next hop has heard packet $p_m$ be $P_m$. Then, the probability, $P_D$, that it can decode its native packet is equal to the probability that it has heard all of the $n - 1$ native packets encoded with its own, i.e.,

$$P_D = P_1 P_2 \ldots P_{n-1} \tag{4}$$

The coding algorithm assures that the decoding probability $P_D$ for all the next hops for a given combination of encoded packets is above the threshold $G$.

## BON

With BON assumptions that each node has positions of all of its neighbouring nodes, and that nodes have fixed locations are made. Algorithm makes coding decisions based solely on the information about the packet's previous and next hop node position, hence, it acts completely independently of all other communication layers, and is completely protocol agnostic.

## The BON Coding Algorithm

The decision making on which packets to encode together can be based on the information already available on the node or the required information needs to be gathered.

The coding process is based on the local bearing (as used in the navigation) of the packet, which is defined on the relay node and depends on the positions of the packet's previous hop and the packet's next hop [2, 3]. Packets that have not travelled at least one hop are not codable, and so a bearing is not defined for them. Let us use the example from Fig. 4a to explain the definition of bearing. A packet $p_{ij}^{(R)}$ has been transmitted from the node $V_i$ to the relay node $V_R$ and its next hop is $V_j$. Locations of nodes $V_i$ and $V_j$ are known and are given in Cartesian coordinate system as $(X_i, Y_i)(X_j, Y_j)$. We define bearing for the packet $p_{ij}^{(R)}$ on the relay node $V_R$ as a unit vector calculated as:

$$\vec{b}_{ij} = \frac{(X_j - X_i, Y_j - Y_i)}{\|(X_j - X_i, Y_j - Y_i)\|} \ , \tag{5}$$

and translated into three dimensional space where also the node elevation ($Z$) is taken into account:

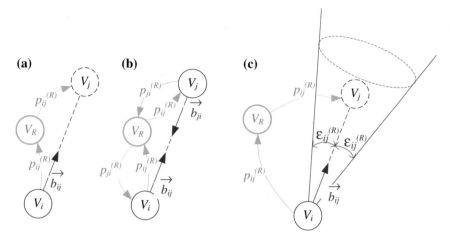

**Fig. 4** Graphical presentation of **a** a bearing, **b** bearings for two packets travelling in the opposite direction and **c** the vicinity for packet $p_{ij}{}^{(R)}$ in the shape of an infinite cone

$$\vec{b}_{ij} = \frac{(X_j - X_i,\ Y_j - Y_i,\ Z_j - Z_i)}{\|(X_j - X_i,\ Y_j - Y_i,\ Z_j - Z_i)\|}\ . \tag{6}$$

In the coding process we are primarily interested in packets that are travelling in opposite directions. Without loss of generality we build on a two-dimensional example presented in Fig. 4b where the packet $p_{ij}{}^{(R)}$ has the same next hop as was the packet $p_{ji}{}^{(R)}$'s previous hop and vice versa (the case of the so-called chain topology). In such a case, the bearings of the two packets are $\vec{b}_{ij} - \vec{b}_{ji} = \vec{0}$. The two packets can be seen as representatives of the two flows that are travelling in the opposite directions. Coding of such packets is an ideal combination as no overhearing between the nodes is needed and the recipient nodes can always decode the encoded packet.

Since such opportunities (i.e. when packets travel in exactly opposite directions) do not provide a sufficient number of coding opportunities, the coding opportunities need to be searched for also amongst other packets. In general, we can assume that the nodes that are located in the *vicinity* of the transmitting node have a better chance of successfully overhearing the packet transmission than nodes that are located far from it. Thus, they also provide a possibly good coding opportunity. With the coding algorithm we are looking for packets ($p_{kl}{}^{(R)}$) that are codeable with the transmitting packet ($p_{ij}{}^{(R)}$) on the relay node. We consider $p_{kl}{}^{(R)}$ to be codable with $p_{ij}{}^{(R)}$, if the $p_{kl}{}^{(R)}$ destination node $V_l$ is in the *vicinity* of the $p_{ij}{}^{(R)}$ source node $V_i$ and vice versa.

**Definition 1** With BON the *vicinity* of the node $V_j$ for packet $p_{ij}{}^{(R)}$ on the relay node $V_R$ is described as an area within the shape of an infinite cone with the apex in the packet source $V_i$, and cone axis with the direction $\vec{b}_{ij}$ and the aperture $2\varepsilon_{ij}{}^{(R)}$.

Example of such an area for a packet $p_{ij}^{(R)}$ that is on the node $V_R$, with the previous hop $V_i$ and the next hop $V_j$ is shown in Fig. 4c, where the cone has the apex in node $V_i$ and the aperture $2\varepsilon_i^{(R)}$. The node $V_j$ lies directly on the cone axes. With $\varepsilon_i^{(R)}$ we define how large the *vicinity* for packet $p_{ij}^{(R)}$ is, hence we refer to it as tolerance angle.

**Definition 2** With BON coding the packets $p_{ij}^{(R)}$ and $p_{kl}^{(R)}$ are codeable on the relay node if previous hop of $p_{ij}^{(R)}$ ($V_i$) is in the *vicinity* of the next hop of $p_{kl}^{(R)}$ ($V_l$) and vice versa, where *vicinity* of nodes for packets $p_{ij}^{(R)}$ and $p_{kl}^{(R)}$ is set according to the Definition 1 with $\varepsilon_{ij}^{(R)}$ and $\varepsilon_{ji}^{(R)}$ accordingly.

Let us illustrate using general two-dimensional examples from Fig. 5a, b where matching and a non-matching packet pairs are presented respectively. In both presented cases the initial situation is the same. Let us point out that the location of the $V_R$ has no impact on the coding decision, and its location is changed only for the reasons of graphical presentation. Packets $p_{ij}^{(R)}$ and $p_{kl}^{(R)}$ are both on the relay node $V_R$. $V_i$ and $V_j$ are the previous and the next hop for $p_{ij}^{(R)}$ and $V_k$ and $V_l$ are the previous and the next hop for $p_{kl}^{(R)}$. $\varepsilon_{ij}^{(R)}$ and $\varepsilon_{kl}^{(R)}$ are tolerance angles that relate to packets' previous hops on the relay $V_R$.

In Fig. 5a we can see that the node $V_i$ is located in the *vicinity* of the node $V_l$ for packet $p_{kl}^{(R)}$ and the similar goes for the node $V_k$, which is in the *vicinity* of the node $V_j$ for packet $p_{ij}^{(R)}$. Since we want that the condition in Definition 2 is met for both situations in the given example, the two packets are codeable.

In Fig. 5b we can see that the node $V_i$ is located in the *vicinity* of the node $V_l$ for packet $p_{kl}^{(R)}$ and the node $V_k$ is outside the *vicinity* of the node $V_j$ for packet $p_{ij}^{(R)}$.

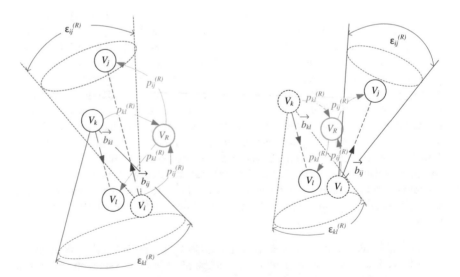

**Fig. 5** Graphical presentation of general coding case for **a** a matching packet pair and **b** of a general coding case for a non-matching coding pair

Since we want that the condition is met for both situations in the given example the two packets are not codeable. Packets $p_{ij}^{(R)}$ and $p_{kl}^{(R)}$ would be codeable if $\varepsilon_{ij}$ was larger. By increasing the tolerance angle we cover a larger area and thus increase probability that packets will meet the conditions. However, by increasing the tolerance angle parameter, the probability that the receiver will not be able to decode the packet is also increased. By reducing the parameter $\varepsilon$ towards zero, the coding opportunities are reduced, but the probability of a successful packet decoding on the receiving nodes is increased.

By further generalization the BON coding procedure allows also coding of multiple packets. Multiple packets are encoded into one encoded packet when all packet pairs and their corresponding nodes are within the *vicinity* area. The higher the $\varepsilon$ values, the higher the possibility of coding multiple packets on the other hand, if at least one $\varepsilon = 0$ only two packets can be encoded. In BON coding the balance between coding opportunities and successful packet decoding is set through the parameter $\varepsilon$, which is adapted automatically on every node. The algorithm handling self-adjustment is presented in Algorithm 1.

- ***Algorithm 1:***      *Update tolerance angle $\varepsilon$.*
-
- 1: **if** ($K_P > K_{PMIN}$) **do**
- 2:    **if** ($K_C > 0$) **do**
- 3:       **if** ($\varepsilon$ increased in the last update) **do**
- 4:          CounterTurnsWithoutCoding = 0;
- 5:          scheduleUpdate(t + $T_{RET}$)
- 6:       **else do**
- 7:          scheduleUpdate(t + $T_{\varepsilon U}$)
- 8:          **if** ($RR \leq RR_{min}$) **do**
- 9:             $\varepsilon$ = increaseEpsilon($\varepsilon$, $\Delta\varepsilon$, $\varepsilon_{max}$)
- 10:          **else if** ($RR \geq RR_{max}$) **do**
- 11:             $\varepsilon$ = decreaseEpsilon($\varepsilon$, $\Delta\varepsilon$)
- 12:          **end if**
- 13:       **end if**
- 14:    **else do**
- 15:       scheduleUpdate(t + $T_{\varepsilon U}$)
- 16:       **if** (CounterTurnsWithoutCoding > No_Coding_Threshold) **do**
- 17:          $\varepsilon$ = increaseEpsilon($\varepsilon$, $\Delta\varepsilon$, $\varepsilon_{max}$)
- 18:       **else do**
- 19:          CounterTurnsWithoutCoding ++
- 20:       **end if**
- 21:    **end if**
- 22: **else do**
- 23:    scheduleUpdate(t + $T_{\varepsilon U}$)
- 24: **end if**

## *Integration into the Communication Stack*

In order to make use of the opportunistic network coding process, several supporting mechanisms are required to be implemented on the nodes. Using the opportunistic network coding, the nodes change the way packets are processed after being received and a new signalization is required. Moreover, the opportunistic network coding requires that the nodes are put in a broadcast receiving mode, while a pseudo-broadcast mechanism is used for sending the packets.

All the nodes in the network listen to all the transmissions and try to overhear as many packets as possible – including the ones that are not addressed to them. All the overheard packets are saved into the packet pool for decoding purposes. A copy of a received packet (i.e., a packet headed to the node) is also stored in the packet pool, while the original is forwarded to the network layer.

### Packet Queues

BON and COPE procedures use different outgoing queues for different packet types i.e.:

- ACK packets. These are standalone acknowledgement messages that are created only when no opportunity has been found to add them as a header to regular outgoing packets.
- Retransmitted (native) packets dedicated queue is used in BON only. Packets are placed in special sub-queue to prevent coding the same packet set together again and thus trying to avoid the possible mistake in the coding process.
- Regular outgoing (native) packets that have arrived from the network layer and have arrived from one of the neighbouring nodes or have their origin on this node. COPE uses this queue also for retransmitted packets, which are placed at the top of the queue and thus processed first.

When there are packets present in one of the output queues, the BON and COPE procedure signal the link layer that the access to the wireless media is required. When the link layer signals that it has successfully gained access to the channel, the queue is selected depending on its priority. If we are dealing with retransmitted or regular queue, the coding procedure is initiated.

### Coding Procedure

The coding process using the FIFO system always takes the packet $p_0$ that is at the head of the output queue (retransmission or regular) and searches for possible coding opportunities with the other packets in the regular outgoing queue. If there are no coding options found, the packet is sent out as-is (Fig. 6).

### Signalization

The two algorithms under the spotlight use two types of signalization, namely acknowledgement messages and reception reports. The acknowledgements in BON and COPE have the same structure, while reception reports are unique to COPE only.

**Fig. 6** A flow chart of the
BON and COPE coding
process

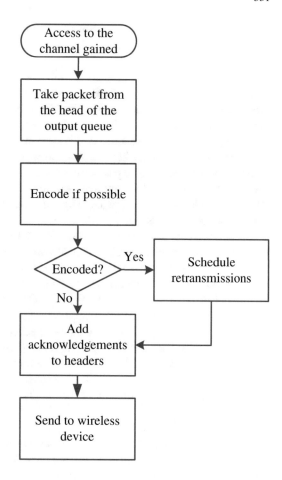

Acknowledgement messages in the network coding layer are used only for native packets that have been received as part of the encoded packet. Acknowledgement messages are generated by the receiving node, designated as a packet's next hop. Please note that we are discussing only the acknowledgement messages in the network coding layer, while acknowledgements in the other layers remain as they were. The acknowledgement message in the network coding layer is required to indicate the success in the packet decoding procedure.

Cumulative acknowledgement in a structure proposed in [1] are used. Acknowledgement massages are sent in a bulk every time an opportunity arises and when there are new messages in the pool. The latest received packet sequence number is used for reference. This is followed by a sequence of Boolean values that also indicates status of the received packets for the packets with lower sequence number. E.g. an entry {A, 50, 01011111} confirms that packets 50, 48, 46-42 have been received and that packets 49 and 47 are still missing.

To make this operate, packet sequence number needs to be recorded. Each node indexes outgoing packet per each neighbour individually and the same holds true for all the received packets.

Acknowledgement messages confirm reception for every received native packet that has been received as part of the encoded packet. Cumulative acknowledgement report messages are broadcasted periodically, every $T_u$ (referred to as acknowledgement packets). If the opportunity arises, the acknowledgement messages are attached to the regular outgoing packets. In this case cumulative acknowledgement can be seen only as an additional header, thus introducing less overhead. In case that the upper limit of acknowledged packets per is reached, the acknowledgement process is triggered and acknowledgement massages are sent out as individual packet. Cumulative acknowledgement messages reduce the overhead compared to the individual acknowledgement messages.

Reception reports are messages used in COPE only. With these nodes inform their neighbours on which packets they have received Also reports are sent out when opportunity arises or in the absence of those periodically. Similar as acknowledgements, also reports are compact. They are formed as headers, where first information on number of reports in the header is given. One report is used for each neighbour with updated status. Second information in the header is address of the neighbour that transmitted the packet, followed by the last packet id received from that node. This is followed by a sequence of Boolean values that also indicates status of the received packets for the packets with lower sequence number. Reports include information on the packet ID and neighbour address because both data are needed to identify the packet in a unique way. More can be read in [1].

**Packet Reception**

In addition to handling acknowledgement packets, the recipient node needs to be able to also handle native and encoded packets, which both can carry additional header that contains acknowledgement massages.

Packet reception process is presented in Fig. 7. Upon packet reception in the network coding module, further actions depend on whether the packet is encoded or native. In the case when an encoded packet (consisting of $M$ native packets) is received, the process checks the packet pool where all the received, sent and overheard packets are stored for decoding purposes. The encoded packet can be decoded if a node has in its packet pool at least $M-1$ packets. The process has to determine whether it has already received $M-1$ of the packets encoded in the encoded packet. If not, the encoded packet cannot be decoded and it is simply dropped. If the node has at least the required $M-1$ packets, i.e., sufficient information, it decodes the encoded packet using these packets with the $XOR$ operations, thus obtaining a native packet that has not been received before. From here on the process is the same as upon receiving a native packet. If the packet is new (i.e. newer received before), its copy is inserted into the packet pool for decoding purposes. It does so for every received native packet, as all the received packets are potentially needed for further decoding purposes. The process checks whether the node is the next hop of the native packet. If so, and if the packet has been a part of the encoded packet, an

**Fig. 7** A flow chart of the
BON and COPE reception
process

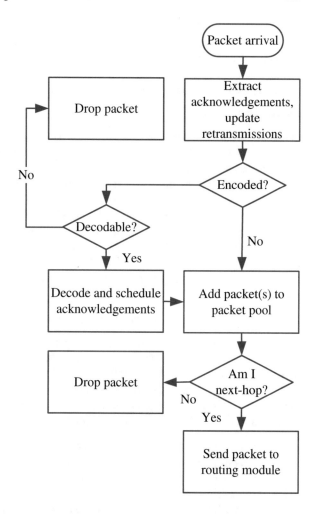

acknowledgement message is scheduled and the packet is sent to the network layer
for further processing.

**Packet Retransmission**

If an ACK message is not received within a predetermined time, the retransmission
event is triggered. A native packet that has been sent out as part of the encoded packet
and has not been acknowledged, is placed in the retransmission output queue.

Packets require retransmissions for two reasons. Packets cannot be decoded on
the recipient node (i) because there are not enough packets in the packet pool, and
(ii) because packets get lost in the transmission/reception procedure.

The packets in the retransmissions queue can be coded again. However, coding
opportunities are searched for only within the packets in the regular output queue.

This mechanism ensures that it is not possible for the same set of packets to be coded again. Hence, the situation where the same set of packets is encoded again is avoided.

With every packet retransmission we want to increase the probability of successful packet decoding on the recipient node. and we can do so by decreasing the parameter $\varepsilon$. Hence, we foresee that with every packet retransmission parameter $\varepsilon$ is decreased.

**Pseudo Broadcast**

With network coding we need to overhear as many transmissions as possible for the network coding mechanisms to seize their full potential. Hence, a natural approach would be to use a broadcast mechanism. As shown in [1] this does not work due to poor reliability and lack of backoff and they propose a pseudo broadcast mechanism which unicasts packets that are meant for the broadcast.

The link-layer destination field is set to the MAC address of one of the intended recipients. Since all the nodes are set in the promiscuous mode, they can overhear packets not addressed to them. When a node receives a packet with a MAC address that is identical to its own, it sends an ACK message to the sender. Regardless of the address of the next hop of the packet, the node sends the packet to the network coding module.

### *Performance Metrics for Network Coding Procedures*

Network coding is a relatively new approach; whose fundaments we are still learning. With opportunistic network coding we may claim the same, as knowledge in what kind of networks and with which traffic distributions we can expect benefits and how does the new layer affect the performance of applications with high QoS demands has not yet been obtained. Depending on what we wanted to show, we adopted different metrics, and on the high level we can group them into two categories:

- Performance metrics that capture the detailed network response on the use of network coding. Corresponding set of results is shown against the network load, which is always measured over longer periods. Network load was selected as the key influence for the network coding performance as it can be measured in a straight-forward manner and there is a strong correlation between performance benefits and load.
- Performance metrics that allow higher level conclusions such as the influence of the network topology on the performance of the network. The main issue in showing such results is in defining a performance metric that is only dependant on the parameter under scope.

Thus, in the following we are presenting some performance metrics, which we found suitable for performance evaluation of opportunistic network coding algorithms. In addition, we also show some illustrative results. As the elementary metric reflecting the quantity of service we can observe the network goodput ($g$), which is the number of useful information bits delivered by the network to a certain destination per unit of time. The goodput in graphs can be shown as a sum of all the goodput on all the network nodes at particular network scenario e.g. particular load $g(i)$ or plain as current goodput on node or network.

**Fig. 8** Goodput (g) with
respect to the network load
for COPE, BON and the case
when coding is not used

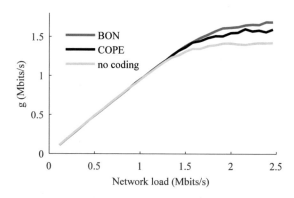

**Fig. 9** Gain (G) with respect
to the network load for
COPE, BON and the case
when coding is not used

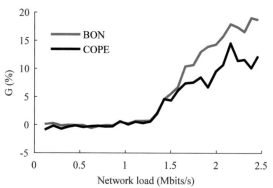

Example of goodput defined as a sum of all the goodput on all the network nodes at particular network load is shown in Fig. 8. Goodput is shown for the two opportunistic coding procedures and reference scenario, where coding is not used.

We further define the gain $G(i)$ in $i$th simulation run as the relative increase of goodput obtained with network coding with respect to goodput without network coding:

$$G(i) = \frac{g_{NC}(i) - g_{no\,coding}(i)}{g_{no\,coding}(i)} 100\% \qquad (7)$$

where gain can be observed for the whole traffic in the network or just on the individual flow level. Similar as goodput also gain can be observed for particular scenarios. Example graph for gain for the same case scenario as presented in Fig. 8 is shown in Fig. 9.

As a typical QoS metric we measure End-to-End Delay and jitter at the application layer. Both ETE delay and jitter are measured for each particular flow separately and for all the flows. Regardless of the case, we can write the following:

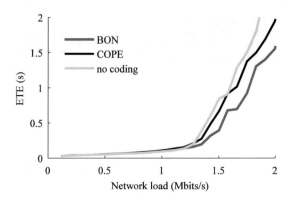

**Fig. 10** End-to-end delay (ETE) with respect to the network load for COPE, BON and the case when coding is not used

$$ETE(i) = \frac{\sum_{n-1}^{K_a(i)} d_n}{K_a(i)} \tag{8}$$

where $d_n$ is the *ETE* delay of the $n$th packet and $K_a(i)$ is the number of packets received in the application layer.

Jitter is the standard deviation from true periodicity of a presumed periodic signal. We measure jitter only for flows with constant interarrival packet rates:

$$jitter(i) = \frac{1}{F(i)} \sum_{f=1}^{F(i)} \frac{1}{K_f(i)} \sum_{n=1}^{K_f(i)} (d_{nf} - ETE_f(i))^2 \tag{9}$$

where $F$ is the number of flows, $K_f$ is the number of packets received in the application layer belonging to the $f$th flow, and $ETE_f$ is *ETE* for $f$th flow. Where we are interested in jitter of only one particular flow, this same equations is used for calculation and $F(i) = 1$.

Typical delay results are presented in Fig. 10. On the figure values for COPE, BON and reference scenario where coding is not used are shown.

Opportunistic network coding performance primarily depends on the quantity of load and also on different parameters. When analysis or evaluation on of the influence of different network and traffic characteristics on the performance efficiency of the network coding needs to be performed we require a load independent metric to quantify the result, in particular, if we want to access the dependency of gain on other parameters (e.g. queue length, packet length, topology, etc.). Thus, we propose *MaxGain*, which is calculated as the average of the 3 highest gain values obtained among results for a given set of parameter values. We first sort vector $G(i)$ where $i = \{1, 2,..., I\}$ and $I$ is the number of simulation runs from the highest to lowest value:

$$G_s = sort\{G(1), G(2), \quad \dots, G(I)\} \tag{10}$$

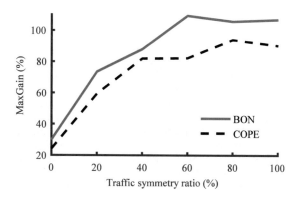

**Fig. 11** MaxGain with respect to the traffic symmetry ratio for COPE and BON

and then the *MaxGain* is:

$$MaxGain = \frac{1}{3}(G_s(1) + G_s(2) + G_s(3)) \tag{11}$$

For example in Fig. 11 we show dependency of opportunistic network coding performance against traffic symmetry ratio. In a network limited number of node pairs were selected. Traffic between the two loads was generated, while the symmetry between the two flows in the load pair was changing.

**Summary**

In this chapter we show how coding theory can be used in existing communication networks. We introduced opportunistic network coding which can be applied to the wireless networks and to multiple unicast streams. The term opportunistic describes the opportunistic nature of this type of coding, as packets are encoded only if the opportunity arises and there are no mechanisms within it to increase the number of coding opportunities.

Different coding approaches have been proposed that cover different network configurations and traffic patterns. The core of all opportunistic network coding approaches is which packets the coding (i.e. relaying) node should code together in such a way that the recipient nodes will be able to decode the encoded packet. In this chapter we focus on design aspects of two typical opportunistic network coding representatives namely BON and COPE. Both are positioned between the network and link layers and are performing network coding on the packet level and can significantly improve the network throughput. As the network coding is a relatively new approach, we are also describing performance metrics, which we found suitable for performance evaluation of network coding algorithms. For the illustration purposes we also show some typical result for the above mentioned algorithms. The results show, that opportunistic network coding can significantly improve the wireless mesh network performance in terms of throughput and delay.

# References

1. S. Katti, H. Rahul, W. Hu, D. Katabi, M. Médard, J. Crowcroft, XORs in the air: practical wireless network coding. IEEE/ACM Trans. Netw. **16**, 497–510 (2008)
2. K. Alic, E. Pertovt, A. Svigelj, Bearing-opportunistic network coding. Int. J. Comput. Commun. Control. **10**, 154–164 (2015)
3. K. Alic, A. Svigelj, A one-hop opportunistic network coding algorithm for wireless mesh networks, Wirel. Netw. (2016) https://doi.org/10.1007/s11276-016-1384-y
4. K. Alic, A. Svigelj, Self-adaptive practical opportunistic network-coding procedure for static wireless mesh networks, Ad Hoc Sens. Wirel. Netw. **36**(1–4), 87–105 (2017)
5. S. Katti, D. Katabi, W. Hu, H. Rahul, M. Medard, The importance of being opportunistic: practical network coding for wireless environments. Presented at the 43rd Allerton conference on communication, control, and computing, 2005

# Coded Random Access

## Čedomir Stefanović and Dejan Vukobratović

**Abstract** This chapter presents an overview of coded slotted ALOHA (CSA), which is a slotted ALOHA-based random access scheme with iterative interference cancellation. The iterative reception algorithm of CSA is analogous to the iterative belief-propagation erasure-decoding, motivating the use of the tools from codes-on-graphs to design and analyze CSA schemes. The asymptotic performance analysis of CSA for the collision channel model is derived using the and-or tree evaluation and instantiated for the case of frameless ALOHA. The and-or tree evaluation is then adapted to the case of block-fading channels and threshold-based reception criterion, which is another frequently used model in wireless systems. Finally, the performance of CSA is assessed in multi access point scenario for the collision channel model, assuming different variants of cooperation among the access points.

## 1 Introduction

The problem of random access arises in communication scenarios in which multiple users (i.e., terminals) have to access the common access point (AP) over the shared communication medium, see Fig. 1, but the activation patterns of the terminals (i.e., instances of when the terminals will initiate the access) and, perhaps, even the number of the accessing terminals are not known a-priori. A typical example can be found in the connection establishment procedure taking place between a mobile phone and a base station in a cellular network, with the purpose of initiating a call. The typical approach in such case is to employ a decentralized algorithm that distributes the

Č. Stefanović (✉)
Department of Electronic Systems, Aalborg University (Copenhagen Campus), Frederikskaj 12, 2450 Copenhagen, Denmark
e-mail: cs@cmi.aau.dk

D. Vukobratović
Faculty of Technical Sciences, Department of Power, Electronics and Communication Engineering, University of Novi Sad,
21000 Novi Sad, Serbia
e-mail: dejanv@uns.ac.rs

© Springer International Publishing AG 2018
M. Greferath et al. (eds.), *Network Coding and Subspace Designs*,
Signals and Communication Technology,
https://doi.org/10.1007/978-3-319-70293-3_13

**Fig. 1** The random access
protocols are used for the
scenarios in which the
a-priori unknown subset of
users attempts access to the
common access point over
shared medium

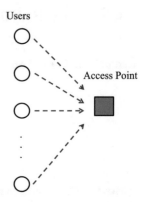

time-instants in which users attempt the access in a random manner; the aim is to
avoid the interference among the users, as it may prevent successful reception of
their access requests by the AP.

Slotted ALOHA (SA) is a prominent example of a random access protocol, used,
e.g., in RFID networks, satellite networks, and as the basis for the connection establishment in cellular access networks. In its original form, SA assumes that the time
is divided in equal-length slots and that users are slot-synchronized, i.e., there is a
certain level of coordination among the users, provided by the AP that broadcasts
time-references. Each accessing user, independently and uncoordinatedly, decides on
the slot-basis with a certain probability whether to transmit. From the AP perspective,
a slot can be an idle (i.e., containing no transmission), a singleton (i.e., containing
a single transmission) or a collision slot (i.e., containing multiple transmissions).
The SA performance is standardly assessed on the collision channel model, which
assumes that (i) transmissions occurring in singleton slots are successfully received
(i.e., decoded) with probability 1, and (ii) any transmission occurring in a collision
slot is not received with probability 1. One of the basic performance metrics of SA is
the expected throughput, denoted by $T$, which is the measure of the efficiency of the
use of the slots. Specifically, the expected throughput can be defined as the average
number of transmissions (i.e., user packets) that can be successfully decoded in a
slot, which, for the collision channel model, is equal to the probability that the slot
is singleton. Denoting the slot access probability by $p_A$, the throughput is computed
as

$$T = \Pr\{\text{slot is singleton}\} = \binom{N}{1} p_A (1 - p_A)^{N-1} \approx N p_A e^{-N p_a}, \quad (1)$$

where $N$ denotes the number of accessing users, and where the approximation holds
as $N \rightarrow \infty$, and $N p_A = \beta$, where $\beta$ is a constant. Obviously, in order to boost the
throughput of SA for the collision channel model, one should maximize the proba-
bility that an observed slot is singleton, which is done by optimizing the slot-access
probability $p_A$. It is easy to verify that, asymptotically, the limit on the maximum
expected throughput is $\frac{1}{e}$ packet/slot, achieved when $p_A = \frac{1}{N}$ (i.e., $\beta = 1$).

**Fig. 2** Frame slotted ALOHA: The users perform access by transmitting in randomly selected slots of the frame

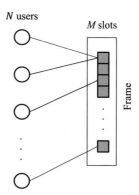

A scheme closely related to SA is the so-called framed slotted ALOHA (FSA), where the time-slots are organized into frames, and the users are both frame- and slot-synchronized. In FSA, a user attempts the access by transmitting in a randomly selected slot in the frame, see Fig. 2. Using similar arguments as in case of SA, it is easy to verify that asymptotic limit on the maximum expected throughput of FSA for the collision channel model is again $\frac{1}{e}$, achieved when the number of the slots in the frame, denoted by $M$, is equal to the number of accessing users $N$.

The collision channel model may serve as a valid approximation only for narrowband, single antenna systems, in which powers of the signals transmitted by users are the same at the point of reception and in which the impact of the noise can be neglected. On the other hand, the collision channel fails to accurately model the cases when the signal powers at the point of reception are unequal, which may happen due to fading and shadowing – the phenomena characteristic for wireless transmissions, and/or when the noise power is substantial relative to the received signal power. In particular, in the former case, the *capture* effect may occur in collision slots, when one of the signals involved in a collision is sufficiently stronger then the interfering signals and the noise combined, and thus becomes successfully decoded; in other words, the existence of the capture effect removes the assumption of collision slots being unusable by default. In the latter case, the noise may prevent receiving a signal in a singleton slot; i.e., the singleton slots may not be usable by default.

A model often used due to its simplicity and analytical tractability is the threshold-based capture model, according to which a signal is successfully received (i.e., captured) in a slot if its signal-to-interference-and-noise ratio (SINR) is above a certain threshold. Specifically, by the threshold-based capture model, the capture occurs if

$$\frac{P_S}{P_N + P_I} \geq b, \tag{2}$$

where $P_S$, $P_N$ and $P_I$ are the signal power, the noise power and the combined power of the interfering signals, respectively, and $b$ is the capture threshold. It is typically assumed that $b$ does not depend on the SINR. Also, note that in the case when $P_I = 0$, which holds for singleton slots, the signal has to capture the slot against the noise.

frame

|  | slot 1 | slot 2 | slot 3 | slot 4 |
|---|---|---|---|---|
| packet 1 | ▓▓▓ | - - - | - - - | ▓▓▓ |
| packet 2 | ☐ | ←- - -→ | ☐ |  |
| packet 3 |  |  |  | ▨▨▨ |

**Fig. 3** Example of FSA with iterative IC: Initially, packet 2 is decoded in singleton slot 3. In the next step, the replica of packet 2 is cancelled from slot 1. Slot 1 now becomes singleton and packet 1 becomes decoded. The replica of packet 1 is cancelled from slot 4, slot 4 becomes singleton and packet 3 becomes decoded. The throughput in this example is 3/4 packet/slot, while the throughput of standard FSA would be 1/4 packet/slot

users        slots

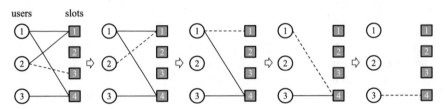

**Fig. 4** Graph representation of the example in Fig. 3: The decoding and the removal of replicas correspond to the removal of the graph edges

## 2    Coded Slotted ALOHA

A substantial improvement of FSA performance can be achieved by modifying the scheme such that (i) a user accesses the shared medium by sending replicas of its packet in multiple slots of the frame, where each replica embeds a pointer to all other replicas, and (ii) when a replica is decoded, it is removed from the slot as well as all other replicas from the corresponding slots using an interference cancellation (IC) algorithm.[1] During this modified reception procedure, some of the slots affected by the IC may provide for the decoding of new packets, propelling new iterations of the IC, etc., as depicted in Fig. 3. The ultimate result is that the overall number of successfully received packets may significantly increase in comparison to FSA without IC, implying better slot utilization, and hence, larger throughputs. The price to pay is that the scheme requires buffering of the (composite) signals received in the slots, as well as additional signal processing in order to perform the IC.

The iterative decoding of users' packets and application of IC can be represented via removal of the edges on the corresponding graph, see Fig. 4. Moreover, for the collision channel model, this graph representation is completely analogous to the graph representation of the iterative belief-propagation of erasure-correcting codes. In other words, the theory and tools of codes-on-graphs can be applied to the analysis and design of SA-based schemes that exploit iterative (i.e., successive) IC, referred

---

[1] Throughout the text it is assumed that the IC is perfect, i.e., the signal affected by the IC is perfectly "erased" from the slot.

**Fig. 5** Bipartite graph representation of the access scheme

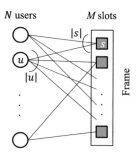

N users                M slots

$|s|$

$u$

$|u|$

Frame

to with an umbrella term Coded Slotted ALOHA (CSA). CSA schemes can asymptotically achieve throughput of 1 packet per slot on the collision channel, which is the ultimate bound for this channel model.

## 2.1 Graph Representation

The contention process of CSA can be represented via a bipartite graph, see Fig. 5, standardly used for the representation of codes-on-graphs. The nodes on the left represent users, the nodes on the right represent slots, and an edge that connects a user node with a slot node corresponds to a transmission of the user packet replica.[2] The number of edges incident to a user node $u$ is called user degree, denoted by $|u|$. Correspondingly, the number of edges incident to a slot node $s$ is called slot degree $|s|$.

In the basic version of frame-based CSA, so-called Irregular Repetition Slotted ALOHA (IRSA), each user randomly and independently selects its degree, i.e., number of replicas to be sent, from a predefined probability distribution $\Lambda_k$, $k = 1, \ldots, M$,

$$\Lambda_k = \Pr\{|u| = k\}, \tag{3}$$

where $\alpha = \sum_{k=1}^{M} k \Lambda_k$ is the average user degree. In the next step, each user $u$ randomly, independently and without repetition chooses $|u|$ out of $M$ slots from the frame, in which it sends replicas of its packet. In this way, the bipartite graph is created.

It is easy to verify the slot degrees are independent and identically distributed (i.i.d.) binomial random variables. In particular,

$$\Omega_l = \Pr\{|s| = l\} = \binom{N}{l} \left(\frac{\beta}{N}\right)^l \left(1 - \frac{\beta}{N}\right)^{N-l}, \tag{4}$$

---

[2]In the rest of text, the terms user/user node, slot/slot node, and edge/replica will be used interchangeably.

where $\beta = \frac{\alpha N}{M}$ is the average slot degree,[3] and $\frac{\beta}{N}$ is the probability that a user chooses a particular slot when transmitting a replica. Letting $M, N \to \infty$, and $M \propto N$, (5) becomes

$$\Omega_l = \frac{\beta^l}{l!} e^{-\beta}, \tag{5}$$

i.e., $|s|$ becomes a Poisson distributed random variable. For the ease of exposition, it is customary to use polynomial notation to denote the user and the slot degree distributions:

$$\Lambda(x) \stackrel{\text{def}}{=} \sum_{k=1}^{M} \Lambda_k x^k, \tag{6}$$

$$\Omega(x) \stackrel{\text{def}}{=} \sum_{l=1}^{N} \Omega_l x^l \approx e^{-\beta(1-x)}, \tag{7}$$

where (7) follows from (5). Strictly speaking, $\Lambda(x)/\Omega(x)$ are referred to as *node-oriented* user/slot degree distributions. There are also *edge-oriented* user/slot degree distributions, which correspond to probabilities that a randomly chosen edge is connected to a user/slot node of a certain degree. Specifically, denote by $\lambda_k$ the probability that an edge is connected to a user node of degree $k$ and by $\omega_l$ the probability that an edge is connected to a slot node of degree $l$. $\lambda_k$ can be computed as

$$\lambda_k = \frac{k \Lambda_k N}{N\alpha} = \frac{k \Lambda_k}{\alpha}, \ k = 1, \dots, M, \tag{8}$$

where $k\Lambda_k N$ is the number of edges incident to user nodes of degree $k$ and $N\alpha$ is the total number of edges. Similarly, $\omega_l$ can be computed as

$$\omega_l = \frac{l\Omega_l}{\beta} \approx \frac{\beta^{l-1}}{(l-1)!} e^{-\beta}, \ l = 1, \dots, N. \tag{9}$$

The edge-oriented user/slot degree distributions are compactly written as

$$\lambda(x) \stackrel{\text{def}}{=} \sum_{k=1}^{M} \lambda_k x^{k-1} = \frac{1}{\alpha} \sum_{k=1}^{M} k \Lambda_k x^{k-1} = \frac{\Lambda'(x)}{\alpha}, \tag{10}$$

$$\omega(x) \stackrel{\text{def}}{=} \sum_{l=1}^{N} N\omega_l x^{l-1} = \frac{1}{\beta} = \sum_{l=1}^{N} l\Omega_l x^{l-1} = \frac{\Omega'(x)}{\beta} \approx e^{-\beta(1-x)}, \tag{11}$$

where $F'(x)$ denotes the derivative of $F(x)$, and where (11) follows from (7).

---

[3] It is easy to verify that $\beta = \sum_{l=1}^{N} l\Omega_l$.

**Fig. 6** The tree
representation of the iterative
reception algorithm of CSA

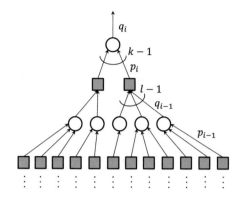

## 2.2 Asymptotic Performance Evaluation

There is an elegant way to perform the asymptotic performance evaluation of IRSA on the collision channel, derived from the so-called and-or tree evaluation. The underlying assumption is that, as $M, N \to \infty$, the bipartite graph of the scheme does not contain loops and can be unfolded into a rooted tree, see Fig. 6. The evaluation models the iterative reception algorithm and is concerned with the iterative updates of the probabilities that: (i) an edge incident to a slot node is *not* removed,[4] denoted by $p_i$, and (ii) the an edge incident to user node is *not* removed, denoted by $q_i$, where $i$ denotes the iteration of the IC. Note that a single iteration consists of two probability updates.

Pick an edge incident to a slot of degree $l$ in the $i$-the iteration of the and-or tree evaluation, and designate it as the reference edge. For the collision channel model, the probability that the reference edge is removed is

$$1 - p_i^{(l)} = (1 - q_{i-1})^{l-1}, \tag{12}$$

i.e., the edge is removed if and only if all the other $l - 1$ edges incident to the slot node have been removed via interference cancellation in previous iteration, which happens with probability $(1 - q_{i-1})^{l-1}$.[5] Averaging over the edge-oriented slot degree distribution yields

$$p_i = \sum_l \omega_l p_i^{(l)} = 1 - \sum_l \omega_l (1 - q_{i-1})^{l-1} = 1 - \omega(1 - q_{i-1}) = 1 - e^{-\beta q_{i-1}}. \tag{13}$$

Similarly, pick an edge user node of degree $k$ in the $i$-the iteration of the and-or tree evaluation, and designate it as the reference edge. The probability that the reference edge is not removed is equal to

---

[4]I.e., the corresponding packet replica is not decoded.

[5]Obviously, the recovery of the user packet in a slot can be modeled as logical 'and' operation.

$$q_i^{(k)} = p_i^{k-1}, \tag{14}$$

i.e., the edge is not removed only if none of the other edges incident to a user node has been removed in the iteration.[6] Averaging over the user edge-oriented degree distribution yields

$$q_i = \sum_k \lambda_k q_i^{(k)} = \sum_k \lambda_k p_i^{k-1} = \lambda(p_i) = \lambda(1 - e^{-\beta q_{i-1}}). \tag{15}$$

The initial value is $q_0 = 1$; i.e., initially, all edges are present in the graph. The output of the evaluation is the probability that a user packet is recovered

$$P_R = 1 - \lim_{i \to \infty} q_i. \tag{16}$$

The throughput can be calculated as

$$T = \frac{N P_R}{M} = G P_R, \tag{17}$$

where $G = \frac{N}{M}$ is the *load* of the scheme.

Evidently, from (15)–(17), the asymptotic performance crucially depends on the choice of $\lambda(x)$ and the average slot degree $\beta$, where $\beta = G\alpha$. On the other hand, both $\lambda(x)$ and the average user degree $\alpha$ depend on $\Lambda(x)$. Therefore, the performance of the scheme ultimately depends on $\Lambda(x)$ and the load $G$. The threshold of the scheme, denoted by $G^*$, is defined as the maximum value of the load for which the following condition holds

$$q > \lambda(1 - e^{-qG\alpha}), \quad \forall q \in (0, 1]. \tag{18}$$

Comparing (18) with (15) reveals that, for given $\Lambda(x)$, $G^*$ is the maximum load for which the probability of not recovering user packet will decrease over the iterations of the reception algorithm. In other words, for given $\Lambda(x)$ and when $G \leq G^*$, the probability of user recovery $P_R$ tends to 1, and, thus, $T$ becomes equal to $G$, see (17). Finally, it can be proven the upper bound on the achievable threshold, denoted by $G_{max}$ for any user degree distribution whose average degree is $\alpha$ is given by

$$G_{max} = 1 - e^{-G_{max}\alpha}. \tag{19}$$

---

[6]I.e., a packet is recovered if any of its replicas is decoded, which can be modeled via logical 'or'.

## 2.3  Frameless ALOHA

Frameless ALOHA is a variant of CSA that has two distinctive features: (i) frame-length is not a-priori defined, but determined on-the-fly such that the instantaneous throughput is maximized, and (ii) like in the original SA framework, each user independently decides whether to transmit its packet replica on the slot basis using a predefined slot access probability.

Assume that the slot access probability $p_A$ is uniform for all users and over all slots

$$p_A = \frac{\beta}{N}, \tag{20}$$

where $N$ denotes the number of users. Assume that all users are synchronized, such that they start contending from the first slot of the frame and denote the current length of the frame as $M$. The probability the a user degree is equal to $k$, $k = 0, \ldots, M$, is

$$\Lambda(k) = \binom{M}{k} p_A^k (1 - p_A)^{M-k} = \binom{M}{k} \left(\frac{\beta}{N}\right)^k \left(1 - \frac{\beta}{N}\right)^{M-k} \approx \frac{\left(\frac{\beta}{G}\right)^k}{k!} e^{-\frac{\beta}{G}}, \tag{21}$$

where $G$ here denotes the current load. Thus, the edge-oriented user degree distribution (10) becomes

$$\lambda(x) = e^{-\frac{\beta}{G}(1-x)}, \tag{22}$$

while (15) evaluates to

$$q_i = e^{-\frac{\beta}{G} e^{-\beta q_{i-1}}}. \tag{23}$$

Asymptotic performance of frameless ALOHA cannot reach the one of IRSA, due to the simplicity of the scheme. In fact, the probability of a user not transmitting at all is

$$\Lambda_0 \approx e^{-\frac{\beta}{G}}, \tag{24}$$

which also implies that a user packet can not be recovered at least with the same probability. In other words, $P_R \geq 1 - \Lambda_0$, and $\Lambda_0 \to 0$ only when $G \to \infty$. However, the key strength of frameless ALOHA is the capability to adapt to the actual evolution of the iterative reception algorithm, attempting to terminate the contention when the instantaneous throughput is maximized. The instantaneous throughput is defined as

$$T_I = \frac{N_R}{M}, \tag{25}$$

**Fig. 7** Frameless ALOHA: An example of the evolution of the instantaneous throughput $T_I$ and the instantaneous fraction of resolved users $F_I$ as the number of the slots in the frame increases $M$

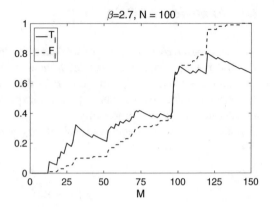

where $N_R$ is the current number of recovered users and $M$ denotes the current frame length. Similarly, the instantaneous fraction of resolved users is defined as

$$F_I = \frac{N_R}{N}. \tag{26}$$

Figure 7 shows an example how $T_I$ and $F_I$ change with $M$ for an instance of frameless ALOHA. For this instance, the optimum frame length that maximizes the throughput is when $M = 120$ and $T_I = 0.8$. An optimal contention-termination criterion has to identify when the maximum $T_I$ has been reached in every instance of the contention, where both the slot-position of the maximum and the value of the maximum vary across the instances.

## 2.4 Performance of CSA on Block Fading Channels with the Threshold-Based Capture Model

Assume that the wireless links between the users and the AP are affected with block-fading and noise, such that the composite signal received in slot $s_m$, denoted by $S_m$ is

$$S_m = \sum_{n \in \mathscr{A}_m} h_{m,n} U_n + Z_m, \quad m = 1, \ldots, M, \tag{27}$$

where $\mathscr{A}_m$ is the set of the indices of the users active in $s_m$, $h_{m,n}$ is the channel coefficient between the user $u_n$ and the AP that describes the impact of the block-fading, $U_n$ is the signal (i.e., packet replica) of user $u_n$, and $Z_m$ is the additive white Gaussian noise. The total received power in $s_m$ is

$$R_m = \sum_{n \in \mathscr{A}_m} ||h_{m,n} U_n||^2 + P_N, \tag{28}$$

where the expected noise power $P_N$ is constant for all slots. A frequent assumption is that the users are able to perform long-term power control, such that the expected powers of their transmissions at the point of reception are the same and equal to some constant $P$. Further, assuming that $h_{m,n}$ are i.i.d. random variables for every $m$ and $n_m$, (28) can be transformed to

$$R_m = \sum_{n \in \mathscr{A}_m} X_n P + P_N, \tag{29}$$

where $X_n$ is the random variable with unit power describing the impact of fading. In the threshold-based model, the capture of signal $U_j$ in slot $s_m$ occurs if

$$\frac{X_j}{\sum_{n \in \mathscr{A}_m, n \neq j}^{k} X_n + P_N/P} \geq b, \tag{30}$$

c.f. (2). Moreover, the use of IC potentially enables the iterative recovery of collided signals *within* $s_m$. Specifically, if (30) is satisfied and $U_j$ recovered, then $U_j$ is removed from $s_m$ (as well as all other slots where its replicas appear) and the condition (30) may become satisfied for some of the signals remaining in $s_m$. In other words, besides *inter-slot* IC, the fading and capture effect also enable *intra-slot* IC; note that the latter is not possible for the collision channel model. In terms of the asymptotic performance evaluation, the possibility of performing intra-slot IC impacts the probability of user recovery (12) in the following way

$$p_i^{(l)} = 1 - \sum_{t=0}^{l-1} C(t) \binom{l-1}{t} q_{i-1}^t (1 - q_{i-1})^{l-t-1}, \tag{31}$$

where $C(t)$ is the probability that the reference edge is removed in slot of degree $l$ where $l - t - 1$ edges have been removed via inter-slot IC in the previous iteration (i.e., $t + 1$ edges remain in the slot).[7] Assuming that $b \geq 1$, i.e., that the receiver is able to decode a single transmission at a time, $C(t)$ can be further decomposed into

$$C(t) = \sum_{r=0}^{t} C(t, r) \tag{32}$$

where $C(t, r)$ is the probability that the reference edge is recovered after $r$ applications of the intra-slot IC.

---

[7]Note that for the collision channel model, it is implicitly assumed that $C(0) = 1$ and $C(t) = 0$, $t = 1, \ldots, l - 1$.

Without loss of generality and the slight abuse of notation, label the edges remaining in slot $s_m$ at the start of the iteration $i$ in the following way: (i) the first $r$ are arranged by their powers in the descending order, (ii) the rest have their powers lower than the power of the edge $r$ but do not feature any particular arrangement, (iii) the reference edge is labeled by $r + 1$, and (iv) the remaining $t - r$ edges are labeled arbitrarily. Then, the probability $C(t, r)$ is equal to

$$C(t, r) = \frac{t!}{(t-r)!} \times$$

$$\times \Pr \left\{ \frac{X_1}{\sum_{j=2}^{t+1} X_j + P_N/P} \geq b; \frac{X_2}{\sum_{j=3}^{t+1} X_j + P_N/P} \geq b; \ldots; \frac{X_{r+1}}{\sum_{j=r+2}^{t+1} X_j + P_N/P} \geq b \right. \tag{33}$$

where $X_i$ is the power of the $i$th edge, and where $\frac{t!}{(t-r)!}$ is the number of realizations in which the power of the reference edge is not among $r$ largest.[8]

If the coefficients $h_{m,n}$ are i.i.d. Rayleigh distributed random variables, which is an often used model in wireless communications, the probability distribution function of $X_j$ is given by

$$p_{X_j}(x) = e^{-x}, \; x \geq 0, \; j = 1, \ldots, t + 1. \tag{34}$$

It can be shown that in this case (33) evaluates to

$$C(t, r) = \frac{t!}{(t-r)!} \frac{e^{-\frac{1}{\gamma}((1+b)^{r+1}-1)}}{(1+b)^{(r+1)\left(t+1-\frac{r+2}{2}\right)}}, \tag{35}$$

where $\gamma = \frac{P}{P_N}$ is the expected signal-to-noise ratio.

Combining (35), (33) and (31) produces

$$p_i^{(l)} = 1 - \sum_{t=0}^{l-1} \sum_{r=0}^{t} \frac{t!}{(t-r)!} \frac{e^{-\frac{1}{\gamma}((1+b)^{r+1}-1)}}{(1+b)^{(r+1)\left(t+1-\frac{r+2}{2}\right)}} \binom{l-1}{t} q_{i-1}^t (1 - q_{i-1})^{l-t-1} \tag{36}$$

$$= 1 - \sum_{t=0}^{l-1} \sum_{r=0}^{t} \frac{l-1!}{(t-r)!(l-t-1)!} \frac{e^{-\frac{1}{\gamma}((1+b)^{r+1}-1)}}{(1+b)^{(r+1)\left(t+1-\frac{r+2}{2}\right)}} q_{i-1}^t (1 - q_{i-1})^{l-t-1}, \tag{37}$$

and averaging over the edge-oriented slot degree distribution yields

_____

[8]It is assumed that the all realizations in terms of the ordering of the powers are a priori equally likely.

$$p_i = \sum_l \omega_l p_i^{(l)} = 1 - \sum_l \frac{\beta^{l-1}}{(l-1)!} p_i^{(l)} \tag{38}$$

$$= 1 - e^{-\beta} \sum_l \beta^{l-1} \sum_{t=0}^{l-1} \sum_{r=0}^{t} \frac{e^{-\frac{1}{\gamma}((1+b)^{r+1}-1)}}{(t-r)!(l-t-1)!} \frac{q_{i-1}^t (1-q_{i-1})^{l-t-1}}{(1+b)^{(r+1)(t+1-\frac{r+2}{2})}}. \tag{39}$$

As an example, for frameless ALOHA with $\beta = 7.21$, capture threshold $b = 1$, and the expected signal-to-noise-ratio $\gamma = 10$, the maximum throughput that can be achieved is 2.37 packet/slot, which is significantly above the absolute upper bound of 1 packet/slot that holds for the collision channel.

## 3 Coded Slotted ALOHA for Multi-AP Scenario

The previous part of the chapter considered the CSA scheme in the single AP scenario. However, in many wireless communication scenarios of interest, a user transmission can be detected at multiple APs. Examples include satellite communication networks, networks of WiFi APs, or dense deployment of cellular base stations in urban areas, including e.g., the cellular small cell networks. One such scenario is illustrated in Fig. 8, where a large number of sensors attempt to upload their packets to a network of APs.

In multi-AP scenarios, as introduced earlier, each AP can exploit the temporal diversity of user's packet replicas via IC across the time slots. However, the CSA scheme can also exploit the spatial diversity, where IC is done across multiple APs in a single time slot. More precisely, due to limited range of user transmissions, a packet replica of a transmitting user may appear as a singleton slot at one of the surrounding APs, while at the same time being a collision slot at other surrounding

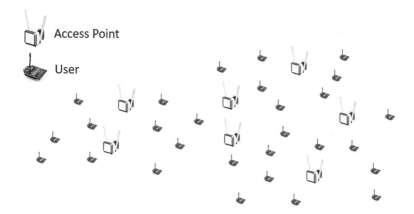

**Fig. 8** Multi-AP communication scenario

APs. In that case, an AP decoding a clean packet replica in a singleton slot can (spatially) cooperate with neighboring APs by sharing the decoded packet replica, thus allowing them to use IC to remove the replica from their collision slots.

In the rest of this chapter, we formalize more precisely the model for multi-AP scenario and provide a brief review of some of the recent results and possible directions for further study.

## 3.1 System Model for CSA in Multi-AP Scenario

We consider a scenario with $N$ contending users and $M$ APs, where both users and APs are placed independently and uniformly at random in a unit-square area. The time domain is slotted and slots are organized into frames containing $\tau$ slots per frame. The system is both frame- and slot-synchronous across all APs and all users. As in CSA, each user transmits packet replicas in one or more randomly selected slots within the frame, where the number of transmitted replicas is governed by the (temporal) degree distribution $\Lambda(x) = \sum_i \Lambda_i x^i$, and $\Lambda_i$ is the probability of sending $i$ replicas. However, unlike in the single-AP model, in the multi-AP model we assume that a user transmission is of limited range, so that the user's signal can be detected only by APs that lie in a circle of radius $r$ centered at the user. The system load is defined as the number of users per AP per slot: $G = N/(M\tau)$.

After users transmit their replicas in a given frame, the decoding process at all the APs is initiated. We consider two decoding models: (i) non-cooperative, where APs do not cooperate during the decoding process, and (ii) cooperative, where APs cooperate by exchanging decoded users' replicas during the decoding process. Note that the cooperative case requires that if two APs exchange information, i.e., if they share common users, then they have to be connected via a backhaul link. After the decoding process is finished, each user is decoded if and only if it is decoded by any of the APs in the network (thus we assume no explicit user-to-AP association). We are interested in the two main performance metrics: (i) the probability that an arbitrary user is recovered, denoted as $P_R$, and (ii) the system throughput $T$ defined as $T = G \cdot P_R$, both of which depend on the system load $G$.

For convenience, it is useful to introduce a graph representation of the multi-AP setup. To this end, we define two different but related graph models. The first model illustrates the connectivity between the users and the APs. Let $\mathcal{U} = \{U_1, U_2, \ldots, U_N\}$ represents the set of user nodes, and $\mathcal{B} = \{B_1, B_2, \ldots, B_M\}$ represent the set of AP nodes. An edge of the graph connects the user node $U_i$ and the AP node $B_j$ iff the $j$th AP is within the range $r$ of the $i$th user. The resulting graph $\mathcal{G} = (\mathcal{U} \cup \mathcal{B}, \mathcal{E}_\mathcal{G})$ is a random bipartite geometric graph. We denote the (spatial) degree distributions of user and AP nodes as $Q(x) = \sum_i Q_i x^i$ and $R(x) = \sum_i R_i x^i$, respectively, where $Q_i$ (resp. $R_i$) is the probability a user (resp. AP) node is of degree $i$.

The second graph model refines the previous spatial model by introducing the temporal dimension. In other words, it expands each AP node into $\tau$ slot nodes observed at the AP during a frame of duration $\tau$. Thus, apart from the set $\mathscr{U}$, we now have the set $\mathscr{S}$ of slot nodes $S_{j,t}$, $1 \le j \le M$, $1 \le t \le \tau$, representing the $t$th time slot at the $j$th AP. An edge of the graph connects the user node $U_i$ and the slot node $S_{j,t}$ if and only if: (i) the $i$th user transmitted a replica in the $t$th time slot, and (ii) the $j$th AP is within the range $r$ from the $i$th user. The resulting graph $\mathscr{H} = \{\mathscr{U} \cup \mathscr{S}, \mathscr{E}_{\mathscr{H}}\}$ is a random bipartite geometric graph. We denote the degree distributions of the user and the slot nodes in $\mathscr{H}$ as $U(x)$ and $S(x)$, and note that they are easily derived from $\Lambda(x)$, $Q(x)$ and $R(x)$.

Figure 9 illustrates an example of the graph representation of the multi-AP system from Fig. 8, for a simple case when the frame contains a single slot, i.e., $\tau = 1$. Note that, in this case, the two graph models $\mathscr{G}$ and $\mathscr{H}$ are equivalent. Figure 10 illustrates an example of the same multi-AP scenario for the frame length equal $\tau = 3$ slots. In this case, Fig. 10 represents the graph $\mathscr{H}$, while the underlying graph $\mathscr{G}$ remains the same as in Fig. 9.

Similarly as in the single-AP model, the graph model $\mathscr{H}$ is useful for formalization of the iterative IC decoding algorithm. We also consider the iterative IC decoding that resembles the iterative graph-peeling erasure decoder for LDPC codes. However, the

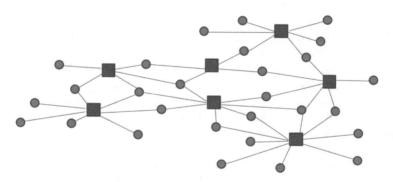

**Fig. 9** Graph representation of the multi-AP model ($\tau = 1$)

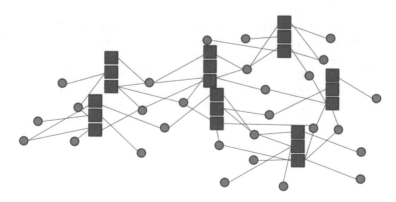

**Fig. 10** Graph representation of the multi-AP model ($\tau = 3$)

IC decoder now operates on the graph $\mathscr{H}$ of the multi-AP model. Due to the fact that this graph is based on the random bipartite geographic graph model, analyzing the decoding performance is considerably more challenging for the multi-AP scenario. Unfortunately, asymptotic analysis based on the density evolution method, which is a standard tool for analyzing the IC decoders on sparse random graphs, does not hold here. This follows from the fact that the probability of appearance of small cycles in the underlying (geometric) graph does not vanish asymptotically with the size of the graph.

In the following subsection, we discuss the performance of the IC decoding over the graph models for the multi-AP scenario. We consider four different decoding scenarios depending on the assumptions if the APs cooperate during the decoding process and if each AP performs temporal IC decoding.

## 3.2 Performance of CSA in Multi-AP Scenario

In this section, we consider the performance of the iterative IC decoder in the multi-AP scenario under the asymptotic setting. We let the number of user nodes $N \to \infty$ and the number of APs $M = M(N) \to \infty$, the number of slots in the frame $\tau \to \infty$ and the transmission range $r \to 0$, under constraints that the average number of APs within the range of a user: $Mr^2\pi \to \delta$, and the load: $N/(M\tau) \to G$, where $\delta$ and $G$ are positive constants. We ignore the edge-effects of users/APs placed close to the edges of the unit-square area as they asymptotically vanish.

In the asymptotic setup, it is easy to show that the spatial degree distribution of users $Q(x)$ is a Poisson distribution with parameter $\delta$. Thus an upper limit on $P_R$ is the probability that a user is not heard by any AP, which equals $P_R \leq 1 - \exp(-\delta)$. For different decoding scenarios, we will be interested in the (asymptotic) user recovery probability $P_R$ and the system throughput $T$. We will be also interested in threshold phenomena involving the system load $G$; namely, the maximum load $G^\star(\delta)$ for which the decoding scheme is still able to achieve the upper bound: $G^\star(\delta) = \sup\{G \geq 0 : P_R \to 1 - e^{-\delta}\}$.

**Case 1: Non-cooperative Decoding Without Temporal IC** - In this model, each AP follows standard SA protocol without IC decoding across the time slots. In addition, APs do not communicate and exchange information with each other. In this case, it is sufficient to observe the decoder operation on a multi-AP graph with a frame length $\tau = 1$ (as in Fig. 9). Note that the decoding process will be able to collect all the users that have at least one edge incident to a degree one AP. However, characterizing the number of such users is a rather challenging task.

- Recovery probability:

$$P_R \to \sum_{k=1}^{\infty} (-1)^{k-1} \frac{\delta^k}{k!} \int_1^4 e^{-a\delta G} d\mu_k(a), \tag{40}$$

where $\mu_k$ is the probability distribution of $\alpha_k$: the area covered by the union of $k$ circles randomly thrown in a circle of radius $r$, normalized by $\pi$. However, although above equation provides exact asymptotic limit, calculating $P_R$ as above is a tedious task, albeit solvable using numerical methods.

- Throughput: Let $\varepsilon = \exp(-\delta)$, then the peak normalized (per AP) throughput is lower-bounded as $T^\star \geq \frac{1}{e}\frac{1-\varepsilon}{\ln(1/\varepsilon)}$.
- Threshold load: $G^\star(\delta) = 0$. In other words, $P_R$ splits away from $1 - \exp(-\delta)$ exactly at $G = 0$.

**Case 2: Cooperative Decoding Without Temporal IC** - Performance improvement over the previous case is obtained if the APs are allowed to share decoded user's replicas. As in the previous case, it is sufficient to observe the decoder operation on a multi-AP graph with $\tau = 1$ (Fig. 9). The cooperative decoding reduces to the iterative graph-peeling erasure decoder on the underlying random bipartite geometric graph $\mathcal{G}$. Unfortunately, as noted earlier, rich asymptotic analysis tools from coding theory do not directly apply to this scenario, due to the fact that very short cycles do not asymptotically vanish.

- Recovery probability:

$$P_R \leq 1 - e^{-\delta} - (1 - e^{-\delta/4})e^{-2\delta}(1 - e^{-G\delta/4}). \tag{41}$$

The above bound follows from analyzing the probability of occurrence of the stopping sets of size two in an underlying random bipartite geometric graph.

- Threshold load: $G^\star(\delta) = 0$. As in the previous case, $P_R$ splits away from $1 - \exp(-\delta)$ at $G = 0$, however, compared to Case 1, the negative slope of decay at $G = 0$ is significantly lower, thus providing better performance of spatial cooperation over non-cooperative case.

The above two cases show that, without temporal IC across the time slots at each AP, it is difficult to exploit the full power of IC decoding due to the structure of the underlying geometric graphs.

**Case 3: Non-cooperative Decoding with Temporal IC** - In this scenario, we observe a system of $M$ APs distributed in a unit-square area that do not cooperate, but independently run the CSA across time slots of a frame. If observed independently, the performance of each AP depends only on the temporal degree distribution $\Lambda(x)$ the users apply to generate replicas. More precisely, for a given $\Lambda(x)$ and the load $H$ (number of users per slot), let $P_R^S(H)$ be the user recovery probability in the asymptotic setup (where the number of users and slots tend to infinity, but their ratio tends to $H$). Then, a threshold load $H^\star$ exists such that, for $H \leq H^\star$, it holds that $P_R^S(H) = 1$. As we present next, using the performance of CSA in the single-AP case, we can express the performance of non-cooperative decoding with temporal IC in the multi-AP case.

- Recovery probability:

$$P_R \leq (1 - e^{-\delta})P_R^S(H = 8e\delta G). \tag{42}$$

- Throughput: Let $\varepsilon = \exp(-\delta)$, then the peak normalized (per AP) throughput is lower-bounded as $T^\star \geq \frac{H^\star}{8e} \frac{1-\varepsilon}{\ln(1/\varepsilon)}$.
- Threshold load: $G^\star(\delta) \geq \frac{1}{8e} \frac{H^\star}{\delta}$. In other words, the recovery probability stays at the maximal possible value $1 - \exp(-\delta)$ at least in the range $G \in [0, \frac{1}{8e} \frac{H^\star}{\delta}]$.

**Case 4: Cooperative Decoding with Temporal IC** - Performance-wise, this case is the most powerful decoding scenario, where the IC decoder operates across the complete spatio-temporal graph $\mathscr{H}$ (such as the one in Fig. 10). However, compared to the Case 3, no stronger results exist for this case (note that the bounds for Case 3 also hold for Case 4). In the following, we differentiate the performance of Case 4 from the previous cases using simulations.

**Numerical results**: In the simulation setup, we set the number of base stations $M = 40$, and the number of slots in the frame to $\tau = 40$. We simulate the recovery probability $P_R$ versus $G = N/(M\tau)$ by varying $N$. We perform Monte Carlo simulations where for each $N$, we generate 30 random placements of users and APs. For the cases with temporal IC, we apply the temporal degree distribution $\Lambda(x) = x^2$ (i.e., each user generates two replicas per frame).

Figure 11 plots the normalized throughput $T(G)$ versus normalized load $G$ for the four decoding cases, assuming $\delta = 6$. We can see that the Case 4 (spatio-temporal cooperation) achieves much higher peak normalized throughput than the remaining three schemes.

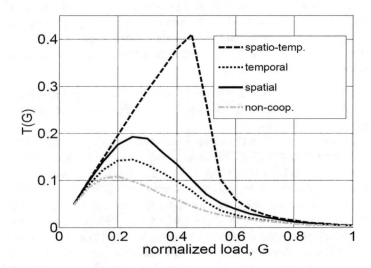

**Fig. 11** Performance comparison of multi-AP decoding schemes

## 3.3 Discussion and Open Problems

In the case of CSA in Multi-AP scenario, a number of open problems arise that yet have to be solved. Note that even the simplest extension to Multi-AP scenario, the case of SA without cooperation and without temporal IC (Case 1) lacks an elegant and closed-form solution for the average system throughput. In the following, we provide several scenarios which we find relevant and which are subject of our current study.

**Different Multi-AP Models**; The Multi-AP model presented in this chapter is mainly motivated by applications in dense cellular systems where APs correspond to small base stations. In this case, the geometric graph model is a reasonable approximation of a user to AP coverage and connectivity. However, in certain scenarios such as satellite networks, all users might be heard at all satellites, although with different channel conditions (e.g., some user-to-AP channels being exposed to harsh fading while others not). This has led to slightly different and analytically more tractable model, wherein underlying graph representation follows random graph model (unlike random geometric graph model).

**Asynchronous System Operation**: In reality, signals transmitted from users will travel different distances to different neighbouring APs, thus arriving at APs with different delay. This delay variability may vary from rather small in dense small cell networks, to very large in multi-AP satellite systems. Thus the assumption that the system is perfectly slot synchronized across all users and all APs might be impossible to achieve in practice. In more realistic model, we might explore the scenario in which users are asynchronous and send their packet replicas in arbitrary time instants which are not aligned to slot boundaries. Such a system may still employ IC decoding and may still provide significant performance improvements over traditional ALOHA. Thus the performance of asynchronous ALOHA in Multi-AP scenario is an interesting and relevant research direction.

**Multi-AP Models with Directional Antennas**: The latest trends in wireless communications advocate shifting the operational communications band to higher frequencies above 6 GHz into the so called mmWave band. In such an environment, to overcome high signal attenuation, typical transmission strategies involve directional antennas in which the transmitted signal is emitted as a focused and narrow-angle beam in a certain direction. An interesting extension to CSA in Multi-AP environment would be to observe how the system throughput changes with the transition towards mmWave communications and narrow-beam antennas. Our preliminary results show that this should help improving the system throughput in ultra-dense deployments since users transmitting in narrow beams could better avoid interfering each other, which in terms of the underlying graph representation means that the occurrence of short cycles could be significantly reduced, leading to better performance of the iterative IC decoding.

# 4 Literature Review

The performance of slotted ALOHA was analyzed in [13]. Framed slotted ALOHA was proposed in [9]. The threshold-based capture model was topic of multitude works, c.f. [8, 17–19].

There is an abundance of literature related to CSA. In the following, some of notable works are briefly mentioned. An accessible overview of CSA is presented in [12]. FSA with iterative IC was originally proposed in [1], assuming that each user sends two replica in randomly selected slots of the frame and showing that the maximum expected throughput on the collision channel is 0.55 packet/slot [1]. IRSA was introduced in [4], where it was also shown how to design user degree distributions whose thresholds are close to 1. The upper bound on the threshold of any user degree distribution with a given average degree was derived in [10]. A generalization of the scheme, in which users contend with segments produced by applying a linear segment-oriented code on the segments of the user packet, is introduced and analyzed in [11].

Frameless ALOHA was inspired by the rateless coding paradigm [5]; it was introduced and its asymptotic performance analyzed in [15]. The optimization of the termination criterion that maximizes expected throughput of the scheme in scenarios with finite number of users was addressed in [14].

Derivation of the and-or tree evaluation for the threshold-based capture model was performed in [16]; the same work also addresses the performance of frameless ALOHA for Rayleigh black-fading channel with the threshold-based capture effect. Design of the user degree distributions for IRSA on Rayleigh black-fading channel with threshold-based capture was assessed in [2].

The and-or tree evaluation was originally conceived in [6].

CSA in Multi-AP scenario is introduced and analyzed in [3]. The exposition in this chapter follow this approach. Somewhat different model and overview of results is also presented in [7].

# References

1. E. Casini, R.D. Gaudenzi, O. del Rio Herrero, Contention resolution diversity slotted ALOHA (CRDSA): an enhanced random access scheme for satellite access packet networks. IEEE Trans. Wirel. Commun. **6**(4), 1408–1419 (2007)
2. F. Clazzer, E. Paolini, I. Membeli, C. Stefanovic, Irregular repetition slotted ALOHA over the Rayleigh block fading channel with capture, in *Proceedings of IEEE ICC 2017. Paris, France* (2017)
3. D. Jakovetic, D. Bajovic, D. Vukobratovic, V. Crnojevic, Cooperative slotted ALOHA for multi-base station systems. IEEE Trans. Commun. **63**(4), 1443–1456 (2015)
4. G. Liva, Graph-based analysis and optimization of contention resolution diversity slotted ALOHA. IEEE Trans. Commun. **59**(2), 477–487 (2011)
5. M. Luby, LT codes, in *Proceedings of 43rd IEEE FOCS. Vancouver, BC, Canada* (2002)
6. M.G. Luby, M. Mitzenmacher, A. Shokrollahi, Analysis of random processes via And-Or tree evaluation, in *Proceedings of 9th ACM-SIAM SODA. San Francisco, CA, USA* (1998)

7. A. Munari, F. Clazzer, G. Liva, Multi receiver ALOHA: a survey and new results, in *Proceedings of IEEE ICC 2015 - MASSAP Workshop. London, UK* (2015)
8. G.D. Nguyen, A. Ephremides, J.E. Wieselthier, On capture in random-access systems, in *Proceedings of IEEE ISIT 2006. Seattle, WA, USA* (2006)
9. H. Okada, Y. Igarashi, Y. Nakanishi, Analysis and application of framed ALOHA channel in satellite packet switching networks - FADRA method. Electron. Commun. Jpn. **60**, 60–72 (1977)
10. E. Paolini, G. Liva, M. Chiani, Graph-based random access for the collision channel without feed-back: capacity bound, in *Proceedings of IEEE Globecom 2011. Houston, TX, USA* (2011)
11. E. Paolini, G. Liva, M. Chiani, Coded slotted ALOHA: a graph-based method for uncoordinated multiple access. IEEE Trans. Inf. Theory **61**(12), 6815–6832 (2015)
12. E. Paolini, C. Stefanovic, G. Liva, P. Popovski, Coded random access: how coding theory helps to build random access protocols. IEEE Commun. Mag. **53**(6), 144–150 (2015)
13. L.G. Roberts, ALOHA packet system with and without slots and capture. SIGCOMM Comput. Commun. Rev. **5**(2), 28–42 (1975)
14. C. Stefanovic, P. Popovski, ALOHA random access that operates as a rateless code. IEEE Trans. Commun. **61**(11), 4653–4662 (2013)
15. C. Stefanovic, P. Popovski, D. Vukobratovic, Frameless ALOHA protocol for wireless networks. IEEE Commun. Lett. **16**(12), 2087–2090 (2012)
16. C. Stefanovic, M. Momoda, P. Popovski, Exploiting capture effect in frameless ALOHA for massive wireless random access, in *Proceedings of IEEE WCNC 2014. Istanbul, Turkey* (2014)
17. A. Zanella, M. Zorzi, Theoretical analysis of the capture probability in wireless systems with multiple packet reception capabilities. IEEE Trans. Commun. **60**(4), 1058–1071 (2012)
18. M. Zorzi, Capture probabilities in random-access mobile communications in the presence of Rician fading. IEEE Trans. Veh. Technol. **46**(1), 96–101 (1997)
19. M. Zorzi, R.R. Rao, Capture and retransmission control in mobile radio. IEEE J. Sel. Areas Commun. **2**(4), 1289–1298 (1994)

# Part IV
# Codes for Distributed Storage Systems

# An Overview of Coding for Distributed Storage Systems

Shiqiu Liu and Frédérique Oggier

**Abstract** This chapter provides a short survey of coding for distributed storage systems. It describes the code design criteria for such codes, emphasizing what makes them different from traditional codes for communication. It then focuses on two large families of codes, regenerating codes and locally repairable codes, including a discussion on how these codes are used in an adversarial setting.

## 1 Distributed Storage Systems and Erasure Codes

A distributed storage system consists of a set of hard drives (disks), or nodes, which is used to store data in a distributed manner: the same file could be stored multiple times, one copy per hard drive, over a set of two, three, or more hard drives, or pieces of the same file could be stored across a set of hard drives. Reasons why one may want to store data in a distributed manner (rather than on a single disk) include

- ease of scale: if the system runs out of memory, one just adds more disks to the storage system,
- reliability: if there is a single disk, and it fails, the stored data is lost, while if there are several disks containing the same data, the data will have chances of surviving, depending on the failure(s), and on how the data is stored.

If there are several disks, but each file is stored only on a single disk, then reliability cannot be achieved, because in the event of this specific disk failure, data will be lost. This is why *redundancy* is needed. The simplest form of redundancy is *replication*. Two or three copies of a file are made, and they are all stored in different disks. The

S. Liu (✉)
Institute of Network Coding-The Chinese University of Hong Kong,
Hong Kong, China
e-mail: sqliu@inc.cuhk.edu.hk

F. Oggier
Division of Mathematical Sciences, School of Physical and Mathematical
Sciences, Nanyang Technological University, Singapore, Singapore
e-mail: frederique@ntu.edu.sg

© Springer International Publishing AG 2018
M. Greferath et al. (eds.), *Network Coding and Subspace Designs*,
Signals and Communication Technology,
https://doi.org/10.1007/978-3-319-70293-3_14

notion of redundancy has been well studied in the context of noisy communication, via coding theory: a signal is transmitted over a noisy channel, because of the noise, part of the signal will be lost, so redundancy is added to the transmitted signal, to help the receiver recover the transmitted signal, despite it being distorted through the communication channel. The process of adding redundancy to the transmitted signal is called "encoding".

We will assume throughout this chapter that data can be modeled as a vector with coefficients in some finite field $\mathbb{F}_q$, for $q$ a prime power. Encoding thus consists of mapping a vector $\mathbf{f} \in \mathbb{F}_q^M$ to a vector $\mathbf{x} \in \mathbb{F}_q^n$, called *codeword*, where $n > M$. The set of codewords is called a *code*. If this mapping is linear, we can write $\mathbf{x} = \mathbf{f}G$ for some $M \times n$ matrix $G$ of rank $M$, and we speak of a *linear code*, with parameters $[n, M]$, where $n$ is the *length* and $M$ the *dimension*. The simplest example of linear code is the repetition code with parameters $[n, 1]$ which maps $\mathbf{f}$ to a vector containing $n$ copies of $\mathbf{f}$: $(\mathbf{f}, \ldots, \mathbf{f})$. The repetition code corresponds to a replication scheme for storage: there are $n$ copies of $\mathbf{f}$, each of them are stored on a different disk, over $n$ disks. This scheme is very reliable: there are $n$ copies of $\mathbf{f}$, therefore $n-1$ disk failures can happen, and still the data will survive. The same is true for communication over a noisy channel: up to $n-1$ transmitted signals can be lost, the intended message will still be communicated reliably. The problem with this scheme is that its rate $M/n$ is poor, respectively its *storage overhead* $n/M$ is high.

The trade-off between reliability and rate is well known in coding theory, and is characterized by the Singleton bound. Reliability is captured by the minimum Hamming distance $d_H$ of a code $C$:

$$d_H(C) = \min_{\mathbf{x} \neq \mathbf{x}'} d(\mathbf{x}, \mathbf{x}')$$

where $d(\mathbf{x}, \mathbf{x}')$ counts in how many coefficients the vectors $\mathbf{x}$ and $\mathbf{x}'$ differ. The minimum Hamming distance tells us that given a codeword in $C$, we can lose up to $d_H(C)$-1 coefficients, and still be able to recover our data. This is because any $\mathbf{x}$ differs in at least $d_H(C)$ coefficients from any other $\mathbf{x}'$. Now the Singleton bound tells us, given $n$ and $M$, what is the maximum possible $d_H$.

**Proposition 1** *Given a code $C$ of size $|C| = q^M$, we have $d_H(C) \leq n - M + 1$.*

*Proof* Take the $q^M$ codewords of $C$, and look at their first $M - 1$ coefficients. Since we can have at most $q^{M-1}$ distinct vectors of length $M - 1$, among the vectors obtained by looking at the first $M - 1$ coefficients of $q^M$ vectors, there must be at least two vectors which are the same by the pigeon hole principle, corresponding to two codewords $\mathbf{x} \neq \mathbf{x}'$ which are equal on their first $M - 1$ coefficients, implying that they can different at most on $n - (M - 1)$ coefficients, which completes the proof.

Codes reaching the Singleton bound are called *maximum distance separable (MDS) codes*. The most famous class of MDS codes is the class of Reed–Solomon codes [33]. A Reed–Solomon code $C$ can be described as

$$C = \{(p(w_1), p(w_2), \ldots, p(w_n)) : p(X) = \sum_{i=0}^{M-1} p_i X^i \in \mathbb{F}_q[X]\} \qquad (1)$$

where $w_1, \ldots, w_n$ are $n$ distinct elements of $\mathbb{F}_q$ (and thus $n \leq q$). From the description, a codeword is made of $n$ evaluations of a polynomial of degree $M - 1$. Polynomial interpolation tells us that we can recover the polynomial as long as we still have the evaluation of $p(X)$ in at least $M$ points, which means that we can lose at most $n - M$ of them, corresponding to a Hamming distance of $n - M + 1$.

Reed–Solomon codes have been used for communications, and for storage applications such as CDs [12]. In the context of distributed storage systems, they are also optimal when it comes to trade reliability and storage overhead. Why then not just adopt Reed–Solomon codes as such[1] instead of researching novel coding strategies for distributed storage systems?

The reason is that coding for distributed storage includes one more dimension, on top of reliability and storage overhead, namely that of *maintenance* or *repairability*. In a communication scenario, a signal is transmitted, and received at a given point of time. Once the signal is received, it must be recovered taking into account lost information due to channel noise, but this degradation is now given, it will not change. This is a main difference with a distributed storage scenario. Suppose a disk has failed at a given point of time, nothing prevents another disk to fail later on, in fact, the failure of one disk may trigger other disk failures. Therefore, without a maintenance mechanism, the data will eventually be lost, because no coding solution can protect against an arbitrarily large amount of failures. What maintenance does is illustrated on an example with two copies of a file **f**. If the first copy is lost because of a disk failure, a disk or node which is still alive make a new copy of **f**, so that the storage system still has two copies overall, despite one failure. In fact, keeping three copies of **f** is somewhat even more prudent, because once the first failure occurred, there is no more protection. Having three copies allows to lose one, start recreating a new copy, and tolerate one more failure in the process of creating a new copy (which may take time). Suppose now a Reed–Solomon codeword is used to encode **f**, each codeword coefficient is stored on a distinct disk, across $n$ different disks. In the event of one failure, recovering the one missing coefficient is asking for a lot of resources: data needs to be transferred from $M$ nodes, the codeword needs to be recomputed, the missing coefficient is then recovered, but at a high cost in terms of both communication and computation costs. If the system were to wait for several failures to happen before starting this repair procedure, the cost would be amortized, but at the risk of losing the data if the system waits for too long.

Research on coding for distributed storage has thus focused on getting codes with

- high reliability (fault tolerance),

---

[1]We are not claiming that Reed–Solomon codes are not used in distributed storage systems, we added "as such" as a way to say that something else has to be taken into account in all cases, even if one wants to use Reed–Solomon codes.

- low storage overhead (if storage overhead is not an issue, replication is a great practical solution),
- good maintenance mechanisms.

What good maintenance means is still under discussion, but in this chapter, we will focus on two aspects, low communication costs in Sect. 2, and a low number of live nodes to be contacted during repair in Sect. 3. Other aspects are also discussed in [23].

## 2   Regenerating Codes

When a node failure happens in a distributed storage system, a maintenance mechanism tries to replace the lost data using data available in the system. This involves downloading data from live nodes, which incurs communication costs, referred to as *repair bandwidth*.

In [4], Dimakis et al. introduced a min-cut max-flow technique from network coding to the setting of distributed storage to optimize the repair bandwidth. The result of their analysis is a trade-off bound between storage and repair bandwidth, and codes achieving this bound are called *regenerating codes*. In [4], the scenario of only one single failure repair at a time is treated. The main idea of regenerating codes is that they could satisfy an MDS-like property (the file can be retrieved by contacting any choice of $k$ nodes, where $k$ is a fixed threshold, not necessarily the code dimension), and in the repair process, many nodes could be contacted (typically all but the failed one) and less symbols downloaded, in turn reducing the cost of repairing one single failure.

### 2.1   Parameters of Regenerating Codes and Min-Cut Bound

Considering a network of $N$ nodes, a file $\mathbf{f}$ of length $M$ stored in $n$ of these $N$ nodes. Each node is assumed to have the same *storage capacity* $\alpha$, that is $\alpha$ symbols are stored in each of the nodes. When a failure occurs, a live node (called newcomer) among these $N$ nodes, which does not yet store data from $\mathbf{f}$, is joining the repair process. The newcomer contacts $d$ live nodes and downloads $\beta$ symbols from each of them, to repair the data lost by the failed node, as shown on Fig. 1. The newcomer thus downloads $\gamma = d\beta$ amount of data, which constitutes the total repair bandwidth. To analyze the trade-off between storage capacity $\alpha$ and repair bandwidth $\beta$, a min-cut bound is computed [4] in an information flow graph, where the data flows from a source $S$, into the storage system, through different nodes when repair happens, to a data collector $DC$ which wants to retrieve the data back.

**Theorem 1** ([4]) *Consider a directed information flow graph, with a source $S$ and a sink as data collector. Every node (apart the source and the sink) has a storage*

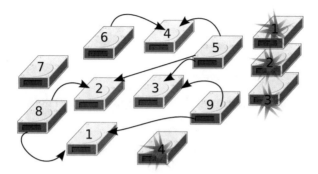

**Fig. 1** Four nodes, labelled from 1 to 4 failed. Four other nodes in the center with the same labels are repairing the lost data by contacting live nodes 5, 6, 8, 9. We have $d = 2$, e.g., node 3 downloads data from nodes 5 and 9. In the original regenerating code case, the four failures will be repaired sequentially. In the collaborative case discussed next, nodes 1, 2, 3, 4 involved in the repair are exchanging data as well

*capacity of $\alpha$. When a failure occurs, a newcomer connects to $d \geq k$ live nodes and obtains $\beta$ symbols from each of them. Suppose that a data collector DC connects to $k$ nodes which were all involved in different phases of repair. Then a min-cut bound between the source $S$ and the data collector DC is given by*

$$mincut(S, DC) \geq \sum_{i=0}^{k-1} \min\{(d-i)\beta, \alpha\}. \tag{2}$$

*Proof* Let $x_i$ denote the newcomer during the $i$th repair, $i \geq 0$. Write $x_i$ as a logical pair $(x_i^{in}, x_i^{out})$ formed by an incoming node, so that an edge between $x_i^{in}$ and $x_i^{out}$ models the storage capacity $\alpha$. Consider a data collector that connects to $k$ output nodes, say $\{x_i^{out} : i = 0, 1, \ldots, k-1\}$. We show that any cut between $S$ and $DC$ in the graph has a capacity that satisfies (2). Since we may assume that outgoing edges of $S$ and incoming edges of DC have infinite capacity, we only need to consider cuts $(U, \overline{U})$, $S \in U$ and $DC \in \overline{U}$, where these infinite capacity edges do not participate. Let $C$ denote the edges in the cut.

Let $(x_0^{in}, x_0^{out})$ be the first repair node. We have two cases: (1) If $x_0^{in} \in U$, then the edge with weight $\alpha$ between $x_0^{in}$ and $x_0^{out}$ must be in the cut $C$. (2) If $x_0^{in} \in \overline{U}$, then the cut contains $d$ edges carrying $\beta$ coefficients. Thus the first node brings a contribution of $c_0 \geq \min\{d\beta, \alpha\}$ to the cut.

Let us consider the second repair node $(x_1^{in}, x_1^{out})$. There are again two cases: (1) $x_1^{in} \in U$, then the edge with weight $\alpha$ is again in the cut. (2) If $x_1^{in} \in \overline{U}$, then at most one of its incoming edges could be from $x_0^{out}$, thus at least $d-1$ edges carrying $\beta$ coefficients are in the cut. The min-cut contribution from the second node is $c_1 \geq \min\{(d-1)\beta, \alpha\}$.

In general, for the $i$th repair node $(x_{i-1}^{in}, x_{i-1}^{out})$, case 1) involves cutting the storage capacity link, while for the second case, there are at most $i - 1$ incoming edges from nodes in $\overline{U}$, thus at least $d - i + 1$ edges carrying $\beta$ coefficients, yielding $c_{i-1} \geq \min\{(d - i + 1)\beta, \alpha\}$.

Summing these contributions leads to (2).

Using the Minimum cut-Maximum flow Theorem, we get that the initial file size $M$ must satisfy

$$M \leq \sum_{i=0}^{k-1} \min\{(d - i)\beta, \alpha\}.$$

Thus the largest that $M$ can get is $M = \sum_{i=0}^{k-1} \min\{(d - i)\beta, \alpha\}$, and codes reaching this bound are called *regenerating codes*. We may now ask to minimize $\alpha = \alpha(d, \gamma)$ subject to the aforementioned constraint on the size of $M$, which can equivalently be stated as a function of $\gamma$ instead of $\beta$: $M = \sum_{i=0}^{k-1} \min\{(1-i/d)\gamma, \alpha\}$. This optimization problem can be easily solved in closed form. This is because $(1-(k-1)/d)\gamma < (1-(k-2)/d)\gamma < \cdots < (1-1/d)\gamma < \gamma$. Since $\alpha$ must belong to any of these intervals, the constraints can be restated removing all the minima, and the minimum value of $\alpha$ is found. Suppose for example that $\alpha \leq (1 - (k - 1)/d)\gamma$, then the constraint becomes $k\alpha = M$, for which $\gamma \geq \alpha\frac{d}{d-(k-1)} = \frac{M}{k}\frac{d}{d-(k-1)}$. Similarly, if $\alpha \geq \gamma$, we get

$$M = \sum_{i=0}^{k-1}(1 - i/d)\gamma = \left(k - \frac{1}{d}\frac{(k - 1)k}{2}\right)\gamma = k\left(\frac{2d - (k - 1)}{2d}\right)\gamma,$$

showing that $\gamma = \frac{M}{k}\frac{2d}{2d-(k-1)}$. Then $\alpha \geq \gamma$ gives $\alpha = \gamma$ for minimum value of $\alpha$. The other regimes in between are obtained similarly, which show that the optimal is a piece wise linear function, which forms a trade-off between $\alpha$ and $\gamma$, as illustrated on Fig. 2 for $k = 4$ and $d = 5$.

The smallest value of $\alpha$, corresponding to

$$(\alpha, \gamma) = \left(\frac{M}{k}, \frac{M}{k}\frac{d}{d - (k - 1)}\right), \tag{3}$$

is indeed characterizing the smallest possible value for $\alpha$, since we have the constraint that any $k$ nodes should be able to retrieve the file $\mathbf{f}$, thus no node can store less than $M/k$ symbols. This point of the trade-off is called the *minimum storage regenerating point (MSR)*. Then the other extreme point

$$(\alpha, \gamma) = \left(\frac{M}{k}\frac{2d}{2d - (k - 1)}, \frac{M}{k}\frac{2d}{2d - (k - 1)}\right), \tag{4}$$

says that $\alpha = \gamma$, which minimizes the repair bandwidth in absolute, irrespectively of $\alpha$. This point is thus called the *minimum repair bandwidth regenerating point (MBR)*.

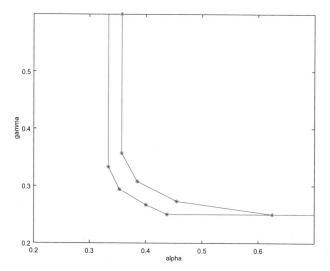

**Fig. 2** Two trade-off curves, with $\alpha$ on the $x$-axis, and $\gamma = \gamma(\alpha)$ on the $y$-axis. The outer line corresponds to $t = 1$ failure repair. The inner one corresponds to $t = 3$ simultaneous repairs using a collaborative strategy. The parameters are set to be $k = 4, d = 5$ and we normalized the file size. We observe that $t = 3$ yields a better trade-off than $t = 1$

It is also clear from this formula that $d = n - 1$ brings most benefits in terms of bandwidth savings.

No matter how many failures occurred in the storage system, the strategy that we just analyzed holds for one failure repair at a time, so multiple failures are handled by performing multiple repairs sequentially.

To repair more than one failure simultaneously, two independent works [14, 38] introduced *collaborative (coordinated, cooperative) regenerating codes*, which allow repairs using the data not only from live nodes, but also from all nodes currently being repaired. This strategy introduces a collaborative phase, on top of the download phase done previously. The analysis generalized the one presented above: a min-cut bound is computed, taking into account the collaboration, and a trade-off for collaborative regenerating codes is obtained, between storage and repair bandwidth, where repair bandwidth takes into account that data is exchanged during collaboration.

**Theorem 2** ([14, 38]) *Consider an information flow graph, where every node has a storage capacity of $\alpha$, and repairs are performed by a group of t nodes, that is the system triggers a repair when t failures occur. Suppose that a data collector DC connects to a subset of k nodes which were all involved in different phases of repairs, where each phase involves a group of $u_i$ nodes, $1 \le u_i \le t$, and $k = \sum_{i=0}^{g-1} u_i$. Then a min-cut bound for collaborative repair is*

$$M \leq \sum_{i=0}^{g-1} u_i \min \left\{ \alpha, \left( d - \sum_{j=0}^{i-1} \right) \beta + (t - u_i)\beta' \right\}, \tag{5}$$

where $\beta'$ is the exchange repair bandwidth, that is each of the repairing nodes exchanges $\beta'$ symbols with every other $t - 1$ repair nodes.

Solving the problem of minimizing $\alpha$ under the constraints given by the above bound now becomes difficult, the interested reader may refer to [40] for the computations. An example of the trade-off curve obtained for $t = 3$ is shown in Fig. 2. It turns out that increasing $t$ actually improves the trade-off between storage and repair bandwidth.

Both scenarios are included as particular cases of *partially collaborative regenerating codes*, proposed in [19]. The idea is to introduce freedom in the extent of collaboration, by allowing repair nodes to exchange data with only a subset of the nodes currently being repaired. A min-cut bound is available, however determining the corresponding trade-off curve is open.

**Theorem 3** ([19]) *For partial collaboration, the repair process of t failures involves t nodes, which will all download $\beta$ amount of data from d live nodes, and exchange $\beta'$ amount of data with a subset of $t - s$ nodes, $1 \leq s \leq t$. Then the min-cut bound is*

$$M \leq \min_{\mathbf{u} \in P} \left( \sum_{i \in I} u_i \min \left\{ \alpha, \left( d - \sum_{j=0}^{i-1} u_j \right) \beta + (t - s + 1 - u_i)\beta' \right\} \right.$$
$$\left. + \sum_{i \in \bar{I}} u_i \min \left\{ \alpha, \left( d - \sum_{j=0}^{i-1} u_j \right) \beta \right\} \right) \tag{6}$$

where $I = \{i, \ t - s + 1 - u_i \geq 0\}$, $\bar{I} = \{i, \ t - s + 1 - u_i < 0\}$ and $P = \{\mathbf{u} = (u_0, \ldots, u_{g-1}), \ 1 \leq u_i \leq t$ and $\sum_{i=0}^{g-1} u_i = k\}$.

When $s = t$, there is no collaboration, and (6) simplifies to (2). When $s = 1$, the collaboration phase involves all the other $t - s = t - 1$ nodes, and we find (5) again.

## 2.2 Functional Versus Exact Repair

The above bounds and trade-offs all address the preservation of data through different repair phases. Suppose a node contains the piece $\mathbf{f}_i^2$ of a file $\mathbf{f}$, and after repair, $a\mathbf{f}_i$ is being stored, for $a \neq 0$. For example, using the alphabet $\mathbb{F}_3 = \{0, 1, 2\}$, a node storing the coefficient 1 fails, is repaired, but the value after repair is now 2, not 1.

---

[2]The same example can be done on a linear combination of $\mathbf{f}_i$.

In terms of recovering $\mathbf{f}_i$, both $\mathbf{f}_i$ and $a\mathbf{f}_i$ play the same role (since $a \neq 0$, $a^{-1}$ exists and computing $a^{-1}a\mathbf{f}_i$ gives back $\mathbf{f}_i$, in the example, $2 \cdot 2$ gives back 1). However, they are not bitwise the same data. Therefore, we distinguish *exact repair*, where lost data is exactly recovered bit by bit, from *functional repair*, where the data itself is not the same, but the information contained in the data is maintained. Exact repair is often preferred, since it is easier to keep track of the data with respect to functional repair, where transformation of the data should also be available. The above bounds however hold for functional repair.

The question thus becomes that of achievability of these trade-offs for exact repair. It is known that the extreme points (MSR and MBR) are achievable, and we will give some example for these two cases in the next subsection. How to characterize the interior points for exact repair remains open in general. We summarize some of the results known so far. It was first shown in [46], using a particular code with parameters $n = 4, k = 3, d = 3$, that the exact repair trade-off lies strictly away from the functional repair trade-off. In [35], it was more generally shown that the exact repair trade-off lies strictly above the functional trade-off for any values $n, k, d$.

From these results, it became clear that one would have to address the question of finding trade-off bounds for exact repair. However, the exact repair trade-off for a fixed $(n, k, d)$ varies with the file size $M$, making it more natural to consider instead the normalized pairs $(\alpha/M, \beta/M)$.[3] In this case, we refer to the *normalized exact repair (ER) trade-off*. In [34, Theorem III.4], an upper bound on the normalized ER trade-off is computed for every choice of parameters $n, k, d$. This upper bound can be combined with the code construction proposed in [36] for $k = 3$ and $d = n - 1$, for any $n$, to fully characterize the normalized ER tradeoff for this set of parameters.

In the introduction, we discussed linear codes, but so far, nothing was mentioned about whether codes are linear or not. This is because the properties of information preservation, or that of data recovery, are independent of the regenerating code being linear. However, linear codes are of course nicer than non-linear ones, because they have been better studied, they come with easier encoding and decoding algorithms. So one may wonder whether adding the constraint of linearity might help in understanding the normalized ER trade-off. In [5], a non-explicit upper bound is given for the general ER trade-off case, which, when applied to the special case of linear codes, gives an explicit upper bound for the ER linear trade-off problem, for any parameter $n, k, d$. In [21], another upper bound is given for the ER linear codes case, which turns out to be optimal for the parameters $k = n - 1, d = n - 1$ for any $n$, but for the limited region where $\beta/M \leq 2\alpha/Mk$. For an improvement of this bound, as well as comparisons between the different bounds discussed, we refer to [34].

---

[3]Since $\gamma = d\beta$, we may consider the trade-off in terms of either $(\alpha, \gamma)$ or $(\alpha, \beta)$.

## 2.3  MSR and MBR Code Constructions

We next present two code constructions, one at MSR point, and one at MBR point.
**Construction of MSR codes**. This construction is actually illustrating how an MDS
code can be used for storage, in the setting of regenerating codes. Consider a file **f**
of length $M = k$ with coefficients in $\mathbb{F}_q$, $q$ a prime power. Set the parameters of the
regenerating code to be

$$d = k, \ n \geq k + 1, \ M = k.$$

Let $G$ be a generator matrix of an $(n, k)$ MDS code over $\mathbb{F}_q$, say that of a Reed–
Solomon code, and denote by $\mathbf{g}_i$ the $i$th column of $G$, $i = 1, \ldots, n$, that is $G =
[\mathbf{g}_1, \ldots, \mathbf{g}_n]$. By the MDS property, any $k$ columns of $G$ form an invertible matrix.
We denote the data file **f** by a $k$-dimensional row vector, $\mathbf{f} = (\mathbf{f}_1, \ldots, \mathbf{f}_k)$. Suppose
the $i$th node store $\mathbf{fg}_i$, that is $\alpha = 1$.
   **Recovery**. The data can be recovered by contacting nodes $i_1, \ldots, i_k$. The $i_l$-th node
contains $\mathbf{fg}_{i_l}$, for $l = 1, \ldots, k$, for a total of $k$ coefficients, by the MDS property, the
data file can be retrieved.
   **Repair phase**. Suppose that the $i$th node has failed. A newcomer connects to
$d = k$ live nodes say $i_1, \ldots, i_d$ and downloads $\mathbf{fg}_{i_l}$ for $l = 1, \ldots, d$ from each of the
nodes, thus $\beta = 1$. By the MDS property, the newcomer can compute **f**.
   The total number of coefficients involved in the repair is $d\beta = k$, that is the repair
bandwidth is $k$, and the construction indeed achieves the MSR point, since

$$(\alpha_{MSR}, \ \gamma_{MSR}) = \left( \frac{M}{k}, \ \frac{M}{k} \frac{d}{d - k + 1} \right) = (1, k).$$

**Construction of MBR codes** [39]. This time, consider a data file **f** of length $M =
k(k + 1)$ with coefficients in $\mathbb{F}_q$. Set the parameters of the regenerating code to be

$$d = k, \ n = k + 1, \ M = k(k + 1).$$

The object **f** can be written as $\mathbf{f} = (\mathbf{f}_1, \ldots, \mathbf{f}_{k+1})$ where $\mathbf{f}_i = (f_{(i-1)k}, \ldots, f_{ik-1})$ has
length $k$, $1 \leq i \leq n = k + 1$.
   Let $\mathbf{g}_1, \ldots, \mathbf{g}_{n-1}$ be $n - 1 = k$ independent column vectors of length $k$ over $\mathbb{F}_q$.
   For $i = 1, 2, \ldots, n$, the content of node $i$ consists of the $k$ coefficients of $\mathbf{f}_i$, and
of the $n - 1$ parity-check coefficients

$$\mathbf{f}_{i+1}\mathbf{g}_1, \ldots, \mathbf{f}_{i+(n-1)}\mathbf{g}_{n-1},$$

computed from other groups $\mathbf{f}_{i+j}, j = 1, \ldots, n-1$. Node $i$ thus stores $\alpha = k+n-1 =
2k$ coefficients of the form $\mathbf{f}_{i+j}\mathbf{g}_j$ where $i + j$ and $j$ are understood modulo $n$.
   **Recovery**. Suppose without loss of generality that a data collector connects to
nodes $1, \ldots, k$, and downloads $\mathbf{f}_1, \mathbf{f}_{1+k}\mathbf{g}_k$ from node $1, \ldots, \mathbf{f}_k, \mathbf{f}_{k+1}\mathbf{g}_1$ from node $k$.
The coefficients of $\mathbf{f}_{k+1}$ can be computed from $\mathbf{f}_{k+1}\mathbf{g}_k, \ldots, \mathbf{f}_{k+1}\mathbf{g}_1$ since $\mathbf{g}_1, \ldots, \mathbf{g}_{n-1}$
are linearly independent.

**Repair process**. Suppose that node $j$ failed. The newcomer downloads $\mathbf{f}_i \cdot \mathbf{g}_{i+(n-j)}$ and $\mathbf{f}_j \cdot \mathbf{g}_{j+(n-i)}$ from node $i$, $i = 1, \ldots, n$, $i \neq j$. Thus the parity-check coefficients stored in node $j$ can be recovered directly from $\mathbf{f}_i \cdot \mathbf{g}_{i+(n-j)}$ with $i = 1, \ldots, n$, $i \neq j$, while $\mathbf{f}_j$ is recovered from $\mathbf{f}_j \cdot \mathbf{g}_{j+(n-i)}$, using that the $\mathbf{g}_j$ involved are linearly independent. The total number of coefficients involved in the repair is $2k$, that is the repair bandwidth is $2k = 2d$, and the construction indeed achieves MBR point, since

$$(\alpha_{MBR}, \gamma_{MBR}) = \left( \frac{M}{k} \frac{2d}{2d - k + 1}, \frac{M}{k} \frac{2d}{2d - k + 1} \right) = (2k, 2k).$$

There have been many works containing code constructions for regenerating codes, the above two examples were meant to give an illustration, and are in no way representative of the richness of the literature on this topic. Maybe the most complete reference is http://storagewiki.ece.utexas.edu/doku.php.

## 2.4 Security

Once constructions and fault tolerance of regenerating codes have been well studied, it became natural to wonder about their performance in the presence of an adversary. The *security* of storage systems using (collaborative) regenerating codes has been considered, both against *passive* adversaries, or *eavesdroppers*, who can only read the data but not modify it, and *active*, or *Byzantine* attackers, who on the contrary are able to corrupt data.

The techniques used in the context of secure regenerating codes are similar to those discussed in the previous section: an information flow graph is considered, this time taking into account the presence of an adversary, and a min-cut bound is computed. We present next one such analysis [28], for the passive adversary case.

As before, a file $\mathbf{f} \in \mathbb{F}_q^M$ is encoded into $n$ pieces $\mathbf{f}_1, \ldots, \mathbf{f}_n$, and each piece is stored at a distinct node. We label the nodes such that the $i$th node $x_i$ stores $\mathbf{f}_i$, $i = 1, \ldots, n$. As in Theorem 1, we represent the node $x_i$ by $(x_i^{in}, x_i^{out})$, with an edge of capacity $\alpha$ in between. Assume that nodes $x_1, \ldots, x_k$ have failed, and have been repaired by the newcomers $x_{n+1}, \ldots, x_{n+k}$. In the presence of a passive eavesdropper, data objects are encrypted for ensuring their confidentiality. We therefore distinguish the secure file $\mathbf{f}^s$ of size $M^s$ from its encrypted version, the file $\mathbf{f}$ of size $M$. Encoding is then done on top of the encrypted file $\mathbf{f}$.

Denote by $\mathbf{f}_\varepsilon$ the encoded data stored in nodes belonging to the set $\varepsilon$. An eavesdropper who can read up to $l$, $l < k$, nodes among all the $n$ storage nodes, possibly at different time instances as the system evolves, accesses the $l$ input nodes in the set $\varepsilon_1 = \{x_{n+1}^{in}, \ldots, x_{n+l}^{in}\}$ while they were being repaired.

**Theorem 4** *Consider a secure object $\mathbf{f}$ of secure file size $M^s$, i.e., $M^s = H(\mathbf{f}^s)$, the entropy of $\mathbf{f}^s$. Suppose that over time, $k$ nodes have failed and been repaired at nodes $x_{n+1}, \ldots, x_{n+k}$. An eavesdropper has accessed the set of nodes $\varepsilon_1 = \{x_{n+1}^{in}, \ldots, x_{n+l}^{in}\}$.*

*Consider a data collector DC that collects data from the $k$ output nodes $\varepsilon = \{x_{n+1}^{out}, \ldots, x_{n+k}^{out}\}$. We have*

$$M^s \leq \sum_{j=l+1}^{k} \min\{(d+i-1)\beta, \alpha\}.$$

*Proof* That $\mathbf{f}^s$ remains confidential despite the knowledge of $\mathbf{f}_{\varepsilon_1}$ is modeled by $H(\mathbf{f}^s | \mathbf{f}_{\varepsilon_1}) = H(\mathbf{f}^s)$. Thus

$$
\begin{align}
M^s = H(\mathbf{f}^s) &= H(\mathbf{f}^s | \mathbf{f}_{\varepsilon_1}) \\
&= H(\mathbf{f}^s | \mathbf{f}_{\varepsilon_1}) - H(\mathbf{f}^s | \mathbf{f}_{\varepsilon}) \tag{7} \\
&= I(\mathbf{f}^s; \mathbf{f}_{\varepsilon \setminus \varepsilon_1} | \mathbf{f}_{\varepsilon_1}) \tag{8} \\
&\leq H(\mathbf{f}_{\varepsilon \setminus \varepsilon_1} | \mathbf{f}_{\varepsilon_1}) \tag{9} \\
&= \sum_{i=l+1}^{k} H(\mathbf{f}_{n+i} | \mathbf{f}_{n+1}, \ldots, \mathbf{f}_{n+i-1}) \tag{10} \\
&\leq \sum_{i=l+1}^{k} \min\{(d-i+1)\beta, \alpha\} \tag{11}
\end{align}
$$

where (7) holds since $|\varepsilon| = k$, which is enough to retrieve $\mathbf{f}$ by definition of $k$. Then since mutual information satisfies $I(X; Y|Z) = H(X|Z) - H(X|Y, Z)$, (8) follows. Alternatively, $I(X; Y|Z) = H(X, Z) - H(Z) - H(X, Y, Z) + H(Y, Z) = H(X, Z) - H(X, Y, Z) + H(Y|Z) \leq H(Y|Z)$, from which (9) is derived, and the Chain rule for entropy $H(X_1, \ldots, X_n|Y) = \sum_{i=1}^{n} H(X_i|X_1, \ldots, X_{i-1}, Y)$ implies (10). Finally, each node has a storage capacity of $\alpha$, for each repair node $H(\mathbf{f}_i) \leq d\beta$, and from the proof of Theorem 1, the repair node $x_{n+i}^{in}$ downloads $\beta$ symbols from each output node $x_{n+1}^{out}, \ldots, x_{n+i-1}^{out}$, yielding (11).

**Corollary 1** *At the minimum storage repair point (MSR), the secure file size $M^s$ is upper bounded by*

$$M^s \leq (k-l)\alpha.$$

**Corollary 2** *At the minimum repair bandwidth point (MBR) with $d = n - 1$, the secure file size $M^s$ is upper bounded by*

$$M^s \leq \sum_{i=l+1}^{k} (n-i)\beta.$$

The above analysis was extended to the collaborative case [22], in particular in the presence of rogue nodes which voluntarily corrupt the data that they transmit during repair. It was shown that the repair bandwidth obtained to secure (information theoretically) collaborative regenerating codes from Byzantine attacks is worse than having no collaboration at all. This is a rational drawback of collaboration, since a single corrupted node can pollute all other honest nodes involved in the repair.

In [37], Shah et al. gave a bound on the secure file size considering a passive eavesdropper model that generalizes the model in [28], in that the eavesdropper can access not only the stored data in nodes that have been repaired, $l_1$ of them from a set $\varepsilon_1$, but also the data which is downloaded during the repair of $l_2$ nodes from the set $\varepsilon_2$. This makes a difference, since more data may be in transit in nodes being repaired than what will actually be stored. Code constructions of secure codes are given both at MBR and MSR points. Rawat et al. [31] tightened the bound given in [37] at MSR point, and also provided secure MSR codes for $d = n - 1$. Finally, Koyluoglu et al. [15] extended the bound in [31] to collaborative regenerating codes. A bound is also available for partial collaboration [20].

**Theorem 5** ([15, 20]) *Consider a secure object* $\mathbf{f}$ *of secure file size* $M^s$, *i.e.,* $M^s = H(\mathbf{f}^s)$. *Consider an eavesdropper who has access to the set of nodes* $\varepsilon_1, \varepsilon_2, |\varepsilon_1| = l_1,$ $|\varepsilon_2| = l_2,$ *then*

$$H(\mathbf{f}^s) \leq \sum_{j \in R} H(\mathbf{f}_j | \mathbf{f}_{n+1}, \ldots, \mathbf{f}_{j-1}, \mathbf{f}_{\varepsilon_1}, \mathbf{d}_{\varepsilon_2}),$$

*with* $R = \{n + 1, \ldots, n + k\} \backslash (\varepsilon_1 \cup \varepsilon_2)$, *and* $\mathbf{d}_{\varepsilon_2}$ *denotes the downloaded data in the set of nodes* $\varepsilon_2$.

As a consequence of the above min-cut bound, we obtain bounds on the secure file size:

**Proposition 2** ([15, Propositions 5 and 9]) *For a collaborative regenerating code at the minimum repair bandwidth point (MBR), with* $\beta = 2\beta'$, $\alpha = \gamma = (2d + t - 1)\beta'$, *and file size* $M = k(2d - k + t)\beta'$, *the secure file size* $M^s$ *is upper bounded by*

$$M^s \leq (k - l_1)(2d + t - k - l_1)\frac{\beta}{2}. \tag{12}$$

*At the minimum storage repair point (MSR), with by* $\beta = \beta'$, $\alpha = (d - k + t)\beta$ *for a file of size* $M = k(d - k + t)\beta$, *the secure file size* $M^s$ *is upper bounded by*

$$M^s \leq \sum_{i=1}^{k - l_1 - l_2} (\alpha - I(\mathbf{f}_i; \mathbf{d}_{i, \varepsilon_2})). \tag{13}$$

*If* $I(\mathbf{f}_i; \mathbf{d}_{i, \varepsilon_2}) \geq \beta' = \beta$, *then (13) becomes*

$$M^s \leq (k - l_1 - l_2)(\alpha - \beta). \tag{14}$$

The generalization of the above two bounds for the partial collaborative case is found in [20]. Code constructions, when available, are found in the respective aforementioned papers. See also e.g. [3, 8, 29].

# 3 Locally Repairable Codes

In the last section, we saw how regenerating codes minimize repair bandwidth by contacting a large number of live nodes (preferably all the $n-1$ live nodes in the case of one failure). One may argue that getting answers from all these nodes in time to ensure an efficient repair may not be easy: other nodes may just be busy reading, writing, or managing their data. Locally repairable codes somehow stand in an opposite corner of the code design spectrum, by focusing on reducing the number $d$ of live nodes to be contacted per repair. For a repetition code, $d = 1$ is achieved. For other codes, we need at least $d \geq 2$.

The idea that repair efficiency improves when a small number of nodes is contacted appears in the construction of Pyramid codes [10]. Pyramid codes contain information symbols, followed by redundancy symbols, and distinguish local and global redundancy symbols, the local ones being those which are computed using a small number of information symbols, while the global ones possibly need all the information symbols to be computed. This can be achieved by taking a Reed–Solomon code, then shortening it, and finally by adding some new symbols obtained by encoding information symbols, part of them set to zero, e.g., to add two local symbols, set the first half and the second half respectively of the information symbols to zero. For example, suppose a file $\mathbf{f} = (f_1, \ldots, f_6)$ is encoded using a Reed–Solomon code. Retain two parity symbols $p_1, p_2$ from it, then set $f_1 = f_2 = f_3 = 0$ and obtain a new Reed–Solomon codeword from which a parity $p_3$ is kept, then alternate and set $f_4 = f_5 = f_6 = 0$ and keep a parity $p_4$ from its Reed–Solomon codeword. This yields a final codeword $(f_1, \ldots, f_6, p_1, p_2, p_3, p_4)$ where $p_1, p_2$ are the global redundancy symbols, and $p_3, p_4$ are the local ones. The motivation for this is to handle *degraded read*, or how to read data when some of the symbols are missing. One solution is to quickly compute the missing symbols of the data to be read using the local redundancy symbols. This in turn implies that the local nodes can be repaired by contacting a small number of live nodes. *Local Reconstruction Codes* form a class of codes for distributed storage systems which optimize Pyramid codes, proposed by the same research group, and which have been incorporated in Windows Azure [11].

In [24], minimizing the repair degree, that is the number of live nodes to be contacted per repair, was proposed as a code design criterion, and *self-repairing* codes were designed, which mimic the encoding of Reed–Solomon codes via polynomial evaluation, but using linearized polynomials. A linearized polynomial $p(X) \in \mathbb{F}_q[X]$ satisfies that $p(w + w') = p(w) + p(w')$ for $w, w' \in \mathbb{F}_q$.

To be consistent with the existing literature, we will use in the following the notation $d$ for the minimum distance of a code, and $k$ for its dimension (while $d$ was the repair degree, and $k$ the number of nodes contacted to retrieve an object, for regenerating codes).

## 3.1 A Singleton-Type Bound on Locality

In [7], the notion of *locality* was introduced. The $i$th coordinate of a codeword is said to have locality $r$, if its value is determined by at most $r$ other coordinates. Codes that have a minimum Hamming distance of $d$ but also have the property that any information coordinate has locality $r$ or less were proposed, and the following Singleton-type bound was given. We state and prove it for linear codes, even though it is known to hold for non-linear codes[4] [6].

**Theorem 6** *Let $C$ be an $[n, k]$ linear code, that is a code of length $n$ and dimension $k$, with minimum distance $d$ and locality $r$. Then*

$$n - k + 1 - d \geq \lfloor \frac{k-1}{r} \rfloor. \tag{15}$$

*Proof* [16] Let $G$ be the generator matrix of $C$, and for a codeword in $C$, each of its coefficients is stored by a distinct node. Choose any $\lfloor \frac{k-1}{r} \rfloor$ nodes, call these the "leaders". Each leader can be written as a linear combination of at most $r$ other nodes, call this set the "set of friends of the leader". Now define $N$ as the set of nodes which is the union of all sets of friends of the leaders (at most $r \lfloor \frac{k-1}{r} \rfloor$ of them) but without the leaders themselves. Then clearly $N$ has at most $r \lfloor \frac{k-1}{r} \rfloor$ elements, thus less than $k$ elements so that the set of columns in $G$ that corresponds to $N$ spans a space of rank $< k$. Since $G$ has full rank it is possible to enlarge $N$ to a set $N$ of more than $k - 1$ columns such that the rank of its corresponding columns equals exactly $k - 1$. Note that because the code has locality $r$, this enlargement operation can be done without involving any of the leaders. Now define $U$ as the union of $N'$ and the set of leaders. Then $U$ has at least $k - 1 + \lfloor \frac{k-1}{r} \rfloor$ nodes but still, because the code has locality $r$, the corresponding columns in $G$ span a space of dimension $< k$. By definition of the minimum distance, all $(k \times \cdot)$-submatrices of $G$ that have rank $< k$ must have less than $n - d$ columns. It therefore follows that $k - 1 + \lfloor \frac{k-1}{r} \rfloor \leq n - d$, which proves the theorem. ∎

The bound (15) has been shown to be tight [7] over a large enough finite field.

The term *locally repairable code (LRC)* was coined in [27] to refer to codes with a prescribed repair degree, or locality. The term is reminiscent of the theoretical computer science terminology "locally decodable" or "locally correctable" codes. Describing possible technical connections among these families of codes is beyond the scope of this survey.

By now, a *locally repairable (or locally recoverable) code* is a linear $[n, k, r]$ code that encode $k$ information symbols into a codeword of length $n$, with the property that for any symbol of the codeword, there exist at most $r$ other symbols such that the value of the symbol can be recovered from them. Then $r$ is called the *repair locality*.

---

[4]In fact, we will not discuss non-linear codes, even though some works have been done on this topic.

For LRC codes, if a symbol is lost due to a node failure, its value can be recovered by contacting at most $r$ other nodes.

MDS codes are actually LRC codes, with locality $r = k$. MDS codes can recover the largest possible number of erased symbols among all $[n, k]$ codes, since their Hamming distance is $n - k + 1$. Thus (15) for MDS codes becomes

$$n - k + 1 - (n - k + 1) \geq \lfloor \frac{k-1}{k} \rfloor$$

and the bound is tight, even though $r$ is high. An $[n, k, r]$ LRC code that achieves the bound (15) with equality is called an optimal LRC code. Pyramid codes are known to yield LRC codes [7] whose information symbols have optimal repair locality. In [31, 41], rank-metric codes are used to construct optimal LRCs, while [45] gave a construction based on Reed–Solomon codes that is optimal for any $1 < r < k$.

### 3.2 Code Constructions

We repeat the disclaimer used for regenerating codes: the literature on locally repairable codes is vast, and we do not pretend to provide an exhaustive list of constructions. Since MDS codes are optimal with repair locality $r = k$, we choose to present a construction from [30], which uses generalized Reed–Muller codes, and achieves $r = 2$ and $r = 3$.

**LRC codes with locality 2.** Let $q$ be a prime. Consider an object $\mathbf{f} = (f_1, \ldots, f_M) \in \mathbb{F}_q^M$ of length $M = m$, and the multivariate polynomial $g$ in $\mathbb{F}_q[X_1, \ldots, X_m]$ of degree 1 given by

$$g(X_1, \ldots, X_m) = \sum_{i=1}^{m} f_i X_i.$$

A codeword is obtained by evaluating $g$ in the points

$$a_i, a_i + t((q-1)a_i + a_j) \in \mathbb{F}_q^m,$$

for $i = 1, \ldots, N$, $1 \leq i < j \leq N$, $2 \leq t \leq 1 + L$, where $N$ and $L$ are designed parameters. Each codeword coefficient is stored in one node. Set $\mathbf{h} = (q-1)a_i + a_j$, then the polynomial is evaluated in the points

$$a_i, a_j, a_i + t\mathbf{h}.$$

Since $t$ goes from 2 to $L + 1$, and with $a_i, a_j$, there are $L + 2$ points on a line, that is

$$a_i, a_i + \mathbf{h}, \ldots, a_i + (L+1)\mathbf{h}.$$

Suppose the node storing $a_i$ fails, the repair can be realized by contacting two other nodes storing the values corresponding to two points on the same line, say the nodes containing $g(a_i + t_1\mathbf{h})$, $g(a_i + t_2\mathbf{h})$ for $L \geq 1$, $t_1 \neq t_2$. Consider $g$ as a polynomial in $t$, then $g$ has degree 1 in $t$, so by polynomial interpolation, the two values $g(a_i + t_1\mathbf{h})$, $g(a_i + t_2\mathbf{h})$ are enough to compute $g(a_i + t\mathbf{h})$, and evaluating it in $t = 0$ yields back $g(a_i)$. This shows that this code has repair locality $r = 2$.

**LRC codes with locality 3.** Similarly, consider a file $\mathbf{f} = (f_1, \ldots, f_M) \in \mathbb{F}_q^M$ of length $M = \binom{m+2}{m}$, and the polynomial $g$ in $\mathbb{F}_q[X_1, \ldots, X_m]$ of degree 2 given by

$$g(X_1, \ldots, X_m) = \sum_{i \in \mathcal{M}} f_i X_1^{\alpha_{i,1}} \cdots X_m^{\alpha_{i,m}},$$

where $\mathcal{M}$ is the index set for monomials in $\mathbb{F}_q[X_1, \ldots, X_m]$ of degree at most 2 arranged in lexical order. A codeword is obtained by evaluating $g$ in the points

$$2a_i, a_i + a_j, 2a_i + t((q-1)a_i + a_j) \in \mathbb{F}_q^m,$$

for $i = 1, \ldots, N$, $1 \leq i < j \leq N$, $3 \leq t \leq 2 + L$, where $N$ and $L$ are designed parameters. Each coefficient in the codeword is stored in one node. Again, set $\mathbf{h} = (q-1)a_i + a_j$, then the polynomial is evaluated in

$$2a_i, a_i + a_j, 2a_j, 2a_i + t\mathbf{h}.$$

Since $t$ goes from 3 to $L + 2$, and with $2a_i$, $a_i + a_j$, $2a_j$, there are $L + 3$ points on a line, that is

$$2a_i, 2a_i + \mathbf{h}, 2a_i + 2\mathbf{h}, \ldots, a_i + (L+2)\mathbf{h}.$$

Suppose the node storing $2a_i$ fails, the repair can be realized by contacting three other nodes storing the values corresponding to three points on the same line, say the nodes containing $g(2a_i + t_1\mathbf{h})$, $g(2a_i + t_2\mathbf{h})$, $g(2a_i + t_3\mathbf{h})$ for $L \geq 1$, $t_1 \neq t_2 \neq t_3$. Consider again $g$ as a polynomial in $t$, which this time has degree at most 2. By polynomial interpolation, three values are enough to reconstruct the polynomial $g(2a_i + t\mathbf{h})$, and $g(a_i)$ is obtained by setting $t = 0$. This shows a repair locality of $r = 3$.

We note three directions that have been pursued in the design of LRC codes:

- To get practical codes, it is often desirable to have a small alphabet $\mathbb{F}_q$, preferably in characteristic 2. In [1], integrated interleaved codes have been studied and showed to be good candidates to obtain LRC codes over $\mathbb{F}_{2^b}$, for small values of $b$. In [44], cyclic codes and subfield subcodes of cyclic codes are considered as LRC codes, where their minimum distance and locality are derived. In particular, the binary case is treated.
- Suppose that one may want different codeword symbols to have different locality. Bounds and constructions, optimal with respect to these bounds, for LRC codes with different localities have been considered in [13, 49].

- The security model presented in Sect. 2.4 for regenerating codes has also been considered for LRCs in [31]. Under the same eavesdropper threat, bounds on the secure file size were computed. For a reference on Byzantine attacks for LRCs, see e.g. [42].

## 3.3 LRC Codes with Multiple Recovering Sets

We have seen above that LRC codes have the local repair property that each symbol can be recovered using at most $r$ other symbols, so a natural question is whether this set of at most $r$ symbols, called *recovering set* is unique. When each symbol has several recovering sets, then we may ask whether these sets are disjoint. These questions are important for the health of the distributed storage system. If the node that fails has a unique recovering set, and some other failures affect this set, then the recovery set cannot be used, and local repair becomes impossible. Therefore it is important to understand the structure of recovering sets.

As an illustration, consider the code constructions of the last subsection with parameters $L = 1$ and $N = 3$. For the construction with locality 3, the evaluating points of the polynomial $g$ are

$$2a_1, 2a_2, 2a_3,$$
$$a_1 + a_2, a_1 + a_3, a_2 + a_3,$$
$$a_2 - a_1, a_3 - a_1, a_3 - a_2,$$

and the points $2a_1, a_1 + a_2, 2a_2, a_2 - a_1$ are actually on the same line. We similarly list the points on the same line:

$$L_1 : 2a_1, a_1 + a_2, 2a_2, a_2 - a_1,$$
$$L_2 : 2a_1, a_1 + a_3, 2a_3, a_3 - a_1,$$
$$L_3 : 2a_2, a_2 + a_3, 2a_3, a_3 - a_2.$$

Suppose a node stores $g(2a_1)$. Since there are two lines $L_1, L_2$ containing $2a_1$, the symbol $g(2a_1)$ has two recovering sets, $a_1 + a_2, 2a_2, a_2 - a_1$ and $a_1 + a_3, 2a_3, a_3 - a_1$, and they do not intersect. This is also true for the points $2a_2$ and $2a_3$. However, for other points, they only have one recovering set.

In [26, 45], LRC codes with disjoint recovering sets are proposed. Tamo and Barg [45] gave two methods to construct LRC codes with multiple recovering sets. One use the concept of orthogonal partition, the second combines several LRC codes into a longer multiple recovering code. In [47], codes providing $\delta - 1$ non-overlapping local repair groups of size no more than $r$ per coordinate were introduced. In [32], recovering sets are studied using the concept of $(r, t)$-availability: a code symbol is said to have $(r, t)$-availability if it can be rebuilt from $t$ disjoint recovering sets, each of size at most $r$.

# 4 Concluding Remarks

This chapter surveyed two main families of codes for distributed storage systems. While we have already mentioned that the references in terms of code constructions are far from exhaustive, we would like to conclude by also listing aspects of the topic which have not been considered.

- Other design criteria, for codes to provide efficient data insertion (e.g. [25]), data update (e.g. [43]), or versioning (e.g. [9]), have not been covered.
- The view point of code design the way it was presented is looking at one snapshot of the network, it is oblivious of the actual system, and does not take into account the fact that a node actually stores many objects (e.g. [2]), which means that data allocation, or management of metainformation is important. There have been works looking at systems aspects and implementation of these codes (e.g. [17]) which we did not discuss either.
- Regarding security aspects, we only described information theoretical approaches, while there are many works discussing cryptographic or system approaches to secure code based distributed storage systems.
- Finally, there is a vast body of work coming from the storage systems and distributed systems community. They started discussing coding for distributed storage systems before the information and coding community did (e.g. [18, 48]), and they also have proposed many interesting code designs, often motivated by considerations closer to working systems.

**Acknowledgements** This work is supported by the MoE Tier-2 grant eCODE: Erasure Codes for Datacenter Environments.

# References

1. M. Blaum, S.R. Hetzler, Integrated interleaved codes as locally recoverable codes: properties and performance. Int. J. Inf. Coding Theory **3**(4), 324 (2016)
2. A. Datta, L. Pamies-Juarez, F. Oggier, A study of the performance of novel storage-centric repairable codes. Springer Comput. **98**(3) (2015)
3. T.K. Dikaliotis, A.G. Dimakis, T. Ho, Security in distributed storage systems by communicating a logarithmic number of bits, in *Proceedings of the 2010 IEEE International Symposium on Information Theory (ISIT)* (2010), pp. 1948–1952
4. A.G. Dimakis, P.B. Godfrey, Y. Wu, M.J. Wainwright, K. Ramchandran, Network coding for distributed storage systems. IEEE Trans. Inf. Theory **56**(9) (2010)
5. I.M. Duursma, Shortened regenerating codes. CoRR (2015), arXiv:1505.00178
6. M. Forbes, S. Yekhanin, On the locality of codeword symbols in non-linear codes. Discret. Math. **324**, 78–84 (2014)
7. P. Gopalan, C. Huang, H. Simitci, S. Yekhanin, On the locality of codeword symbols. IEEE Trans. Inf. Theory **58**(11), 6925–6934 (2011)
8. Y.S. Han, R. Zheng, W.H. Mow, Exact regenerating codes for Byzantine fault tolerance in distributed storage, in *Proceedings of IEEE INFOCOM* (2012), pp. 2498–2506
9. J. Harshan, F. Oggier, A. Datta, Sparsity exploiting erasure coding for distributed storage of versioned data. Springer Comput. (2016)

10. C. Huang, M, Chen, J. Li, Pyramid codes: flexible schemes to trade space for access efficiency in reliable data storage systems, in *Sixth IEEE International Symposium on Network Computing and Applications, July 2007*, pp. 79–86
11. C. Huang, H. Simitci, Y. Xu, A. Ogus, B. Calder, P. Gopalan, J. Lin, S. Yekhanin, Erasure coding in windows Azure storage, in *2012 USENIX Annual Technical Conference*, 12–15 June 2012
12. K.A.S. Immink, *Codes for Mass Data Storage Systems*, Second fully revised edition (Shannon Foundation Publishers, Eindhoven, 2004), http://www.turing-machines.com/pdf/codes_for_mass_data2.pdf. ISBN 90-74249-27-2
13. S. Kadhe, A. Sprintson, Codes with unequal locality, in *IEEE International Symposium on Information Theory (ISIT)* (2016), pp. 435–439
14. A.-M. Kermarrec, N. Le Scouarnec, G. Straub, Repairing multiple failures with coordinated and adaptive regenerating codes, in *The 2011 International Symposium on Network Coding (NetCod 2011)*
15. O.O. Koyluoglu, A.S. Rawat, S. Vishwanath, Secure cooperative regenerating codes for distributed storage systems. IEEE Trans. Inf. Theory **60**(9), 5228–5244 (2014), arXiv:1210.3664
16. M. Kuijper, D. Napp, Erasure codes with simplex locality, in *21st International Symposium on Mathematical Theory of Networks and Systems*, Groningen, The Netherlands (2014)
17. M. Li, P.P.C. Lee, STAIR codes: a general family of erasure codes for tolerating device and sector failures in practical storage systems, in *Proceedings of the 12th USENIX Conference on File and Storage Technologies (FAST'14)*, Santa Clara, CA (2014)
18. W.K. Lin, D.M. Chiu, Y.B. Lee, Erasure code replication revisited, P2P 2004
19. S. Liu, F. Oggier, On storage codes allowing partially collaborative repairs, in *IEEE International Symposium on Information Theory (ISIT) 2014*, pp. 2440–2444
20. S. Liu, F. Oggier, Partially collaborative storage codes in the presence of an eavesdropper. Int. J. Inf. Coding Theory **3**(3), 177–196 (2016)
21. S. Mohajer, R. Tandon, New bounds on the (n, k, d) storage systems with exact repair, in *IEEE International Symposium on Information Theory, ISIT 2015, Hong Kong, China, 14–19 June 2015*, pp. 2056–2060
22. F. Oggier, A. Datta, Byzantine fault tolerance of regenerating codes, in *The 11th IEEE International Conference on Peer-to-Peer Computing (P2P 2011)*, Kyoto, arXiv:1106.2275
23. F. Oggier, A. Datta, Coding techniques for repairability in networked distributed storage systems, *Foundations and Trends in Communications and Information Theory* (Now Publishers, Breda, 2013)
24. F. Oggier, A. Datta, Self-repairing homomorphic codes for distributed storage systems, INFOCOM 2011
25. L. Pamies-Juarez, A. Datta, F. Oggier, RapidRAID: pipelined erasure codes for fast data archival in distributed storage systems, arXiv:1207.6744
26. L. Pamies-Juarez, H.D.L. Hollmann, F. Oggier, Locally repairable codes with multiple repair alternatives, in *IEEE International Symposium on Information Theory (ISIT)* (2013), pp. 892–896
27. D.S. Papailiopoulos, A.G. Dimakis, Locally repairable codes, in *IEEE International Symposium on Information Theory (ISIT)* (2012), arXiv:1206.3804
28. S. Pawar, S. El Rouayheb, K. Ramchandran, Securing dynamic distributed storage systems against eavesdropping and adversarial attacks. IEEE Trans. Inf. Theory (Spec. Issue Facet. Coding Theory Algorithm Netw.) **57**(9) (2011)
29. K.V. Rashmi, N.B. Shah, K. Ramchandran, P.V. Kumar, Regenerating codes for errors and erasures in distributed storage, in *Proceedings of the 2012 IEEE International Symposium on Information Theory (ISIT)* (2012), pp. 1202–1206
30. A.S. Rawat, S. Vishwanath, On locality in distributed storage systems, in *IEEE Information Theory Workshop (ITW)* (2012), pp. 497–501
31. A.S. Rawat, O.O. Koyluoglu, N. Silberstein, S. Vishwanath, Optimal locally repairable and secure codes for distributed storage systems. IEEE Trans. Inf. Theory **60**(1), 212–236 (2014)

32. A.S. Rawat, D.S. Papailiopoulos, A.G. Dimakis, S. Vishwanath, Locality and availability in distributed storage. IEEE Trans. Inf. Theory **62**(8), 4481–4493 (2016)
33. I. Reed, G. Solomon, Polynomial codes over certain finite fields. J. Soc. Ind. Appl. Math. **8**, 300–304 (1960)
34. B. Sasidharan, N. Prakash, M.N. Krishnan, M. Vajha, K. Senthoor, P.V. Kumar, Outer bounds on the storage-repair bandwidth tradeoff of exact-repair regenerating codes. Int. J. Inf. Coding Theory **3**(4), 255–298 (2016)
35. B. Sasidharan, K. Senthoor, P.V. Kumar, An improved outer bound on the storage repair-bandwidth tradeoff of exact-repair regenerating codes, in *IEEE International Symposium on Information Theory, ISIT 2014*, pp. 2430–2434
36. K. Senthoor, B. Sasidharan, P.V. Kumar, Improved layered regenerating codes characterizing the exact-repair storage-repair bandwidth tradeoff for certain parameter sets, in *2015 IEEE Information Theory Workshop, ITW 2015*, Jerusalem, Israel, 26 April–1 May 2015, p. 15
37. N.B. Shah, K.V. Rashmi, P.V. Kumar, Information theoretically secure regenerating codes for distributed storage, in *Proceedings of IEEE Globecom* (2011)
38. K.W. Shum, Cooperative regenerating codes for distributed storage systems, in *The International Conference on Communications (ICC 2011)*, Kyoto, arXiv:1101.5257
39. K.W. Shum, Y. Hu, Exact minimum-repair-bandwidth cooperative regenerating codes for distributed storage systems, in *The International Symposium on Information Theory (ISIT 2011)*, Saint-Petersburg
40. K.W. Shum, Y. Hu, Cooperative regenerating codes. IEEE Trans. Inf. Theory **59**(11), 7229–7258 (2013)
41. N. Silberstein, A.S. Rawat, O.O. Koyluoglu, S. Vishwanath, Optimal locally repairable codes via rank-metric codes, in *IEEE International Symposium on Information Theory (ISIT)* (2013)
42. N. Silberstein, A.S. Rawat, S. Vishwanath, Error-correcting regenerating and locally repairable codes via rank-metric codes. IEEE Trans. Inf. Theory **61**(11), 5765–5778 (2015)
43. A. Singh Rawat, S. Vishwanat, A. Bhowmick, E. Soljanin, Update efficient codes for distributed storage, in *IEEE International Symposium on Information Theory (ISIT)* (2011)
44. I. Tamo, A. Barg, S. Goparaju, R. Calderbank, Cyclic LRC codes, binary LRC codes, and upper bounds on the distance of cyclic clodes. Int. J. Inf. Coding Theory **3**(4), 32 (2016)
45. I. Tamo, A. Barg, A family of optimal locally recoverable codes. IEEE Trans. Inf. Theory **60**(8), 4661–4676 (2014)
46. C. Tian, Characterizing the rate region of the (4,3,3) exact-repair regenerating codes. IEEE J. Sel. Areas Commun. **32**(5), 967–975 (2014)
47. A. Wang, Z. Zhang, Repair locality with multiple erasure tolerance. IEEE Trans. Inf. Theory **60**(11), 6979–6987 (2014)
48. H. Weatherspoon, J. Kubiatowicz, Erasure coding vs replication: a quantitative comparison, *Peer-to-Peer Systems*, LNCS (2002)
49. A. Zeh, E. Yaakobi, Bound and constructions of codes with multiple localities, in *IEEE International Symposium on Information Theory (ISIT)* (2016), pp. 641–644

# Matroid Theory and Storage Codes: Bounds and Constructions

Ragnar Freij-Hollanti, Camilla Hollanti and Thomas Westerbäck

**Abstract** Recent research on distributed storage systems (DSSs) has revealed interesting connections between matroid theory and locally repairable codes (LRCs). The goal of this chapter is to introduce the reader to matroids and polymatroids, and illustrate their relation to distributed storage systems. While many of the results are rather technical in nature, effort is made to increase accessibility via simple examples. The chapter embeds all the essential features of LRCs, namely locality, availability, and hierarchy alongside with related generalised Singleton bounds.

## 1 Introduction to Locally Repairable Codes

In this chapter, we will discuss the theoretical foundations of *locally repairable codes* (LRCs), which were introduced in chapter "An Overview of Coding for Distributed Storage Systems". While our main interest is in the codes and their applicability for distributed storage systems, significant parts of our machinery comes from matroid theory. We will develop this theory to the extent that is needed for the applications, and leave some additional pointers to interpretations in terms of graphs and projective geometries.

The need for large-scale data storage is continuously increasing. Within the past few years, *distributed storage systems* (DSSs) have revolutionised our tradi-

The authors gratefully acknowledge the financial support from the Academy of Finland (grants #276031 and #303819), as well as the support from the COST Action IC1104.

R. Freij-Hollanti (✉) · C. Hollanti · T. Westerbäck
Department of Mathematics and Systems Analysis, Aalto University,
P.O. Box 11100, 00076 Aalto, Finland
e-mail: ragnar.freij@aalto.fi

C. Hollanti
e-mail: camilla.hollanti@aalto.fi

T. Westerbäck
e-mail: thomas.westerback@aalto.fi

© Springer International Publishing AG 2018
M. Greferath et al. (eds.), *Network Coding and Subspace Designs*,
Signals and Communication Technology,
https://doi.org/10.1007/978-3-319-70293-3_15

tional ways of storing, securing, and accessing data. Storage node failure is a frequent obstacle in large-scale DSSs, making repair efficiency an important objective. A bottle-neck for repair efficiency, measured by the notion of *locality* [27], is the number of contacted nodes needed for repair. The key objects of study in this paper are *locally repairable codes* (LRCs), which are, informally speaking, storage systems where a small number of failing nodes can be recovered by boundedly many other (close-by) nodes. Repair-efficient LRCs are already in use for large-scale DSSs used by, for example, Facebook and Windows Azure Storage [40].

Another desired attribute, measured by the notion of *availability* [31], is the property of having multiple alternative ways to repair nodes or access files. This is particularly relevant for nodes containing so-called hot data that is frequently and simultaneously accessed by many users. Moreover, as failures are often spatially correlated, it is valuable to have each node repairable at several different *scales*. This means that if a node fails simultaneously with the set of nodes that should normally be used for repairing it, then there still exists a larger set of helper nodes that can be used to recover the lost data. This property is captured by the notion of *hierarchy* [10, 32] in the storage system.

Network coding techniques for large-scale DSSs were considered in [7]. Since then, a plethora of research on DSSs with a focus on linear LRCs and various localities has been carried out, see [13, 27, 30, 34, 39] among many others. Availability for linear LRCs was defined in [31]. The notion of hierarchical locality was first studied in [32], where bounds for the global minimum distance were also obtained.

Let us denote by $(n, k, d, r, \delta, t)$, respectively, the code length, dimension, global minimum distance, locality, local minimum distance, and availability. Bold-faced parameters $(\mathbf{n}, \mathbf{k}, \mathbf{d}, \mathbf{t})$ will be used in the sequel to refer to hierarchical locality and availability. It was shown in [40] that the $(r, \delta = 2)$-locality of a linear LRC is a matroid invariant. The connection between matroid theory and linear LRCs was examined in more detail in [47]. In addition, the parameters $(n, k, d, r, \delta)$ for linear LRCs were generalised to matroids, and new results for both matroids and linear LRCs were given therein. Even more generally, the parameters $(n, k, d, r, \delta, t)$ were generalised to polymatroids, and new results for polymatroids, matroids and both linear and nonlinear LRCs over arbitrary finite alphabets were derived in [46]. Similar methods can be used to bound parameters of *batch codes* [49], as discussed in chapter "Batch and PIR Codes and Their Connections to Locally Repairable Codes". For more background on batch codes, see e.g. [16, 23]. Moreover, as certain specific LRCs and batch codes[1] belong to the class of *private information retrieval (PIR)* codes as defined in [9, Definition 4], the related LRC and batch code bounds also hold for those PIR codes. See Sect. 5.3 and chapter "Batch and PIR Codes and Their Connections to Locally Repairable Codes" for more discussion.

The main purpose of this chapter is to give an overview of the connection between matroid theory and linear LRCs with availability and hierarchy, using examples for improved clarity of the technical results. In particular, we are focusing on how the

---

[1]To this end, we need to make specific assumptions on the locality and availability of the LRC [9, Theorem 21], which also implies restrictions on the query structure of the batch code.

parameters of a LRC can be analysed using the *lattice of cyclic flats* of an associated matroid, and on a construction derived from matroid theory that provides us with linear LRCs. The matroidal results on LRCs reviewed here are mostly taken from [10, 46, 47].

The rest of this chapter is organised as follows. In Sects. 1.1 and 1.2, we introduce distributed storage systems and how they can be constructed by using linear codes. In particular, we consider locally repairable codes with availability. Section 2 gives a brief introduction to the concepts and features related to matroids relevant to LRCs. In Sect. 3, we summarise the state-of-the-art generalised Singleton bounds on the code parameters for linear codes, as well as discuss existence of Singleton-optimal linear codes and matroids. Section 4 reviews some explicit (linear) code constructions. In Sect. 5, we go beyond linear codes and consider polymatroids and related generalised Singleton bounds, which are then valid for all LRCs over any finite alphabet, and also imply bounds for PIR codes when restricted to systematic linear codes. Section 6 concludes the chapter and discusses some open problems. Further results in matroid theory and especially their representability are given in appendix for the interested reader. The following notation will be used throughout the paper:

| | |
|---|---|
| $\mathbb{F}$ | : a field; |
| $\mathbb{F}_q$ | : the finite field of prime power size $q$; |
| $E$ | : a finite set; |
| $G$ | : a matrix over $\mathbb{F}$ with columns indexed by $E$; |
| $G(X)$ | : the matrix obtained from $G$ by restricting to the columns indexed by $X$, where $X \subseteq E$; |
| $C(G)$ | : the vector space generated by the columns of $G$; |
| $R(G)$ | : the vector space generated by the rows of $G$; |
| $C$ | : linear code $C = R(G)$ over $\mathbb{F}$ generated by $G$; |
| $C_X$ | : the *punctured* code of $C$ on $X$, i.e., $C_X = R(G(X))$, where $X \subseteq E$; |
| $2^E$ | : the collection of all subsets of a finite set $E$; |
| $[j]$ | : the set $\{1, 2, \ldots, j\}$ for a positive integer $j$; |
| $n, k, d, r, \delta, t, h$ | : code length, dimension, minimum distance, locality, failure tolerance, availability, hierarchy, respectively; |
| $[n, k, d], (n, k, d)$ | : parameters of a linear/general code, respectively; |
| $(n, k, \ldots)_i$ | : parameter values when we consider information symbols, e.g., information symbol locality; |
| $(n, k, \ldots)_a$ | : parameter values when we consider all code symbols, e.g., all symbol locality; |
| $(n, k, \ldots)_s$ | : parameter values when we consider systematic code symbols, e.g., systematic symbol locality; |
| $(\mathbf{n, k, d, t})$ | : parameter values for different hierarchy levels. |

*Remark 1* Here, $d$ denotes the minimum (Hamming) distance of the code, rather than the number of nodes that have to be contacted for repair, as is commonplace in

the theory of regenerating codes. In chapter "An Overview of Coding for Distributed Storage Systems", the minimum distance of the code was denoted by $d_H$.

The motivation to study punctured codes arises from hierarchical locality; the locality parameters at the different hierarchy levels correspond to the global parameters of the related punctured codes. The puncturing operation on codes corresponds to the so-called restriction (or deletion) operation on matroids.

We also point out that $G(E) = G$ and $C_E = C$. We will often index a matrix $G$ by $[n]$, where $n$ is the number of columns in $G$.

## 1.1 Distributed Storage Systems from Linear Codes

A linear code $C$ can be used to obtain a DSS, where every coordinate in $C$ represents a storage node in the DSS, and every point in $C$ represents a stored data item. While one often assumes that the data items are field elements in their own right, no such assumption is necessary. However, if $C$ is a code over the field $\mathbb{F}$ and the data items are elements in an alphabet $\mathbb{A}$, then we must be able to form the linear combinations $f_1 a_1 + f_2 a_2$ for $f_1, f_2$ in $\mathbb{F}$ and $a_1, a_2$ in $\mathbb{A}$. Moreover, if we know the scalar $f$, we must be able to read off $a$ from $fa$. This is achieved if $\mathbb{A} \cong \mathbb{F}^\alpha$ is a vector space over $\mathbb{F}$, wherefore we must have $|\mathbb{A}| \geq |\mathbb{F}|$. Thus, the length of the data items must be at least the number of symbols needed to represent a field element. In particular if the data items are measured in, e.g., kilobytes, then we are restricted to work over fields of size not larger than about $2^{8000}$. Beside this strict upper bound on the field size, the complexity of operations also makes small field sizes — ideally even binary fields — naturally desirable.

*Example 1* Let $C$ be the linear code generated by the following matrix $G$ over $\mathbb{F}_3$:

$$G = \begin{array}{c} \phantom{G=}\begin{array}{ccccccccc} 1 & 2 & 3 & 4 & 5 & 6 & 7 & 8 & 9 \end{array} \\ \begin{bmatrix} 1 & 0 & 0 & 0 & 1 & 1 & 1 & 1 & 1 \\ 0 & 1 & 0 & 0 & 1 & 0 & 1 & 2 & 2 \\ 0 & 0 & 1 & 0 & 0 & 1 & 1 & 0 & 0 \\ 0 & 0 & 0 & 1 & 0 & 0 & 0 & 1 & 2 \end{bmatrix} \end{array}$$

Then, $C$ corresponds to a 9 node storage system, storing four files $(a, b, c, d)$, each of which is an element in $\mathbb{F}_3^\alpha$. In this system, node 1 stores $a$, node 5 stores $a + b$, node 9 stores $a + 2b + 2d$, and so on.

Two very basic properties of any DSS are that every node can be repaired by some other nodes and that every node contains some information.[2] We therefore give the following definition.

---

[2]We remark that if one takes into account queueing theoretic aspects, then data allocation may become less trivial (some nodes may be empty). Such aspects are discussed, especially in a wireless setting, in chapters "Opportunistic Network Coding" and "Coded Random Access". However, these considerations are out of the scope of this chapter.

**Definition 1** A linear $[n, k, d]$-code $C$ over a field is a *non-degenerate storage code* if $d \geq 2$ and there is no zero column in a generator matrix of $C$.

The first example of a storage code, and the motivating example behind the notion of locality, is the notion of a *maximum distance separable (MDS) code*. It has several different definitions in the literature, here we list a few of them.

**Definition 2** The following properties are equivalent for linear $[n, k, d]$ storage codes:

(i) $n = k + d - 1$.
(ii) The stored data can be retrieved from any $k$ nodes in the storage system.
(iii) To repair any single erased node in the storage system, one needs to contact $k$ other nodes.

A code that satisfies one (and therefore all) of the above properties is called an $MDS$ code.

By the Singleton bound, $n \geq k + d - 1$ holds for any storage code, so by property (i), MDS codes are "optimal" in the sense that they have minimal length for given storage capacity and error tolerance. However, (iii) is clearly an unfavourable property in terms of erasure correction. This is the motivation behind constructing codes with small $n - k - d$, where individual node failures can still be corrected "locally".

## 1.2 Linear Locally Repairable Codes with Availability

The very broad class of linear LRCs will be defined next. It is worth noting that, contrary to what the terminology would suggest, a LRC is not a novel kind of code, but rather the "locality parameters" $(r, \delta)$ can be defined for any code. What we call a LRC is then only a code that is specifically designed with the parameters $(r, \delta)$ in mind. While the locality parameters can be understood directly in terms of the storage system, it is more instructive from a coding theoretic point of view to understand them via punctured codes. Then, the punctured codes will correspond exactly to the restrictions to "locality sets", which can be used to locally repair a small number of node failures within the locality set.

**Definition 3** Let $G$ be a matrix over $\mathbb{F}$ indexed by $E$ and $C$ the linear code generated by $G$. Then, for $X \subseteq E$, $C_X$ is a linear $[n_X, k_X, d_X]$-code where

$$n_X = |X|,$$
$$k_X = \mathrm{rank}(G(X)),$$
$$d_X = \min\{|Y| : Y \subseteq X \text{ and } k_{X \setminus Y} < k_X\}.$$

Alternatively, one can define the minimum distance $d_X$ as the smallest support of a non-zero codeword in $C_X = \mathrm{R}(G(X))$. We use Definition 3, as it has the advantage of not depending on the linearity of the code.

*Example 2*  Consider the storage code $C$ from Example 1. Let $Y_1 = \{1, 2, 3, 5, 6, 7\}$, $X_1 = \{1, 2, 5\}$ and $X_2 = \{2, 6, 7\}$. Then $C_{Y_1}$, $C_{X_1}$ and $C_{X_2}$ are storage codes with

$$[n_{Y_1}, k_{Y_1}, d_{Y_1}] = [6, 3, 3],$$
$$[n_{X_1}, k_{X_1}, d_{X_1}] = [3, 2, 2],$$
$$[n_{X_2}, k_{X_2}, d_{X_2}] = [3, 2, 2].$$

The parameter $d_X$ is the minimum (Hamming) distance of $C_X$. We say that $C$ is an $[n, k, d]$-code with $[n, k, d] = [n_E, k_E, d_E]$.

We choose the following definition for *general* $(n, k, d, r, \delta, t)$-LRCs (i.e., both linear and nonlinear), which we will compare to known results for *linear* LRCs.

**Definition 4**  An $(n, k, d)$-*code* $C$ over $A$ is a nonempty subset $C$ of $A^n$, where $A$ is a finite set of size $s$, $k = \log_s(|C|)$, and $d$ the minimum (Hamming) distance of the code. For $X = \{i_1, \ldots, i_m\} \subseteq E$, the puncturing $C_X$ is defined as

$$C_X = \{(c_{i_1}, \ldots, c_{i_m}) : c \in C\}.$$

The code $C$ is *non-degenerate*, if $d \geq 2$ and $|C_{\{i\}}| > 1$ for all coordinates $i \in [n]$.

**Definition 5**  A *locally repairable code* over $A$ is a non-degenerate $(n, k, d)$-code $C$. A coordinate $x \in [n]$ of $C$ has *locality* $(r, \delta)$ and *availability* $t$ if there are $t$ subsets $R_1, \ldots, R_t$ of $[n]$, called repair sets of $x$, such that for $i, j \in [t]$

$$(i) \ x \in R_i,$$
$$(ii) \ |R_i| \leq r + \delta - 1,$$
$$(iii) \ d(C_{R_i}) \geq \delta,$$
$$(iv) \ i \neq j \ \Rightarrow \ R_i \cap R_j = \{x\}.$$

If every element $x \in X \subseteq E$ has availability with parameters $(n, k, d, r, \delta, t)$ in $C$, then we say that the set $X$ has $(n, k, d, r, \delta, t)$-availability in $C$. We will often talk about codes with $(n, k, d, r, \delta)$-locality, by which we mean a code that has $(n, k, d, r, \delta, 1)$-availability, so that symbols are not required to be included in more than one repair set. If the other parameters are clear from the context, we may shortly say that $X$ has "locality $(r, \delta)$" or "availability $t$", along the lines of the above definition.

An *information set* of a linear $[n, k, d]$-code $C$ is defined as a set $X \subseteq E$ such that $k_X = |X| = k$. Hence, $X$ is an information set of $C$ if and only if there is a generator matrix $G$ of $C$ such that $G(X)$ equals the identity matrix, i.e., $C$ is systematic in the coordinate positions indexed by $X$ when generated by $G$. In terms of storage systems, this means that the nodes in $X$ together store all the information of the DSS.

*Example 3*  Two examples of an information set of the linear code $C$ generated by $G$ in Example 1 are $\{1, 2, 3, 4\}$ and $\{1, 2, 6, 8\}$.

More formally we define:

**Definition 6** Let $C$ be an $(n, k, d)$-code and $X$ a subset of $[n]$. Then $X$ is an *information set* of $C$ if $\log_s(|C_X|) = k$ and $\log_s(C_Y) < k$ for all $Y \subsetneq X$. Further, $X$ is *systematic* if $k$ is an integer, $|X| = k$ and $C_X = A^k$. Also, $X$ is an *all-coordinate set* if $X = [n]$.

**Definition 7** A *systematic-symbol*, *information-symbol*, and *all-symbol* LRC, respectively, is an $(n, k, d)$-LRC with a systematic, information set and all-coordinate set $X$, such that every coordinate in $X$ has locality $(r, \delta)$ and availability $t$. These are denoted by

$$(n, k, d, r, \delta, t)_s\text{-LRC},\quad (n, k, d, r, \delta, t)_i\text{-LRC},\quad \text{and}\quad (n, k, d, r, \delta, t)_a\text{-LRC},$$

respectively. Further, when availability is not considered ($t = 1$), we get natural notions of $(n, k, d, r, \delta)_s$, $(n, k, d, r, \delta)_i$, and $(n, k, d, r, \delta)_a$-LRCs.

## 2  Introduction to Matroids

Matroids were first introduced by Whitney in 1935, to capture and generalise the notion of linear dependence in purely combinatorial terms [48]. Indeed, the combinatorial setting is general enough to also capture many other notions of dependence occurring in mathematics, such as cycles or incidences in a graph, non-transversality of algebraic varieties, or algebraic dependence of field extensions. Although the original motivation comes from linear algebra, we will see that a lot of matroid terminology comes from graph theory and projective geometry. More details about these aspects of matroid theory are relegated to the appendix.

### 2.1  Definitions

We begin by presenting two equivalent definitions of a matroid.

**Definition 8** *(Rank function)* A *(finite) matroid* $M = (\rho, E)$ is a finite set $E$ together with a *rank function* $\rho : 2^E \rightarrow \mathbb{Z}$ such that for all subsets $X, Y \subseteq E$

$$(R.1)\ 0 \le \rho(X) \le |X|,$$
$$(R.2)\ X \subseteq Y \implies \rho(X) \le \rho(Y),$$
$$(R.3)\ \rho(X) + \rho(Y) \ge \rho(X \cup Y) + \rho(X \cap Y).$$

An alternative but equivalent definition of a matroid is the following.

**Definition 9** *(Independent sets)* A *(finite) matroid* $M = (\mathscr{I}, E)$ is a finite set $E$ and a collection of subsets $\mathscr{I} \subseteq 2^E$ such that

(*I*.1) $\emptyset \in \mathscr{I}$,

(*I*.2) $Y \in \mathscr{I}, X \subseteq Y \Rightarrow X \in \mathscr{I}$,

(*I*.3) For all pairs $X, Y \in \mathscr{I}$ with $|X| < |Y|$, there exists
$y \in Y \setminus X$ such that $X \cup \{y\} \in \mathscr{I}$.

The subsets in $\mathscr{I}$ are the *independent sets* of the matroid.

The rank function $\rho$ and the independents sets $\mathscr{I}$ of a matroid on a ground set $E$ are linked as follows: For $X \subseteq E$,

$$\rho(X) = \max\{|Y| : Y \subseteq X \text{ and } Y \in \mathscr{I}\},$$

and $X \in \mathscr{I}$ if and only if $\rho(X) = |X|$. It is an easy (but not trivial) exercise to show that the two definitions are equivalent under this correspondence. Another frequently used definition is in terms of the set of *bases* for a matroid, which are the maximal independent sets. Further, we will also use the *nullity function* $\eta : 2^E \to \mathbb{Z}$, where $\eta(X) = |X| - \rho(X)$ for $X \subseteq E$.

Any matrix $G$ over a field $\mathbb{F}$ generates a matroid $M_G = (\rho, E)$, where $E$ is the set of columns of $G$, and $\rho(I)$ is the rank over $\mathbb{F}$ of the induced matrix $G(I)$ for $I \subseteq E$. Consequently, $I \in \mathscr{I}$ precisely when $I$ is a linearly independent set of vectors. It is straightforward to check that this is a matroid according to Definition 8. As elementary row operations preserve the row space $R(G(I))$ for all $I \subseteq E$, it follows that row equivalent matrices generate the same matroid.

Two matroids $M_1 = (\rho_1, E_1)$ and $M_2 = (\rho_2, E_2)$ are *isomorphic* if there exists a bijection $\psi : E_1 \to E_2$ such that $\rho_2(\psi(X)) = \rho_1(X)$ for all subsets $X \subseteq E_1$.

**Definition 10** A matroid that is isomorphic to $M_G$ for some matrix $G$ over $\mathbb{F}$ is said to be *representable* over $\mathbb{F}$. We also say that such a matroid is $\mathbb{F}$-*linear*.

Two trivial matroids are the zero matroid where $\rho(X) = 0$ for each set $X \subseteq E$, and the one where $\rho(X) = |X|$ for all $X \subseteq E$. These correspond to all-zeros matrices and invertible $n \times n$-matrices respectively. The first non-trivial example of a matroid is the following:

**Definition 11** The *uniform matroid* $U_n^k = (\rho, [n])$, where $[n] = \{1, 2, \ldots, n\}$, is given by the rank function $\rho(X) = \min\{|X|, k\}$ for $X \subseteq [n]$.

The following straightforward observation gives yet another characterisation of MDS codes.

**Proposition 1** *$G$ is the generator matrix of an $[n, k, n - k + 1]$-MDS code if and only if $M_G$ is the uniform matroid $U_n^k$.*

## 2.2 Matroid Operations

For explicit constructions of matroids, as well as for analysing their structure, a few elementary operations are useful. Here, we will define these in terms of the rank function, but observe that they can equally well be formulated in terms of independent sets. The effect of these operations on the representability of the matroid is discussed in appendix. In addition to the operations listed here, two other very important matroid operations are dualisation and contraction. As these are not explicitly used here to understand locally repairable codes, we leave their definition to appendix.

**Definition 12** The *direct sum* of two matroids $M = (\rho_M, E_M)$ and $N = (\rho_N, E_N)$ is

$$M \oplus N = (\tau, E_M \sqcup E_N),$$

where $\sqcup$ denotes the disjoint union, and $\tau : 2^{E_M \sqcup E_N} \to \mathbb{Z}$ is defined by $\tau(X) = \rho_M(X \cap E_M) + \rho(X \cap E_N)$.

Thus, all dependent sets from $M$ and $N$ remain dependent in $M \oplus N$, whereas there is no dependence between elements in $M$ and elements in N. If $M$ and $N$ are graphical matroids,[3] then $M \oplus N$ is graphical, and obtained from the disjoint union of the graphs associated to $M$ and $N$.

**Definition 13** The *restriction* of $M = (\rho, E)$ to a subset $X \subseteq E$ is the matroid $M_{|X} = (\rho_{|X}, X)$, where

$$\rho_{|X}(Y) = \rho(Y), \text{ for } Y \subseteq X. \tag{1}$$

Obviously, for any matroid $M$ with underlying set $E$, we have $M_{|E}=M$. The restriction operation is also often referred to as *deletion* of $E \setminus X$, especially if $E \setminus X$ is a singleton. Given a matrix $G$ that represents $M$, the submatrix $G(X)$ represents $M_{|X}$.

**Definition 14** The truncation of a matroid $M = (\rho, E)$ at rank $k \leq \rho(E)$ is $M_k = (\rho', E)$, where $\rho'(X) = \min\{\rho(X), k\}$.

Geometrically, the truncation of a matroid corresponds to projecting a point configuration onto a *generic* $k$-dimensional space. However, this does not imply that truncations of $\mathbb{F}$-linear matroids are necessarily $\mathbb{F}$-linear, as it may be the case that there exists no $k$-space that is in general position relative to the given point configuration. However, it is easy to see that $M_k$ is always representable over some field extension of $\mathbb{F}$. In fact, via a probabilistic construction, one sees that the field extension can be chosen to have size at most $q\binom{n}{k}$ [17].

The *relaxation* is the elementary operation that is most difficult to describe in terms of rank functions. It is designed to destroy representability of matroids, and corresponds to selecting a hyperplane in the point configuration, and perturbing it so that its points are no longer coplanar. To prepare for the definition, we say that a *circuit* is a dependent set, all of whose subsets are independent. For any nonuniform matroid $M = (\rho, E)$, there are circuits of rank $\rho(E) - 1$. This is seen by taking any dependent set of rank $\rho(E) - 1$, and deleting elements successively in such a way that the rank does not decrease. We mention that a matroid that has no circuits of

---

[3]See appendix for the definition of graphical matroids.

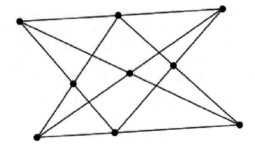

**Fig. 1** The non-Pappus matroid, which is not representable over any field. If there were a line between the three points in the middle, then the figure would illustrate the matroid $M_G$ in Example 4. Relaxation of the circuit $\{4, 5, 6\}$ corresponds to deletion of this line in the figure

rank $< \rho(E) - 1$ is called a *paving matroid*. It is conjectured that asymptotically (in the size) almost all matroids are paving [24]. Recent research shows that this is true at least on a "logarithmic scale" [28].

**Definition 15** Let $M = (\mathscr{I}, E)$ be a matroid with rank function $\rho$, and let $C$ be a circuit of rank $\rho(E) - 1$. The *relaxation* of $M$ at $C$ is the matroid $(\mathscr{I} \cup \{C\}, E)$.

*Example 4* The first example of a matroid constructed by relaxation is the *non-Pappus matroid* of Fig. 1. This is constructed by relaxing the circuit $\{4, 5, 6\}$ from the representable matroid $M_G$, where

$$
G = \begin{array}{c} \begin{array}{ccccccccc} 1 & 2 & 3 & 4 & 5 & 6 & 7 & 8 & 9 \end{array} \\ \left[ \begin{array}{ccc|ccc|c|c|c} 1 & 0 & -1 & 1/2 & 0 & -1/2 & 1 & 0 & -1 \\ 1 & 1 & 1 & 0 & 0 & 0 & -1 & -1 & -1 \\ 1 & 1 & 1 & 1 & 1 & 1 & 1 & 1 & 1 \end{array} \right] \end{array}
$$

and can be defined over any field of odd (or zero) characteristic, other than $\mathbb{F}_3$.

## 2.3 Matroid Invariants of Codes

There is a straightforward connection between linear codes and matroids. Indeed, let $C$ be a linear code generated by a matrix $G$. Then $C$ is associated with the matroid $M_G = (\rho, E)$. As two different generator matrices of $C$ have the same row space, they will generate the same matroid. Therefore, without any inconsistency, we denote the associated linear matroid of $C$ by $M_C = (\rho_C, E)$. In general, there are many different codes $C \neq C'$ with the same matroid structure $M_C = M_{C'}$. In the appendix, we will see how this phenomenon can be interpreted as a stratification of the Grassmannian over a finite field.

A property of linear codes that depends only on the matroid structure of the code is called *matroid invariant*. For example, the collection of information sets and the parameters $[n, k, d]$ of a code are matroid invariant properties. This is the content of the following easy proposition.

**Proposition 2** *Let $C$ be a linear $[n, k, d]$-code and $X \subseteq E$. Then for $M_C = (\rho_C, E)$,*

$(i)\ n_X = |X|,$

$(ii)\ k_X = \rho_C(X),$

$(iii)\ d_X = \min\{|Y| : Y \subseteq X, \rho_C(X \setminus Y) < \rho_C(X)\},$

$(iv)\ X$ *is an information set of $C$* $\Longleftrightarrow$
$X$ *is a basis of $M_C$* $\Longleftrightarrow$ $\rho_C(X) = |X| = k.$

In addition to the parameters $[n, k, d]$ of a linear code $C$, we are also interested in the length, rank and minimum distance of the punctured codes, since these correspond to the locality parameters at the different hierarchy levels, which we will discuss in more detail in Sect. 5.

A punctured code can be analysed using matroid restrictions, since $M_{C|X} = M_C|X$ for every coordinate subset $X$. Thus, the parameters $[n_X, k_X, d_X]$ of $C_X$ are also matroid invariant properties for $C$.

*Example 5* Let $C$ denote the $[n, k, d]$-code generated by the matrix $G$ given in Example 1. Then $[n, k, d] = [9, 4, 3]$, where the value of $d$ arises from the fact that $\rho_C([9] \setminus \{i, j\}) = 4$ for $i, j = 1, 2, \ldots, 9$, and $\rho_C([9] \setminus \{4, 8, 9\}) = 3$. Two information sets of $C$ are $\{1, 2, 3, 4\}$ and $\{1, 2, 6, 8\}$, as we already saw before in Example 3.

It is rather easy to see that two different linear codes can have the same associated matroid. As a consequence, not every property of a linear code is matroid invariant. An example of a code invariant that is not matroid invariant is the covering radius [5, 37]. Indeed, an $[n, k, d]$-MDS code, i.e., a realisation of the uniform matroid $U_n^k$, generically has covering radius $d - 1 = n - k$, yet there exist MDS codes with lower covering radii. An explicit example is given in [5].

## 2.4 The Lattice of Cyclic Flats

One matroid invariant that has singled out as essential for describing the repairability of storage codes is the lattice of cyclic flats. To define this, remember that $X \subseteq E$ is a *circuit* in $M = (\mathscr{I}, E)$ if $X$ is dependent, but all proper subsets of $X$ are independent. A *cyclic set* is a (possibly empty) union of circuits. Equivalently, $X$ is *cyclic* if for every $x \in X$

$$\rho(X \setminus \{x\}) = \rho(X).$$

Let us define the operation $\mathrm{cyc} : 2^E \to 2^E$ by

$$\mathrm{cyc}(X) = \{x \in X : \rho(X \setminus \{x\}) = \rho(X)\}.$$

Then $X$ is cyclic if and only if $\mathrm{cyc}(X) = X$. We refer to cyc as the *cyclic core* operator.

Dually, we define the *closure* of $X$ to be

$$cl(X) = \{y \in E : \rho(X \cup \{y\}) = \rho(X)\},$$

and notice that $X \subseteq cl(X)$ by definition. We say that $X$ is a *flat* if $cl(X) = X$. Therefore, $X$ is a *cyclic flat* if

$$\rho(X \setminus \{x\}) = \rho(X) \text{ and } \rho(X \cup \{y\}) > \rho(X)$$

for all $x \in X$ and $y \in E \setminus X$. The set of flats, cyclic sets, and cyclic flats of $M$ are denoted by $\mathscr{F}(M)$, $\mathscr{U}(M)$, and $\mathscr{Z}(M)$, respectively.

It is not entirely obvious that the set of cyclic flats is nonempty. However, it follows from the matroid axioms that the closure operator cl preserves cyclicity, and that the cyclic core operator cyc preserves flatness. Thus we can consider cl and cyc as maps

$$cl : \begin{cases} 2^E \to \mathscr{F}(M) \\ \mathscr{U}(M) \to \mathscr{Z}(M) \end{cases},$$

and

$$cyc : \begin{cases} 2^E \to \mathscr{U}(M) \\ \mathscr{F}(M) \to \mathscr{Z}(M) \end{cases}.$$

In particular, for any set $X \subseteq E$, we have $cyc \circ cl(X) \in \mathscr{Z}(M)$ and $cl \circ cyc(X) \in \mathscr{Z}(M)$.

Let $M[G] = (\rho, E)$ be a linear matroid, generated by $G$. Then $X \subseteq E$ is a cyclic flat if and only if the following two conditions are satisfied

$$(i) \ C(G(X)) \cap C(G(E \setminus X)) = \mathbf{0}$$
$$(ii) \ x \in X \Rightarrow C(G(X \setminus \{x\})) = C(G(X)).$$

In terms of storage codes, a cyclic flat is thus a set $X \subseteq E$ of storage nodes such that every node in $X$ can be repaired by the other nodes in $X$, whereas no node outside $X$ can be repaired by $X$. This observation shows the relevance of cyclic flats for storage applications. The strength of using $\mathscr{Z}(M)$ as a theoretical tool comes from its additional *lattice structure*, which we will discuss next.

A collection of sets $\mathscr{P} \subseteq 2^E$ ordered by inclusion defines a partially ordered set (poset) $(\mathscr{P}, \subseteq)$. Let $X$ and $Y$ denote two elements of $\mathscr{P}$. $Z$ is the *join* $X \vee Y$ if it is the unique maximal element in $\{W \in \mathscr{P} : X \subseteq W, Y \subseteq W\}$. Dually, $Z$ is the *meet* $X \wedge Y$ if it is the unique minimal element in $\{W \in \mathscr{P} : X \supseteq W, Y \supseteq W\}$.

A pair of elements in an arbitrary poset does not need to have a join or a meet. If $(\mathscr{P}, \subseteq)$ is a poset such that every pair of elements in $\mathscr{P}$ has a join and a meet, then $\mathscr{P}$ is called a *lattice*. The bottom and top elements of a finite lattice $(\mathscr{P}, \subseteq)$ always exist, and are denoted by $1_{\mathscr{P}} = \bigvee_{X \in \mathscr{P}} X$ and $0_{\mathscr{P}} = \bigwedge_{X \in \mathscr{P}} X$, respectively.

Two basic properties of cyclic flats of a matroid are given in the following proposition.

**Proposition 3** ([4]) *Let $M = (\rho, E)$ be a matroid and $\mathscr{Z}$ the collection of cyclic flats of M. Then,*

*(i)* $\rho(X) = \min\{\rho(F) + |X \setminus F| : F \in \mathscr{Z}\}$, *for* $X \subseteq E$,
*(ii)* $(\mathscr{Z}, \subseteq)$ *is a lattice,* $X \vee Y = \mathrm{cl}(X \cup Y)$ *and*
$X \wedge Y = \mathrm{cyc}(X \cap Y)$ *for* $X, Y \in \mathscr{Z}$.

Proposition 3 (i) shows that a matroid is uniquely determined by its cyclic flats and their ranks.

*Example 6* Let $M_C = (\rho_C, E)$ be the matroid associated to the linear code $C$ generated by the matrix $G$ given in Example 1. The lattice of cyclic flats $(\mathscr{Z}, \subseteq)$ of $M_C$ is given in the figure below, where the cyclic flat and its rank are given at each node.

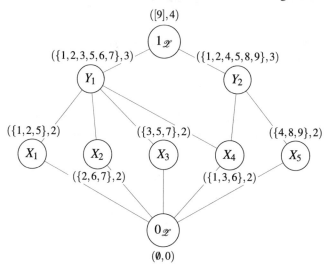

An axiom scheme for matroids via cyclic flats and their ranks was independently given in [4, 35]. This gives a compact way to construct matroids with prescribed local parameters, which we have exploited in [47].

**Theorem 1** (see [4] Theorem 3.2 and [35]) *Let $\mathscr{Z} \subseteq 2^E$ and let $\rho$ be a function $\rho : \mathscr{Z} \to \mathbb{Z}$. There is a matroid M on E for which $\mathscr{Z}$ is the set of cyclic flats and $\rho$ is the rank function restricted to the sets in $\mathscr{Z}$, if and only if*

(Z0) $\mathscr{Z}$ *is a lattice under inclusion,*
(Z1) $\rho(0_{\mathscr{Z}}) = 0$,
(Z2) $X, Y \in \mathscr{Z}$ *and* $X \subsetneq Y \Rightarrow$
$\quad 0 < \rho(Y) - \rho(X) < |Y| - |X|$,
(Z3) $X, Y \in \mathscr{Z} \Rightarrow \rho(X) + \rho(Y) \geq$
$\quad \rho(X \vee Y) + \rho(X \wedge Y) + |(X \cap Y) \setminus (X \wedge Y)|$.

For a linear $[n, k, d]$-code $C$ with $M_C = (\rho_C, E)$ and $\mathscr{L} = \mathscr{L}(M_C)$, and for a coordinate $x$, we have

$$(i)\ d \geq 2 \iff 1_{\mathscr{L}} = E,$$
$$(ii)\ C_{\{x\}} \neq \{0_{\mathbb{F}}\} \text{ for every } x \in E \iff 0_{\mathscr{L}} = \emptyset.$$

Hence, by Definition 1, we can describe non-degeneracy in terms of the lattice of cyclic flats, as follows.

**Proposition 4** *Let $C$ be a linear $[n, k, d]$-code and $\mathscr{L}$ denote the collection of cyclic flats of the matroid $M_C = (\rho_C, E)$. Then $C$ is a non-degenerate storage code if and only if $0_{\mathscr{L}} = \emptyset$ and $1_{\mathscr{L}} = E$.*

**Proposition 5** *Let $C$ be a non-degenerate storage code and $M_C = (\rho_C, E)$. Then, for $X \subseteq E$, $C_X$ is a non-degenerate storage code if and only if $X$ is a cyclic set of $M_C$.*

As cyclic sets correspond to non-degenerate subcodes, and hence to systems where every symbol is stored with redundancy, we will use these as our "repair sets". Therefore, we want to determine from the lattice of cyclic flats, whether a set is cyclic or not, which we achieve through the following theorem.

**Theorem 2** *Let $M = (\rho, E)$ be a matroid with $0_{\mathscr{L}} = \emptyset$ and $1_{\mathscr{L}} = E$ where $\mathscr{L} = \mathscr{L}(M)$. Then, for any $X \subseteq E$, $X \in \mathscr{U}(M)$ if and only if the cyclic flat*

$$F^X = \bigwedge \{F \in \mathscr{L} : X \subseteq F\}$$

*is such that*

$$\rho(F) + |X \setminus F| > \rho(F^X).$$

*for all $F \subsetneq F^X$ in $\mathscr{L}$.*

If this is indeed the case, then it is easy to verify that $F^X$ as defined in Theorem 2 is indeed the closure $\text{cl}(X)$ as defined earlier. In order to analyze the parameters $[n_X, k_X, d_X]$ of a punctured code $C_X$, we will use the lattice of cyclic flats of $M_{C|X}$.

**Theorem 3** *Let $M = (\rho, E)$ be a matroid with $0_{\mathscr{L}} = \emptyset$ and $1_{\mathscr{L}} = E$ where $\mathscr{L} = \mathscr{L}(M)$. Then, for $X \in \mathscr{U}(M)$,*

$$(i)\ \mathscr{L}(M|X) = \{X \cap F \in \mathscr{U}(M) : F \in \mathscr{L}, F \subseteq F^X\},$$
$$(ii)\ Y \in \mathscr{L}(M|X) \Rightarrow \rho_{|X}(Y) = \rho(F^Y).$$

We remark that if $X$ is a cyclic flat of a matroid $M$, then $\mathscr{L}(M|X) = \{F \in \mathscr{L}(M) : F \subseteq X\}$.

*Example 7* Let $\mathscr{L} = \mathscr{L}(M_C)$ be the lattice of cyclic flats given in Example 6, where $M_C = (\rho_C, E)$ is the matroid associated to the linear LRC $C$ generated by the matrix

$G$ given in Example 1. Then, $F^X = F^Y = Y_1$ for $X = \{1, 2, 3, 7\}$ and $Y = \{1, 2, 3\}$. Further, $X$ is a cyclic set but $Y$ is not a cyclic set. The lattice of cyclic flats $(\mathscr{Z}_X, \subseteq)$ for $M_{C|X} = M_C|X$ is shown in the following figure.

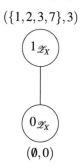

$$(\{1,2,3,7\},3)$$

$$1_{\mathscr{Z}_X}$$

$$0_{\mathscr{Z}_X}$$

$$(\emptyset, 0)$$

The very simple structure of $\mathscr{Z}_X$ shows that $C_X$ has the very favourable property of being an MDS code. Indeed, the following proposition is immediate from the definitions of the involved concepts.

**Proposition 6** *Let $C$ be a linear code of length $n$ and rank $k$. The following are equivalent:*

(i) *$C$ is an $[n, k, n - k + 1]$-MDS code.*
(ii) *$M_C$ is the uniform matroid $U_n^k$.*
(iii) *$\mathscr{Z} = \mathscr{Z}(M_C)$ is the two element lattice with $1_{\mathscr{Z}} = [n]$ and $0_{\mathscr{Z}} = \emptyset$.*

For linear LRCs we are also interested in when a coordinate set is an information set, or equivalently, if it is a basis for the matroid. This property is determined by the cyclic flats as follows.

**Proposition 7** *Let $C$ be a linear $[n, k, d]$-code with $0_{\mathscr{Z}} = \emptyset$ and $1_{\mathscr{Z}} = E$ where $\mathscr{Z}$ is the collection of cyclic flats of the matroid $M_C = (\rho_C, E)$. Then, for any $X \subseteq E$, $X$ is an information set of $C$ if and only if the following two conditions are satisfied,*

$$(i) \; |X| = \rho_C(1_{\mathscr{Z}}),$$
$$(ii) \; |X \cap F| \leq \rho_C(F) \text{ for every } F \in \mathscr{Z}.$$

*Example 8* Let $C$ be the linear $[n, k, d]$-code generated by the matrix $G$ given in Example 1. Then, by the lattice of cyclic flats for $M_C$ given in Example 6, $C$ is a linear LRC with all-symbol $(2, 2)$-locality. We notice that, by Proposition 7, $\{1, 2, 3, 4\}$ is an information set of $C$. This follows as it is not contained in either of $Y_1$ and $Y_2$, while all its subsets are. On the other hand, $\{1, 2, 8, 9\}$ is not an information set of $C$, as it is itself a subset of $Y_2$.

The parameters $[n, k, d]$ of a linear LRC $C$ and $[n_X, k_X, d_X]$ of a punctured code $C_X$ that is a non-degenerate storage code can now be determined by the lattice of cyclic flats as follows.

**Theorem 4** *Let C be a linear $[n, k, d]$-LRC, where $\mathscr{L} = \mathscr{L}(M_C)$ for the matroid $M_C = (\rho_C, E)$. Then, for any $X \in \mathscr{U}(M_C)$, $C_X$ is a linear $[n_X, k_X, d_X]$-LRC with*

$$(i) \; n_X = |X|,$$
$$(ii) \; k_X = \rho(F^X),$$
$$(iii) \; d_X = n_X - k_X + 1 - \max\{\eta(Y) : Y \in \mathscr{L}(M_C|X) \setminus X\}.$$

*Example 9* Let $\mathscr{L} = \mathscr{L}(M_C)$ be the lattice of cyclic flats given in Example 6, where $M_C = (\rho_C, E)$ is the matroid associated to the linear $[n, k, d]$-LRC $C$ generated by the matrix $G$ given in Example 1. Then by Example 7, $X = \{1, 2, 3, 7\}$ is a cyclic set, and by Theorem 4 $C_X$ is a linear $[n_X, k_X, d_X]$-LRC with parameters $n_X = 4, k_X = 3$ and $d_X = 4 - 3 + 1 - 0 = 2$. Moreover, $n = n_E = 9, k = k_E = 4$ and $d = d_E = 9 - 4 + 1 - 3 = 3$.

## 3 Singleton-Type Bounds

Many important properties of a linear code $C$ are due to its matroid structure, which is captured by the matroid $M_C$. By the results in [10, 47], matroid theory seems to be particularly suitable for proving Singleton-type bounds for linear LRCs and nonexistence results of Singleton-optimal linear LRCs for certain parameters.

Even though matroids can be associated to linear codes to capture the key properties for linear LRCs, this cannot be done in general for nonlinear codes. Fortunately, by using some key properties of entropy, any code (either linear and nonlinear) $C$ can be associated with a polymatroid $P_C$ so that $P_C$ captures the key properties of the code when it is used as a LRC. A polymatroid is a generalisation of a matroid. For any linear code $C$ the associated polymatroid $P_C$ and matroid $M_C$ are the same object. We will briefly discuss the connection between polymatroids and codes in Sect. 5. Singleton-type bounds for polymatroids were derived in [46], and polymatroid theory for its part seems to be particularly suitable for proving such bounds for general LRCs. In Sect. 5 we will also review Singleton-type bounds for polymatroids and general codes with availability and hierarchy.

### 3.1 Singleton-Type Bounds for Matroids and Linear LRCs

Matroid theory provides a unified way to understand and connect several different branches of mathematics, for example linear algebra, graph theory, combinatorics, geometry, topology and optimisation theory. Hence, a theorem proven for matroids gives "for free" theorems for many different objects related to matroid theory. As described earlier, the key parameters $(n, k, r, d, \delta, t)$ of a linear LRCs $C$ are matroid properties of the matroid $M_C$ and can therefore be defined for matroids in general.

**Definition 16** Let $M = (\rho, E)$ be a matroid and $X \subseteq E$. Then

(i) $n_X = |X|$,
(ii) $k_X = \rho(X)$,
(iii) $d_X = \min\{|Y| : Y \subseteq X, \rho(X \setminus Y) < \rho(X)\}$,
(iv) $X$ is an information set of $M$ if $\rho(X) = k_E$ and $\rho(Y) < k_E$ for all $Y \subsetneq X$,
(v) $M$ is non-degenerate if $\rho(x) > 0$ for all $x \in E$ and $d_E \geq 2$.

Further, $n = n_E$, $k = k_E$ and $d = d_E$, and the definitions of repair sets, $(r, \delta)$-locality and $t$-availability for elements $x \in E$, as well as the concepts of $(n, k, d, r, \delta, t)_i$-matroids and $(n, k, d, r, \delta, t)_a$-matroids are directly carried over from Definitions 5 and 7.

Before stating Theorem 5 below, it is not at all clear that the Singleton-type bounds, already proven for linear LRCs, also hold for matroids in general. Especially, one could doubt this generality of the bound because of the wide connection between matroids and a variety of different mathematical objects, as well as for the sake of the recently proven result, stated in Theorem 13 later on, that almost all matroids are nonrepresentable. However, Theorem 5 gives a Singleton-type bound that holds for matroids in general. This implies, as special cases, the same bound on linear LRCs and other objects related to matroids, e.g., graphs, almost affine LRCs, and transversals. For the bound to make sense for various objects, a description of the parameters $(n, k, d, r, \delta)$ has to be given for the objects in question. To give an example, for a graph,

- $n$ equals the number of edges,
- $k$ equals the difference of the number of vertices and the number of connected components,
- $d$ is the smallest number of edges in an edge-cut (i.e., a set of edges whose removal increases the number of connected components in the graph).

Recall that the Singleton bound [36] states that for any linear $[n, k, d]$-code we have

$$d \leq n - k + 1. \tag{2}$$

In what follows, we state generalised versions of this bound, accounting for the various parameters relevant for storage systems. We start with the general one for matroids.

**Theorem 5** ([47] Singleton-type bound for matroids) *Let $M = (\rho, E)$ be an $(n, k, d, r, \delta)_i$-matroid. Then*

$$d \leq n - k + 1 - \left( \left\lceil \frac{k}{r} \right\rceil - 1 \right)(\delta - 1). \tag{3}$$

Theorem 5 was stated for all-symbol locality in [47]. However, the proof given in [47] implies also information-symbol locality. As an illustration on how matroid theory and cyclic flats can be useful for proving Singleton-type of bounds we will here give a proof of Theorem 5.

*Proof* We know from Theorem 4 that $d = n - k + 1 - \max\{\eta(Z) : Z \in \mathscr{Z} \setminus E\}$. Hence to prove the theorem we need to show that there exists a cyclic flat $Z \neq E$ in $M$ with $\eta(Z) \geq \left(\left\lceil \frac{k}{r} \right\rceil - 1\right)(\delta - 1)$.

Let $B$ be an information set of $M$, i.e, $B$ is a basis of $M$, such that $M$ is an $(n, k, d, r, \delta)_i$-matroid. For $x \in B$ let $R_x$ denote the repair set of $x$. Since $R_x$ is a cyclic set of we obtain that $Z_x = \mathrm{cl}(R_x)$ is a cyclic flat of $M$ with

$$\rho(Z_x) = \rho(R_x) \leq r \text{ and } \eta(Z_x) \geq \eta(R_x) \geq d_{R_x} - 1 \geq \delta - 1.$$

As $\rho(B) = k$ we can choose a subset of cyclic flats $\{Z_1, \ldots, Z_m\} \subseteq \{Z_x : x \in B\}$ such that we obtain a chain of cyclic flats

$$\emptyset = Y_0 \subsetneq Y_1 \subsetneq \cdots \subsetneq Y_m = E$$

with $Y_i = Y_{i-1} \vee Z_i$ for $i = 1, \ldots, m$. Since $\rho(Y_0) = \eta(Y_0) = 0$ and $\rho(Y_m) = k$, the theorem will now be proved if we can prove that $\rho(Y_i) - \rho(Y_{i-1}) \leq r$ and $\eta(Y_i) - \eta(Y_{i-1}) \geq \delta - 1$ for $i = 1, \ldots, m$.

First, by the use of Axiom (R.3) and Proposition 3,

$$\rho(Y_i) = \rho(\mathrm{cl}(Y_{i-1} \cup Z_i)) = \rho(Y_{i-1} \cup Z_i) \leq \rho(Y_{i-1}) + \rho(Z_i) - \rho(Y_{i-1} \cap Z_i)$$
$$\leq \rho(Y_{i-1}) + r.$$

Second, by Axiom (R.3), $\eta(X) + \eta(Y) \leq \eta(X \cap Y) + \eta(X \cup Y)$ for $X, Y \subseteq E$. Further, we observe that, $\mathrm{cyc}(Y_{i-1} \cap Z_i)$ and $Z_i$ are cyclic flats of $M|Z_i$ and that $\mathrm{cyc}(Y_{i-1} \cap Z_i) \subsetneq Z_i$. Hence,

$$\eta(Y_i) = \eta(\mathrm{cl}(Y_{i-1} \cup Z_i)) \geq \eta(Y_{i-1} \cup Z_i) \geq \eta(Y_{i-1}) + \eta(Z_i) - \eta(Y_{i-1} \cap Z_i)$$
$$= \eta(Y_{i-1}) + \eta(Z_i) - \eta(\mathrm{cyc}(Y_{i-1} \cap Z_i))$$
$$= \eta(Y_{i-1}) + |Z_i| - \rho(Z_i) - \eta(\mathrm{cyc}(Y_{i-1} \cap Z_i)) \geq \eta(Y_{i-1}) + d_{Z_i} - 1$$
$$\geq \eta(Y_{i-1}) + \delta - 1.$$

That $d_{Z_i} \geq \delta$ follows from the fact that $Z_i = \mathrm{cl}(R_x)$ for some $x \in B$ and therefore

$$d_{Z_i} = d_{\mathrm{cl}(R_x)} = \min\{|Y| : Y \subseteq \mathrm{cl}(R_x), \rho(\mathrm{cl}(R_x) \setminus Y) \leq \rho(\mathrm{cl}(R_x))\}$$
$$\geq \min\{|Y| : Y \subseteq R_x, \rho(R_x \setminus Y) \leq \rho(R_x)\}$$
$$= d_{R_x}.$$

$\square$

The Singleton bound given in (2) was generalised by Gopalan et al. in [13] as follows. A linear $(n, k, d, r)_i$-LRC satisfies

$$d \leq n - k + 1 - \left( \left\lceil \frac{k}{r} \right\rceil - 1 \right). \tag{4}$$

The bound (4) shows that there is a penalty for requiring locality. That is, the smaller the locality $r$ the smaller the upper bound on $d$. By the definition of LRCs, any linear $[n, k, d]$-code with locality $r$ is also a linear $[n, k, d]$-code with locality $k$. Hence, by letting the locality be $k$, the bound (4) implies (2).

The bound (4) was generalised in [30] as follows. A linear $(n, k, d, r, \delta)_i$-LRC satisfies

$$d \leq n - k + 1 - \left( \left\lceil \frac{k}{r} \right\rceil - 1 \right) (\delta - 1). \tag{5}$$

The bound (5) again shows that there is a penalty on the upper bound for $d$ depending on the size of the local distance $\delta$. This is, the bigger the local distance $\delta$ the smaller the upper bound on the global distance $d$. However, we must remark that any linear $(n, k, d, r, \delta)_i$-LRC satisfies $d \geq \delta$, and this property also holds more generally for matroids [47]. The bound (4) follows from the bound (5) by letting $\delta = 2$.

A bound including availability was proven in [44]. This bound states that a linear $(n, k, d, r, t)_i$-LRC satisfies

$$d \leq n - k + 1 - \left( \left\lceil \frac{t(k-1)+1}{t(r-1)+1} \right\rceil - 1 \right). \tag{6}$$

Again, the bound (4) follows from (6) above by letting $t = 1$.

The bounds (3)–(6) are stated assuming information-symbol locality. However, since every matroid contains an information set, this implies that the bound is also valid under the stronger assumption of all-symbol locality.

## 3.2 Stronger Bounds for Certain Parameter Values

A linear LRC, or more generally a matroid, that achieves any of the Singleton-type bounds given above will henceforth be called *Singleton-optimal*.

Any $(n, k, d, r, \delta)_i$-matroid $M$ satisfies that $\delta \leq d$. Hence, by the bound (5), $k \leq n - (\delta - 1) \lceil \frac{k}{r} \rceil$ for $M$. Thus, regardless of the global minimum distance $d$, any $(n, k, d, r, \delta)$-LRC with either information or all-symbol locality, has parameters $n, k, r, \delta$ in the set

$$P(n, k, r, \delta) = \left\{ (n, k, r, \delta) \in \mathbb{Z}^4 : 2 \leq \delta \text{ and } 0 < r \leq k \leq n - (\delta - 1) \left\lceil \frac{k}{r} \right\rceil \right\}.$$
(7)

A very natural question to ask then is for which parameters $(n, k, r, \delta) \in P(n, k, r, \delta)$ there exists a Singleton-optimal matroid or linear LRC, regarding both information-tion and all-symbol locality. We remark that existence results on Singleton-optimal linear LRCs imply existence results on Singleton-optimal matroids. Conversely, nonexistence results on Singleton-optimal matroids implies nonexistence results on Singleton-optimal linear LRCs.

When considering information-symbol locality it is known that the upper bound for $d$ given in (5) is achieved for all parameters $(n, k, r, \delta) \in P(n, k, r, \delta)$ by linear LRCs over sufficient large fields. This follows from [14], where a new class of codes called pyramid codes was given. Using this class of codes, Singleton-optimal linear $(n, k, d, r, \delta)_i$-LRCs can be constructed for all parameters in $P(n, k, r, \delta)$.

It is well known that Singleton-optimal linear $(n, k, d, r, \delta)_a$-LRCs exist when $r = k$. Namely, the LRCs in these cases are linear $[n, k, n - k + 1]$ MDS-codes. However, existence or nonexistence results when $r < k$ are in general not that easy to obtain. In [38], existence and nonexistence results on Singleton-optimal linear $(n, k, d, r, \delta)_a$-LRCs were examined. Such results were given for certain regions of parameters, leaving other regions for which the answer of existence or nonexistence of Singleton-optimal linear LRCs is not known. The results on nonexistence were extended to matroids in [47]. All the parameter regions for the nonexistence of Singleton-optimal linear LRCs in [47] were also regions of parameters for the nonexistence of Singleton optimal matroids for all-symbol locality. Further, more regions of parameters for nonexistence of Singleton-optimal matroids with all-symbol locality were given in [47]. This implies new regions of parameters for nonexistence of Singleton-optimal linear LRCs with all-symbol locality.

The nonexistence results for Singleton-optimal matroids were proven via the following structure result in [47]. Before we state the theorem we need the concept of nontrivial unions. Let $M = (\rho, E)$ be a matroid with repair sets $\{R_x\}_{x \in E}$. For $Y \subseteq E$, we say that

$$R_Y = \bigcup_{x \in Y} R_x \quad \text{and} \quad R_Y \text{ is a } \textit{nontrivial union} \text{ if } R_x \not\subseteq R_{Y \setminus \{x\}} \text{ for every } x \in Y.$$

**Theorem 6** ([47] Structure theorem for Singleton-optimal matroids) *Let $M = (\rho, E)$ be an $(n, k, d, r, \delta)_a$-matroid with $r < k$, repair sets $\{R_x\}_{x \in E}$ and*

$$d = n - k + 1 - \left( \left\lceil \frac{k}{r} \right\rceil - 1 \right) (\delta - 1).$$

*Then, the following properties must be satisfied by the collection of repair sets and the lattice of cyclic flats $\mathscr{Z}$ of $M$:*

(i) $0_{\mathscr{L}} = \emptyset$,

(ii) *for each $x \in E$,*

  a) *$R_x$ is an atom of $\mathscr{L}$,*

   *(i.e., $R_x \in \mathscr{L}$, $0_{\mathscr{L}} < R_x$ and $\nexists Z \in \mathscr{L}$ such that $0_{\mathscr{L}} < Z < R_x$),*

  b) $\eta(R_x) = \delta - 1$,

(iii) *for each $Y \subseteq E$ with $R_Y$ being a nontrivial union,*

  c) $|Y| < \lceil \frac{k}{r} \rceil$  $\Rightarrow$  $R_Y \in \mathscr{L}$,

  d) $|Y| < \lceil \frac{k}{r} \rceil$  $\Rightarrow$  $\rho(R_Y) = |R_Y| - |Y|(\delta - 1)$,

  e) $|Y| \leq \lceil \frac{k}{r} \rceil$  $\Rightarrow$  $|R_x \cap (R_{Y \setminus \{x\}})| \leq |R_x| - \delta$, *for each $x \in Y$,*

  f) $|Y| \geq \lceil \frac{k}{r} \rceil$  $\Rightarrow$  $\{Z \in \mathscr{L} : Z \supseteq R_Y\} = 1_{\mathscr{L}} = E$.

Conditions (i) and (ii) in the structure theorem above for Singleton-optimal matroids show that each repair set $R_x$ must correspond to a uniform matroid with $|R_x|$ elements and rank $|R_x| - (\delta - 1)$. Further, condition (iii) gives structural properties on nontrivial unions of repair sets. This can be viewed as structural conditions on how nontrivial unions of uniform matroids need to be glued together in a Singleton-optimal matroid, with the uniform matroids corresponding to repair sets. For Singleton-optimal linear $(n, k, d, r, \delta)_a$-LRCs, the property of the repair sets being uniform matroids corresponds to the repair sets being linear $[|R_x|, |R_x| - (\delta - 1), \delta]$-MDS codes. We remark that structure theorems when $r | k$ for Singleton-optimal linear $(n, k, d, r, \delta)_a$-LRCs and the special case of $(n, k, d, r, \delta = 2)_a$-LRCs are given in [13, 18], respectively. These theorems show that the local repair sets correspond to linear $[r + \delta - 1, r, \delta]$-MDS codes that are mutually disjoint. This result is a special case of Theorem 6.

*Example 10* By Theorem 1, the poset with its associated subsets of $E = [16]$ and rank of these subsets in the figure below defines the set of cyclic flats $\mathscr{L}$ and the rank function restricted to the sets in $\mathscr{L}$ of a matroid $M$ on $E$. From Theorem 4, we obtain that $(n, k, d) = (16, 7, 6)$ for the matroid $M$. Choosing repair sets $R_1 = \cdots = R_5 = X_1$, $R_6 = \cdots = R_9 = X_2$, $R_{10} = \cdots = R_{13} = X_3$ and $R_{14} = \cdots = R_{16} = X_4$, we obtain that $M$ is an $(n = 16, k = 7, d = 6, r = 3, \delta = 3)_a$-matroid. It can easily be checked that all the properties (i)–(iii) are satisfied by the matroid $M$ and the chosen repair sets. Further, we also have that $M$ is Singleton-optimal as

$$n - k + 1 - \left( \left\lceil \frac{k}{r} \right\rceil - 1 \right) (\delta - 1) = 6 = d.$$

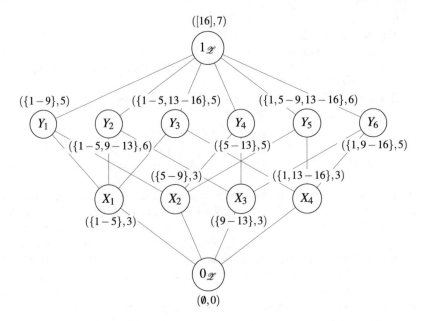

## 4    Code Constructions

### 4.1    Constructions of $(n, k, d, r, \delta)_a$-Matroids via Cyclic Flats

In [47], a construction of a broad class of linear $(n, k, d, r, \delta)_a$-LRCs is given via matroid theory. This is generalised in [10, 46] to account for availability and hierarchy, respectively.

*A construction of $(n, k, d, r, \delta)_a$-matroids via cyclic flats:*

Let $F_1, \ldots, F_m$ be a collection of finite sets and $E = \bigcup_{i=1}^m F_i$. Assign a function $\rho : \{F_i\} \cup \{E\} \to \mathbb{Z}$ satisfying

$$
\begin{aligned}
&(i)\ \ 0 < \rho(F_i) < |F_i| \text{ for } i \in [m],\\
&(ii)\ \ \rho(F_i) < \rho(E) \text{ for } i \in [m],\\
&(iii)\ \ \rho(E) \le |E| - \sum_{i \in [m]} \eta(F_i),\\
&(iv)\ \ j \in [m] \Rightarrow |F_{[m]\setminus\{j\}} \cap F_j| < \rho(F_j),
\end{aligned}
\tag{8}
$$

where

$$
\eta(F_i) = |F_i| - \rho(F_i) \text{ for } i \in [m] \quad \text{and} \quad F_I = \bigcup_{i \in I} F_i \text{ for } I \subseteq [m].
$$

Extend $\rho$ to $\{F_I\} \to \mathbb{Z}$ by

$$\rho(F_I) = \min\left\{|F_I| - \sum_{i \in I} \eta(F_i), \rho(E)\right\} \tag{9}$$

and let $\mathcal{Z}$ be the following collection of subsets of $E$,

$$\mathcal{Z} = \{F_I : I \subseteq [m] \text{ and } \rho(F_I) < \rho(E)\} \cup E. \tag{10}$$

**Theorem 7** ([47] Construction of $(n, k, d, r, \delta)_a$-matroids) *Let* $F_1, \ldots, F_m$, *be a collection of finite sets with* $E = \bigcup_{i=1}^{m} F_i$ *and* $\rho : \{F_i\}_{i \in [m]} \to \mathbb{Z}$ *satisfying (8). Then the pair* $(\rho, \mathcal{Z})$, *defined in (9) and (10), defines a* $(n, k, d, r, \delta)_a$-matroid $M(F_1, \ldots, F_m; \rho)$ *on* $E$ *for which* $\mathcal{Z}$ *is the collection of cyclic flats,* $\rho$ *is the rank function restricted to the cyclic flats in* $\mathcal{Z}$, $F_1, \ldots, F_m$, *are the repair sets and*

> (i) $n = |E|$,
> (ii) $k = \rho(E)$,
> (iii) $d = n - k + 1 - \max\{\sum_{i \in I} \eta(F_i) : F_I \in \mathcal{Z} \setminus E\}$,
> (iv) $\delta = 1 + \min\{\eta(F_i) : i \in [m]\}$,
> (v) $r = \max\{\rho(F_i) : i \in m\}$.

That $M(F_1, \ldots, F_m; \rho)$ defines a matroid follows from a proof given in [47] that the pair $(\rho, \mathcal{Z})$ satisfies the axiomatic scheme of matroids via cyclic flats and their ranks stated in Theorem 1. The correctness of the parameters $(n, k, d, r, \delta)$ when $F_1, \ldots, F_m$ are considered as the repair sets also follows from [47].

We remark, that the matroids constructed in Theorem 7 satisfy, for all unions of repair sets $F_I$ with $\rho(F_I) < \rho(E)$, that

> (i) $F_I$ is a cyclic flat,
> (ii) the nullity $\eta(F_I)$ of $F_I$ is as small as possible. $\qquad(11)$

Properties (i) and (ii) above are trivially seen to be fulfilled by uniform matroids $U_{k,n}$, where $\mathcal{Z} = \{\emptyset, E\}$, $\rho(\emptyset) = 0$ and $\rho(E) = k$. However, uniform matroids cannot be constructed by Theorem 7, since all constructed matroids by this theorem have $r < k$ and uniform matroids have $r = k$. Though both uniform matroids and the matroids constructed in Theorem 7 satisfy properties (i) and (ii) in (11), we will consider them in terms of a class of matroids $\mathcal{M}$, defined as follows:

$$\mathcal{M} = \{M = M(F_1, \ldots, F_m; \rho) : M \text{ is constructed in Theorem 7}\} \cup \{U_{k,n}\}. \tag{12}$$

By the structure Theorem 6, the properties (i) and (ii) in (11) are necessary (but not sufficient) for Singleton-optimal $(n, k, d, r, \delta)_a$-matroids.

*Example 11* Let $E = [12]$ and let $F_1 = \{1, \ldots, 4\}$, $F_2 = \{3, \ldots, 6\}$, $F_3 = \{7, \ldots, 10\}$, $F_4 = \{10, \ldots, 12\}$ with $\rho(F_1) = \rho(F_2) = \rho(F_3) = 3$, $\rho(F_4) = 2$, and

$\rho(E) = 7$. Then, by Theorem 7, $M(F_1, \ldots, F_m; \rho)$ is an $(n, k, d, r, \delta)_a$-matroid over $E$ with $(n, k, d, r, \delta) = (12, 7, 3, 3, 2)$ and the following lattice of cyclic flats and their ranks.

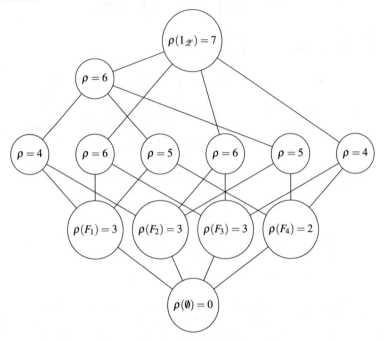

Further, the matroid is not Singleton-optimal since

$$d = 3 < n - k + 1 - \left( \left\lceil \frac{k}{r} \right\rceil - 1 \right) (\delta - 1) = 4.$$

## 4.2 A Matroidal Construction of Linear all Symbol LRCs

As will be explained below, all matroids constructed in Theorem 6 are contained in a class of matroids called gammoids. These matroids are linear, which especially implies that all $(n, k, d, r, \delta)_a$-matroids constructed by Theorem 6 are matroids associated with linear $(n, k, d, r, \delta)_a$-LRCs.

**Definition 17** Any (finite) directed graph $\Gamma = (V, D)$ and vertex subsets $E, T \subseteq V$ define a *gammoid* $M(\Gamma, E, T)$, where $M(\Gamma, E, T) = (\mathscr{I}, E)$ is a the matroid with

$$\mathscr{I} = \{X \subseteq E : \exists \text{ a set of } |X| \text{ vertex-disjoint paths from } X \text{ to } T\}.$$

**Theorem 8** ([20]) *Every gammoid $M(\Gamma, E, T)$ is $\mathbb{F}_q$-linear for all prime powers $q \geq 2^{|E|}$.*

In [47], it is proven that the matroids constructed in Theorem 7 are indeed gammoids, and hence representable. This is achieved by explicitly constructing a triple $(\Gamma, E, T)$ whose associated matroid is $M(F_1, \ldots, F_m; \rho)$. The details of the construction are left to Theorem 14 in the appendix. The essence of the argument is to construct a graph of depth three, whose sources correspond to the ground set of the matroid, and whose middle layer corresponds to the repair sets, with multiplicities to reflect the ranks of the repair sets.

*Example 12* The following directed graph $\Gamma = (V = E \cup H \cup T, D)$ is constructed in Theorem 14 from the matroid $M(F_1, \ldots, F_m, E; k; \rho)$ given in Example 11.

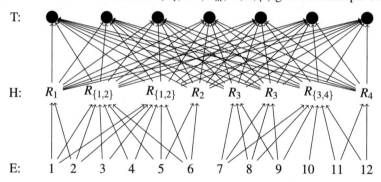

In general it is extremely hard to prove that a matroid is linear (or the converse). There is no known deterministic algorithm to solve this problem in general. However, by combining the results given in Theorems 7–14, we obtain the following result.

**Theorem 9** ([47] A matroidal construction of $(n, k, d, r, \delta)_a$-LRCs) *For every $(n, k, d, r, \delta)_a$-matroid $M(F_1, \ldots, F_m, E; k; \rho)$ given by Theorem 7 and every prime power $q \geq 2^{|E|}$ there is a linear $(n, k, d, r, \delta)_a$-LRC $C$ over $\mathbb{F}_q$ with repair sets $F_1, \ldots, F_m$ such that $M(F_1, \ldots, F_m, E; k; \rho) = M_C$.*

*Example 13* The $(12, 7, 3, 3, 2)_a$-matroid $M(F_1, \ldots, F_m, E; k; \rho)$ given in Example 11 equals the matroid $M_C = M[G]$, where $G$ equals the following matrix over $\mathbb{F}_5$:

|  | 1 | 2 | 3 | 4 | 5 | 6 | 7 | 8 | 9 | 10 | 11 | 12 |
|---|---|---|---|---|---|---|---|---|---|---|---|---|
|  | 1 | 1 | 0 | 0 | 0 | 0 | 0 | 0 | 0 | 0 | 1 | 3 |
|  | 0 | 2 | 1 | 0 | 0 | 1 | 0 | 0 | 0 | 0 | 1 | 3 |
| $G =$ | 0 | 3 | 0 | 1 | 0 | 3 | 0 | 0 | 0 | 0 | 1 | 3 |
|  | 0 | 0 | 0 | 0 | 1 | 2 | 0 | 0 | 0 | 0 | 1 | 3 |
|  | 0 | 0 | 0 | 0 | 0 | 0 | 1 | 0 | 1 | 0 | 1 | 3 |
|  | 0 | 0 | 0 | 0 | 0 | 0 | 0 | 1 | 2 | 0 | 1 | 3 |
|  | 0 | 0 | 0 | 0 | 0 | 0 | 0 | 0 | 3 | 1 | 1 | 1 |

Hence, the code $C$ generated by the rows of $G$ is a linear $(12, 7, 3, 3, 2)$-LRC over $\mathbb{F}_5$ with repair sets $F_1 = \{1, 2, 3, 4\}$, $F_2 = \{3, 4, 5, 6\}$, $F_1 = \{7, 8, 9, 10\}$ and $F_1 = \{10, 11, 12\}$.

Note that the bound $q \geq 2^{|E|}$ given in Theorem 9 is a very rough bound. There are many matroids $M(F_1, \ldots, F_m, E; k; \rho) = M_C$ for linear LRCs $C$ over $\mathbb{F}_q$ where $q \ll 2^{|E|}$. In Example 13, for instance, we constructed a code over $\mathbb{F}_5$, while the field size predicted by Theorem 9 was $2^{12} = 4096 \gg 5$. To construct an explicit linear $(n, k, d, r, \delta)_a$-LRC from a matroid $M(F_1, \ldots, F_m, E; k; \rho)$, one can use the directed graph representation of the matroid given in Theorem 14, together with results on how to construct a generator matrix from this representation [20].

As we saw earlier, it is known that there exists a Singleton-optimal linear $(n, k, d, r, \delta)_i$-LRC for all parameters $(n, k, r, \delta) \in P(n, k, r, \delta)$ (cf. (7) for a definition of $P(n, k, r, \delta)$). Further, it is also known that if $r = k$, then all Singleton-optimal linear LRCs are linear $[n, k, n - k + 1]$-MDS codes. In [38] existence and nonexistence of Singleton-optimal linear $(n, k, d, r, \delta)_a$-LRCs were examined. The parameter regions for existence given in [38] were both obtained and extended in [47] by the construction of linear LRCs via matroid theory given in Theorem 9. Hence, the results in [47] about nonexistence and existence of Singleton-optimal linear $(n, k, d, r, \delta)_a$-LRCs settled large portions of the parameter regions left open in [38] leaving open only a minor subregion. Some improvements of the results in [38] were also given for $\delta = 2$ in [45] via integer programming techniques.

For $(n, k, r, \delta) \in P(n, k, r, \delta)$ it is also very natural to ask what is the maximal value of $d$ for which there exist an $(n, k, d, r, \delta)_a$-matroid or a linear $(n, k, d, r, \delta)_a$-LRC. We denote this maximal value by $d_{max}(n, k, r, \delta)$. In [47] it was proven that

$$d_{max}(n, k, r, \delta) \geq n - k + 1 - \left\lceil \frac{k}{r} \right\rceil (\delta - 1)$$

for linear LRCs. For matroids, this result is straightforward, as a matroid with $d = n - k + 1 - \left\lceil \frac{k}{r} \right\rceil (\delta - 1)$ can be constructed as a truncation of the direct sum of $\left\lceil \frac{n}{r+\delta-1} \right\rceil$ uniform matroids of size $s_i \leq r + \delta - 1$ and rank $s_i - \delta + 1 \leq r$. As representability (over some field) is preserved under direct sums and truncation, the result follows for linear LRCs. However, with this straightforward argument, and with the bound on the field size of truncated matroids from [17], the field size required could be as large as

$$(r + \delta - 1) \cdot \prod_{i=k}^{r \cdot \left\lceil \frac{n}{r+\delta-1} \right\rceil} \binom{n}{i}.$$

Significant work is needed in order to bound the field size even in this special case.

This result was improved in [47] and further in [29]. Also, the parameter region of Singleton-optimal linear $(n, k, d, r, \delta)$-LRCs was also extended in [29]. The existence of Singleton-optimal linear LRCs obtained by the matroidal construction described here depends mainly on the relation between the parameters $a$ and $b$ where $a = \left\lceil \frac{k}{r} \right\rceil r - k$ and $b = \left\lceil \frac{n}{r+\delta-1} \right\rceil (r + \delta - 1) - n$. Thus, we can easily get Singleton-optimal linear LRCs for all possible coding rates.

## 4.3   Random Codes

An alternative way to design $(n, k, d, r, \delta)$-LRCs with prescribed parameters is by exploiting the fact that independence is a generic property for $r$- and $k$-tuples of vectors over large fields. This allows us to use randomness to generate $(n, k, d, r, \delta)$-LRCs in a straightforward way, once the matroid structure of the code is prescribed. This is the key element in [8]. As opposed to in the gammoid construction from the last section, we will now consider the field size $q$ to be fixed but large. Indeed, a sufficiently large field will be $\mathbb{F}_q$ with

$$q > (r\delta)^{4^r r} \binom{n + (r\delta)^{(r-1)4^r}}{k - 1}.$$

For given $(n, k, r, \delta)$, we will construct $(n, k, d, r, \delta)$-LRCs where

$$d \geq n - k + 1 - \left\lceil \frac{k}{r} \right\rceil (\delta - 1).$$

Comparing this to the generalised Singleton bound (5), we notice that the codes we construct are "almost Singleton-optimal".

The underlying matroid will again be a truncation of

$$\bigsqcup_{i=1}^{\lceil \frac{n}{r+\delta-1} \rceil} U_{s_i}^{s_i - \delta - 1}.$$

However, rather than first representing this direct sum, which has rank

$$\sum_i (s_i - \delta + 1) = n - \lceil \frac{n}{r + \delta - 1} \rceil (\delta - 1),$$

we will immediately represent its truncation as an $n \times k$ matrix. The random construction proceeds as follows. Divide the columns $[n]$ into locality sets $S_i$ of size $s_i$. For each $S_i$, we first generate the $r_i = s_i - \delta + 1$ first columns uniformly at random from the ambient space $\mathbb{F}^k$. This gives us an $r_i \times k$-matrix $G_i$. After this, we draw $\delta - 1$ vectors from $\mathbb{F}^{r_i}$, and premultiply these by $G_i$. The resulting $r_i + \delta - 1 = s_i$ vectors will be in the linear span of the $r_i$ first vectors, and so have rank $\leq r_i$ as a point configuration in $\mathbb{F}^k$. We arrange the $s_i$ vectors into a matrix $G'_i$ of rank $\leq r_i$ in $\mathbb{F}^{s_i \times k}$. Let $A_i$ be the event that all $r_i$-tuples of columns in $G_i$ are linearly independent. It is easy to see that, if the field size grows to infinity, the probability of $A_i$ tends to one.

Juxtaposing the matrices $G'_i$ for $i = 1, \ldots, \lceil \frac{n}{r+\delta-1} \rceil$, we obtain a generator matrix $G$ for a code of length $n$. Let $B$ be the event that $G$ has full rank. Again, assuming the field size is large enough, the probability of $B$ can be arbitrarily close to one. Now, the random matrix $G$ generates an $(n, k, d, r, \delta)$-LRC if all the events $A_1, \ldots, A_{\lceil \frac{n}{r+\delta-1} \rceil}$, $B$ simultaneously occur. A simple first moment estimate shows that, if

$$q > (r\delta)^{r4^r} \binom{n + (r\delta)^{(r-1)4^r}}{k - 1},$$

then the probability of this is positive, so there exists an $(n, k, d, r, \delta)$-LRC.

## 4.4   Constructing LRCs as Evaluation Codes

As suggested in the previous sections, there are several assumptions that can be made in order to give more explicit code constructions for optimal LRCs. Next, we will follow [39, 40] in assuming that $n$ is divisible by $r + \delta - 1$ and $k$ is divisible by $r$. Then, an optimal LRC with $d = n - k - (\frac{k}{r} - 1)(\delta - 1) + 1$ exists for any choice of $k$. We will also assume that $n = q$ is a prime power, although this assumption can easily be removed at the price of a more technical description of the code.

We will construct a Singleton-optimal code in this case as an evaluation code, generalising the construction of MDS codes as Reed-Solomon codes. The main philosophy goes back to [40], but due to a technical obstacle, [40] still required exponential field size. This technicality was overcome by the construction in [39], which we will present next. Evaluation codes have a multitude of favourable properties, not least that the field size can often be taken to be much smaller than in naïve random code constructions. Moreover, the multiplicative structure used for designing evaluation codes can also be exploited when one needs to do computations with the codes in question.

Let $A$ be a subgroup of $\mathbb{F}_q$ of size $r + \delta - 1$ and let $g = \prod_{i \in A}(x - i)$ be the polynomial of degree $r + \delta - 1$ that vanishes on $A$. We will construct a storage code whose nodes are the elements of $\mathbb{F}_q$ and whose locality sets are the cosets of $A$. Thus, there are $\frac{n}{r+\delta-1}$ locality sets, each of size $r + \delta - 1$. The codewords will be the evaluations over $\mathbb{F}_q$ of polynomials of a certain form. As the rank of the code that we are designing is $k = r \cdot \frac{k}{r}$, we can write the messages as a $r \times \frac{k}{r}$ matrix

$$a = \begin{pmatrix} a_{0,0} & \cdots & a_{r-1,0} \\ \vdots & \ddots & \vdots \\ a_{0,\frac{k}{r}-1} & \cdots & a_{r-1,\frac{k}{r}-1} \end{pmatrix}$$

over $\mathbb{F}_q$. Now consider the polynomial function

$$f_a = \left(1 \; g(x) \; g(x)^2 \; \ldots \; g(x)^{\frac{k}{r}-1}\right) \cdot \begin{pmatrix} a_{0,0} & \cdots & a_{r-1,0} \\ \vdots & \ddots & \vdots \\ a_{0,\frac{k}{r}-1} & \cdots & a_{r-1,\frac{k}{r}-1} \end{pmatrix} \cdot \begin{pmatrix} 1 \\ x \\ x^2 \\ \vdots \\ x^{r-1} \end{pmatrix}.$$

Consider the code

$$C = \{f_a(x) \; : \; x \in \mathbb{F}_q, \; a \in \mathbb{F}_q^{r \times \frac{k}{r}}\}.$$

By design, $f_a$ has degree

$$\deg f_a \le (r + \delta - 1)\left(\frac{k}{r} - 1\right) + r - 1 = k - 1 + (\delta - 1)\left(\frac{k}{r} - 1\right),$$

and can therefore be computed for every point in $\mathbb{F}_q$ by evaluation on any $k + (\delta - 1)(\frac{k}{r} - 1)$ points. Therefore, the code $C$ protects against

$$d - 1 = n - k - (\delta - 1)\left(\frac{k}{r} - 1\right) + 1$$

errors. It remains to see that it has locality $(r, \delta)$.

To this end, note that the row vector

$$\left(1 \; g(x) \; g(x)^2 \; \ldots \; g(x)^{\frac{k}{r}-1}\right) \cdot \begin{pmatrix} a_{0,0} & \cdots & a_{r-1,0} \\ \vdots & \ddots & \vdots \\ a_{0,\frac{k}{r}-1} & \cdots & a_{r-1,\frac{k}{r}-1} \end{pmatrix}$$

of polynomials is constant over the subgroup $A \subseteq \mathbb{F}_q$ and thus on all of its cosets by construction of $g$. It follows that when restricted to any such coset, the function $f_a$ is a polynomial of degree $\le r - 1$, and so can be extrapolated to all points in the coset from any $r$ such evaluation points. This proves the $(r, \delta)$-locality.

As discussed, this construction depends on a collection of assumptions on the divisibility of parameters that are needed for the rather rigid algebraic structures to work. Some of these assumptions can be relaxed, using more elaborate evaluation codes, such as algebraic geometry codes over curves and surfaces [2, 3]. While this field of research is still very much developing, it seems that the rigidity of the algebraic machinery makes it less suitable for generalisations of the LRC concept, for example when different nodes are allowed to have different localities.

# 5 Beyond Linear Storage Codes

In this section we will introduce the notion of hierarchical codes, which are natural generalisations of locally repairable codes. After this, we will briefly describe the connection between $(n, k, d, r, \delta, t)$-LRCs and polymatroids given in [46].

## 5.1 Hierarchical Codes

**Definition 18** Let $h \geq 1$ be an integer, and let

$$(\mathbf{n}, \mathbf{k}, \mathbf{d}, \mathbf{t}) = [(n_1, k_1, d_1, t_1), \ldots, (n_h, k_h, d_h, t_h)]$$

be a $h$-tuple of integer 4-tuples, where $k_i \geq 1$, $n_i, d_i \geq 2$, and $t_i \geq 1$ for $1 \leq i \leq h$. Then, a coordinate $x$ of a linear $[n, k, d] = [n_0, k_0, d_0]$-LRC $C$ indexed by $E$ has $h$-level hierarchical availability $(\mathbf{n}, \mathbf{k}, \mathbf{d}, \mathbf{t})$ if there are $t_1$ coordinate sets $X_1, \ldots, X_{t_1} \subseteq E$ such that

$(i)$ $x \in X_i$ for $i \in [t_1]$,
$(ii)$ $i, j \in [t_1], i \neq j \Rightarrow X_i \cap X_j = \{x\}$,
$(iii)$ $n_{X_i} \leq n_1$, $k_{X_i} = k_1$ and $d_{X_i} \geq d_1$ for the punctured $[n_{X_i}, k_{X_i}, d_{X_i}]$-code $C_{X_i}$, for $i \in [t_1]$,
$(iv)$ for $i \in [t_1]$, $x$ has $(h-1)$-level hierarchical availability $[(n_2, k_2, d_2, t_2), \ldots, (n_h, k_h, d_h, t_h)]$ in $C_{X_i}$.

The code $C$ above as well as all the related subcodes $C_{X_i}$ should be non-degenerate. For consistency of the definition, we say that any symbol in a non-degenerate storage code has 0-level hierarchical availability.

*Example 14* Let $C$ be the code generated by the matrix $G$ in Example 1 and let $x = 2$. Then $x$ has 2-level hierarchical availability

$$(\mathbf{n}, \mathbf{k}, \mathbf{d}, \mathbf{t}) = [(6, 3, 3, 1), (3, 2, 2, 2)].$$

This follows from Example 2 where $C_{Y_1}$ implies the $(6, 3, 3, 1)$-availability, and the $(3, 2, 2, 2)$-availability is implied by $C_{X_1}$ and $C_{X_2}$.

The most general Singleton bound for matroids with hierarchy in the case $t = 1$ are the following given in [10, 32]:

$$d_i(M) \leq n_i - k_i + 1 - \sum_{j > i}(d_j - d_{j+1}) \left( \left\lceil \frac{k_i}{k_j} \right\rceil - 1 \right),$$

where we say $d_{h+1} = 1$.

## 5.2 General Codes from Polymatroids

**Definition 19** Let $E$ be a finite set. A pair $P = (\rho, E)$ is a (finite) *polymatroid* on $E$ with a *set function* $\rho : 2^E \to \mathbb{R}$ if $\rho$ satisfies the following three conditions for all $X, Y \subseteq E$:

$$(R1)\ \rho(\emptyset) = 0\,,$$
$$(R2)\ X \subseteq Y \Rightarrow \rho(X) \le \rho(Y)\,,$$
$$(R3)\ \rho(X) + \rho(Y) \ge \rho(X \cup Y) + \rho(X \cap Y)\,.$$

Note that a *matroid* is a polymatroid which additionally satisfies the following two conditions for all $X \subseteq E$:

$$(R4)\ \rho(X) \in \mathbb{Z}\,,$$
$$(R5)\ \rho(X) \le |X|\,.$$

Using the joint entropy and a result given in [11] one can associate the following polymatroid to every code.

**Definition 20** Let $C$ be an $(n, k)$-code over some alphabet $A$ of size $s$. Then $P_C = (\rho_C, [n])$ is the polymatroid on $[n]$ with the set function $\rho_C : 2^{[n]} \to \mathbb{R}$ where

$$\rho_C(X) = \sum_{z_X \in C_X} \frac{|\{c \in C : c_X = z_X\}|}{|C|} \log_s \left( \frac{|C|}{|\{c \in C : c_X = z_X\}|} \right)$$

and $\rho_C(\emptyset) = 0$.

We remark that for linear codes $M_C = P_C$. Using the above definition of $P_C$, one can now prove the following useful properties.

**Proposition 8** *Let $C$ be an $(n, k)$-code over $A$ with $|A| = s$. Then for the polymatroid $P_C = (\rho_C, [n])$ and any subsets $X, Y \subseteq [n]$,*

$$(i)\ P_C(X) \le |X|,$$
$$(ii)\ |C_{X \cup Y}| > |C_X| \iff \rho_C(X \cup Y) > \rho_C(X),$$
$$(iii)\ |C| = s^{\rho_C([n])},$$
$$(iv)\ |C|/|A^n| = s^{\rho_C([n]) - n}.$$

We remark that, even though $|C| = s^{\rho_C([n])}$ for nonlinear codes and $|C_X| = s^{\rho_C(X)}$ for all $X$ for linear codes, it is not true in general that $|C_X| = s^{\rho_C(X)}$ for $X \subsetneq [n]$ for nonlinear codes. This stems from the fact that, for non-linear codes, the uniform distribution over the code does not necessarily map to the uniform distribution under coordinate projection.

After scaling the rank function of a finite polymatroid $P = (\rho, E)$ by a constant $c$ such that $c\rho(X) \le |X|$ for all $X \subseteq E$, we obtain a polymatroid satisfying axiom (R5). We will assume that such a scaling has been performed, so that all polymatroids satisfy axiom (R5).

We are now ready to define a cyclic flat of a polymatroid $P = (\rho, E)$, namely $X \subseteq E$ is a *cyclic flat* if

$$\rho(X \cup \{e\}) > \rho(X) \text{ for all } e \in E \setminus X \text{ and } \rho(X) - \rho(X \setminus \{x\}) < 1 \text{ for all } x \in X.$$

Let $P = (\rho, E)$ be a polymatroid and $X \subseteq E$. The restriction of $P$ to $X$ is the polymatroid $P|X = (\rho_{|X}, X)$ where $\rho_{|X}(Y) = \rho(Y)$ for $Y \subseteq X$. We can now define the distance of $P|X$ as

$$d(P|X) = \min\{|Y| : Y \subseteq X, \rho_{|X}(X \setminus Y) < \rho_{|X}(X)\}.$$

Let $\mathscr{Z}$ denote the family of cyclic flats of the polymatroid $P$. Assuming that $E \in \mathscr{Z}$, we can define the parameters $n, k, d$ of $P$ via the cyclic flats and their ranks, namely

$$n = |E|, k = \rho(E) \text{ and } d = \lfloor n - k + 1 - \max\{|X| - \rho(X) : X \in \mathscr{Z} \setminus E\} \rfloor.$$

The definitions of $(n, k, d, r, \delta, t)_i$ and $(n, k, d, r, \delta, t)_a$-polymatroids are carried over directly from Definition 16. In addition, the parameters $(n, k, d, r, \delta, t)_i$ and $(n, k, d, r, \delta, t)_a$ of a LRC $C$ are the same as the corresponding parameters for $P_C$. Using the cyclic flats and similar methods as for matroids, Singleton-type bounds can be proven for polymatroids in general, which then imply bounds on all objects related to polymatroids, e.g., matroids, linear and nonlinear LRCs, and hypergraphs. This is the content of the next section.

## 5.3 Singleton-Type Bounds for Polymatroids and General LRCs

It is not clear whether the Singleton-type bounds given for linear LRCs in (2)–(6) also hold for general LRCs — in general the upper bound on $d$ might have to be larger. As we will describe briefly in Sect. 5, any general LRC can be associated with a polymatroid that captures the key properties of the LRC. Using this connection we are able to define the $(n, k, d, r, \delta, t)$-parameters and information-symbol, systematic-symbol, and all-symbol locality sets for polymatroids in general.

   The class of polymatroids is much bigger than the class of the polymatroids arising from general LRCs. Hence, it is also not clear whether the Singleton-type bounds given in (2)–(6) also hold for polymatroids in general. However, from [46], we obtain a Singleton-type bound for polymatroids in Theorem 10 below. This theorem shows that all the Singleton-type bounds given in (2)–(6) are polymatroid properties. Further, the polymatroid result also extends all these bounds by including all the parameters $(n, k, d, r, \delta, t)$ at the same time.

The methods used to prove the Singleton-type bound given for polymatroids in Theorem 10 are similar to those used for proving the Singleton-type bound for matroids in Theorem 5. Especially, the notion of cyclic flats is generalised to polymatroids and used as the key tool in the proof. However, some obstacles occur since we are dealing with real-valued rank functions in the case of polymatroids instead of integer-valued rank functions, which was the case for matroids. As a direct consequence of Theorem 10, the Singleton-type bounds given in (2)–(6) are valid for all objects associated to polymatroids.

**Theorem 10** ([46] Singleton-type bound for polymatroids) *Let $P = (\rho, E)$ be an information-set $(n, k, d, r, \delta, t)_i$-polymatroid. Then*

$$d \le n - \lceil k \rceil + 1 - \left( \left\lceil \frac{t(\lceil k \rceil - 1) + 1}{t(r - 1) + 1} \right\rceil - 1 \right) (\delta - 1). \tag{13}$$

Theorem 10 is stated for information-symbol locality. This implies that the bound (13) is also valid for systematic-symbol and all-symbol locality. Hence, as a direct corollary, the bounds (2)–(13) hold for information-symbol, systematic-symbol, and all-symbol locality for all objects associated to polymatroids, e.g., entropy functions, general LRCs, hypergraphs, matroids, linear LRCs, graphs and many more. If we restrict to systematic linear codes, then the bound also holds for PIR codes [9, Definition 4]. The connection is not as straightforward in the nonlinear case, since the definitions of a repair group are then slightly different for LRCs (as defined here) and PIR codes, while coinciding in the linear case.

The bound (4), for all-symbol LRCs (as subsets of size $|B|^K$ of $B^{\alpha n}$, where $B$ is a finite set, $A = B^\alpha$ is the alphabet, and $\alpha$ and $K$ are integers), follows from a result given in [27]. The bound (5), for all-symbol LRCs (as a linear subspace of $\mathbb{F}_q^{\alpha n}$ with the alphabet $A = \mathbb{F}_q^\alpha$), is given in [34]. This result is slightly improved for information-symbol locality in [18]. The bound (6), for $(n, k, d, r, t)_s$-LRCs where $k$ is a positive integer, follows from a result given in [31]. The following bound for $(n, k, d, r, t)_a$-LRCs with integral $k$ was given in [41],

$$d \le n - k + 1 - \sum_{i=1}^{t} \left\lfloor \frac{k - 1}{r^i} \right\rfloor.$$

One parameter which has not been included above is the alphabet size. Small alphabet sizes are important in many applications because of implementation and efficiency reasons. The bound (14) below takes the alphabet size into account, but is only inductively formulated. Before stating this bound we introduce the following notation:

$$k_{\text{opt}}^{(q)}(n, d) = \max\{k : C \text{ is an } (n, k, d)\text{-code over an alphabet of size } q\}.$$

By [6], an all-symbol $(n, k, d, r)$-LRC over a finite alphabet $A$ of size $q$ satisfies

$$k \leq \min_{s \in \mathbb{Z}_+} (sr + k_{\text{opt}}^{(q)} (n - s(r + 1), d)). \tag{14}$$

It is a hard open problem in classical coding theory to obtain a value for the parameter $k_{\text{opt}}^{(s)}(n, d)$ for linear codes. This problem seems to be even harder for codes in general. However, by using other known bounds, such as the Plotkin bound or Greismer bound, it is possible to give an explicit value for $k_{\text{opt}}^{(s)}(n, d)$ for some classes of parameters $(s, n, d)$. This has been done for example in [33].

We remark that when considering nonlinear LRCs, some extra care has to be taken in terms of how to define the concepts associated with the LRCs. Two equivalent definitions in the linear case may differ in the nonlinear case. In this chapter, we have chosen to consider $\delta$ as a parameter for the local distance of the repair sets, *i.e.*, any node in a repair set $R$ can be repaired by any other $|R| - \delta + 1$ nodes of $R$. The condition used in [6, 27, 31, 41] is for $\delta = 2$ only assuming that a specific node in a repair set $R$ can be repaired by the rest of the nodes of $R$. It is not assumed that any node in $R$ can be repaired by the other nodes of $R$, i.e., that the local distance is 2. A Singleton bound using the weaker condition of guaranteeing only repair of one node in each repair set implies directly that the same upper bound on $d$ is true for the case with local distance 2.

## 6   Conclusions and Further Research

We have shown how viewing storage codes from a matroidal perspective helps our understanding of local repairability, both for constructions and for fundamental bounds. However, many central problems about linear LRCs boil down to notoriously hard representability problems in matroid theory.

A famous conjecture, with several consequences for many mathematical objects, is the so called *MDS-conjecture*. This conjecture states that, for a given finite field $\mathbb{F}_q$ and a given $k$, every $[n, k, d]$-MDS code over $\mathbb{F}_q$ has $n \leq q + 1$, unless in some special cases. Currently, the conjecture is known to hold only if $q$ is a prime [1]. Linear Singleton-optimal LRCs may be seen as a generalisation of linear MDS codes. An interesting problem would therefore be to consider an upper bound on $n$ for linear Singleton-optimal LRCs over a certain field size $q$ with fixed parameters $(k, r, \delta, t)$. In this setting, a sufficiently good upper bound on $n$ would be a good result.

Instead of fixing the Singleton-optimality and trying to optimise the field size, we could also fix the field $\mathbb{F}_q$, and try to optimise the locality parameters. This would give us bounds on the form

$$d \leq n - k + 1 - \left( \left\lceil \frac{k}{r} \right\rceil - 1 \right) (\delta - 1) - p(q, n, k, r, \delta),$$

where the dependence on the field size $q$ is isolated to a "penalty" term $p(q, n, k, r, \delta)$. Partial results in this direction are given by the Cadambe–Mazumdar bound [6], and LRC versions of the Griesmer and Plotkin bounds [33]. However, the optimality of these bounds is only known for certain ranges of parameters. Further research in this direction is definitely needed, but seems to lead away from the most obvious uses of matroid theory.

Finally, it would be interesting to characterise all Singleton-optimal LRCs up to matroid isomorphism. The constructions discussed in this paper appear to be rather rigid, and unique up to shifting a few "slack" elements between different locality sets. However, it appears to be difficult to prove that all Singleton-optimal matroids must have this form. Once a complete characterisation of Singleton-optimal matroids has been obtained, this could also be taken as a starting point for possibly finding Singleton-optimal nonlinear codes in the parameter regimes where no Singleton-optimal linear codes exist.

# Appendix: More About Matroid Theory

A matroid realisation of an $\mathbb{F}$-linear matroid $M$ has two geometric interpretations. Firstly, we may think of a matrix representing $M$ as a collection of $n$ column vectors in $\mathbb{F}^k$. As the matroid structure is invariant under row operations, or in other words under change of basis in $\mathbb{F}^k$, we tend to think of $M$ as a configuration of $n$ points in abstract projective $k$-space.

The second interpretation comes from studying the row space of the matrix, as an embedding of $\mathbb{F}^k$ into $\mathbb{F}^n$. Row operations correspond to a change of basis in $\mathbb{F}^k$, and hence every matroid representation can be thought of as a k-dimensional subspace of $\mathbb{F}^n$. In other words, a matroid representation is a point in the Grassmannian $\mathrm{Gr}(n, k; \mathbb{F})$, and $\mathrm{Gr}(n, k; \mathbb{F})$ has a stratification as a union of realisation spaces $R(M)$, where $M$ ranges over all $\mathbb{F}$-representable matroids of size $n$ and rank $k$. This perspective allows a matroidal perspective also on the subspace codes discussed in chapters "Codes Endowed with the Rank Metric"–"Generalizing Subspace Codes to Flag Codes Using Group Actions", where the codewords themselves are matroid representations. However, so far this perspective has not brought any new insights to the topic.

Another instance where matroids appear naturally in mathematics is graph theory. Let $\Gamma$ be a finite graph with edge set $E$. We obtain a matroid $M_\Gamma = (\mathscr{I}, E)$, where $I \subseteq E$ is independent if the subgraph $\Gamma_I \subseteq \Gamma$ induced on $I \subseteq E$ is a forest, i.e., has no cycles. A matroid that is isomorphic to $M_\Gamma$ for some graph $\Gamma$ is said to be a *graphical* matroid.

*Example 15* The matrix $G$ and the graph $\Gamma$ given below generate the same matroid, regardless of the field over which $G$ is defined.

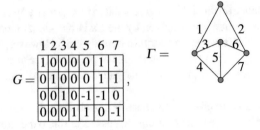

$$G = \begin{array}{c} \begin{array}{ccccccc} 1 & 2 & 3 & 4 & 5 & 6 & 7 \end{array} \\ \begin{array}{|c|c|c|c|c|c|c|} \hline 1 & 0 & 0 & 0 & 0 & 1 & 1 \\ \hline 0 & 1 & 0 & 0 & 0 & 1 & 1 \\ \hline 0 & 0 & 1 & 0 & -1 & -1 & 0 \\ \hline 0 & 0 & 0 & 1 & 1 & 0 & -1 \\ \hline \end{array} \end{array},$$

Some examples of independent sets in $G$ and $\Gamma$ are $\{3, 4, 6\}$, $\{1, 2, 3, 5\}$, $\{2, 3, 4, 6\}$. The set $X = \{5, 6, 7\}$ is dependent in $M_\Gamma$ as these edges form a cycle, and it is dependent in $M_G$ as the submatrix

$$G(X) = \begin{array}{c} \begin{array}{ccc} 5 & 6 & 7 \end{array} \\ \begin{array}{|c|c|c|} \hline 0 & 1 & 1 \\ \hline 0 & 1 & 1 \\ \hline -1 & -1 & 0 \\ \hline 1 & 0 & -1 \\ \hline \end{array} \end{array}$$

has linearly dependent columns.

Indeed, graphical matroids are representable over any field $\mathbb{F}$. To see this, for a graph $\Gamma$ with edge set $E$, we will construct a matrix $G(\Gamma)$ over $\mathbb{F}$ with column set $E$ as follows. Choose an arbitrary spanning forest $T \subseteq E$ in $\Gamma$, and index the rows of $G(\Gamma)$ by $T$. Thus $G(\Gamma)$ is a $T \times E$-matrix. Choose an arbitrary orientation for each edge in the graph. For $e \in T \subseteq E$ and $uv \in E$, the entry in position $(e, \{uv\})$ is 1 (respectively $-1$) if $e$ is traversed forward (respectively backward) in the unique path from $u$ to $v$ in the spanning forest $T$. In particular, the submatrix $G(\Gamma)(T)$ is an identity matrix. It is straightforward to check that the independent sets in $G(\Gamma)$ are exactly the noncyclic sets in $\Gamma$.

*Example 16* The matrix $G$ in Example 15 is $G(\Gamma)$ where $\Gamma$ is the graph in the same example, and the spanning forest $T$ is chosen to be $\{1, 2, 3, 4\}$.

The restriction to $X \subseteq E$ of a graphical matroid $M_\Gamma$ is obtained by the subgraph of $\Gamma$ containing precisely the edges in $X$.

A third example of matroids occurring naturally in mathematics are *algebraic matroids* [22]. These are associated to field extensions $\mathbb{F} : K$ together with a finite point sets $E \subseteq K$, where the independent sets are those $I \subseteq E$ that are algebraically independent over $\mathbb{F}$. In particular, elements that are algebraic over $\mathbb{F}$ have rank zero, and in general $\rho(I)$ is the transcendence degree of the field extension $\mathbb{F}(I) : \mathbb{F}$.

It is rather easy to see that every $\mathbb{F}$-linear matroid is also algebraic over $\mathbb{F}$. Indeed, let $X_1, \ldots, X_k$ be indeterminates, and let

$$g : \mathbb{F}^k \to \mathbb{F}(X_1, \ldots, X_k)$$

**Fig. 2** The Vamos matroid
of size 8 and rank 4, which is
not algebraically
representable

be given by $e_i \mapsto X_i$ for $i = 1, \ldots, k$. Then $J \subseteq E$ is linearly independent over $\mathbb{F}$ if and only if $\{g(j) : j \in J\}$ is algebraically independent over $\mathbb{F}$. Over fields of characteristic zero the converse also holds, so that all algebraic matroids have a linear representation. However, in positive characteristic there exist algebraic matroids that are not linearly representable. For example, the non-Pappus matroid of Example 4 is algebraically representable over $\mathbb{F}_4$, although it is not linearly representable over any field [21]. The smallest example of a matroid that is not algebraic over any field is the Vamos matroid, in Fig. 2 [15].

**Definition 21** The *dual* of $M = (\rho, E)$ is $M^* = (\rho^*, E)$, where

$$\rho^*(X) = |X| + \rho(E \setminus X) - \rho(E).$$

The definition of the dual matroid lies in the heart of matroid theory, and has profound interpretations. In geometric terms, let $M$ be represented by a $k$-dimensional subspace $V$ of $\mathbb{F}^n$. Then, the matroid dual $M^*$ is represented by the orthogonal complement $V^\perp \subseteq \mathbb{F}^n$. Surprisingly and seemingly unrelatedly, if $\Gamma$ is a planar graph and $M = M_\Gamma$ is a graphical matroid, then $M^* = M_{\bar{\Gamma}}$, where $\bar{\Gamma}$ is the planar dual of $\Gamma$. Moreover, the dual $M_\Gamma^*$ of a graphical matroid is graphical if and only if $\Gamma$ is planar.

**Definition 22** The *contraction* of $X \subseteq E$ in the matroid $M = (\rho, E)$ is $M/X = (\rho', M \setminus X)$, where $\rho'(Y) = \rho(Y \cup X) - \rho(X)$.

Contraction is the dual operation of deletion, in the sense that $M/X = (M_{|E \setminus X}^*)^*$. The terminology comes from graphical matroids, where contraction of the edge $e \in E$ corresponds to deleting $e$ and identifying its endpoints in the graph. Notice that it follows directly from submodularity of the rank function that $\rho_{M/X}(Y) \leq \rho_{M|E \setminus X}(Y)$ for every $Y \subseteq E \setminus X$. In terms of subspace representations, contraction of $e \in E$ corresponds to intersecting the subspace that represents $M$ with the hyperplane $\{x_e = 0\}$.

As matroids are used as an abstraction for linear codes, it would be desirable to have a way to go back from matroids to codes, namely to determine whether a given matroid is representable, and when it is, to find such a representation. Unfortunately, there is no simple criterion to determine representability [25, 43]. However, there are a plethora of sufficient criteria to prove nonrepresentability, both over a given field and over fields in general. In recent years, these methods have been used to prove two long-standing conjectures, that we will discuss in sections Rota's Conjecture and Most Matroids are Nonrepresentable respectively.

## Rota's Conjecture

While there is no simple criterion to determine linear representability, the situation is much more promising if we consider representations over a fixed field. It has been known since 1958, that there is a simple criterion for when a matroid is binary representable.

**Theorem 11** ([42]) *Let $M = (\rho, E)$ be a matroid. The following two conditions are equivalent.*

1. *$M$ is linearly representable over $\mathbb{F}_2$.*
2. *There are no sets $X \subseteq Y \subseteq E$ such that $M|Y/X$ is isomorphic to the uniform matroid $U_4^2$.*

In essence, this means that the only obstruction that needs to be overcome in order to be representable over the binary alphabet, is that no more than three nonzero points can fit in the same plane. For further reference, we say that a *minor* of the matroid $M = (\rho, E)$ is a matroid of the form $M|Y/X$, for $X \subseteq Y \subseteq E$. Clearly, if $M$ is representable over $\mathbb{F}$, then so is all its minors. Let $L(\mathbb{F})$ be the class of matroids that are not representable over $\mathbb{F}$, but such that all of their minors are $\mathbb{F}$-representable. Then the class of $\mathbb{F}$-representable matroids can be written as the class of matroids that does not contain any matroid from $L(\mathbb{F})$ as a minor. Gian-Carlo Rota conjectured in 1970 that $L(\mathbb{F})$ is a finite set for all finite fields $\mathbb{F}$. A proof of this conjecture was announced by Geelen, Gerards and Whittle in 2014, but the details of the proof still remain to written up [12].

**Theorem 12** *For any finite field $\mathbb{F}$, there is a finite set $L(\mathbb{F})$ of matroids such that any matroid $M$ is representable if and only if it contains no element from $L(\mathbb{F})$ as a minor.*

Since the 1970s, it has been known that a matroid is representable over $\mathbb{F}_3$ if and only if it avoids the uniform matroids $U_5^2$, $U_5^3$, the Fano plane $P^2(\mathbb{F}_2)$, and its dual $P^2(\mathbb{F}_2)^*$ as minors. The list $L(\mathbb{F}_4)$ has seven elements, and was given explicitly in 2000. For larger fields, the explicit list is not known, and there is little hope to even find useful bounds on its size.

## Most Matroids are Nonrepresentable

For a fixed finite field $\mathbb{F}$, it follows rather immediately from the minor-avoiding description in the last section that the fraction of $n$-symbol matroids that is $\mathbb{F}$-representable goes to zero as $n \to \infty$. It has long been a folklore conjecture that this is true even when representations over arbitrary fields are allowed. However, it was only in 2016 that a verifiable proof of this claim was announced [26].

**Theorem 13**
$$\lim_{n \to \infty} \frac{\#linear\ matroids\ on\ n\ elements}{\#matroids\ on\ n\ elements} = 0.$$

The proof is via estimates of the denominator and enumerator of the expression in Theorem 13 separately. Indeed, it is shown in [19] that the number of matroids on $n$ nodes is at least $\Omega(2^{(2-\varepsilon)^n})$ for every $\varepsilon > 0$. The proof of Theorem 13 thus boiled down to proving that the number of representable matroids is $O(2^{n^3})$. This is in turn achieved by bounding the number of so called *zero-patterns* of polynomials.

## *Gammoid Construction of Singleton-Optimal LRCs*

For completeness, we end this appendix with a theorem that explicitly presents the matroids constructed in Theorem 7 as gammoids. As discussed in Sect. 4.2, this proves the existence of Singleton-optimal linear LRCs whenever a set system satisfying (8) exists.

**Theorem 14** ([47], $M(F_1, \ldots, F_m, E; k; \rho)$-matroids are gammoids) *Let $M(F_1, \ldots, F_m; \rho)$ be a matroid given by Theorem 7 and define $s : E \to 2^{[m]}$ where $s(x) = \{i \in [m] : x \in F_i\}$. Then $M(F_1, \ldots, F_m, E; k; \rho)$ is equal to the gammoid $M(\Gamma, E, T)$, where $\Gamma = (V, D)$ is the directed graph with*

> *(i)* $V = E \cup H \cup T$ *where $E, H, T$ are pairwise disjoint,*
> *(ii)* $T = [k]$,
> *(iii)* $H$ *equals the union of the pairwise disjoint sets*
> $H_1, \ldots, H_m, H_{\geq 2}$, *where*
> $|H_i| = \rho(F_i) - |\{x \in F_i : |s(x)| \geq 2\}|$ *for $i \in [m]$,*
> $H_{\geq 2} = \{h_y : y \in E, |s(y)| \geq 2\}$,
> *(iv)* $D = D_1 \cup D_2 \cup D_3$, *where*
> $D_1 = \bigcup_{i \in [m]} \{(\overrightarrow{x, y}) : x \in E, s(x) = \{i\}, y \in H_i\}$,
> $D_2 = \{(\overrightarrow{x, h_y}) : x \in E, h_y \in H_{\geq 2}, s(x) \subseteq s(y)\}$,
> $D_3 = \{(\overrightarrow{x, y}) : x \in H, y \in T\}$.

## References

1. S. Ball, On sets of vectors of a finite vector space in which every subset of basis size is a basis. J. Eur. Math. Soc. **14**, 733–748 (2012)
2. A. Barg, I. Tamo, S. Vlăduţ, Locally recoverable codes on algebraic curves (2016), arXiv:1603.08876
3. A. Barg, K. Haymaker, E. Howe, G. Matthews, A. Várilly-Alvarado, Locally recoverable codes from algebraic curves and surfaces (2017), arXiv:1701.05212
4. J.E. Bonin, A. de Mier, The lattice of cyclic flats of a matroid. Ann. Comb. **12**, 155–170 (2008)

5.  T. Britz, C.G. Rutherford, Covering radii are not matroid invariants. Discret. Math. **296**, 117–120 (2005)
6.  V. Cadambe, A. Mazumdar, An upper bound on the size of locally recoverable codes, in *International Symposium on Network Coding* (2013), pp. 1–5
7.  A. Dimakis, P.B. Godfrey, Y. Wu, M.J. Wainwright, K. Ramchandran, Network coding for distributed storage systems. IEEE Trans. Inf. Theory **56**(9), 4539–4551 (2010)
8.  T. Ernvall, T. Westerbäck, R. Freij-Hollanti, C. Hollanti, Constructions and properties of linear locally repairable codes. IEEE Trans. Inf. Theory **62**, 5296–5315 (2016)
9.  A. Fazeli, A. Vardy, E. Yaakobi, PIR with low storage overhead: coding instead of replication (2015), arXiv:1505.06241
10. R. Freij-Hollanti, T. Westerbäck, C. Hollanti, Locally repairable codes with availability and hierarchy: matroid theory via examples, in *International Zürich Seminar on Communications* (IEEE/ETH, 2016), pp. 45–49
11. S. Fujishige, Polymatroidal dependence structure of a set of random variables. Inf. Control **39**(1), 55–72 (1978)
12. J. Geelen, B. Gerards, G. Whittle, Solving Rota's conjecture. Not. Am. Math. Soc. **61**, 736–743 (2014)
13. P. Gopalan, C. Huang, H. Simitci, S. Yekhanin, On the locality of codeword symbols. IEEE Trans. Inf. Theory **58**(11), 6925–6934 (2012)
14. C. Huang, M. Chen, J. Lin, Pyramid codes: flexible schemes to trade space for access efficiency in reliable data storage systems, in *International Symposium on Network Computation and Applications* (IEEE, 2007), pp. 79–86
15. A. Ingleton, R. Main, Non-algebraic matroids exist. Bull. Lond. Math. Soc. **7**, 144–146 (1975)
16. Y. Ishai, E. Kushilevitz, R. Ostrovsky, A. Sahai, Batch codes and their applications, in *The 36th ACM Symposium on Theory of Computing (STOC)* (2004)
17. R. Jurrius, R. Pellikaan, Truncation formulas for invariant polynomials of matroids and geometric lattices. Math. Comput. Sci. **6**, 121–133 (2012)
18. G.M. Kamath, N. Prakash, V. Lalitha, P.V. Kumar, Codes with local regeneration and erasure correction. IEEE Trans. Inf. Theory **60**(8), 4637–4660 (2014)
19. D. Knuth, The asymmetric number of geometries. J. Comb. Theory Ser. A **16**, 398–400 (1974)
20. B. Lindström, On the vector representations of induced matroids. Bull. Lond. Math. Soc. **5**, 85–90 (1973)
21. B. Lindström, On $p$-polynomial representations of projective geometries in algebraic combinatorial geometries. Math. Scand. **63**, 36–42 (1988)
22. B. Lindström, On algebraic matroids. Discret. Math. **111**, 357–359 (1993)
23. H. Lipmaa, V. Skachek, Linear batch codes, in *The 4th International Castle Meeting on Coding Theory and Applications (4ICMCTA)* (2015)
24. D. Mayhew, M. Newman, D. Welsh, G. Whittle, On the asymptotic proportion of connected matroids. Eur. J. Comb. **32**(6), 882–890 (2011)
25. D. Mayhew, M. Newman, G. Whittle, Yes, the missing axiom of matroid theory is lost forever (2015), arXiv:1412.8399
26. P. Nelson, Almost all matroids are non-representable, arXiv:1605.04288
27. D. Papailiopoulos, A. Dimakis, Locally repairable codes, in *International Symposium on Information Theory* (IEEE, 2012), pp. 2771–2775
28. R. Pendavingh, J. van der Pol, On the number of matroids compared to the number of sparse paving matroids. Electron. J. Comb. **22** (2015), 17pp
29. A. Pöllänen, T. Westerbäck, R. Freij-Hollanti, C. Hollanti, Improved singleton-type bounds for locally repairable codes, in *International Symposium on Information Theory* (IEEE, 2016), pp. 1586–1590
30. N. Prakash, G.M. Kamath, V. Lalitha, P.V. Kumar, Optimal linear codes with a local-error-correction property, in *International Symposium on Information Theory* (IEEE, 2012), pp. 2776–2780
31. A.S. Rawat, D. Papailiopoulos, A. Dimakis, S. Vishwanath, Locality and availability in distributed storage (2014), arXiv:1402.2011v1

32. B. Sasidharan, G.K. Agarwal, P.V. Kumar, Codes with hierarchical locality (2015), arXiv:1501.06683v1
33. N. Silberstein, A. Zeh, Optimal binary locally repairable codes via anticodes (2015), arXiv:1501.07114v1
34. N. Silberstein, A.S. Rawat, O. Koyluoglu, S. Vishwanath, Optimal locally repairable codes via rank-metric codes, in *International Symposium on Information Theory* (IEEE, 2013), pp. 1819–1823
35. J.A. Sims, Some problems in matroid theory. Ph.D. thesis, Oxford University (1980)
36. R.C. Singleton, Maximum distance q-nary codes. IEEE Trans. Inf. Theory **10**(2), 116–118 (1964)
37. A. Skorobogatov, Linear codes, strata of Grassmannians, and the problems of Segre, in *International Workshop on Coding Theory and Algebraic Geometry* (1992), pp. 210–223
38. W. Song, C. Yuen, S.H. Dau, T.J. Li, Optimal locally repairable linear codes. IEEE J. Sel. Areas Commun. **32**(5), 1019–1036 (2014)
39. I. Tamo, A. Barg, A family of optimal locally recoverable codes. IEEE Trans. Inf. Theory **60**(8), 4661–4676 (2014)
40. I. Tamo, D. Papailiopoulos, A. Dimakis, Optimal locally repairable codes and connections to matroid theory, in *International Symposium on Information Theory* (IEEE, 2013), pp. 1814–1818
41. I. Tamo, A. Barg, A. Frolov, Bounds on the parameters of locally recoverable codes. IEEE Trans. Inf. Theory **62**(6), 3070–3083 (2016)
42. W. Tutte, A homotopy theorem for matroids, I, II. Trans. Am. Math. Soc. **88**, 148–178 (1958)
43. P. Vámos, The missing axiom of matroid theory is lost forever. J. Lond. Math. Soc. **18**, 403–408 (1978)
44. A. Wang, Z. Zhang, Repair locality with multiple erasure tolerance. IEEE Trans. Inf. Theory **60**(11), 6979–6987 (2014)
45. A. Wang, Z. Zhang, An integer programming-based bound for locally repairable codes. IEEE Trans. Inf. Theory **61**(10), 5280–5294 (2015)
46. T. Westerbäck, R. Freij, C. Hollanti, Applications of polymatroid theory to distributed storage systems, in *Allerton Conference on Communication, Control, and Computing* (2015), pp. 231–237
47. T. Westerbäck, R. Freij-Hollanti, T. Ernvall, C. Hollanti, On the combinatorics of locally repairable codes via matroid theory. IEEE Trans. Inf. Theory **62**, 5296–5315 (2016)
48. H. Whitney, On the abstract properties of linear dependence. Am. J. Math. **57**, 509–533 (1935)
49. H. Zhang, V. Skachek, Bounds for batch codes with restricted query size, in *IEEE International Symposium on Information Theory (ISIT)* (2016)

# Batch and PIR Codes and Their Connections to Locally Repairable Codes

**Vitaly Skachek**

**Abstract** Two related families of codes are studied: batch codes and codes for private information retrieval. These two families can be viewed as natural generalizations of locally repairable codes, which were extensively studied in the context of coding for fault tolerance in distributed data storage systems. Bounds on the parameters of the codes, as well as basic constructions, are presented. Connections between different code families are discussed.

## 1 Introduction

In this chapter, we discuss two related families of codes: batch codes and codes for private information retrieval (PIR codes). These two families can be viewed as natural generalizations of locally repairable codes, which were extensively studied in the context of coding for fault tolerance in distributed data storage systems.

Batch codes were first presented in [15], where it was suggested to use them for load balancing in the multi-server distributed data storage systems. It was also suggested in [15] to use these codes in private information retrieval. A number of constructions of batch codes were presented therein. Later, the authors of [33] proposed to use so-called "switch codes" for facilitating the routing of data in the network switches. It turns out, however, that switch codes are a special case of batch codes.

Coding schemes for PIR were studied in [12]. The authors showed that a family of codes, which is a relaxed version of batch codes, can be employed in classical linear PIR schemes in order to reduce the redundant information stored in a distributed server system. This relaxed version of the batch codes is termed PIR codes.

In these schemes, typically a distributed data storage system is considered. The coded words are written across the block of disks (servers), where each disk stores a single symbol (or a group of symbols). The reading of data is done by accessing

V. Skachek (✉)
Institute of Computer Science, University of Tartu, 50409 Tartu, Estonia
e-mail: vitaly.skachek@ut.ee

© Springer International Publishing AG 2018

M. Greferath et al. (eds.), *Network Coding and Subspace Designs*,
Signals and Communication Technology,
https://doi.org/10.1007/978-3-319-70293-3_16

a small number of disks. Mathematically, this can be equivalently represented by the assumption that each information symbol depends on a small number of other symbols. However, the type of requested queries varies in different code models. Thus, in PIR codes several copies of the same information symbols are requested, while in batch codes combinations of different symbols are also possible.

In this chapter, we mathematically define the corresponding families of codes, and study their properties. We derive bounds on the parameters of such codes, and show some basic constructions. We also show relations between different families of codes. In Sect. 3, we introduce various models of locally repairable codes. In Sect. 4, we define batch codes. In Sect. 5, we discuss properties of linear batch codes. In Sect. 6, we introduce codes for private information retrieval. In Sect. 7, we study connections between locally repairable and batch/PIR codes. In Sects. 8 and 9, we present bounds on the parameters of various families of codes. In Sect. 10, we pose some open questions. For the sake of completeness, we introduce all necessary notations and definitions, which are used in the sequel.

## 2 General Settings

Throughout this chapter, we denote by $\mathbb{N}$ the set of nonnegative integer numbers. For $n \in \mathbb{N}$, define $[n] \triangleq \{1, 2, \ldots, n\}$. Let $\mathbf{e}_i$ be the row vector having one at position $i$ and zeros elsewhere (the length of vectors will be clear from the context).

Let $\Sigma$ be a finite alphabet. Let $\mathbf{x} = (x_1, x_2, \ldots, x_k) \in \Sigma^k$ be an information vector. The code is a set of coded vectors

$$\left\{ \mathbf{y} = (y_1, y_2, \ldots, y_n) = \mathscr{C}(\mathbf{x}) \; : \; \mathbf{x} \in \Sigma^k \right\} \subseteq \Sigma^n \; ,$$

where $\mathscr{C} : \Sigma^k \to \Sigma^n$ is a bijection, for some $n \in \mathbb{N}$. By slightly abusing the notation, sometimes we denote the above set by $\mathscr{C}$.

Let $\mathbb{F} = \mathbb{F}_q$ be a finite field with $q$ elements, where $q$ is a prime power. If $\mathscr{C} : \mathbb{F}^k \to \mathbb{F}^n$ is a linear mapping, then $\mathscr{C}$ is a linear $[n, k, d]$ code over $\mathbb{F}$. Here, $d$ is the minimum Hamming distance of $\mathscr{C}$. In that case, the encoding can be viewed as a multiplication by a $k \times n$ generator matrix $\mathbf{G}$ over $\mathbb{F}$ of an information vector $\mathbf{x}$,

$$\mathbf{y} = \mathbf{x} \cdot \mathbf{G} \; . \tag{1}$$

The rate of the code is defined as $\mathscr{R} \triangleq k/n$.

## 3 Codes with Locality and Availability

Codes with locality were proposed for use in the distributed data storage systems [11]. In such systems, the data is stored in many disks (servers), and these servers may fail from time to time. It is possible to use erasure-correcting codes, where parts of a codeword are stored in different servers. A failure of one server can be viewed as an erasure of a symbol or of a group of symbols. In order to repair the erased symbols, there is a need to bring a few other symbols from other servers. In general, it would be beneficial to minimize the traffic in the network. For an overview of coding for distributed storage systems the reader is referred to chapter "An Overview of Coding for Distributed Storage Systems". We continue by recalling the definition of *locally repairable codes* (LRCs).

**Definition 1** The code $\mathscr{C}$ has locality $r \geq 1$, if for any $\mathbf{y} \in \mathscr{C}$, any symbol in $\mathbf{y}$ can be recovered by using *at most r* other symbols of $\mathbf{y}$.

Codes with low locality were extensively discussed, for example, in [11]. A somewhat similar family of codes, known as *one-step majority-logic decodable codes*, was investigated in the classical literature [18]. Recently, the bounds on the parameters of LRCs were derived in [14]. It was shown therein that the parameters of a linear $[n, k, d]$ code with locality $r$ over $\mathbb{F}$ satisfy[1]:

$$n \geq k + d + \left\lceil \frac{k}{r} \right\rceil - 2 . \tag{2}$$

This bound can be viewed as a refinement of the classical Singleton bound, where $\lceil \frac{k}{r} \rceil - 1$ is an additive penalty for locality of the code, when compared to the classical Singleton bound. The proof is done by iterative expurgating of the code, and by taking into account that there are dependencies between sets of $r + 1$ symbols, which consist of an arbitrary symbol in $\mathbf{y}$ and its recovery set. The bound in (2) is tight. In fact, several known constructions attain it with equality (see, for example, [14, 26, 30]).

Assume that the linear code $\mathscr{C}$ is systematic, i.e. the matrix $\mathbf{G}$ contains a $k \times k$ identity submatrix $\mathbf{I}$. Then, the symbols of $\mathbf{y}$ corresponding to $\mathbf{I}$ are called *information symbols*. It is possible to require recoverability of information symbols only (from sets of size at most $r$). In that case, the code is said to have *locality of information symbols*. Otherwise, if *all* symbols of $\mathbf{y}$ are recoverable from small sets, the code has *locality of all symbols*.

The above model was extended to codes with *locality and availability* in [24].

**Definition 2** The code $\mathscr{C}$ has locality $r \geq 1$ and availability $\delta \geq 1$, if for any $\mathbf{y} \in \mathscr{C}$, any symbol in $\mathbf{y}$ can be reconstructed by using any of $\delta$ disjoint sets of symbols, each set is of size at most $r$.

In [34], the authors consider linear codes with locality $r$ and availability $\delta$ of all symbols. They derive the following bound on the parameters of the code:

$$n \geq k + d + \left\lceil \frac{\delta(k - 1) + 1}{\delta(r - 1) + 1} \right\rceil - 2 . \tag{3}$$

---

[1]This result is proven in Theorem 6, Chap. "An Overview of Coding for Distributed Storage Systems".

In [24], systematic codes (linear or non-linear) are considered, with locality $r$ and availability $\delta$ of information symbols. The authors show the bound analogues to (3) for that case. In particular, we observe that when availability $\delta = 1$, i.e. there is only one recovery set for each symbol, then (3) coincides with (2). The proof technique in both cases is based on the idea similar to that of [14].

Another related model is considered in [23]. In that model, several *different* symbols are recovered from a small set of recovery symbols. By building on the ideas in the previous works, the authors derive a variation of the bound (3) for the model under consideration. Other related works, for example, include [6, 13, 19, 21, 29, 35, 36] and the references therein.[2]

## 4  Batch Codes

Batch codes were first presented in the cryptographic community in [15]. In that work, the authors proposed to use batch codes for load balancing in the distributed systems, as well as for private information retrieval. The authors of [15] have also presented a few constructions of various families of batch codes. Those constructions were based on recursive application of simple batch codes (so-called "sub-cube codes"), on classical Reed-Muller codes, on locally decodable codes, and others.

The following definition is based on [15].

**Definition 3** Let $\Sigma$ be a finite alphabet. We say that $\mathscr{C}$ is an $(k, n, t, M, \tau)_\Sigma$ *batch code* over a finite alphabet $\Sigma$ if it encodes any string $\mathbf{x} = (x_1, x_2, \ldots, x_k) \in \Sigma^k$ into $M$ strings (buckets) of total length $n$ over $\Sigma$, namely $\mathbf{y}_1, \mathbf{y}_2, \ldots, \mathbf{y}_M$, such that for each $t$-tuple (batch) of (not necessarily distinct) indices $i_1, i_2, \ldots, i_t \in [k]$, the symbols $x_{i_1}, x_{i_2}, \ldots, x_{i_t}$ can be retrieved by reading at most $\tau$ symbols from each bucket.

More formally, by following the presentation in [31], we can state an equivalent definition.

**Definition 4** An $(k, n, t, M, \tau)_\Sigma$ batch code $\mathscr{C}$ over a finite alphabet $\Sigma$ is defined by an encoding mapping $\mathscr{C} : \Sigma^k \rightarrow (\Sigma^*)^M$ (there are $M$ buckets in total), and a decoding mapping $\mathscr{D} : \Sigma^n \times [k]^t \rightarrow \Sigma^t$, such that

1. The total length of all buckets is $n$;
2. For any $\mathbf{x} \in \Sigma^k$ and

$$i_1, i_2, \ldots, i_t \in [k] , \tag{4}$$

$$\mathscr{D}(\mathscr{C}(\mathbf{x}), i_1, i_2, \ldots, i_t) = (x_{i_1}, x_{i_2}, \ldots, x_{i_t}) ,$$

and $\mathscr{D}$ depends only on $\tau$ symbols in each bucket in $\mathscr{C}(\mathbf{x})$.

---

[2]In particular, the use of matroids in bounding the parameters of various families of codes with locality is thoroughly treated in chapter "Matroid Theory and Storage Codes: Bounds and Constructions".

In particular, an interesting case for consideration is when $\tau = 1$, namely only at most one symbol is read from each bucket. If the requested information symbols $(x_{i_1}, x_{i_2}, \ldots, x_{i_t})$ can be reconstructed from the data read by $t$ different users independently (i.e., the symbol $x_{i_\ell}$ is reconstructed by the user $\ell$, $\ell = 1, 2, \ldots, t$, respectively), and the sets of the symbols read by these $t$ users are all disjoint, such a model is called a *multiset* batch code.

In the sequel, we only consider multiset batch codes, and therefore we usually omit the word "multiset" for convenience.

An important special case of batch codes is defined as follows.

**Definition 5** ([15]) A *primitive* batch code is a batch code, where each bucket contains exactly one symbol. In particular, $n = M$.

Following the work [15], a number of subsequent papers have studied *combinatorial* batch codes. In combinatorial batch codes, a number of replicas of the information symbols are stored in different positions in the codeword. Usually, the symbols are associated with servers according to some optimal or sub-optimal combinatorial objects, such as block designs. Combinatorial batch codes were studied, for example, in [2, 4, 5, 27, 28].

## 5 Linear Batch Codes

In what follows, we consider a special case of primitive multiset batch codes with $n = M$ and $\tau = 1$. Under these conditions, each symbol can be viewed as a separate bucket, and only one reading per bucket is allowed.

We assume that the information and the coded symbols are taken from the finite field $\mathbb{F} = \mathbb{F}_q$, where $q$ is a prime power. Additionally, we assume that the encoding mapping $\mathscr{C} : \mathbb{F}^k \to \mathbb{F}^n$ is linear over $\mathbb{F}$, and therefore the code $\mathscr{C}$ is a linear $[n, k]$ code over $\mathbb{F}$. In that case, $\mathscr{C}$ falls under the linear coding framework defined in (1). We also refer to the parameter $t$ as the size of a query of the code. The batch code with the parameters $n$, $k$ and $t$ over $\mathbb{F}_q$ is denoted as $[n, k, t]_q$-batch code (or simply $[n, k, t]$-batch code) in the sequel.

This framework was first considered in [17], and the similarities with locally repairable codes were mentioned. The main difference between these two families, however, is that the supported query types are different. In batch codes we are interested in reconstruction of the information symbols in $\mathbf{x}$, while in locally repairable codes the coded symbols in $\mathbf{y}$ are to be recovered.

The following simple result was established in [17, Theorem 1].

**Theorem 1** *Let* $\mathscr{C}$ *be an* $[n, k, t]_q$ *batch code. It is possible to retrieve* $x_{i_1}, x_{i_2}, \ldots, x_{i_t}$ *by* $t$ *different users in the primitive multiset batch code model (where the symbol* $x_{i_\ell}$ *is retrieved by the user* $\ell$, $\ell = 1, 2, \ldots, t$, *respectively) if and only if there exist* $t$ *non-intersecting sets* $T_1, T_2, \ldots, T_t$ *of indices of columns in the generator matrix* $\mathbf{G}$, *and for each* $T_\ell$, $1 \le \ell \le t$, *there exists a linear combination of columns of* $\mathbf{G}$ *indexed by that set, which equals to the column vector* $e_{i_\ell}^T$, *for all* $\ell \in [t]$.

The reader can find the proof of this theorem in [17]. Next, we show examples that further illustrate this concept.

*Example 1* ([15]) Consider the following binary $2 \times 3$ generator matrix of a batch code $\mathscr{C}$ given as

$$\mathbf{G} = \begin{pmatrix} 1 & 0 & 1 \\ 0 & 1 & 1 \end{pmatrix}.$$

The corresponding code is a sub-cube code constructed in [15, Sect. 3.2]. By using this code, the information symbols $(x_1, x_2)$ are encoded into $(y_1, y_2, y_3) = (x_1, x_2, x_1 + x_2)$.

Assume that the query contains two different symbols $(x_1, x_2)$. Then, we can retrieve these symbols directly by using the following equations:

$$\begin{cases} x_1 = y_1 \\ x_2 = y_2 \end{cases}.$$

Alternatively, assume that the query contains two copies of the same symbol, for example $(x_1, x_1)$. Then, we can retrieve these symbols by using the following equations:

$$\begin{cases} x_1 = y_1 \\ x_1 = y_2 + y_3 \end{cases}.$$

Similarly, $(x_2, x_2)$ can be retrieved. We conclude that $\mathscr{C}$ is a $[3, 2, 2]_2$ batch code.

*Example 2* ([17]) Pick the following binary $4 \times 9$ generator matrix of a batch code $\mathscr{C}$ given as

$$\mathbf{G} = \begin{pmatrix} 1 & 0 & 1 & 0 & 0 & 0 & 1 & 0 & 1 \\ 0 & 1 & 1 & 0 & 0 & 0 & 0 & 1 & 1 \\ 0 & 0 & 0 & 1 & 0 & 1 & 1 & 0 & 1 \\ 0 & 0 & 0 & 0 & 1 & 1 & 0 & 1 & 1 \end{pmatrix}.$$

The corresponding code is a second-order sub-cube code constructed as in [15, Sect. 3.2].

Assume that the query contains the information symbols $(x_1, x_1, x_2, x_2)$. Then, we can retrieve these symbols using the following equations:

$$\begin{cases} x_1 = y_1 \\ x_1 = y_2 + y_3 \\ x_2 = y_5 + y_8 \\ x_2 = y_4 + y_6 + y_7 + y_9 \end{cases}.$$

It can be verified in a similar manner that any 4-tuple $(x_{i_1}, x_{i_2}, x_{i_3}, x_{i_4})$, where $i_1, i_2, i_3, i_4 \in [4]$, can be retrieved by using the symbols of $\mathbf{y}$, by using each symbol at most once. We conclude that $\mathscr{C}$ is a $[9, 4, 4]_2$ batch code.

*Example 3* ([32]) Pick the following binary $3 \times 7$ generator matrix of a batch code $\mathscr{C}$ given as

$$G = \begin{pmatrix} 1 & 0 & 0 & 1 & 1 & 0 & 1 \\ 0 & 1 & 0 & 1 & 0 & 1 & 1 \\ 0 & 0 & 1 & 0 & 1 & 1 & 1 \end{pmatrix} .$$

The corresponding code is a binary $[7, 3, 4]$ classical error-correcting simplex code.

Assume that the query contains the information symbols $(x_1, x_1, x_2, x_2)$. Then, we can retrieve these symbols using the following equations:

$$\begin{cases} x_1 = y_1 \\ x_1 = y_2 + y_4 \\ x_2 = y_3 + y_6 \\ x_2 = y_5 + y_7 \end{cases} .$$

It can be verified in a similar manner that any 4-tuple $(x_{i_1}, x_{i_2}, x_{i_3}, x_{i_4})$, where $i_1, i_2, i_3, i_4 \in [4]$, can be retrieved by using the symbols of $\mathbf{y}$, by using each symbol at most once. We conclude that $\mathscr{C}$ is a $[7, 3, 4]_2$-batch code. Moreover, it was shown in [32] that all queries can be satisfied when each user reads at most $r = 2$ symbols from $\mathbf{y}$.

Constructions of linear batch codes using graphs without short cycles were presented in [10]. A family of codes, related to batch codes, and corresponding to the case $t = k$, termed *switch codes*, was studied in [8, 32, 33]. It was suggested in [33] to use such codes for efficient routing of data in the network switches.

The following property of batch codes was observed in [17] for binary linear codes, and later generalized to nonbinary (and also to non-linear) codes in [40].

**Theorem 2** *Let $\mathscr{C}$ be an $[n, k, t]_q$-batch code. Then, the minimum Hamming distance of $\mathscr{C}$ is at least $t$.*

*Proof* Let $\mathbf{y}_1 = \mathscr{C}(\mathbf{x}_1)$ and $\mathbf{y}_2 = \mathscr{C}(\mathbf{x}_2)$ be two codewords of $\mathscr{C}$, and $\mathbf{x}_1 \neq \mathbf{x}_2$. Then, $\mathbf{x}_1$ and $\mathbf{x}_2$ differ in at least one symbol, i.e. $(\mathbf{x}_1)_\ell \neq (\mathbf{x}_2)_\ell$, for some $1 \leq \ell \leq k$. Consider the query $(\underbrace{x_\ell, x_\ell, \ldots, x_\ell}_{t})$. The $i$-th copy of $x_\ell$ is recovered from the set of symbols indexed by the set $T_i$, $1 \leq i \leq t$. Since $\mathbf{x}_1$ and $\mathbf{x}_2$ differ in the $\ell$-th symbol, the codewords $\mathbf{y}_1$ and $\mathbf{y}_2$ should differ in at least one symbol in each $T_i$, $1 \leq i \leq t$. The sets $T_i$ are all disjoint, and therefore $\mathbf{y}_1$ and $\mathbf{y}_2$ differ in at least $t$ symbols.     □

It follows that any $[n, k, t]_q$-batch code is in particular a classical $[n, k, \geq t]_q$ error-correcting code, and a variety of classical bounds, such as Singleton bound, Hamming bound, Plotkin bound, Griesmer bound, Johnson bound, Elias-Bassalygo bound, are all applicable to batch codes (when the minimum distance $d$ is replaced by the query size $t$).

*Example 4* Let $m > 1$ be an integer. A binary simplex $[2^m - 1, m, 2^{m-1}]$ code $\mathscr{C}$ is defined by its generator matrix

$$\mathbf{G} = \left( \mathbf{g}_1 \,|\, \mathbf{g}_2 \,|\, \cdots \,|\, \mathbf{g}_{2^m-1} \right) \,,$$

where $\mathbf{g}_i$ are all possible different binary nonzero column vectors of length $m$, $i = 1, 2, \ldots, 2^m - 1$ [25, Problem 2.18].

For a classical error-correcting $[n, k, d]$ code over $\mathbb{F}_q$, the Plotkin bound is defined as follows [20, Theorem 2.2.29]:

$$\text{if } qd > (q - 1)n, \text{ then } q^k \leq \left\lfloor \frac{qd}{qd - (q - 1)n} \right\rfloor .$$

It is straightforward to see that, as an error-correcting code, the binary simplex code as above (with $q = 2$) attains the Plotkin bound with equality [25, Problem 2.18] for all $m \geq 2$.

As it was shown in [32], the code $\mathscr{C}$ is a $[2^m - 1, m, 2^{m-1}]_2$ batch code, with $t = 2^{m-1}$. Therefore, by Theorem 2, it attains the corresponding Plotkin-based bound

$$q^k \leq \left\lfloor \frac{qt}{qt - (q - 1)n} \right\rfloor$$

with equality, and therefore it is a Plotkin-optimal batch code.

In [38], a variation of batch codes *with restricted size of reconstruction sets* is defined. These codes are batch codes as in Definition 3 with an additional property that every queried information symbol $x_i$ is reconstructed from *at most* $r \geq 1$ symbols of $\mathbf{y}$. This additional property can be viewed as analogous to locality of the LRCs. Small size of reconstruction sets allows for recovering the requested data symbol from a small number $r$ of servers, thus reducing the traffic and the load in the system.

For example, the binary simplex code $\mathscr{C}$ in the previous example was shown in [32] to have the size of reconstruction sets of at most $r = 2$.

## 6   Codes for Private Information Retrieval

The topic of private information retrieval (PIR) protocols has been a subject of a lot of research over the last two decades [9]. In the PIR scenario, the database is stored in a number of servers in a distributed manner. The user is interested in reading an item from the database without revealing to any server what item was read. In the classical approach, the data is replicated, and the replicas are stored in a number of different servers. The user accesses some of these servers, such that no server learns what data the user is interested in (it is assumed that the servers do not collude).

A novel approach to PIR is based on coding, and it was studied, for example in [1, 7, 16]. More specifically, assume that $\mathbf{x} = (x_1, x_2, \ldots, x_k)$ is an information vector, which is encoded into $\mathscr{C}(\mathbf{x}) = \mathbf{y} = (y_1, y_2, \ldots, y_n)$. The symbols of $\mathbf{y}$ are stored in different servers in a distributed manner.

In [7], the authors show that there is a fundamental trade-off between download communication complexity of the protocol (the number of symbols or bits downloaded by the user from the database) and the storage overhead (the number of redundant symbols or bits stored in the database). Later, the authors of [12] show that it is possible to emulate many of the existing PIR protocols by using a code of length $n$ that approaches $(1 + \varepsilon)k$ for vanishing $\varepsilon$ (for sufficiently large $k$). This approach leads to PIR schemes with storage data rate arbitrarily close to 1. The authors define a special class of codes, which allows for such efficient PIR protocols.

**Definition 6** ([12]) An $k \times n$ binary matrix $\mathbf{G}$ has property $\mathscr{A}_t$ if for all $i \in [k]$, there exist $t$ disjoint sets of columns of $\mathbf{G}$ that add up to $\mathbf{e}_i$. A binary linear $[n, k]$ code $\mathscr{C}$ is called a $t$-*server PIR code* (or, simply, PIR code) if there exists a generator matrix $\mathbf{G}$ for $\mathscr{C}$ with property $\mathscr{A}_t$.

The batch codes turn out to be a special case of PIR codes, with a difference that PIR codes support only queries of type $(x_i, x_i, \ldots, x_i)$, $i \in [k]$, while batch codes support queries of a more general form $(x_{i_1}, x_{i_2}, \ldots, x_{i_t})$, possibly for different indices $i_1, i_2, \ldots, i_t \in [k]$. It follows that batch codes can be used as PIR codes.

Since for PIR codes (as well as for batch codes), the code rate $\mathscr{R}$ approaches 1 [15, 22] for large values of $k$, it is more appropriate to talk about redundancy (as a function of $k$), rather than about the code rate. This is in contrast to PIR/batch codes with restricted size of reconstruction sets, where the asymptotic loss of code rate takes place. The redundancy of the codes will be defined and analyzed in the following sections.

Constructions of PIR codes were presented very recently, for example, in [3, 12, 39].

## 7 Connections Between Batch/PIR Codes and General LRCs

As it was mentioned above, there are two types of LRCs considered in the literature: LRCs with locality of information symbols and LRCs with locality of all symbols.

In order to preserve the information symbols in the coded word, a code with locality of information symbols has a systematic encoding matrix $\mathbf{G} = [\mathbf{I}|\mathbf{A}]$ for some matrix $\mathbf{A}$. Consider LRCs with locality $r$ and availability $\delta = t - 1$ of information symbols. Then, each information symbol $y_i$, $1 \leq i \leq k$, in $\mathbf{y}$, can be recovered from $\delta$ disjoint sets $T_i$ of symbols, $|T_i| \leq r$. Such a code can also be viewed as a PIR code that supports any query of $t$ copies of an information symbol with locality $r$ (including one copy of the information symbol in the systematic part). Generally, it does not follow that such a code is a batch code, since there is no guarantee that mixed queries of different information symbols are supported by disjoint reconstruction sets.

On the other hand, a systematic batch or PIR code with restricted size of reconstruction sets allows to recover any query of $t$ information symbols with recovery sets of size $r$. Since in the systematic case, the information symbols are a part of a coded word $\mathbf{y}$, it follows that this code is an LRC with locality $r$ and availability $\delta = t - 1$ of information symbols.

We obtain the following corollary (see also Theorem 21 in [12]).

**Corollary 1** *A linear systematic code $\mathscr{C}$ is an LRC with locality $r$ and availability $\delta = t - 1$ of information symbols if and only if $\mathscr{C}$ is a PIR code that supports queries of size $t$ with size of reconstruction sets at most $r$.*

It follows that the bounds derived for the parameters of the LRCs with *locality and availability of information symbols* can be applied also to systematic batch or PIR codes.[3]

On the other hand, for the non-systematic case, there is no simple known connection between linear batch codes with restricted size of reconstruction sets and LRCs with availability, as it is illustrated in the following examples.

*Example 5* Let $\mathbf{G}$ be a $k \times (3k)$ generator matrix of a linear binary code $\mathscr{C}$ defined as follows:

$$\mathbf{G} = \begin{pmatrix}
1\,0\ldots0\,1 & 1\,0\ldots0\,1 & 1\,0\ldots0\,1 \\
0\,1\ldots0\,1 & 0\,1\ldots0\,1 & 0\,1\ldots0\,1 \\
0\,0\,\ddots\,0\,1 & 0\,0\,\ddots\,0\,1 & 0\,0\,\ddots\,0\,1 \\
0\,0\ldots1\,1 & 0\,0\ldots1\,1 & 0\,0\ldots1\,1 \\
0\,0\ldots0\,1 & 0\,0\ldots0\,1 & 0\,0\ldots0\,1
\end{pmatrix}.$$

Specifically, the binary information vector $\mathbf{x} = (x_1, x_2, \ldots, x_k)$ is encoded into the codeword $\mathbf{y}$, which consists of three copies of the same sub-vector,

$$\mathbf{y} = (y_1, y_2, \ldots, y_{3k})$$
$$= (x_1, x_2, \ldots, \sum_{i=1}^{k} x_i, \ x_1, x_2, \ldots, \sum_{i=1}^{k} x_i, \ x_1, x_2, \ldots, \sum_{i=1}^{k} x_i).$$

The code $\mathscr{C}$, when viewed as an LRC, has locality $r = 1$ and availability $\delta = 2$, since every symbol in $\mathbf{y}$ can be recovered from a single symbol in $\mathbf{y}$, and there are 2 different recovery sets.

On the other hand, the code $\mathscr{C}$, when viewed as a batch or PIR code, must have a maximal size of reconstruction sets at least $k$, since it is impossible to recover a single information symbol $x_k$ from less than $k$ coded symbols in $\mathbf{y}$.

---

[3]Please note that generally it does not follow here that the bounds for LRCs with *locality of all symbols* are applicable to systematic batch or PIR codes.

The following example shows that batch or PIR code with small size of reconstruction sets is not necessarily LRC with all symbols locality, even in the systematic case (the example can be also modified to a non-systematic case).

*Example 6* Take **G** to be a binary $2\kappa \times (3\kappa + 1)$ matrix, where $\kappa$ is some integer, as follows:

$$
\mathbf{G} = \begin{pmatrix}
1\,1\,0 & 0\,0\,0 & \cdots & 0\,0\,0 & 1 \\
0\,1\,1 & 0\,0\,0 & \cdots & 0\,0\,0 & 1 \\
\hline
0\,0\,0 & 1\,1\,0 & \cdots & 0\,0\,0 & 1 \\
0\,0\,0 & 0\,1\,1 & \cdots & 0\,0\,0 & 1 \\
\hline
\vdots\;\vdots\;\vdots & \vdots\;\vdots\;\vdots & \ddots & \vdots\;\vdots\;\vdots & \vdots \\
\hline
0\,0\,0 & 0\,0\,0 & \cdots & 1\,1\,0 & 1 \\
0\,0\,0 & 0\,0\,0 & \cdots & 0\,1\,1 & 1
\end{pmatrix}.
$$

Here, $k = 2\kappa$. This matrix **G** is a diagonal block matrix, where each block is a $2 \times 3$ generator matrix of a basic sub-cube code in [15], with an additional all-ones column. Let $\mathscr{C}$ be a binary linear code generated by this matrix.

The code $\mathscr{C}$, when viewed as a batch code, supports any two queries of the form $(x_i, x_j)$ $(1 \le i \le k, 1 \le j \le k)$, with size of reconstruction sets at most 2. Therefore, $\mathscr{C}$ is a batch code with $r = 2$, $t = 2$. In particular, it is a PIR code with $r = 2$ and $t = 2$.

On the other hand, the code $\mathscr{C}$, when viewed as an LRC, has locality of at least $\kappa$, since in order to recover $y_n = \sum_{i=1}^{k} x_k$, one needs to combine at least $\kappa$ other symbols of **y**.

As we see, this batch code with $r = 2$ and $t = 2$ has locality at least $k/2 = \kappa$, when used as an LRC with all symbols locality.

## 8 Bounds on the Parameters of PIR and Batch Codes with Unrestricted Size of Reconstruction Sets

In [31], the systematic batch codes with *unrestricted* size of reconstruction sets are considered. The codes under consideration have restriction on the value of $t$ (typically, it is a small constant), yet there is no restriction on $r$, so we can assume that $r = n$.

Define the parameters $\mathscr{B}(k, t)$ and $\mathscr{P}(k, t)$ to be the shortest length $n$ of any linear systematic batch and PIR code, respectively, for given values of $k$ and $t$. Define the *optimal redundancy* of systematic batch and PIR codes, respectively, as

$$
r_B(k, t) \triangleq \mathscr{B}(k, t) - k \quad \text{and} \quad r_P(k, t) \triangleq \mathscr{P}(k, t) - k .
$$

It is known [31] that for any fixed $t$,

$$
\lim_{k \to \infty} \frac{\mathscr{B}(k, t)}{k} = 1 .
$$

In a case of switch codes, $k = t$, it is shown in [33] that $\mathscr{B}(k, k) = O\left(k^2/\log(k)\right)$.
A constructive proof showing that $r_P(k, t) = t \cdot \sqrt{k}(1 + o(1))$ was given in [12]. As
for the lower bound on $r_P(k, t)$, it was recently shown in [22] (see also [37]) that
for a fixed $t \geq 3$, $r_P(k, t) = \Omega(k)$, thus establishing an asymptotic behavior for the
redundancy of the PIR codes.

Since every batch code is also a PIR code, it follows that $\mathscr{B}(k, t) \geq \mathscr{P}(k, t)$, and
$r_B(k, t) \geq r_P(k, t)$. The relations between $\mathscr{B}(k, t)$ and $\mathscr{P}(k, t)$ for specific choices
of $t$ were extensively studied in [31]. Thus, for example, it was shown that $\mathscr{B}(k, t) = \mathscr{P}(k, t)$ for $1 \leq t \leq 4$, while for $5 \leq t \leq 7$,

$$r_B(k, t) \leq r_P(k, t) + 2\lceil \log(k) \rceil \cdot r_P(k/2, t - 2) .$$

It is quite straightforward to verify that $r_B(k, 1) = 0$ and $r_B(k, 2) = 1$ for any $k$. It
was additionally shown in [31] that

$$r_B(k, t) = O(\sqrt{k}) \text{ for } t = 3, 4 ,$$
$$r_B(k, t) = O(\sqrt{k} \cdot \log(k)) \text{ for } 5 \leq t \leq 7 .$$

The following more general result was proven in [31].

**Theorem 3** *For all values of $k$ and $t$, it holds*

$$r_B(k, t) \leq r_P(k, t) + \left\lfloor \frac{t}{2} \right\rfloor \left\lceil \frac{\log \binom{k}{\lfloor t/2 \rfloor}}{-\log \left( 1 - \frac{\lfloor t/2 \rfloor!}{\lfloor t/2 \rfloor^{\lfloor t/2 \rfloor}} \right)} \right\rceil \cdot r_P \left( \left\lceil \frac{k}{\lfloor t/2 \rfloor} \right\rceil, t - 2 \right) .$$

In particular, it follows from Theorem 3 that for any fixed $t$,

$$r_B(k, t) = O\left( \sqrt{k} \cdot \log(k) \right) .$$

## 9 Bounds on the Parameters of PIR and Batch Codes with Restricted Size of Reconstruction Sets

In [38], the authors study linear batch codes with restricted size of reconstruction
sets. They aim at refining the Singleton bound for that case by using ideas in [14] and
subsequent works. Note, however, that these ideas cannot be applied directly, because
in LRCs there are dependencies between different coded symbols, and expurgation of
the code in the proof of the bound (2) (and similar bounds) uses these dependencies.
Therefore, the authors of [38] consider a query of $t$ copies of the same symbol (for
example, $(x_1, x_1, \ldots, x_1)$), and show that the symbols in different reconstruction
sets possess certain dependencies. By using this property, they apply an expurgation
technique similar to that of [14, 23, 24, 34], and obtain the following relation on

the parameters of batch codes with size of reconstruction sets restricted to $r$. The proof actually only assumes property $\mathscr{A}_t$, and therefore it is directly applicable to PIR codes as well.

**Theorem 4** ([38]) *Let $\mathscr{C}$ be a linear $[n, k, t]_q$-batch code (or PIR code) with the size of reconstruction sets restricted to $r$. Then, it holds:*

$$n \geq k + d + (t - 1) \left( \left\lceil \frac{k}{rt - t + 1} \right\rceil - 1 \right) - 1 . \tag{5}$$

Now, observe that if the $[n, k, t]_q$-batch code (or PIR code) allows for reconstruction of any batch of $t$ symbols, then it also allows for reconstruction of any batch of $\beta$ symbols, $1 \leq \beta \leq t$. Therefore, expression (5) in Theorem 4 can be adjusted as follows:

$$n \geq k + d + \max_{1 \leq \beta \leq t, \beta \in \mathbb{N}} \left\{ (\beta - 1) \left( \left\lceil \frac{k}{r\beta - \beta + 1} \right\rceil - 1 \right) \right\} - 1 . \tag{6}$$

If the code is systematic, then there is always a reconstruction set of size 1 for one of the queried symbols. In that case, the last expression can be rewritten as:

$$n \geq k + d + \max_{2 \leq \beta \leq t, \beta \in \mathbb{N}} \left\{ (\beta - 1) \left( \left\lceil \frac{k}{r\beta - \beta - r + 2} \right\rceil - 1 \right) \right\} - 1 . \tag{7}$$

The reader can refer to [38] for the full proofs.

*Example 7* ([38]) Take $r = 2$ and $t = \beta = 2$. Then, the bound in (7) is attained with equality by the linear systematic codes of minimum distance 2, defined as follows:

- $y_i = x_i$ for $1 \leq i \leq k$, and $y_j = x_{2(j-k)-1} + x_{2(j-k)}$ for $k + 1 \leq j \leq k + k/2$, when $k$ is even,
- $y_i = x_i$ for $1 \leq i \leq k$, $y_j = x_{2(j-k)-1} + x_{2(j-k)}$ for $k + 1 \leq j \leq k + (k - 1)/2$, and $y_{k+(k+1)/2} = x_k$, when $k$ is odd.

In that case, $d = 2$, and we obtain

$$n = k + k/2 \qquad \text{if } k \text{ is even} ,$$
$$n = k + (k + 1)/2 \quad \text{if } k \text{ is odd} .$$

In both cases, the bound (7) is attained with equality for all $k \geq 1$.

*Example 8* Consider the code $\mathscr{C}$ in Example 3, which was studied in [32]. As discussed, $\mathscr{C}$ is a linear $[7, 3, 4]_2$-batch code, with the size of reconstruction sets at most $r = 2$. Its minimum Hamming distance is $d = 4$. We pick $\beta = 2$, and observe that the right-hand side of Eq. (7) can be re-written as

$$3 + 4 + (2 - 1) \left( \left\lceil \frac{3}{2 \cdot 2 - 2 - 2 + 2} \right\rceil - 1 \right) - 1 = 7 \,,$$

and therefore the bound in (7) is attained with equality for the choice $\beta = 2$. The code $\mathscr{C}$ in Example 3 is optimal with respect to that bound. We note, however, that general simplex codes (of larger length) do not attain (7) with equality.

A slight improvement to the above bounds for both batch and PIR codes can be obtained, if one considers simultaneously reconstruction sets for, say, two queried batches $(x_1, x_1, \ldots, x_1)$ and $(x_2, x_2, \ldots, x_2)$, and studies intersections of their reconstruction sets. The analysis along those lines was done in [38], and the following result was derived.

Assume that
$$k \geq 2(rt - t + 1) + 1 \,. \tag{8}$$

Denote

$$\mathbb{A} = \mathbb{A}(k, r, d, \beta, \varepsilon) \triangleq k + d + (\beta - 1) \left( \left\lceil \frac{k + \varepsilon}{r\beta - \beta + 1} \right\rceil - 1 \right) - 1 \,,$$

$$\mathbb{B} = \mathbb{B}(k, r, d, \beta, \lambda) \triangleq k + d + (\beta - 1) \left( \left\lceil \frac{k + \lambda}{r\beta - \beta + 1} \right\rceil - 1 \right) - 1 \,,$$

$$\mathbb{C} = \mathbb{C}(k, r, \beta, \lambda, \varepsilon) \triangleq (r\beta - \lambda + 1)k - \binom{k}{2}(\varepsilon - 1) \,.$$

**Theorem 5** ([38]) *Let $\mathscr{C}$ be a linear $[n, k, t]$-batch code over $\mathbb{F}$ with the minimum distance $d$ and size of reconstruction sets at most $t$. Then,*

$$n \geq \max_{\beta \in \mathbb{N} \cap \left[1, \min\left\{t, \left\lfloor \frac{k-3}{2(r-1)} \right\rfloor\right\}\right]} \left\{ \max_{\varepsilon, \lambda \in \mathbb{N} \cap [1, r\beta - \beta]} \left\{ \min\{\mathbb{A}, \mathbb{B}, \mathbb{C}\} \right\} \right\} \,. \tag{9}$$

## 10 Open Questions

Below, we list some open questions related to batch and PIR codes.

1. Derive tighter bounds on the length or redundancy of batch and PIR codes, in particular, for small alphabet size, for large values of $t$, or for bounded values of $r$.
2. Construct new optimal or sub-optimal batch and PIR codes.
3. Do non-linear batch (or PIR) codes have better parameters than their best linear counterparts?
4. Do non-systematic batch (or PIR) codes have better parameters than their best systematic counterparts?
5. Propose batch and PIR codes that allow for efficient reconstruction algorithms.

**Acknowledgements** The material in this chapter has benefited a lot from discussions of the author with his students and colleagues, including Venkatesan Guruswami, Camilla Hollanti, Helger Lipmaa, Sushanta Paudyal, Eldho Thomas, Alexander Vardy, Hui Zhang and Jens Zumbrägel. This work is supported in part by the grants PUT405 and IUT2-1 from the Estonian Research Council and by the EU COST Action IC1104.

# References

1. D. Augot, F. Levy-Dit-Vehel, A. Shikfa, A storage-efficient and robust private information retrieval scheme allowing few servers (2014), arXiv:1412.5012
2. S. Bhattacharya, S. Ruj, B. Roy, Combinatorial batch codes: a lower bound and optimal constructions. Adv. Math. Commun. **6**(2), 165–174 (2012)
3. S.R. Blackburn, T. Etzion, PIR array codes with optimal PIR rates (2016), arXiv:1609.07070
4. R.A. Brualdi, K. Kiernan, S.A. Meyer, M.W. Schroeder, Combinatorial batch codes and transversal matroids. Adv. Math. Commun. **4**(3), 419–431 (2010)
5. C. Bujtás, Z. Tuza, Batch codes and their applications. Electron. Notes Discret. Math. **38**, 201–206 (2011)
6. V. Cadambe, A. Mazumdar, An upper bound on the size of locally recoverable codes, in *Proceedings International Symposium on Network Coding (NetCod)* (2013), pp. 1–5
7. T.H. Chan, S. Ho, H. Yamamoto, Private information retrieval for coded storage (2014), arXiv:1410.5489
8. Y.M. Chee, F. Gao, S.T.H. Teo, H. Zhang, Combinatorial systematic switch codes, in *Proceedings IEEE International Symposium on Information Theory (ISIT), Hong Kong, China* (2015), pp. 241–245
9. B. Chor, E. Kushilevitz, O. Goldreich, M. Sudan, Private information retrieval, in *Proceedings 36-th IEEE Symposium on Foundations of Computer Science (FOCS)* (1995), pp. 41–50
10. A.G. Dimakis, A. Gál, A.S. Rawat, Z. Song, Batch codes through dense graphs without short cycles (2014), arXiv:1410.2920
11. A.G. Dimakis, K. Ramchandran, Y. Wu, C. Suh, A survey on network codes for distributed storage. Proc. IEEE 99(3) (2011)
12. A. Fazeli, A. Vardy, E. Yaakobi, PIR with low storage overhead: coding instead of replication (2015), arXiv:1505.06241
13. M. Forbes, S. Yekhanin, On the locality of codeword sysmbols in non-linear codes. Discret. Math. **324**, 78–84 (2014)
14. P. Gopalan, C. Huang, H. Simitchi, S. Yekhanin, On the locality of codeword symbols. IEEE Trans. Inform. Theory **58**(11), 6925–6934 (2012)
15. Y. Ishai, E. Kushilevitz, R. Ostrovsky, A. Sahai, Batch codes and their applications, in *Proceedings of the 36th ACM Symposium on Theory of Computing (STOC), June 2004, Chicago, IL* (2004)
16. S. Kopparty, S. Saraf, S. Yekhanin, High-rate code with sublinear-time decoding, in *Proceedings of the 43rd Annual ACM Symposium on Theory of Computing (STOC), New York, NY* (2011), pp. 167–176
17. H. Lipmaa, V. Skachek, Linear batch codes, in *Proceedings 4th International Castle Meeting on Coding Theory and Applications, Palmela, Portugal, September 2014* (2014), arXiv:1404.2796
18. J.L. Massey, Threshold decoding, Technical report TR-410, MIT (1963)
19. S. Paudyal, Multi-symbol locally repairable codes, Master's thesis, University of Tartu, June 2015 (2015)
20. R. Pellikaan, X.-W. Wu, S. Bulygin, R. Jurrius, Error-correcting codes, http://www.win.tue.nl/~ruudp/courses/2WC09/2WC09-book.pdf
21. N. Prakash, V. Lalitha, P.V. Kumar, Codes with locality for two erasures, in *Proceedings IEEE International Symposium on Information Theory (ISIT), June-July 2014* (2014), pp. 1962–1966

22. S. Rao, A. Vardy, Lower bound on the redundancy of PIR codes (2016), arXiv:1605.01869
23. A.S. Rawat, A. Mazumdar, S. Vishwanath, Cooperative local repair in distributed storage. EURASIP J. Adv. Signal Process. (2015)
24. A.S. Rawat, D.S. Papailiopoulos, A.G. Dimakis, S. Vishwanath, Locality and availability in distributed storage. IEEE Trans. Inf. Theory **62**(8), 4481–4493 (2016)
25. R.M. Roth, *Introduction to Coding Theory* (Cambridge University Press, Cambridge, 2006)
26. N. Silberstein, A.S. Rawat, O.O. Koyluoglu, S. Vishwanath, Optimal locally repairable codes via rank-metric codes, in *Proceedings IEEE International Symposium on Information Theory (ISIT), Istanbul, Turkey* (2013), pp. 1819–1823
27. N. Silberstein, A. Gál, Optimal combinatorial batch codes based on block designs. Des. Codes Cryptogr. **78**(2), 409–424 (2016)
28. D. Stinson, R. Wei, M. Paterson, Combinatorial batch codes. Adv. Math. Commun. **3**(1), 13–17 (2009)
29. I. Tamo, A. Barg, Bounds on locally recoverable codes with multiple recovering sets, in *Proceedings IEEE International Symposium on Information Theory (ISIT), Honolulu, HI, June-July 2014* (2014), pp. 691–695
30. I. Tamo, A. Barg, A family of optimal locally recoverable codes. IEEE Trans. Inf. Theory **60**(8), 4661–4676 (2014)
31. A. Vardy, E. Yaakobi, Constructions of batch codes with near-optimal redundancy, in *Proceedings IEEE International Symposium on Information Theory (ISIT), Barcelona, Spain* (2016), pp. 1197–1201
32. Z. Wang, H.M. Kiah, Y. Cassuto, Optimal binary switch codes with small query size, in *Proceedings IEEE International Symposium on Information Theory (ISIT), Hong Kong, China* (2015), pp. 636–640
33. Z. Wang, O. Shaked, Y. Cassuto, J. Bruck, Codes for network switches, in *Proceedings IEEE International Symposium on Information Theory (ISIT), Istanbul, Turkey* (2013), pp. 1057–1061
34. A. Wang, Z. Zhang, Repair locality with multiple erasure tolerance. IEEE Trans. Inf. Theory **60**(11), 6979–6987 (2014)
35. T. Westerbäck, R. Freij, C. Hollanti, Applications of polymatroid theory to distributed storage systems, in *Proceedings 53rd Allerton Conference on Communication, Control, and Computing, Allerton, IL, USA, September-October 2015* (2015), pp. 231–237
36. T. Westerbäck, R. Freij-Hollanti, T. Ernvall, C. Hollanti, On the combinatorics of locally repairable codes via matroid theory. IEEE Trans. Inf. Theory **62**(10), 5296–5315 (2016)
37. M. Wootters, Linear codes with disjoint repair groups, unpublished manuscript (2016)
38. H. Zhang, V. Skachek, Bounds for batch codes with restricted query size, in *Proceedings IEEE International Symposium on Information Theory (ISIT), Barcelona, Spain* (2016), pp. 1192–1196
39. Y. Zhang, X. Wang, H. Wei, G. Ge, On private information retrieval array codes (2016), arXiv:1609.09167
40. J. Zumbrägel, V. Skachek, On bounds for batch codes, in *Algebraic Combinatorics and Applications (ALCOMA), Kloster Banz, Germany* (2015)

Printed in the United States
By Bookmasters